MAM 2006:
MARKOV ANNIVERSARY MEETING

Published by **Boson Books**
Raleigh, North Carolina

ISBN 1-932482-34-2 (print edition)
ISBN 1-932482-35-0 (online ebook edition)

An imprint of **C&M Online Media, Inc.**
3905 Meadow Field Lane
Raleigh, North Carolina 27606 USA

Cover art by A.N. Langville, D.F. McAllister and W.J. Stewart

Design by Once Removed

www.bosonbooks.com
cm@cmonline.com

MAM 2006:
Markov Anniversary Meeting

An International Conference
to celebrate the
150th Anniversary of the birth of
A.A. Markov

June 12-14
Charleston, South Carolina
USA

Edited by
Amy N. Langville
William J. Stewart

An imprint of C & M Online Media
Raleigh, North Carolina USA

Contents

Preface

The Markov Anniversary Meeting, June 12-14, 2006 was the fifth in a series of conferences devoted to the Numerical Solution of Markov Chains (NSMC). The prior four NSMC conferences were held in 1990, 1995, 1999, and 2003. The first two conferences were held in Raleigh, North Carolina, the third moved overseas to Zarazoga, Spain, and the fourth was held in Urbana-Champaign, Illinois.

This fifth and most recent Markov meeting occurred sooner than the usual 4-5 year schedule in order to celebrate two very important events: the 150th anniversary of Andrei A. Markov's birth and the 100th anniversary of his work on his chains. This special Markov Anniversary Meeting was held at the College of Charleston in historic Charleston, SC. The scope of the meeting was expanded to incorporate Markov work beyond the usual NSMC topics. Thus, this proceedings contains several papers describing the diverse applications of Markov's chains in diverse fields from web search and telecommunications to population and language modeling.

We were pleased to receive a wide variety of quality papers for presentation. There was, rightfully, much anticipation for the keynote talk given by Eugene Seneta the life, work, and contemporaries of A. A. Markov. His work as a dedicated mathematician and historian fills the need for a complete paper about Markov. With the historical stage set, we turned to present uses and extensions of Markov's chains. Papers discuss Markov reward processes and Markov bounds, while another group of papers cover the use of chains in queueing theory. A third group of papers examines the techniques for analyzing enormous chains with the help of Kronecker algebra. Such techniques connect nicely with one of today's most famous applications of Markov chains, the PageRank chain used by Google to rank webpages. Several invited speakers discussed the mathematics behind PageRank and other web search techniques. The proceedings conclude with a section devoted to applications of Markov chains.

We are proud to present a body of work that represents a continuation of A. A. Markov's work over a century ago. Of course, the resulting proceedings is the product of many contributors. First, we thank this year's Program Committee members, many of whom held multiple roles as publicists, promoters, contributing authors, reviewers, and attendees. Next, we thank all submitting authors and participants. Finally, we thank Nancy and David McAllister of Boson Books for their help preparing this timely compendium.

Amy N. Langville and William J. Stewart.
Charleston, South Carolina and Raleigh, North Carolina, USA.
June 2006

Program Committee

Michele Benzi,	USA	Peter King,	UK
Dario Bini,	Italy	William Knottenbelt,	UK
Gianfranco Ciardo,	USA	Udo Krieger,	Germany
Tugrul Dayar,	Turkey	Beatrice Meini,	Italy
Susanna Donatelli,	Italy	Carl D. Meyer,	USA
Iain Duff,	UK & France	Andrew Miner,	USA
Jean-Michel Fourneau,	France	Harry G. Perros,	USA
Winfried Grassmann,	Canada	Bernard Philippe,	France
Reinhard German,	Germany	Bill Sanders,	USA
Gunter Haring,	Austria	Roger B. Sidje,	Australia
Pete Harrison,	UK	Evgenia Smirni,	USA
Boudewijn Haverkort,	The Netherlands	Pete Stewart,	USA
Graham Horton,	Germany	Daniel Szyld,	USA
Ilse Ipsen,	USA	Kishor S. Trivedi,	USA
Joost-Pieter Katoen,	Germany	Jean-Marc Vincent,	France
Peter Kemper,	Germany		

MARKOV AND THE CREATION OF MARKOV CHAINS.

EUGENE SENETA*

Abstract.
We describe the life, times and legacy of Andrei Andreevich Markov (1856 -1922), and his writings on what became known as Markov chains. One focus is on his first paper [27] of 1906 on this topic, which already contains important contractivity principles embodied in the Markov - Dobrushin coefficient of ergodicity, which in fact makes an explicit appearance in that paper. The contractivity principles are shown directly to underpin a number of results of the later theory. The coefficient is especially useful as a condition number in measuring the effect of perturbation of a stochastic matrix on the stationary distribution (sensitivity analysis). Some recent work in this direction is reviewed from the standpoint of the paper [53], presented at the first of the present series of conferences [63].

Key words. biography, Markov chain, contractivity, stochastic matrix, coefficient of ergodicity, perturbation, condition number, Google matrix, Dobrushin, Besicovitch, Uspensky, Romanovsky.

AMS subject classifications. 01A55, 01A60, 60-03, 60J10, 60E15, 15A51, 15A12

1. Introduction. Andrei Andreevich Markov was born June 14th (June 2nd, old style), 1856, in Ryazan, Imperial Russia, and died on July 20, 1922, in Petrograd, which was – before the Revolution, and is now again – called Sankt Peterburg (St. Petersburg).

In his academic life, totally associated with St. Petersburg University and the Imperial Academy of Science, he excelled in three mathematical areas: the theory of numbers, mathematical analysis, and probability theory. What are now called Markov chains first appear in his work in a paper of 1906 [27], when Markov was 50 years old. It is the 150th anniversary of his birth, and the 100th anniversary of the appearance of this paper that we celebrate at the Markov Anniversary Meeting, Charleston, South Carolina, June 12 - 14, 2006.

Markov's writings on chains occur within his interest in probability theory. On the departure in 1883 of his mentor, Pafnuty Lvovich Chebyshev (1821 - 1894) from the university, Markov took over the teaching of the course on probability theory and continued to teach it yearly, even in his capacity of a Privat-Dozent (lecturer) after his own retirement from the university as Emeritus Professor in 1905.

His papers on Markov chains utilize the theory of determinants (of finite square matrices), and focus heavily on what are in effect finite stochastic matrices. However, explicit formulation and treatment in terms of matrix multiplication, properties of powers of stochastic matrices, and more generally of inhomogeneous products of stochastic matrices, and of associated spectral theory, are somewhat hidden, even though striking results, rediscovered by other authors many years later, follow from ideas in [27]. Our mathematical focus is an exploration of the contractivity ideas of that paper in the context of finite stochastic matrices , and specifically of the structure and usage of the Markov-Dobrushin coefficient of ergodicity.

Markov's motivation in writing the Markov chain papers was to show that the two classical theorems of probability theory, the Weak Law of Large Numbers and the Central Limit Theorem, could be extended to sums of dependent random variables. Thus he worked very much in terms of probabilistic quantities such as moments and

---
*School of Mathematics and Statistics, University of Sydney, NSW 2006, Australia.(eseneta@maths.usyd.edu.au).

expectations, and particularly with positive matrices. The underlying matrix properties of general non-negative stochastic matrices, such as irreducibility, periodicity, stationary (invariant) vector, and asymptotic behavior of powers, which determine the nature of the probabilistic behavior, were not clearly in evidence.

The theory of finite non-negative matrices was beginning to emerge only contemporaneously with Markov's [27], [29] first papers on Markov chains, with the work of Perron [44] and Frobenius [14], [15]. The connection between the two directions, Markov and Perron-Frobenius is probably due to von Mises [40]. The theory of finite Markov chains was then developed from this standpoint in the treatises of Fréchet [13] and Romanovsky [45] on homogeneous finite Markov chains.

In our own times, the heavily influential book on finite homogeneous Markov chains has been that of Kemeny and Snell [23], which, while heavily matrix theoretic in its operations, avoids any mention of spectral theory, and in its discussion of ergodicity is closest in spirit to Markov's original memoir [27].

The issues raised in the preceding paragraphs have been discussed in more detail, especially with respect to coefficients of ergodicity and inhomogeneous products, in the author's paper [48], and the author's book [49]. This book of 1981 has been reissued in paperback form in February, 2006, with an (incomplete) additional bibliography on coefficients of ergodicity. More recently (1996) the paper [57] of the author, written for statisticians, explores some of what might have entered the content of the present paper. However, the technical emphasis there is probabilistic, including Section 5 ("Techniques of Markov's 1906 Paper") and Section 6 ("The Ergodicity Coefficient"). We shall recast some of this into matrix analytic form, and proceed in a generally different direction in our exploration of the consequences of [27].

2. Biographical Notes. In an anniversary paper , it is appropriate to give some details on Markov's life.

Markov's father was Andrei Grigorievich Markov (1823 - ?). In Russian usage the second name is a patronymic. Thus our Markov, baptized in the Russian Orthodox Church as Andrei, became Andrei Andreevich Markov. A.A. Markov's father, on completion of his studies in a theological seminary in 1844, entered the administration of the Ryazan Guberniia (a guberniia (governorship) was an administrative region). He eventually rose to a senior position as counsellor, becoming noted for his directness, honesty, and uncompromising nature, qualities reflected later in his son Andrei Andreevich. His diligence in unmasking financial corruption was not to the taste of his superiors, and he was eventually asked to retire. Consequently he became a para-legal clerk, and found the legal profession much to his taste. He was reputed to be a gambler, and an inveterate card player. Markov family lore has it that he once lost all his family assets to a card-sharp; but the loss was later reinstated.

Andrei Grigorievich married Nadezhda Petrovna Fedorova (1829 - ?) early in 1847. She was the daughter of a guberniia official. They had 6 children: Piotr (1849 - ?), Yevgeniia (1850 - 1920), Pavel (1852, died in childhood), Maria (1854 - 1875), our Andrei (1856 - 1922) and Mikhail (1859 - ?). Andrei Grigorievich was married twice. His second wife, Anna Iosifovna, was also the daughter of an official. They had three children: Vladimir (1871 - 1897), Lidia (1873 - 1942) and Ekaterina (1875 - ?).

Andrei Andreevich's half-brother, Vladimir Andreevich, was on the way to eminence as mathematician at St. Petersburg University in the area of number theory, but died early. He figures significantly in modern Russian mathematical historiography [25]. Number theory was one of the areas in which Andrei Andreevich excelled, and in which he seems to have influenced the young J.V. Uspensky (of whom more shortly), in a direction akin to Vladimir's.

In the early 1860's Andrei Grigorievich moved with his family from Ryazan to St. Petersburg. He became steward to the estate of Ekaterina Aleksandrovna Valvatieva, a widow, who had two daughters, Maria and Elizaveta. Maria Ivanovna Valvatieva (1860 - 1942) was to become Andrei Andreevich's wife in 1883, when Maria's mother finally assented to the marriage, until then judging Andrei Andreevich's prospects insufficient. At the time of the marriage he was Privat-Dozent at the university, close to defending his Doctoral dissertation, with the prospect of a Professorship.

Andrei Andreevich was a sickly child. In childhood he had a bone disorder; one leg wouldn't straighten at the knee and he had to walk on crutches. He determinedly learned to dispense with the crutches during games by hopping on one leg. In St. Petersburg an eminent surgeon straightened the leg, allowing him to walk normally, although he retained a slight limp all his life. This did not stop him from taking long hikes of which he was fond, sometimes saying "While you can walk, you know you're alive". Thus he was not a "cripple", as oral tradition has it.

His carefree childhood ended in 1866 when, at age 10, he was placed into the 5th St. Petersburg Gimnaziia (High School), which was then on the outskirts of St. Petersburg. (His younger half-brother Vladimir was also a student there.) Markov was not a particularly good student in high school except in mathematics, with a strong interest also in the burning social issues of the time. On occasion he revealed a rather rebellious and uncompromising nature. This was to manifest itself later in numerous clashes with academic colleagues and with the tsarist regime itself. Nevertheless, even in high school he established contact with the St. Petersburg University's mathematics professors A.N. Korkin and E.I. Zolotarev through his precocious, though as it turned out, not new, method of solving differential equations. Finally, on graduation from high school in 1874, he entered in that year the physico-mathematical faculty of St. Petersburg University. In 1877 he received a gold medal and on completion of his studies in 1878 was retained by the university to prepare for a career as an academic. His first mathematical papers appeared in 1879.

His Master's and Doctoral dissertations (defended in 1880 and 1885 respectively) were in number theory. He began lecturing in 1880 as Privat-Dozent, and as already mentioned, in probability theory in 1883.

3. Probability and chain dependence. The stream of Markov's publications in probability was initially motivated by inadequacies in Chebyshev's treatment of the Central Limit Problem in 1887, and begins with a letter to his friend A.V. Vasiliev (1853 - 1929), which Vasiliev published in the *Izvestiia (Bulletin)* of Kazan University's Physico-Mathematical Society. The Weak Law of Large Numbers (WLLN) and the Central Limit Theorem were the focal probabilistic issues of the times.

The paper [27] in which a Markov chain, as a stochastically dependent sequence for which the WLLN holds, first appeared in Markov's writings, was likewise published in the *Izvestiia*. The paper is motivated by the need to construct a counterexample to what Markov interpreted [50], [57] as a claim in 1902 of P.A. Nekrasov (1853 - 1924) that pairwise independence of summands was necessary as well as sufficient for the WLLN to hold. From showing that the WLLN held in the presence of chain dependence, it was an obvious step to investigate that refinement of the WLLN which is the Central Limit Theorem. There followed a stream of papers in this direction [29], [30], [32], now in more high profile journals, but unfortunately omitting any further use of the contractivity methodology of the original 1906 paper. in particular omitting the ergodic coefficient. It should be mentioned that in correspondence [41]

with the St. Petersburg statistician A.A. Chuprov (1874 - 1926), initiated by fiery
post-cards in late 1910 by Markov subsequent to Chuprov's mentioning Markov's
arch-enemy P.A. Nekrasov in a positive light, Markov was made aware of earlier work
on special kinds of Markov chains by Ernst Heinrich Bruns (1848 - 1919), later called
"Markov-Bruns Chains" by Romanovsky [45]. Although the publication of Bruns 's
book [7] in 1906 is contemporaneous with [27], in the preface Bruns claims that his
book arises out of his lectures of the preceding 25 years. Bruns's own methodology,
like Markov's, is direct, that is: not matrix-based. Immediately on becoming aware
due to Chuprov, of Bruns's work, Markov [34] produced a paper on "Markov-Bruns"
chains. This paper together with another stimulated by correspondence with Chuprov
were both presented to the Imperial St. Petersburg Academy of Science on the same
day (19 January, 1911, o.s.). Markov was not particularly well-read on the relevant
probabilistic literature, and indeed appears not to have been conversant with the fact
that the Bernoulli-Laplace urn model, of much earlier provenance, could be cast in the
form of a homogeneous Markov chain until, apparently, [35]. When these early models
of homogeneous Markov chains are cast in transition matrix form, the transition
matrices all have zero entries. Markov, as we have mentioned, did not completely
resolve the matrix structural issues (reducibility, periodicity) which can arise out of
such forms of stochastic matrix. His (probabilistic) methodology was strongly focused
on the Method of Moments in the guise of conditional and absolute expectations, and
double probability generating functions. These functions are, indeed, closely linked
to the determinant [46], [58] and spectral theory of stochastic matrices, and thus
necessarily interact with the positioning of zeros in the transition matrix. We omit a
more detailed discussion of them from this paper.

4. Markov's academic progress and later years. In 1886 he was appointed
Extraordinary (Associate) Professor in the Department of Pure Mathematics, and
later in that year elected Adjunct of the Imperial St. Petersburg Academy of Science,
at the proposal of Chebyshev. On the 30 January (o.s.), 1890, he was elected Extraordinary (Associate) Academician, in place of V. Ya. Buniakovsky who had died in
1889. Competing for this vacant position was Sofiia Kovalevskaya (Sonia Kowalewski),
whose work Markov continued, apparently mistakenly, to attack in characteristically
volatile and unreflective fashion even after her death in January, 1891. Markov was
promoted to Ordinary (full) Professor in 1893, and elected Ordinary (full) Academician in 1896. His mentor Chebyshev had died in 1894. The year 1903 saw the birth
of his son, also Andrei, and thus also Andrei Andreevich Markov (1903 - 1979), who
himself was to become an eminent mathematician, and Corresponding Member of the
Academy of Sciences of the U.S.S.R. The identical 3-part name with his father has
sometimes confused western writers producing photographs.

1900 saw the publication of the first edition of Markov's textbook *Ischislenie
Veroiatnostei (The Calculus of Probabilities)*. The second edition appeared in Russian
in 1908 and was translated into German by Heinrich Liebmann in 1912 as *Wahrscheinlichkeitsrechnung*, which became well-known in the West. This edition already contains Markov's (Tail) Inequality, a simple and more direct approach to the Bienaymé-
Chebyshev Inequality in probability theory. The third edition of 1913, substantially
expanded, and with a portrait of Jacob Bernoulli, was timed to appear in the year of
the 200th anniversary of Jacob Bernoulli's WLLN. Markov organized a Commemora-
tive Meeting of the Academy of Science in honor of the anniversary. Other speakers
were Vasiliev and Chuprov. The fourth edition of 1924 [36] was posthumous, and
published in Moscow in the early years of the Soviet era. It is again much expanded

and, of the Russian-language editions, now the most readily available.

The last years of Markov's life coincided with a stormy period of history. The revolution in Russia of February 1917 saw the fall of the monarchy, with the abdication of Tsar Nicholas II and the establishment of the Provisional Government in Russia. The October Revolution (25 October (o.s.); 7th November) of 1917 resulted in Bolshevik seizure of power. The name St. Petersburg, used till 1914 and the beginning of World War I, was changed to Petrograd which was used till 1924, the year of Lenin's death. It then became Leningrad until the demise of the Soviet Union.

The name Imperial Academy of Sciences in 1917 became the Russian Academy of Sciences until 1925, and then the Academy of Sciences of the U.S.S.R. In 1934 the Academy was transferred from Leningrad to Moscow, which had become the capital city after the Bolshevik seizure of power.

5. Markov's legacy. Tributes to Markov soon after his death appeared in the 1923 *Izvestiia* of the Russian Academy of Science, by Ya.V.Uspensky (= J.V. Uspensky) [65] and A.S. Bezikovich (= A.S. Besicovitch) [5]. Besicovitch [6] also wrote a biographical sketch for the posthumous edition of Markov's monograph [36] .

The name of Abram Samoilovitch Besicovitch (1891 - 1970) was to become very well known in the mathematical world. Besicovitch (this was the transliteration which he used from the Russian of his name) graduated in 1912 from the St. Petersburg University, where one of his teachers had been A.A. Markov, and Besicovitch's first paper published in 1915 in the then renamed Petrograd was in probability theory, a new proof of the classical (independent summands) Central Limit Theorem very much in the Chebyshev/Markov tradition. He left Leningrad (as Petrograd had become) illegally in 1924 by crossing the then-nearby border with Finland under cover of darkness. He took up a position at the University of Cambridge, England, in 1927 and was elected Fellow of the Royal Society in 1934, and to the Rouse Ball Chair of Mathematics in 1950, receiving many honors for his mathematical research, which was primarily in real variable analysis and measure theory, specifically almost periodic functions, geometry of plane sets, and Hausdorff measure. His *oeuvre* after his first paper contains very few papers on probabilistic topics, but he did influence a number of mathematicians during his life in England who eventually made very significant contributions to probabilistic topics, including Markov chains. In particular: P.A.P. Moran, I.J. Good and S.T. Taylor. Burkill [8] wrote Besicovitch's biographical essay/obituary.

Yakov Viktorovich Uspensky (1883 - 1947) seems to have been Markov's colleague at St. Petersburg University, and would have taken courses from him, being about 30 years old in 1913, when Markov stopped teaching probability at the University (Besicovitch thus about 22 years old). Uspensky's early work in Russia was in number theory, and gets considerable attention in the chapter on this topic in [25]. There seems to have been influence by the work on number theory of A.A. Markov. Uspensky's Master's thesis in this area was published in St. Petersburg in 1910. According to Markov [3] pp.19–20, Uspensky in May 1913 was Privat-Dozent at St. Petersburg University, and translated the celebrated 4th part of Jacob Bernoulli's *Ars Conjectandi* from Latin into Russian, for the 200th anniversary celebrations of Bernoulli's WLLN. A note in [3], p.73, identifies Uspensky as Academician of the Russian Academy of Science from 1921. His election to the Academy was supported by Markov, V.A. Steklov and A.N. Krylov, who give an account of his publications in the Academy's *Izvestiia*, Ser. 6, 15(1921), pp. 4–5. At this time Uspensky was (full) Professor at the University.

Uspensky's apparently last paper in Russian was published in 1924 in the *Doklady ANSSSR* and is on a probabilistic topic. He appears to have left Leningrad at about this time, and made his way to the United States. His first paper in English, according to Math. Sci. Net., appeared in the *American Mathematical Monthly* in 1927 and was also on a probabilistic topic. A note in [3], p. 167, says he worked in the United States from 1929. In the United States he used the English version James of Yakov (which is, more accurately translated, as Jacob). Although he continued to write in several areas, and gained considerable distinction, it is largely for his book of 1937, *Introduction to Mathematical Probability*, written as Professor of Mathematics at Stanford University, and based on his lectures there, that he is best known. The book [66] discusses only two-state Markov chains within its chapter *Fundamental Limit Theorems*. It is certainly heavily influenced by the work of the St. Petersburg School of Probability, and specifically by Markov, on the Central Limit Problem. Uspensky's book seems to have brought analytical probability, in the St. Petersburg tradition, to the United States, where it remained a primary probabilistic source until the appearance of W. Feller's *An Introduction to Probability Theory and Its Applications* in 1951. Feller's book contains a great deal on Markov chains, specifically the case of a denumerable number of states, for which a matrix/spectral approach is not adequate, and renewal theoretic arguments are employed.

We have already mentioned the book of Fréchet [13], which was the first monograph on finite Markov chains from a matrix standpoint. The matrix method for finite Markov chains was subsequently exposited very much from Markov's post-1906 standpoint, in monograph form in Russian, by Romanovsky [45]. It reappeared in English translation by the author of this paper in 1970.

Vsevolod Ivanovich Romanovsky (1879 - 1954), born in the town that became known as Alma Ata, received his secondary eduction in an academic high school ("Reelschule") in Tashkent. He completed his studies at St. Petersburg University in 1906, where he was retained to prepare for an academic career. Then he completed his Master's degree examinations in 1908, at which time he returned to teach mathematics at his old high school. From 1911 to 1915 he was at first Privat-Dozent and then Professor at Warsaw University (at the time part of Poland was still part of the Russian Empire). This university, as a Russian institution, was closed down, and for a year or so from 1915 he worked at Don University at Rostov-on-the-Don, and returned to Tashkent in 1917. From its beginning stages in 1918, till his death he was heavily involved in teaching and research in mathematics, and in administration at what became Tashkent State University (initially called Central Asian University), and with the organization in 1943 and functioning of the Academy of Science of the Uzbek S.S.R. In the early period of his research he worked on differential equations, algebraic questions, and (as expected from his student days at St. Petersburg) on number theory. His later research activities were very largely devoted to the theory and applications of probability theory and mathematical statistics. In spite of his geographical distance from the main academic centers of the Soviet Union, he managed to keep in touch with and publish on statistical topics in the important western European statistical and mathematical journals, such as *Biometrika, Metron, Comptes Rendus Acad. Sci. Paris, Rend. del Circ. Mat. di Palermo,* on issues of mathematics close in spirit to that of the English Biometric School of Karl Pearson, but using the probabilistic methodology of the St. Petersburg School. There was a fundamental paper on finite Markov chains in *Acta Mathematica* in 1936, presumably in imitation of A.A. Markov's French-language publication in this journal in 1910. Romanovsky's geographical isolation within the Soviet Union seems to have helped him maintain a scientific activity in mathematical statistics and its applications when it was being severely attacked in the (European) Soviet centers. The distance from St. Petersburg-Petrograd-Leningrad on the other hand, would have worked against any personal contact with A.A. Markov in the last decade or so of Markov's life. Romanovsky's most important scientific work was on finite Markov chains (it began in 1928), and on their generalization. His *magnum opus* of 1949 on this topic [45], however, was algebraically intricate, and received little attention in comparison with the theory of denumerable chains developed by Kolmogorov from the 1930's.

In the English-speaking world, finite homogeneous Markov chain theory was re-born with Kemeny and Snell's book, as we have mentioned in our Introduction.

6. Some sources on Markov's life and work. For some time the best source on Markov, and the present author's primary source, on his life, and his publications, in number theory and probability theory, has been [37], a Russian-language book of about 720 pages. The part entitled Probability Theory includes reprinting of 7 of Markov's papers on Markov chains, including [27]. There are several important appendices, in particular an extensive biography by his son, and a survey of Markov senior's writings on number theory and probability by Yu. V. Linnik, N.A. Sapogov and V.N. Timofeev. There are additionally commentaries on the individual papers, the ones on the Markov chain papers are written by Sapogov. His commentary does not encompass all the important ideas in [27] , but makes the important point that the strict positivity of all transition probabilities in Markov's exposition can be relaxed to assuming a strictly positive column, a fact which had already been noticed by S.N. Bernstein (see [4]), and which has played an important part in the theory of ergodicity coefficients. A finite stochastic matrix with a strictly positive column has been called a "Markov" matrix. Also in [37] is a very complete and detailed listing, by year, of all of Markov's publications; of his lithographed course lectures; of literature about Markov; and a name index. The paper published in French in *Acta Mathematica* **33** (1910) 87-104 as "Recherches sur un cas remarquable d'épreuves dépendantes" is not included, but is of course readily available. It is essentially encompassed by the Russian-language articles [29], [30]. The 1908 paper [30] is included in [37].

Recently in a privately printed book [62] some of the contents of [37] have become available in English translation: [28], [31], [33] with Sapogov's commentaries; the sketch by Linnik *et al.*, ; and the biography of his father by A.A. Markov Jr. [38].

The biography by A.A. Markov Jr. is, understandably, written in the Soviet polit-ical spirit of the times. For example, P.A. Nekrasov is painted in a very negative way. In the last *glasnost* years of the Soviet Union a more balanced as well as considerably extended biographical study (on the basis of family documents and recollections, and archival documents from a number of archives) was prepared by Grodzensky [16].

Grodzensky's account consists of 5 chapters, contains a number of photographs, and features Markov's activity as a chess player. There are 3 appendices, respectively on number theory, mathematical analysis, and probability theory (this last by B.V. Gnedenko). There is a listing of Markov's publications, which concludes with [37], and a list of 210 references.

In English a recent biographical sketch is given by the author [59]. Markov's mo-tivation for initiating his study of Markov chains, and his interaction with Nekrasov, are encompassed in the author's [50], [57], [60]. Sheynin [61] gives an introductory bio-graphical sketch of Markov, and is largely concerned with describing with the content of Markov's probability monograph [36].

Finally, it is appropriate to mention the study by Basharin, Langville and Naumov [2] prepared for an earlier conference in this series. The present paper (until a late draft) was written without examining [2], There is, in the event, relatively little overlap. On the technical side the emphasis in [2] is on the probabilistic aspects, as is the case in [57], which it cites. The presence of the Markov family photographs in [2] is very welcome. The reader is encouraged to read both [57], [2] in conjunction with the present paper.

7. Contractivity principles in Markov's reasoning. In Sections 7.1 to 7.3 we present what may be extracted in essence from Markov (1906), specifically its Section 5.

7.1. Markov's Contraction Inequality. Lemma 7.1 below, states what we call Markov's Contraction Inequality, which is sometimes inappropriately attributed to Paz [42], We explain this circumstance in Section 7.3. .

LEMMA 7.1. *If $\delta = \{\delta_s\}, \mathbf{w} = \{w_s\}$, are real-valued column N-vectors, and $\delta^T \mathbf{1} = 0$, then*

$$(7.1) \qquad |\delta^T \mathbf{w}| \leq (\max w_s - \min w_s) \frac{1}{2} \sum_{s=1}^{N} |\delta_s|$$

$$= \max_{h,h'} |w_h - w_{h'}| \frac{1}{2} \sum_{s=1}^{N} |\delta_s|$$

Proof. Let $E = \{s; \delta_s \geq 0\}$, $F = \{s; \delta_s < 0\}$. Then

$$(7.2) \qquad \sum_{s \epsilon E} \delta_s = -\sum_{s \epsilon F} \delta_s = \frac{1}{2} \sum_{s=1}^{N} |\delta_s|.$$

Also

$$v = \delta^T \mathbf{w} = \sum_{s=1}^{N} \delta_s w_s = \sum_{s \epsilon E} \delta_s w_s + \sum_{s \epsilon F} \delta_s w_s \leq (\max w_s) \sum_{s \epsilon E} \delta_s + (\min w_s) \sum_{s \epsilon F} \delta_s$$

$$= (\max w_s - \min w_s) \sum_{s \epsilon E} \delta_s. \qquad \square$$

7.2. Contractive property of a stochastic matrix.. The following lemma expresses the averaging property of a stochastic matrix.

LEMMA 7.2. *Let $\mathbf{P} = \{p_{ij}\}, i, j = 1, \cdots, N$ be a stochastic matrix, so that $\mathbf{P} \geq 0, \mathbf{P1} = \mathbf{1}$. Let $\mathbf{w} = \{w_i\}$ be a real-valued column N-vector, and put*

$$(7.3) \qquad\qquad\qquad \mathbf{v} = \mathbf{P}\,\mathbf{w}$$

Then, writing $\mathbf{v} = \{v_i\}$,

$$(7.4) \qquad\qquad \max_{h,h'} |v_h - v_{h'}| \leq H \max_{j,j'} |w_j - w_{j'}|$$

where

$$(7.5) \qquad H = \frac{1}{2} \max_{i,j} \sum_{s=1}^{N} |p_{is} - p_{js}|,$$

so $0 \le H \le 1$.

Proof. From (7.3)

$$(7.6) \qquad v_i - v_j = \sum_{s=1}^{N} (p_{is} - p_{js}) w_s,$$

and since $\sum_{s=1}^{N} (p_{is} - p_{js}) = 0$ by stochasticity of $\mathbf{P}$, we may apply (7.1) to (7.6) to obtain

$$|v_i - v_j| \le (\max w_s - \min w_s) \frac{1}{2} \sum_{s=1}^{N} |p_{is} - p_{js}|$$

$$(7.7) \qquad = (\max w_s - \min w_s) H$$

from (7.5), whence (7.4) follows. □

LEMMA 7.3. *Putting* $\mathbf{P}^{n-1} = \{p_{sr}^{(n-1)}\}, n \ge 1$, *with* $\mathbf{P}^0 = \mathbf{I}$ *(the unit matrix),*

$$(7.8) \qquad \max_{h,h'} |p_{hr}^{(n)} - p_{h'r}^{(n)}| \le H^n, \; n \ge 0.$$

Proof. Since $\mathbf{P}^n = \mathbf{P}\,\mathbf{P}^{n-1}, n \ge 1$, putting $w_s = p_{sr}^{(n-1)}$ for fixed r , and $s = 1, \cdots, N$, from (7.3) and (7.4):

$$(7.9) \qquad \max_{h,h'} |p_{hr}^{(n)} - p_{h'r}^{(n)}| \le H \max_{j,j'} |p_{jr}^{(n-1)} - p_{j'r}^{(n-1)}|$$

so by iterating (7.9) back, (7.8) obtains. □

If $\mathbf{P} > \mathbf{0}$ i.e. all entries are positive, as Markov effectively assumes, it is clear that $H < 1$ from the expression (7.5) for H ; and this is also clearly true if $\mathbf{P}$ has a strictly positive column (i.e. is a "Markov" matrix).

When $H < 1$, (7.9) implies that as $n \to \infty$ all rows of $\mathbf{P}^n$ tend to coincidence (this property was later called "weak ergodicity"). □

LEMMA 7.4. *For fixed* r, $\max_h p_{hr}^{(n)}$ *is non-increasing with increasing* n; *and* $\min_h p_{hr}^{(n)}$ *is non-decreasing with increasing* n, *so (since both sequences are bounded) both have limits as* $n \to \infty$. *When* $H < 1$, *all rows of* $\mathbf{P}^n$ *tend to the same limiting probability vector.*

Proof. Using the notation of Lemma 7.2, since

$$\mathbf{v} = \mathbf{P}\mathbf{w}, \quad v_i = \sum_{j=1}^{N} p_{ij} w_j \le (\max w_j) \sum_{j=1}^{N} p_{ij}$$

so max $v_i \leq$ max w_j. Similarly min $v_i \geq$ min w_j.

Putting, for fixed r, $w_s = p_{sr}^{(n-1)}$, $v_i = p_{ir}^{(n)}$ the respective monotonicities follow. Now from (7.7),

$$v_i - v_j \leq (\max w_s - \min w_s)H$$

so the coincidence in the limit as $n - \infty$ of both the maximal and minimal of sequences follows when $H < 1$. $\square$

The property of all rows of $\mathbf{P}^n$ actually tending to the same limiting probability distribution came to be called "strong ergodicity".

Notice that the argument of Lemma 7.4 uses the "backward" form: $\mathbf{P}^n = \mathbf{P}\,\mathbf{P}^{n-1}$; and obtains ergodicity of a finite homogeneous Markov chain, it would seem, at geometric rate of convergence providing $H < 1$, without use of Perron-Frobenius theory of non-negative matrices. We explore the consequences of these remarks in our Section 8.

The notation "H" of (7.5) is actually Markov's [27], and the expression (7.5) implicitly appears in this paper. The form of H has been ascribed to Dobrushin [12]. We think [57] it appropriate to call it the Markov-Dobrushin coefficient of ergodicity.

7.3. Attribution. A great deal of theory for stochastic matrices/Markov chains can be developed from the inequality (7.1). It remains true if $\mathbf{w}$ is replaced by an N-vector $\mathbf{z} = \{z_i\}$ each of whose elements may be real or complex, so that

$$(7.10) \qquad |\delta^T \mathbf{z}| \leq \max_{h,h'} |z_h - z_{h'}| \frac{1}{2} \sum_{s=1}^{N} |\delta_s|$$

This inequality (7.10) for finite N is due to Alpin and Gabassov [1], where it is proved by induction on N . It follows also from a problem, given without solution in Paz ([42], p.73, Problem 16), and restated in [48], p.583, where (7.10) is derived from it. The inequality (7.10) does not appear in [42]. To rectify the question of attribution further, since (7.1) plays a crucial role in the perturbation (sensitivity) theory of stochastic matrices to be discussed below, and (7.10) plays a crucial role in spectral bounding theory, we restate verbatim Paz's problem:

> **16.** Prove that for any vector ξ such that $\| \xi \| < \infty$ and $\sum \xi_i = 0$ can be expressed in the form $\xi = \sum_{i=1}^{\infty} \zeta_i$ where $\zeta_i = (\zeta_{ij})$ vectors have only two non-zero entries, $\| \zeta_i \| < \infty$, $\sum_j \zeta_{ij} = 0$, and $\| \xi \| = \sum \| \zeta_i \|$.

The norm used in the above is $\| \cdot \|_1$.

The following is, with small notational changes, Lemma 2.4 of [49], p.62. Here $\mathbf{f}_k$ denotes the vector with unity in the kth position, and zeros elsewhere.

LEMMA 7.5.   *Suppose $\delta \in \Re^N, N \geq 2, \delta^T \mathbf{1} = 0, \delta \neq \mathbf{0}$. Then for a suitable set $I = I(\delta)$ of ordered pairs of indices (i,j), $i,j = 1, \cdots, N$,*

$$\delta = \sum_{(i,j) \in I} \left(\frac{\eta_{ij}}{2} \right) \gamma(i,j)$$

where $\eta_{i,j} > 0$ and $\sum_{(i,j) \in I} \eta_{i,j} = \| \delta \|_1$, and $\gamma(i,j) = \mathbf{f}_i - \mathbf{f}_j$. $\square$

A proof of this lemma, by induction on N is given on [49], p.63. This can be worked up into a constructive proof. (7.10) (and (7.1)) follow immediately.

Lemma 7.1 clearly remains valid, with essentially the proof given, for real valued vectors $\delta, \mathbf{w}$, of countably infinite length, providing $\| \delta^T \|_1 = \sum |\delta_i| < \infty, \delta^T \mathbf{1} = 0, |w_i| < K < \infty$, in which case

$$(7.11) \qquad |\delta^T \mathbf{w}| \leq (\sup_s w_s - \inf w_s) \frac{1}{2} \| \delta^T \|_1$$

$$= \sup_{h,h'} |w_h - w_{h'}| \frac{1}{2} \| \delta^T \|_1.$$

We propose that the name Markov's Contraction Inequality be used for both (7.1) and (7.10), although the name Lemma PS, as used in the body of Kirkland, Neumann and Shader [24] is a reasonable compromise. "Paz's Lemma" is not an appropriate name. This is no reflection on A. Paz's excellent [42], Chapter II on finite and countably infinite, homogeneous and inhomogeneous, Markov chains with emphasis on the countably infinite and inhomogeneous, using what we have called the Markov-Dobrushin coefficient (7.5).

For subsequent sections, we need to show the dependence of H on $\mathbf{P}$ explicitly. so we change the notation to more recent usage at (8.1).

8. Some direct consequences of Markov's contractivity principles. With little extra effort, Lemmas 7.1 - 7.4 may be used to obtain direct results, which in qualitative nature are as good as known results using more elaborate (albeit related) superstructure. We give two examples.

8.1. Weak and strong ergodicity of inhomogeneous products. For an $N \times N$ stochastic matrix $\mathbf{P} = \{p_{ij}\}$, write

$$(8.1) \qquad \tau_1(\mathbf{P}) = \frac{1}{2} \max_{i,j} \sum_{k=1}^N |p_{ik} - p_{jk}|,$$

$$(8.2) \qquad \Delta(\mathbf{P}) = \frac{1}{2} \sum_{k=1}^N \max_{i,j} |p_{ik} - p_{jk}|.$$

Notice that

$$\tau_1(\mathbf{P}) \leq \Delta(\mathbf{P}),$$

and that $\Delta(\mathbf{P}) = 0$ (whenever $\tau_1(\mathbf{P}) = 0$, both zero values expressing equality of all rows of $\mathbf{P}$. We have noted that $0 \leq \tau_1(\mathbf{P}) \leq 1$ always, but it is possible for $\Delta(\mathbf{P}) > 1$. For example, if $\mathbf{P} = I$, $\tau_1(\mathbf{P}) = 1$, but $\Delta(\mathbf{P}) = N/2$.

Now let $\mathbf{P_1} = \{p_{ij}(1)\}$ and $\mathbf{P_2} = \{p_{ij}(2)\}$ be $N \times N$ stochastic matrices, and put $\mathbf{U} = \{u_{ij}\} = \mathbf{P_2 P_1}$. From (7.4)

$$\max_{h,h'} |u_{hj} - u_{h'j}| \leq \tau_1(\mathbf{P_2}) \max_{i,i'} |p_{ij}(1) - p_{i'j}(1)|$$

for fixed j, so that

$$(8.3) \qquad \Delta(\mathbf{U}) = \Delta(\mathbf{P_2 P_1}) \leq \tau_1(\mathbf{P_2}) \Delta(\mathbf{P_1}).$$

Now put

$$(8.4) \qquad \mathbf{U_{p,r}} = \{u_{ij}^{(p,r)}\} = \mathbf{H_{p+r}} \cdots \mathbf{H_{p+2}}\,\mathbf{H_{p+1}}$$

where $\{\mathbf{H_i}\}$, $\mathbf{i} \geq 1$ are $N \times N$ stochastic matrices.
Since

$$(8.5) \qquad \mathbf{U_{p,r}} = \mathbf{H_{p+r}}\,\mathbf{U_{p,r-1}}$$

as in Lemma 7.4, as $r \to \infty$:

$$(8.6) \qquad \max_{h} u_{jj}^{(p,r)} \downarrow \overline{u}_j^{(p)},\ \min_{h} u_{hj}^{(p,r)} \uparrow \underline{u}_j^{(p)}.$$

for fixed j, p for some limit quantities $\overline{u}_j^{(p)}$, $\underline{u}_j^{(p)}$. Further, from (8.3) and (8.4)

$$(8.7) \qquad \Delta(\mathbf{U_{p,r}}) \leq \tau_1(\mathbf{H_{p+r}})\,\Delta(\mathbf{U_{p,r-1}})$$

and iterating (8.7)

$$(8.8) \qquad \Delta(\mathbf{U_{p,r}}) \leq \prod_{s=1}^{r} \tau_1(\mathbf{H_{p+s}})(N/2)$$

since $\Delta(\mathbf{I}) = N/2$.

Now, weak ergodicity (for fixed p) is said to obtain if the rows of $\mathbf{U_{p,r}}$ tend to equality as $r \to \infty$; that is, if and only if $\Delta(\mathbf{U_{p,r}}) \to 0$ as $r \to \infty$. From (8.7) and (8.5) we see that weak ergodicity holds for backwards products (8.4) if and only if strong ergodicity (all rows tending to the same probability vector) holds. This result occurs in Chatterjee and Seneta [9] and is discussed in [49], Section 4.6. .

We see that the proof follows very much from the "backward" multiplication structure inherent in Lemma 7.4 .

Further, if we form successive an inhomogeneous matrix products $\mathbf{T_{p,r}}$ stochastic matrices $\{\mathbf{H_i}\}$, $\mathbf{i} \geq 1$, in any order, for fixed $\mathbf{r}$,

$$(8.9) \qquad \Delta(\mathbf{T_{p,r}}) \leq \prod_{\mathbf{s=1}}^{\mathbf{r}} \tau_1(\mathbf{H_{p+s}})(N/2),$$

so a sufficient condition weak ergodicity of the sequence $\mathbf{T_{p,r}}$ is

$$\sum_{\mathbf{s=1}}^{\infty} \{1 - \tau_1(\mathbf{H_{p+s}})\} = \infty,$$

where $\mathbf{H_{p+s}}$, $\mathbf{s} = 1, 2, \cdots, \mathbf{r}$ now simply labels the order of selection of the matrices which go to form $\mathbf{T_{p+r}}$, irrespective of where each new matrix is placed in going from $\mathbf{T_{p,r}}$ to $\mathbf{T_{p,r+1}}$.

Notice that we *have not* used here the submultiplicative property

(8.10) $$\tau_1(\mathbf{P_2}\,\mathbf{P_1}) \leq \tau_1(\mathbf{P_2})\,\tau_1(\mathbf{P_1})$$

for stochastic matrices $\mathbf{P_1}$, $\mathbf{P_2}$. The submultiplicative property is derived, for finite or infinite compatible stochastic matrices, using in effect direct ideas very similar to Markov's contractivity arguments by Isaacson and Madsen [22] as Lemma V.2.3, pp. 143 - 146; and by Iosifescu [21], as Theorem 1.11, pp. 58-59.

We take this opportunity to mention the author's paper [55] which shows that a condition expressed in terms of Birkhoff's coefficient of ergodicity implies a ratio limit property, as well as weak ergodicity, for inhomogeneous products of infinite stochastic matrices. The condition generalizes a classical condition of Kolmogorov. The Birkhoff coefficient $\tau_B(\mathbf{P}) \geq \tau_1(\mathbf{P})$ for a stochastic $\mathbf{P}$ [49], Theorem 3.13.

8.2. Rate of convergence to ergodicity. The Google matrix. Markov's argument embodied in Lemmas 7.3 and 7.4 gives a geometric rate H to equalization of rows as embodied in (7.7), providing $H \equiv \tau_1(\mathbf{P}) < 1$. This is easily extended to the more conventional concept of convergence at geometric rate. For a less direct proof, see the author's [54].

Theorem 8.1. Suppose $\mathbf{P}$ is $(N \times N)$ stochastic, with $H < 1$, and suppose $\pi^T = \{\pi_r\}$ is the common probability distribution to which each row of $\mathbf{P}^n$ converges as $n \to \infty$. Then for fixed r

$$|p_{ir}^{(n)} - \pi_r| \leq H^n,\ n \geq 0.$$

Proof: Since by Lemma 7.4

$$\mathbf{P}^n \to \mathbf{1}\,\pi^T$$

where $\pi \geq \mathbf{0}$, $\pi^T\mathbf{1} = 1$, it follows that $\pi^T\mathbf{P}^n = \pi^T$, $n \geq 0$, so

$$(\mathbf{I} - \mathbf{1}\,\pi^T)\mathbf{P}^n = \mathbf{P}^n - \mathbf{1}\,\pi^T$$

and

$$(\mathbf{I} - \mathbf{1}\,\pi^T)\mathbf{1} = \mathbf{0}.$$

Now

$$\mathbf{I} - \mathbf{1}\,\pi^T = \{\delta_{ij} - \pi_j\}$$

where δ_{ij} is the Kronecker delta. We see that $\sum_j(\delta_{ij} - \pi_j) = 0$. Hence fixing i , the vector $\delta_i = \{\delta_{ij} - \pi_i\}_{j=1}^N$ satisfies $\delta_i^T\mathbf{1} = 0$.

Thus fixing i, r and using (7.1)

$$|p_{ir}^{(n)} - \pi_r| = |\sum_j\{\delta_{ij} - \pi_j\}\,p_{jr}^{(n)}|$$

$$\leq \left(\frac{1}{2}\sum_s|\delta_{is} - \pi_s|\right)\left(\max_s p_{sr}^{(n)} - \min_s p_{sr}^{(n)}\right)$$

$$= \left(\frac{1}{2}\sum_s|\delta_{is} - \pi_s|\right)\max_{k,t}|p_{kr}^{(n)} - p_{tr}^{(n)}|$$

$$\leq \left(\frac{1}{2}\sum_s|\delta_{is} - \pi_s|\right)H^n$$

from Lemma 7.3 ,

$$\leq \frac{1}{2} \left(\sum_s |\delta_{is}| + |\pi_s| \right) H^n$$

$$= \frac{1}{2} 2H^n \leq H^n.$$

This completes the proof. □

The *Google matrix* ([26], p.5) $\mathbf{P}$ is of form

$$\mathbf{P} = \alpha \mathbf{S} + (1 - \alpha) \mathbf{1} \mathbf{v}^T$$

where $0 < \alpha < 1$ and $\mathbf{v}^T > \mathbf{0}^T$ is a probability vector, and both can be arbitrarily chosen. $\mathbf{S}$ is a stochastic matrix. From (8.1)

$$H = \tau_1(\mathbf{P}) = \alpha \tau_1(\mathbf{S}) \leq \alpha$$

since $\tau_1(\mathbf{S}) \leq 1$. By Theorem 8.1, the rate of convergence to the limit distribution vector π^T by the power method is rapid [26], even with a small $(1 - \alpha)$. Using the relations $\pi^T = \pi^T \mathbf{P}$, $\pi^T \mathbf{1} = 1$ it follows immediately that

$$\pi^T = (1 - \alpha) \mathbf{v}^T (\mathbf{I} - \alpha \mathbf{S})^{-1}.$$

These results are not new.

9. Measuring sensitivity under perturbation..

9.1. The setting. Norms and bounds. For $\mathbf{x}^T \in \Re^N$, if $\| \cdot \|$ is a vector norm on $\Re^N$, then the corresponding matrix norm for an $(N \times N)$ matrix $\mathbf{B} = \{b_{ij}\}$ is defined by

$$\| \mathbf{B} \| = \sup\{\mathbf{x}^T, \| \mathbf{x}^T \| = 1 : \| \mathbf{x}^T \mathbf{B} \|\}.$$

We focus on the l_p norms on $\Re^N$, where $\| \mathbf{x}^T \|_p = \left(\sum_{i=1}^N |x_i|^p \right)^{1/p}$ in the cases $p = 1, \infty$, where $\| \mathbf{x}^T \|_\infty = \max_i |x_i|$.

Then

(9.1) $$\| \mathbf{B} \|_1 = \max_i \sum_{j=1}^N |b_{ij}|, \ \| \mathbf{B} \|_\infty = \max_j \sum_{i=1}^N |b_{ij}|.$$

Dobrushin [12] showed that for an $(N \times N)$ stochastic matrix $\mathbf{P}$,

(9.2) $$\tau_1(\mathbf{P}) = \sup\{\delta^T, \| \delta^T \|_1 = 1, \delta^T \mathbf{1} = 0 : \| \delta^T \mathbf{P} \|_1\}$$

in the context of a Central Limit Theorem for non-homogeneous Markov chains (inhomogeneous products of stochastic matrices). The Central Limit Theorem direction was Markov's main concern, but, as we have noted, he never seems to have used the coefficient $\tau_1(\mathbf{P})$ in this setting. The submultiplicative property (8.10) follows trivially from (9.2).

Extending (9.2) to any $(N \times N)$ matrix $\mathbf{B} = \{b_{ij}\}$ we may define a Markov-Dobrushin-type coefficient of ergodicity more generally by

$$(9.3) \qquad \tau_1(\mathbf{B}) = \sup\{\delta^T, \| \delta^T \|_1 = 1, \ \delta^T \mathbf{1} = 0 : \ \| \delta^T \mathbf{B} \|_1\}$$

whence [51]

$$(9.4) \qquad \tau_1(\mathbf{B}) = \frac{1}{2} \max_{i,j} \sum_{s=1}^{N} |b_{is} - b_{js}|.$$

Suppose $\mathbf{P} = \{p_{ij}\}$ is an $N \times N$ stochastic matrix containing a single irreducible set of indices, so that there is a unique stationary distribution vector $\pi^T = \{\pi_i\}$, $[\pi^T(\mathbf{I} - \mathbf{P}) = \mathbf{0}^T, \pi^T \mathbf{1} = 1]$. Let $\overline{\mathbf{P}}$ be any other $(N \times N)$ stochastic matrix with this structure (the irreducible sets need not coincide), and $\overline{\pi}^T = \{\overline{\pi}_i\}$ its unique stationary distribution vector. Under the assumption on $\mathbf{P}$ the corresponding *fundamental matrix* [23] $\mathbf{Z}$ exists, where $\mathbf{Z} = (I - \mathbf{P} + \mathbf{1}\pi^T)^{-1}$. Set $\mathbf{E} = \{e_{ij}\} = \overline{\mathbf{P}} - \mathbf{P}$. Suppose that there exists an $N \times N$ matrix $\mathbf{C} = \{c_{ij}\}$ such that:

$$(9.5) \qquad \overline{\pi}^T - \pi^T = \overline{\pi}^T \mathbf{E} \mathbf{C}$$

THEOREM 9.1. *Under our prior conditions on P and $\overline{P}$, and assuming (9.5) holds:*

$$(9.6) \qquad \| \overline{\pi}^T - \pi^T \|_1 \leq \tau_1(\mathbf{C}) \| \mathbf{E} \|_1$$

$$(9.7) \qquad \| \overline{\pi}^T - \pi^T \|_\infty \leq \mathcal{T}(\mathbf{C}) \| E \|_1$$

where

$$(9.8) \qquad \mathcal{T}(\mathbf{C}) = \frac{1}{2} \max_j \left(\max_{k,k'} |c_{kj} - c_{k'j}| \right).$$

Proof. (9.6) follows by imitating the last part of the proof of [53] Theorem 2, using (9.2).

For the jth column vector of $\mathbf{C}$ write $\mathbf{C}_{\cdot j}$, and for the kth row of $\mathbf{E}$ write $\mathbf{E}_{k\cdot}^T$.

From (9.5)

$$\overline{\pi}_j - \pi_j = \sum_k \overline{\pi}_k (\mathbf{EC})_{kj}$$

so

$$|\overline{\pi}_j - \pi_j| \leq \max_k |(\mathbf{EC})_{kj}|$$
$$= \max_k |\mathbf{E}_{k\cdot}^T \mathbf{C}_{\cdot j}|$$

so by Markov's Contraction Inequality (7.1) since $\mathbf{E}_k^T \mathbf{1} = 0$,

$$\leq \max_k \left(\max_{h,h'} |c_{hj} - c_{h'j}| \left(\frac{1}{2} \sum_s |e_{ks}| \right) \right)$$

$$= \frac{1}{2} \left(\max_{h,h'} |c_{hj} - c_{h'j}| \left(\max_k \sum_s |e_{ks}| \right) \right)$$

so

(9.9) $$\qquad |\bar{\pi}_j - \pi_j| \leq \frac{1}{2} \left(\max_{h,h'} |c_{hj} - c_{h'j}| \right) \| \mathbf{E} \|_1$$

where

$$\| \mathbf{E} \|_1 = \max_k \sum_s |e_{ks}|.$$

Thus from (9.9)

(9.10) $$\qquad \max_j |\pi_j - \pi_j| = \| \bar{\pi}^T - \pi^T \|_\infty \leq \mathcal{T}(\mathbf{C}) \| E \|_1.$$

This completes the argument.□

The result (9.6) was obtained by the author [53] Theorem 2, in the case $\mathbf{C} = \mathbf{C}(\mathbf{u}, \mathbf{v})$ where:

(9.11) $$\begin{aligned} \mathbf{C}(\mathbf{u}, \mathbf{v}) &= (\mathbf{I} - \mathbf{P} + \mathbf{1}\mathbf{u}^T)^{-1} - \mathbf{1}\mathbf{v}^T \\ &= (\mathbf{Z}^{-1} + \mathbf{1}(\mathbf{u} - \pi)^T)^{-1} - \mathbf{1}\mathbf{v}^T \\ &= \mathbf{Z} - \frac{\mathbf{1}(\mathbf{u} - \pi)^T \mathbf{Z}}{\mathbf{u}^T \mathbf{1}} - \mathbf{1}\mathbf{v}^T \end{aligned}$$

since $\mathbf{Z}\mathbf{1} = \mathbf{1}$, for any (real) $\mathbf{v}$, and any (real $\mathbf{u}$) such that $\mathbf{u}^T \mathbf{1} \neq 0$, using Bartlett's Identity. Notice that $\mathbf{C}(\pi, \pi) = \mathbf{A}^\sharp = \mathbf{Z} - \mathbf{1}\pi^T$, the *group generalized inverse* [39] $\mathbf{A}^\sharp$ of $\mathbf{A} = \mathbf{I} - \mathbf{P}$; while $\mathbf{C}(\pi, \mathbf{0}) = \mathbf{Z}$.

In fact it is shown in [53] that $\tau_1(\mathbf{C}(\mathbf{u}, \mathbf{v})) = \tau_1(\mathbf{A}^\sharp) = \tau_1(\mathbf{Z})$.

The steps in the proof of (9.7) are due to Kirkland, Neumann and Shader [24], Theorem 2.2, in the case $\mathbf{C} = \mathbf{A}^\sharp$, The point of imitating the steps here is primarily to show that, in the guise of "Lemma PS", Markov's Contraction Inequality is the central ingredient.

Cho and Meyer [11] have shown that the bound on the right of (9.7) in a different guise also occurs in 1984 in Haviv and Van der Heyden [17], and later, in Cho and Meyer [10]. Haviv and Van der Heyden [17] also used Lemma 7.1, and Hunter [20], in the proof of his Theorem 3.2, ascribes the result to these authors.

9.2. Measuring sensitivity. The relative effect on π^T of the perturbation $\mathbf{E}$ to $\mathbf{P}$ is measured in a natural way by the quantity

(9.12) $$\frac{\| \bar{\pi}^T - \pi^T \| / \| \pi^T \|}{\| \mathbf{E} \| / \| \mathbf{P} \|}.$$

From (9.6), using $\| \cdot \|_1$, and taking $\mathbf{C} = \mathbf{A}^\sharp$ in (9.5), we see since $\| \mathbf{P} \|_1 = 1$ that (9.12) $\leq \tau_1(\mathbf{A}^\sharp)$, so $\tau_1(\mathbf{A}^\sharp)$ is a natural condition number to measure the relative sensitivity of π^T to perturbation of $\mathbf{P}$.

The foundation paper from which sensitivity theory developed is [47].

Cho and Meyer [11] survey various condition numbers $\kappa_l, l = 1, \cdots, 8$ which have occurred in the literature which satisfy

$$\| \bar{\pi}^T - \pi^T \|_p \leq \kappa_l \| \mathbf{E} \|_q$$

where $(p, q) = (1, 1)$ or $(\infty, 1)$ depending on l. In this sense, in particular from their Theorem 3 for $(p, q) = (1, 1)$, $\kappa_6 = \tau_1(\mathbf{A}^\sharp)$, and $\kappa_3 = \kappa_8 = \mathcal{T}(\mathbf{A}^\sharp)$ are condition numbers.

However, one might argue that, inasmuch as a condition number should bound (9.12), the same norm should be used for numerator and denominator of the left-hand side of (9.12). This is not the case in expressing (9.7) in form (9.12).

In their Remark 4.1, Cho and Meyer [11], p.148, point out, in order to obtain a fair comparison between the bounding tightness of $\kappa_6 = \tau_1(\mathbf{A}^\sharp)$ and $\kappa_3 \equiv \kappa_8 = \mathcal{T}(\mathbf{A}^\sharp)$ that $(\bar{\pi}^T - \pi^T)\mathbf{1} = 0$, so $\| \bar{\pi}^T - \pi^T \|_\infty \leq (1/2)\| \bar{\pi}^T - \pi \|_1$. Hence

$$\kappa_3 = \mathcal{T}(\mathbf{A}^\sharp) \leq \frac{1}{2}\tau_1(\mathbf{A}^\sharp) = \frac{1}{2}\kappa_6,$$

from which they conclude that κ_3 is the tighter condition number.

However, one might argue that from (9.9) which underlies (9.7), that

$$\sum_j |\bar{\pi}_j - \pi_j| \leq \frac{1}{2} \sum_j \left(\max_{h,h'} |c_{hj} - c_{h'j}| \right) \max_k \| \mathbf{E}_k^T \|_1,$$

and since from (9.4)

$$\tau_1(\mathbf{C}) \leq \Delta(\mathbf{C}) \equiv \frac{1}{2} \sum_j \left(\max_{h,h'} |c_{hj} - c_{h'j}| \right),$$

defining $\Delta(\mathbf{C})$ in analogy to (8.2), it follows that the consequent bound on $\| \bar{\pi}^T - \pi \|_1$ is not as tight as when using $\tau_1(\mathbf{A}^\sharp)$.

In their role as condition numbers, $\tau_1(\mathbf{A}^\sharp)$ and $\mathcal{T}(\mathbf{A}^\sharp)$ are not really directly comparable in regard to size, since different versions of the norm $\| \bar{\pi}^T - \pi \|$ are involved.

9.3. Recent related results. The discussion of Sections 9.1 and 9.2 has revolved around (9.5). Hunter [19] Theorems 2.1 - 2.2 has obtained this equality by using the general form $\mathbf{G}$ of the g-inverse of $\mathbf{I} - \mathbf{P}$:

(9.13) $$\mathbf{G} = [\mathbf{I} - \mathbf{P} + \mathbf{t}\mathbf{u}^T]^{-1} - \mathbf{1}\mathbf{v}^T + \mathbf{g}\pi^T$$

for any real $\mathbf{t}, \mathbf{v}, \mathbf{u}, \mathbf{g}$ satisfying $\pi^T\mathbf{t} \neq 0$, $\mathbf{u}^T\mathbf{1} \neq 0$, by showing

(9.14) $$\bar{\pi}^T - \pi^T = \bar{\pi}^T\mathbf{E}\mathbf{G}(\mathbf{I} - \mathbf{1}\pi^T)$$

and that $\mathbf{E}\mathbf{C}(\mathbf{u}, \mathbf{v})$ is a special case of $\mathbf{E}\mathbf{G}(\mathbf{I} - \mathbf{1}\pi^T)$.

In fact, we can see immediately from (9.13) that $\mathbf{C}(\mathbf{u}, \mathbf{v})$ itself is a special case of the g-inverse G .

Equation (9.14) leads to a bound of general appearance :

$$(9.15) \qquad \parallel \bar{\pi}^T - \pi^T \parallel_1 \leq \tau_1(\mathbf{G}(\mathbf{I} - \mathbf{1}\pi^T)) \parallel \mathbf{E} \parallel_1.$$

It would seem plausible that $\tau_1(\mathbf{G}(\mathbf{I} - \mathbf{1}\pi^T))$ may give a tighter bound, for some parameter vectors, than $\tau_1(\mathbf{A}^\sharp) = \tau_1(\mathbf{Z})$, but the author has shown, by generalizing the argument in [53] and using (9.3) or (9.4), that in fact all these values are the same, namely $\tau_1(\mathbf{A}^\sharp)$.

In another paper Hunter [20], Corollary 5.1.1, derives the bound

$$(9.16) \qquad \parallel \bar{\pi}^T - \bar{\pi}^T \parallel_1 \leq \text{tr}(\mathbf{A}^\sharp) \parallel \mathbf{E} \parallel_1.$$

This bound is not as strict as (9.6), with $\mathbf{C} = \tau_1(\mathbf{A}^\sharp)$, since [56], p.165, (10), states that

$$(9.17) \qquad \tau_1(\mathbf{A}^\sharp) \leq \text{tr}(\mathbf{A}^\sharp).$$

Work on these issues by J. Hunter and the author is in progress. For the time being, the bound [53]:

$$(9.18) \qquad \parallel \bar{\pi}^T - \pi^T \parallel_1 \leq \tau_1(\mathbf{A}^\sharp) \parallel \mathbf{E} \parallel_1$$

remains sharp for the norm used.

REFERENCES

[1] Yu. A. ALPIN AND N.Z. GABASSOV, *A remark on the problem of localization of the eigenvalues of real matrices (in Russian)*, Izv. Vyssh. Uchebn. Zaved. Matematika, 11(1976), pp. 98–100.

[2] G.P. BASHARIN, A.N. LANGVILLE AND V.A. NAUMOV, *The life and work of A.A. Markov*, Linear Algebra Appl., 386(2004), pp. 3–26.[This was in the Special Issue, pp. 1- 408, on the *Conference on the Numerical Solution of Markov Chains*, 2003.]

[3] J. BERNOULLI, *O Zakone Bol'shikh Chisel (On the Law of Large Numbers)*. Nauka, Moscow, 1986.

[4] S. N. BERNSTEIN, *Sobranie Sochinenii: Tom IV [Collected Works, Vol. IV], Teoriia Veroiatnostei i Matematicheskaia Statistika [1911–1946]*, Nauka, Moscow, 1964.

[5] A.S. BEZIKOVICH [BESICOVITCH], *A. A. Markov's contributions to probability theory (in Russian)*, Izvestiia Rossiiskoi Akademii Nauk, 17(1923), pp. 45–52.

[6] A.S. BEZIKOVICH [BESICOVITCH], *Biograficheskii Ocherk. [Biographical Sketch]*, [36], pp.III – XIV.

[7] H. BRUNS, *Wahrscheinlichkeitsrechnung und Kollektivmasslehre*, Teubner, Leipzig, 1906. [See esp. Achtzehnte Vorlesung].

[8] J.C. BURKILL, *Abram Samoilovitch Besicovitch. 1891-1970*, Biographical Memoirs of Fellows of the Royal Society, 17(1971), pp. 1–16.

[9] S. CHATTERJEE AND E. SENETA, *Towards consensus: some convergence theorems on repeated averaging*, J. Appl. Prob., 14(1977), pp. 89–97.

[10] G.E. CHO AND C.D. MEYER, *Markov chain sensitivity measured by mean first passage time*, Linear Algebra Appl., 316(2000), pp. 21–28.

[11] G.E. CHO AND C.D. MEYER, *Comparison of perturbation bounds for the stationary distribution of a Markov chain*, Linear Algebra Appl., 335(2001), pp. 137–150.

[12] R.L. DOBRUSHIN, *Central limit theorem for non-stationary Markov chains, I and II*, Theory Prob. Appl., 1(1956), pp. 65–80, 329–83 [English translation].

[13] M. FRÉCHET, *Recherches théoriques modernes sur le calcul des probabilités. Secnd livre. Méthode des fonctions arbitraires. Théorie des événements en chaîne dans le cas d'un nombre fini d'états possibles.* Gauthier-Villars, Paris, 1938.

[14] G. FROBENIUS, *Über Matrizen aus positiven Elementen*, S.-B. Preuss. Akad. Wiss.,(Berlin), (1908), pp. 471–476; (1909), pp. 514–518.

[15] G. FROBENIUS, *Über Matrizen aus nicht negativen Elementen*, S.-B. Preuss. Akad. Wiss.,(Berlin), (1912), pp. 456–477.

[16] S. IA. GRODZENSKY, *Andrei Andreevich Markov 1856–1922.* Nauka, Moscow, 1987.

[17] M. HAVIV AND L. VAN DER HEYDEN, *Perturbation bounds for the stationary probabilities of a finite Markov chain*, Adv. Appl. Prob., 16(1984), pp. 804–818.

[18] C.C. HEYDE AND E. SENETA, EDS., *Statisticians of the Centuries*, Springer, New York, 2001.

[19] J.J. HUNTER, *Stationary distributions and mean first passage times of perturbed Markov chains*, Linear Algebra Appl., 410(2005), pp. 217–243.

[20] J.J. HUNTER, *Mixing times with applications to perturbed Markov chains*, LInear Algebra Appl., (2006) to appear.

[21] M. IOSIFESCU, *Finite Markov Processes and Their Applications*, Wiley, Chichester; Editura Tehnica, Bucharest, 1980.

[22] D.L. ISAACSON AND R.W. MADSEN, *Markov Chains*, Wiley, New York, 1976.

[23] J.G. KEMENY AND J.L. SNELL, *Finite Markov Chain*, Van Nostrand, Princeton N.J., 1960.

[24] S.J. KIRKLAND. M. NEUMANN AND B.L. SHADER, *Applications of Paz's inequality to perturbation bounds for Markov chains*, Linear Algebra Appl., 268(1998), pp. 183–196.

[25] A.N. KOLMOGOROV AND A. P. YUSHKEVICH, EDS., *Mathematics in the 19th Century; Mathematical Logic, Algebra, Number Theory, Probability Theory*, Birkhäuser, Basel, Boston, 1992.

[26] A.N. LANGVILLE AND C.D. MEYER, *The use of linear algebra by web search engines*, IMAGE, 33(2004), pp. 2–5.

[27] A.A. MARKOV, *Extension of the law of large numbers to dependent quantities (in Russian)*, Izvestiia Fiz.-Matem. Obsch. Kazan Univ., (2nd Ser.), 15(1906), pp. 135–156 [Also [37], pp. 339–61].

[28] A.A. MARKOV, *The extension of the law of large numbers onto quantities depending on each other*, [62], pp. 143–158. Translation into English of [27].

[29] A.A. MARKOV, *Investigation of a notable instance of dependent trials (in Russian)*, Izvestiia Akad. Nauk (St. Petersburg), (6th Ser.), 1(3)(1907), pp.61-80. [Also in [37], pp. 363–397.]

[30] A.A. MARKOV, *Extension of limit theorems of the calculus of probabilities to sums of quantities associated into a chain(in Russian)*, Zapiski Akad. Nauk (St. Petersburg), Fiz.-Mat. Otd., (7th Ser.). 22(9)(1908), 29 pp. [Also [37], pp. 363–397.]

[31] A.A. MARKOV, *The extension of the limit theorems of the calculus of probability onto a sum of magnitudes connected into a chain*, [62], pp. 159–178. Transalation into English of [30].

[32] A.A. MARKOV, *Investigation of the general case of trials associated into a chain [in Russian]*, Zapiski. Akad. Nauk (St. Petersburg) Fiz.-Matem. Otd., (8th Ser.) 25(3)(1910), 33 pp. [Also in [37]. pp. 465–507.]

[33] A.A. MARKOV, *An investigation of the general case of trials connected into a chain*, [62], pp. 181–203. Translation into English of [32] .

[34] A.A. MARKOV, *On dependent trials not forming a true chain (in Russian)*, Izvestiia Akad. Nauk (St. Petersburg), (6th Ser.), 5(2)(1911), pp.113—126. [Also in [37], pp. 399–416.]

[35] A.A. MARKOV, *On a problem of Laplace (in Russian)*, Izvestiia Akad. Nauk (St. Petersburg), (6th Ser.), 9(2)(1915), pp. 87–104. [Also in [37], pp. 549–571.]

[36] A.A. MARKOV, *Ischislenie Veroiatnostei (Calculus of Probabilities)*, Fourth (posthumous) edn., Gosudarstvennoie Izdatelstvo, Moscow, 1924.

[37] A.A. MARKOV, *Izbrannie Trudy (Selected Works)*, A.N.S.S.S.R., Leningrad, 1951.

[38] A.A. MARKOV [JR.], *The Biography of A. A. Markov.* In [62], pp. 242-256. Translation into English of [37], pp. 599–613.

[39] C.D. MEYER, *The role of the generalized inverse in the theory of finite Markov chains*, SIAM Rev., 17(1975), pp. 443–464.

[40] R. VON MISES, *Wahscheinlichkeitsrechnung*, Fr. Deuticke, Leipzig and Vienna,1931.

[41] KH.O. ONDAR (ED.), *The Correspondence Between A.A. Markov and A.A. Chuprov on the Theory of Probability and Mathematical Statistics*, [Translated by C. & M. Stein], Springer, New York. 1981.

[42] A. PAZ, *Introduction to Probabilistic Automata*, Academic Press, New York,1971.

[43] U. Paŭn, *A class of ergodicity coefficients, and applications*, Math. Reports (Bucharest), 4(2002), pp. 225–232.

[44] O. Perron, *Zur Theorie der Matrices*, Math. Ann., 64(1907), pp. 248–263.

[45] V.I. Romanovsky, *Diskretnie Tsepi Markova*, Gostekhizdat, Moscow, 1949. [English translation by E. Seneta, *Discrete Markov Chains*, Wolters-Noordhoff, Groningen, 1970].

[46] H. Schneider, *The concepts of irreducuibility and full indecomposability of a matrix in the works of Frobenius, König and Markov*, Linear Algebra Appl., 18(1977), pp. 139–162.

[47] P.J. Schweitzer, *Perturbation theory and finite Markov chains*, J. Appl. Prob., 5(1968), pp. 401–413.

[48] E. Seneta, *Coefficients of ergodicity: structure and applications.* Adv. Appl. Prob., 11(1979), pp.576–590.

[49] E. Seneta, *Non-negative Matrices and Markov Chains*, Second edn., Springer, New York, 1981, 2006(revised printing).

[50] E. Seneta, *The central limit problem and linear least squares in pre-revolutionary Russia: The background*, Mathematical Scientist, 9(1984), pp. 37–77.

[51] E. Seneta, *Explicit forms for ergodicity coefficients and spectrum localization*, Linear Algebra. Appl., 60(1984), pp. 187–197.

[52] E. Seneta, *Perturbation of the stationary distribution measured by ergodicity coefficients*, Adv. Appl. Prob., 20(1988), pp. 228–230.

[53] E. Seneta, *Sensitivity analysis, ergodicity coefficients, and rank-one updates for finite Markov chains.* In [63], pp. 121–129.

[54] E. Seneta, *Applications of ergodicity coefficients to homogeneous Markov chains*, Contemporary Mathematics, 149(1993), pp. 189–199.

[55] E. Seneta, *Ergodicity for products of infinite stochastic matrices*, J. Theoret. Probab., 6(1993), pp. 345–362.

[56] E. Seneta, *Sensitivity of finite Markov chains under perturbation*, Statist. Probab. Lett., 17(1993), pp. 163 – 168.

[57] E. Seneta, *Markov and the birth of chain dependence theory*, International Statistical Review, 64(1996), pp. 255–263.

[58] E. Seneta, *Complementation in stochastic matrices and the GTH algorithm*, SIAM J. Matrix Anal. Appl., 19(1998), pp. 556–563.

[59] E. Seneta, *Andrei Andreevich Markov*, In [18], pp.243–247.

[60] E. Seneta, *Statistical regularity and free will: L.A.J. Quetelet and P.A. Nekrasov.* International Statistical Review, 71(2003), pp. 319–334.

[61] O.B.Sheynin, *A.A. Markov's work on probability.* Archive Hist. Exact Sci., 39(1989), pp. 337–77.

[62] O.B.Sheynin, ED., *Probability and Statistics. Russian Papers.*, Selected and Translated by Oscar Sheynin. NG Verlag, Berlin, 2004.

[63] W.J. Stewart, ED., *Numerical Solution of Markov Chains*, Marcel Dekker, New York, 1991.

[64] C.P. Tan. *Coefficients of ergodicity with respect to vector norms.* J. Appl. Prob., 20(1983), pp. 277–287.

[65] J.V. Uspensky, *A sketch of the scientific work of A. A. Markov (in Russian)*, Izvestiia Rossiskoi Akademii Nauk, 17(1923), pp. 19–34.

[66] J.V. Uspensky, *Introduction to Mathematical Probability*, McGraw-Hill, New York, 1937.

A KRYLOV-BASED FINITE STATE PROJECTION ALGORITHM FOR SOLVING THE CHEMICAL MASTER EQUATION ARISING IN THE DISCRETE MODELLING OF BIOLOGICAL SYSTEMS

KEVIN BURRAGE[‡¶], MARKUS HEGLAND[§¶], SHEV MACNAMARA[‡], AND ROGER B. SIDJE[‡†]

Abstract. Biochemical reactions underlying genetic regulation are often modelled as a continuous-time, discrete-state, Markov process, and the evolution of the associated probability density is described by the so-called chemical master equation (CME). However the CME is typically difficult to solve, since the state-space involved can be very large or even countably infinite. Recently a finite state projection method (FSP) that truncates the state-space was suggested and shown to be effective in an example of a model of the Pap-pili epigenetic switch. However in this example, both the model and the final time at which the solution was computed, were relatively small. Presented here is a Krylov FSP algorithm based on a combination of state-space truncation and inexact matrix-vector product routines. This allows larger-scale models to be studied and solutions for larger final times to be computed in a realistic execution time. Additionally the new method computes the solution at intermediate times at virtually no extra cost, since it is derived from Krylov-type methods for computing matrix exponentials. For the purpose of comparison the new algorithm is applied to the model of the Pap-pili epigenetic switch, where the original FSP was first demonstrated. Also the method is applied to a more sophisticated model of regulated transcription. Numerical results indicate that the new approach is significantly faster and extendable to larger biological models.

1. Introduction. Computational and mathematical models of cellular processes promise great benefits in important fields such as molecular biology and medicine. Increasingly researchers are feeling the need to incorporate the fundamentally discrete and stochastic nature of the biochemical processes into the mathematical models that are intended to represent them [1, 3]. This has led to the formulation of models for genetic networks as continuous-time, discrete state, Markov processes [22, 8], giving rise to the so-called chemical master equation (CME) that governs the evolution of the associated probability density. While promising many insights, the CME is difficult to solve, especially as the dimension of the model grows.

The size of the model is closely related to the possible numbers of each species in it. If the biological system being modelled had N species, all of which were known to be bounded in number by s, then the size of the model – as measured by the number of states – would be bounded by s^N. This rough estimate gives an indication of how the size of the problem grows. Typical models of interest may have s about a few thousand and N a few dozen. This explosion in size is known as the *curse of dimensionality*.

It should be noted furthermore that, for many kinds of macromolecules in a cell there is no convenient *a priori* bound s on the number of copies that may be produced, so in principle at least, the corresponding models must be capable of handling a countably infinite set of states. In reality of course a cell may only ever consist of a finite number of molecules but the implication is that the state-space of the models can be extremely large. For example, a crude bound for the number of proteins or other species in the cell may be obtained by considering bounds on the cell's mass.

[†]Corresponding author.

[‡]Advanced Computational Modelling Centre, Department of Mathematics, University of Queensland, Brisbane, QLD 4072, Australia. {kb, shev, rbs}@maths.uq.edu.au.

[§]Centre for Mathematics and its Application, Mathematical Sciences Institute, Australian National University, Canberra, ACT 0200, Australia. markus.hegland@anu.edu.au.

[¶]ARC Centre in Bioinformatics, University of Queensland, Brisbane, QLD 4072, Australia.

It is estimated that a bacterial cell weighs about $1 \times 10^{-12} g$ [6], and that about 15% of this is attributable to proteins [12]. If a protein being modelled weighed about 1×10^6 daltons (1 dalton = $1.660\,538\,73 \times 10^{-24} g$ — this unit equals 1 divided by Avogadro's number), then a simple calculation shows that the cell could contain $s \approx 90\,000$ proteins. This simplified calculation is based on estimates that will vary, and does not take into account, for example, that a cell contains many different kinds of proteins. It also does not cap the number of copies of a particular kind of protein in a cell but it does show that they can be potentially quite large. In contrast some key molecules exist in small numbers. It is known, for example, that when a λ-phage virus has infected an *Escherichia coli* (*E.coli*) bacterial cell, it will remain in the dormant phase while there are around 200 molecules or more per cell of the lambda repressor protein, but that if this level drops to around, say, 50 copies per cell, then the phage will enter lysis, replicating itself many times before lysing its host cell and searching for new hosts to infect [14, 12]. These observations illustrate that although the numbers of species present will be finite, the bounds may vary a lot and are not known in advance, so models must be very large and potentially countably infinite, creating a demand for numerical methods capable of handling them. The λ-phage decision pathway also provides one example (of many) in biology where small numbers of molecules are crucial to cell dynamics, further emphasising the need for discrete and stochastic methods to model them.

Mainly because the chemical master equation is so difficult to solve for such large models, many studies resort to simulation, most notably the Gillespie stochastic simulation algorithm (SSA) [7] is used. The SSA allows the simulation of a continuous-time, discrete state, Markov chain. It is often applied to models of genetic regulatory networks and so gives detailed information about possible paths through the chain as a sequence of biochemical reactions. The SSA and related variants such as the τ-leap method [9, 20] are prized when such detailed path information is of interest. In a reasonably detailed model of the biochemical processes involved in the previously mentioned infection of an *E.coli* bacterial cell by the λ-phage virus, there may be many dozens of species to be tracked, perhaps each varying in numbers from only a few to as much as a few thousand, so that including every possible configuration would require an enormous state-space. Presently the SSA may be used to study such a model [2, 21], whereas solving the corresponding CME lies out of reach computationally. However some important questions are more easily answered from the CME approach. What fraction of infected cells will lyse after a given time period, according to the λ-phage model, for instance? If the CME were tractable this question would be easily and accurately answered. Instead one resorts to a kind of sampling procedure where the SSA must be run many times, and this results in only an estimate with an associated confidence interval. Such examples emphasise the need for the ongoing research for scalable methods for biological applications.

The rest of this paper is structured into five sections. First, the way that a chemical system is modelled as a Markov process will be explained, from which a description of how the chemical master equation arises naturally follows. The article then focuses on methods for the numerical solution of this equation and briefly reviews a recently suggested CME solver known as the Finite State Projection algorithm. An improved version of this method is then suggested and described. Numerical results comparing the two methods and indicating the utility of the new approach are then presented. This section includes application to three models arising in systems biology. Finally a brief discussion is given, identifying potential areas for future work.

2. Stochastic models for biochemical systems.

Some biochemical systems have been successfully modelled by ordinary differential equations that represent the state by the concentration of each chemical species, and that take rate constants as parameters [5, p.15]. Such ODE models are not always appropriate because by their nature they are deterministic and continuous, while chemical systems are not. In reality there can only be an integer number of copies of each kind of chemical species so chemical systems are discrete, but the continuous approximation is a good one when all species are present in large numbers. However when there are small numbers of some molecular species, as is the case for many key regulatory elements in biological systems (recall the λ-phage example mentioned earlier), the continuous approximation breaks down and fluctuations due to apparent randomness in the system play a critical role in determining its behaviour. Thus a discrete and stochastic framework, appropriate for modelling such systems, has been developed and this is briefly described below. It has been observed that the mean behaviour of this underlying stochastic process is what would typically be captured by the ODE model, so the two frameworks are consistent in this sense.

2.1. Chemical kinetics as Markov processes.

Based on strong theoretical grounds it has been shown that a chemically reacting system that is both well-stirred and at thermal equilibrium may be modelled by a Markov process and this has been found to be in good agreement with experimental evidence. A formal derivation of this result may be found in the seminal work by Gillespie [8], for example. The evolution of the probability density associated with the Markov process is governed by the chemical master equation. Although originally developed to model simple chemical systems, the same framework has since been successfully applied to biochemical systems [2], despite the fact that some of the underlying assumptions are no longer strictly satisfied – for example in a living cell, the mixture is far from being perfectly homogeneous.

In this paper a biochemical system consists of $N \geq 1$ different kinds of chemical species $\{S_1, \ldots, S_N\}$, interacting via $M \geq 1$ possible chemical reaction channels $\{R_1, \ldots, R_M\}$. It is assumed that the mixture is homogeneous and that it is at thermal equilibrium. Additionally, it is assumed that the volume of the cell, where the reactions are taking place, is fixed. Relaxing this assumption can be accommodated into the mathematical framework, but it is more difficult computationally.

The system is modelled as a temporally homogeneous, continuous-time, discrete-state, Markov process. The state of the system is defined by the number of copies of each different kind of chemical species. Thus the state $\boldsymbol{x} \equiv (x_1, \ldots, x_N)^T$ is a vector of non-negative integers (negative populations are not physically meaningful and so not part of the models here) where x_i is the number of copies of species S_i. Note that the state depends on time but this is suppressed in the notation for the moment. The length of the vector is equal to the number of different kinds of species that may react in the system and this may be thought of as the dimension of the system. Each possible configuration of the system defines a distinct vector and so must be interpreted as a state in the Markov chain, thus defining the state-space. Hence the state-space for this class of models may always be identified with just the N-dimensional lattice of points with non-negative integer coordinates. It should now be evident where the estimate, made in the introduction, of the model size came from.

Transitions between states occur when (and only when) a reaction occurs. Typically there will be many kinds of reactions (for example transcription and translation) that can occur in the system. Associated with each reaction R_j is a *stoichiometric vector* $\boldsymbol{\nu}_j$, of the same dimension as the state vector, that defines the way the state changes when the reaction occurs; if the system is in state x and reaction j occurs, then the system transitions to state $x + \nu_j$. The components of the stoichiometric vector are integers, recording the increase or decrease in the number of copies of each species after the associated reaction occurs. So $\nu_j \equiv (\nu_{1j}, \ldots, \nu_{Nj})^T$ where ν_{ij} records the amount by which species i changes when reaction j occurs. For example, suppose that three chemical species in the system, S_u, S_v and S_w undergo the chemical reaction

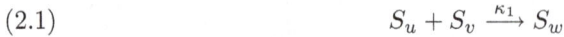

$$(2.1) \qquad\qquad S_u + S_v \xrightarrow{\kappa_1} S_w$$

The associated stoichiometric vector has -1 in the uth and v^{th} components, and $+1$ in the wth component and is zero elsewhere. In general, for a chemical reaction channel of the form

$$(2.2) \qquad\qquad c_1 S_1 + \ldots c_N S_N \xrightarrow{\kappa} b_1 S_1 + \ldots b_N S_N$$

where κ is the reaction rate constant and the c_i and b_i are, respectively, the coefficients of the *reactants* and *products*, the associated stiochiometric vector is $\nu = (b_1 - c_1, \ldots, b_N - c_N)^T$.

Typically in biological applications only a handful of stiochiometric forms arise and in particular, the following forms appear in the examples given in this paper. As well as the reaction (2.1) just mentioned there is decay,

$$(2.3) \qquad\qquad S_u \xrightarrow{\kappa_2} \emptyset$$

where an instance of a species S_u is lost. The associated stoichiometric vector $\nu = (0, \ldots, 0, -1, 0, \ldots, 0)^T$ has -1 in the uth component and is zero elsewhere. Dimerization is another common reaction in which two molecules of the same species, called monomers in this context, combine to form a dimer. The chemical reaction is written as

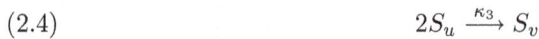

$$(2.4) \qquad\qquad 2S_u \xrightarrow{\kappa_3} S_v$$

where species S_u is the monomer and S_v is the dimer. The associated stoichiometric vector $\nu = (0, \ldots, 0, -2, 0, \ldots, 0, +1, 0, \ldots, 0)^T$ has -2 in the uth component, and $+1$ in the vth component and is zero elsewhere.

It remains only to specify the probability of different kinds of transitions between states, i.e. the relative likelihood of the various chemical reactions. Associated with each state is a set of M *propensities*, $\alpha_1(x), \ldots, \alpha_M(x)$ that determine the relative chance of each reaction channel occurring if the system is in state x. The propensities are defined by the requirement that, given $x(t) = x$, $\alpha_j(x)dt$ is the probability of reaction j occurring in the next infinitesimal time interval $[t, t+dt)$, where the dependence on time has now been made explicit. The propensities are state-dependent and have a functional form defined by the stoichiometry of the reaction. For a chemical reaction with the general form (2.2), the propensity function, when in state x, is a product of binomial terms:

$$(2.5) \qquad\qquad \alpha(x) = \kappa \binom{x_1}{c_1} \binom{x_2}{c_2} \ldots \binom{x_N}{c_N}.$$

Note that (2.5) does not depend on the products. For example the chemical reaction (2.1) considered earlier has an associated propensity of the form $\kappa_1 x_u x_v$, when in state

x. The propensity for a decay reaction is proportional to the number of molecules of the species, so that reaction (2.3) has an associated propensity of $\kappa_2 x_u$, when in state x. Dimerization (2.4) has an associated propensity of the form $\kappa_3 \binom{x_u}{2}$. Values of the reaction rate constants κ must be obtained experimentally but because this is difficult they are often not available or not known to sufficient accuracy, so that biological models are often formulated with incomplete data. Note that in general not all reactions will be possible in a particular state. In particular, if one or more of the components of $x + \nu_j$ is negative then reaction j can not occur from x. For example, the state would have to include at least one each of the chemical species S_u and S_v in order for reaction (2.1) above to occur. This leads to the notion of reachability in a Markov chain, which is used as part of the Finite State Projection algorithm, to be described shortly.

2.2. The chemical master equation. Having defined the Markov chain we may now consider the probability density associated with it. Let the probability of being in state x at time t be denoted by $P(x; t)$ (note that this would depend on an initial condition but this dependence has been suppressed here) and consider the way that this changes over time. It can be shown that for each state x, the previous description of the model implies that this probability satisfies the following discrete, parabolic, partial differential equation,

$$\frac{\partial P(x; t)}{\partial t} = \sum_{j=1}^{M} \alpha_j(x - \nu_j) P(x - \nu_j; t) - P(x; t) \sum_{j=1}^{M} \alpha_j(x)$$

where there are M reaction channels and ν_j are the stoichiometric vectors. This is actually what is known as the chemical master equation. It may be written in an equivalent matrix-vector form so that the evolution of the probability density $p(t)$ (which is a vector of probabilities $P(x; t)$, indexed by the states x) is described by a system of linear, constant coefficient, ordinary differential equations:

$$\dot{p}(t) = Ap(t)$$

where the matrix $A = [a_{ij}]$ is populated by the propensities and represents the *infinitesimal generator* of the Markov process as defined in [19], for example. The rows and columns of the matrix are indexed by the states, so the states have now been implicitly enumerated. With this in mind, for $i \neq j$, the a_{ij} entry of the matrix gives the propensity for the system to transition to state i, given that it is in state j. Thus the a_{ij}, for $i \neq j$, are nonnegative. In order to be a valid infinitesimal generator, it is then required that the diagonal terms be defined by $a_{jj} := -\sum_i a_{ij}$, which means the matrix has zero column sum and so probability is conserved. Given an initial density $p(0)$, the solution at time t is the familiar matrix-exponential function:

$$(2.6) \qquad\qquad p(t) = \exp(tA)p(0)$$

where the exponential is usually defined via a convergent Taylor series

$$\exp(tA) = I + \sum_{n=1}^{\infty} \frac{(tA)^n}{n!}.$$

The numerical solution of (2.6), for the special class of matrices arising in biological applications, is thus the focus of this paper.

For the class of models involving only one species (such as the Pap-pili example), the matrix A has the block-tridiagonal form:

$$\begin{pmatrix} S_0 & D_0 & & & & \text{\Large 0} \\ P_0 & S_1 & D_1 & & & \\ & P_1 & S_2 & D_2 & & \\ & & P_2 & \ddots & \ddots & \\ \text{\Large 0} & & & & \ddots & \ddots \end{pmatrix}$$

where the S_i blocks along the diagonal correspond to *switching* between different genetic configurations, the D_i blocks above the diagonal correspond to *decay*, and the P_i blocks correspond to *production* of more copies of that species. Usually the entries in the sequence of blocks tend to increase only linearly or quadratically, because as already mentioned only a handful of forms of the propensities are encountered in biological applications.

In the case of multiple species, the matrix has a more complicated structure involving a sum of terms of the form $I \otimes T_i \otimes I$ where all the T_i are block-tridiagonal and the I have different sizes. In the terminology of stochastic processes the production of species is referred to as a birth process and the decay as a death process.

In fact the structure of the matrix above for the simple case of one species bears some resemblance to a matrix representing a quasi-birth-death (QBD) process [11, 19] that is usually formulated with constant blocks along the diagonals. However results given there are not directly applicable here because the blocks are not constant along the diagonals. Moreover these results are not generalizable to the case of multiple species.

3. The FSP Algorithm. Due to the fact that the models may involve huge state-spaces and in principle at least, may even involve a potentially countably infinite set of states, the FSP algorithm [13] proposed to *truncate* the system to a finite subsystem that captures enough of the information in the model while remaining tractable and is justified by the accompanying approximation theorems. This also seems intuitive from a biological point of view since, as already observed, a cell will only ever have finite numbers of species.

In the FSP algorithm therefore, the matrix in (2.6) is replaced by A_k where

$$A = \left(\begin{array}{c|c} A_k & * \\ \hline * & * \end{array} \right)$$

i.e. A_k is a $k \times k$ submatrix of the true operator A. The states indexed by $\{1, \ldots, k\}$ then form the *finite state projection*, which will be denoted by X_k. The FSP algorithm then takes

$$(3.1) \qquad p(t_f) \approx \exp(t_f A_k) p_k(0)$$

which is (2.6) evaluated with this approximate operator at the desired final time t_f. Here we have used the subscript k to denote the truncation just described and note that a similar truncation is applied to the initial distribution. Munsky and Khammash [13] then consider the column sum of this approximate solution,

$$(3.2) \qquad \Gamma_k = \mathbb{1}^T \exp(t_f A_k) p_k(0)$$

where $\mathbb{1} = (1, ..., 1)^T$ with appropriate length. Normally the exact solution (2.6) would be a proper probability vector with unit column sum, however due to the truncation the sum Γ_k may be less than one, because in the approximate system, the probability sum condition is no longer conserved. Munsky and Khammash [13] showed that as k is increased, Γ_k increases too, so that the approximation is gradually improved. Additionally it is shown in [13, Theorem 2.2] that if

$$(3.3) \qquad \qquad \Gamma_k \geq 1 - \epsilon$$

for some prespecified tolerance ϵ, then the approximate solution (3.1) is within ϵ of the true solution (2.6). More precisely, in the component-wise sense (with vectors padded as needed), if $\Gamma_k \geq 1 - \epsilon$ then we have

$$\begin{pmatrix} \exp(t_f \boldsymbol{A}_k)\boldsymbol{p}_k(0) \\ \boldsymbol{0} \end{pmatrix} \leq \boldsymbol{p}(t_f) \leq \begin{pmatrix} \exp(t_f \boldsymbol{A}_k)\boldsymbol{p}_k(0) \\ \boldsymbol{0} \end{pmatrix} + \epsilon \mathbb{1}.$$

This is the basis of the FSP algorithm that is outlined in Algorithm 1. Note that $\boldsymbol{X}_0$ is used for the set of states forming the initial projection, $\boldsymbol{X}_k$ for the projection at the kth step, $\boldsymbol{A}_k$ for the corresponding approximating matrix and $\boldsymbol{p}_k(0)$ for the corresponding approximate initial distribution.

In the original example the state-space projection is expanded simply by increasing k. More generally the FSP allows expanding the states in a way that respects the reachability of the model so that, depending on the way the states are enumerated, this may mean that the principal submatrix of the true operator is not simply the intersection of the first k rows and columns.

ALGORITHM 1: $\mathrm{FSP}(\boldsymbol{A}, \boldsymbol{p}_0(0), t_f, \epsilon, \boldsymbol{X}_0)$

$\boldsymbol{A}_0 := submatrix(\boldsymbol{X}_0)$;
$\Gamma_0 := \mathbb{1}^T \exp(t_f \boldsymbol{A}_0)\boldsymbol{p}_0(0)$;
for $k := 1, 2, ...$ **until** $\Gamma_k \geq 1 - \epsilon$ **do**
$\qquad \boldsymbol{X}_k := expand(\boldsymbol{X}_{k-1})$;
$\qquad \boldsymbol{A}_k := submatrix(\boldsymbol{X}_k)$;
$\qquad \Gamma_k := \mathbb{1}^T \exp(t_f \boldsymbol{A}_k)\boldsymbol{p}_k(0)$;
endfor
return $\exp(t_f \boldsymbol{A}_k)\boldsymbol{p}_k(0)$;

In summary, given a fixed final time t_f and a starting state, the FSP algorithm gradually expands the projection around the initial state, via $N-$step reachability, until the likelihood of leaving during the prescribed time interval becomes arbitrarily small. In other words, expansion of the projection continues until, with suitably high probability, the first exit time is less than the final time. This leads to a large sparse problem that involves a principal submatrix of the full infinitesimal generator $\boldsymbol{A}$, gradually increasing in size, that is ever closer to a valid infinitesimal generator for a Markov process. Some of the columns of this approximating matrix will no longer have zero sum, (the sum will be slightly negative) and this means that in the approximating system, probability is no longer conserved, but gradually lost with increasing time. Empirically it is observed that as the state-space is expanded to satisfy the FSP criteria (3.3), the matrix is converging to a Markov matrix, and the eigenvalue with real component closest to zero, tends to zero. This observation makes a connection between the eigenspectra of the truncated matrix and the FSP stopping criteria.

The original FSP algorithm as put into practice in Munsky and Khammash [13] has some drawbacks. It is implemented via MATLAB's *expm* function, and separately exponentiates a matrix before multiplying it by the initial vector. This is not viable for very large matrices as will typically arise in realistic biological models. Furthermore, one is often interested in the evolution of the probability density over time, and so requires many solutions at intermediate time points as well as at the final time, and the original FSP implementation does not provide for this. Even if a more efficient implementation were used, the algorithm first finds the correct projection size by exponentiating a sequence of matrices with the same final time t_f. This is inefficient compared to the implementation suggested in the next section, where these steps are combined. Also it will be seen that the support changes over time. Closely tracking this support leads to an inexact matrix-vector product that makes the new method far more efficient.

4. The Krylov FSP Algorithm. Repeatedly expanding the finite state projection and computing the associated matrix exponentials is expensive. Understanding where to cut this off is one of the keys to a more efficient algorithm, since overshooting means wasted computation (where a smaller projection would still have done the job with the desired accuracy), whereas undershooting means that the whole process must be repeated, including expanding the state-space and again computing a matrix exponential at the final time t_f. However expansion of the state-space and evaluation of the exponential may be performed concurrently, making the process optimal in the sense that the exponential would have to be evaluated anyway and our new approach would be at least as fast. This is one of the improvements made to the original FSP. The other improvement is the inexact matrix-vector product. We detail each in turn.

4.1. Krylov-based exponential. The Krylov FSP outlined in Algorithm 2 is based around Roger Sidje's Expokit [16, 17], available from [15]. As the truncated matrix is never quite a valid infinitesimal generator for a Markov process, the actual Expokit routines involved are the general ones, namely *dgexpv.f* or *expv.m*. Either flavor converts the problem of exponentiating a large sparse matrix to that of exponentiating a small, dense matrix in the Krylov subspace. The dimension m of the Krylov subspace is typically small, and $m = 40$ was used in this implementation (it may be changed adaptively during the FSP process but this is not considered here). The Krylov approximation to $\exp(\tau A)v$ being used is

$$(4.1) \qquad\qquad \beta V_{m+1} \exp(\tau \overline{\overline{H}}_{m+1}) e_1$$

where β is the 2-norm of v, e_1 is the first unit basis vector, and V_{m+1} and $\overline{H}_{m+1}$ are the orthonormal basis and upper Hessenberg matrix resulting from the well-known Arnoldi process. The smaller exponential $\exp(\tau \overline{H}_{m+1})$ is computed via the irreducible Padé approximation, together with scaling and squaring. Another special feature of the Krylov method is that it is *matrix-free*, i.e., the matrix A (or its submatrices) need not be formed explicitly because the method interacts only through matrix-vector products Av, making it possible to deal with very large problems. These aspects should be kept in mind when reading our description. Refer to [16, 17] for other details.

4.2. Embedded exponential. Rather than simply being a mere substitution of MATLAB's *expm* in the original FSP algorithm with the Krylov-based variant as one may think at first, there is actually a deeper improvement to be stressed in our new solver. Unlike the original FSP algorithm that repeatedly computes $\exp(t_f A_k)p_k(0)$ with the *same* t_f, until A_k is sufficiently large, our new solver uses the embedded scheme (with vectors padded with zeros to be of consistent sizes as appropriate)

$$(4.2) \qquad \boldsymbol{p}(t_f) \approx \exp(\tau_K \boldsymbol{A}_K) \dots \exp(\tau_0 \boldsymbol{A}_0) \boldsymbol{p}(0)$$

where the $\{\tau_k\}$ are step-sizes and

$$t_f = \sum_{k=0}^{K} \tau_k$$

with K denoting the total number of steps needed. Literally, our improved FSP scheme (4.2) is evaluated from right to left, harnessing the built-in step-by-step integration procedure of Expokit, with the special feature that the matrix changes between these *internal* integration steps. So the solution at the final time t_f is arrived at via a sequence of solutions at intermediate times

$$0 \equiv t_0 < t_1 < \dots < t_K < t_{K+1} \equiv t_f, \qquad \tau_k = t_{k+1} - t_k$$

satisfying the recurrence scheme

$$(4.3) \qquad \boldsymbol{p}(t_{k+1}) = \boldsymbol{p}(t_k + \tau_k) = \exp(\tau_k \boldsymbol{A}_k) \boldsymbol{p}(t_k), \quad k = 0, 1, \dots, K.$$

Practically, at each step, the current operand vector is padded as needed and the exponential operator is effectively evaluated with the Krylov approximation (4.1), with V_{m+1} and $\bar{H}_{m+1}$ resulting from the current Arnoldi process for the stage. The code inherits Expokit's automatic step-size control to select step-sizes that achieve numerical accuracy – so that the Krylov approximation is a good enough approximation for $\boldsymbol{p}(t_{k+1})$. However, this does not cater for the probability sum condition (not even an exact evaluation of (4.3) would provide this guarantee). The reason is that if the suggested time-step is too long, the system may evolve into states not yet accounted for in the current finite state projection. Therefore at each stage, the step-size may have to be reduced to keep the potential loss of the probability mass sufficiently small. Thus we further compute Γ_k given in (3.2) as if the present time t_k was actually the final time, and a FSP-like criteria (3.3) is checked. More precisely we require that

$$(4.4) \qquad \Gamma_{k+1} = \mathbb{1}^T \exp(\tau_k \boldsymbol{A}_k) \boldsymbol{p}(t_k) \geq 1 - \epsilon \frac{t_{k+1}}{t_f}.$$

If this fails, then the time-step τ_k is halved and the criteria is re-evaluated. This halving is repeated until the criteria is met. Observe that the ratio of times monotonically increases from zero to one, although perhaps a different control strategy could be experimented with in future work. Note that re-computing Γ_k for a smaller time step is not expensive since it involves only recomputing a small matrix exponential in the Krylov subspace (via the Padé approximation, combined with scaling and squaring) and in particular it does not require the relatively expensive Arnoldi process to be repeated.

There is a converse situation to the above. Indeed it is also possible for a time-step to satisfy the FSP criteria (4.4) but be too big to achieve numerical accuracy. In this case, a normal step-by-step integration can pursue its course without the need to expand the subsystem $\boldsymbol{A}_k$ and incur the associated cost of a larger system – recall from (4.2) that each time-step brings us closer to t_f anyway. Hence a further ingredient in our algorithm is that it only expands the FSP projection if the previous time-step has been halved, i.e., if the previous step did not initially satisfy (4.4). Ideally the projection would be expanded via reachability. In our present implementation we merely double the size of the projection at each expansion. The process continues until the sum of the smaller steps reaches the final time. See Algorithm 2.

ALGORITHM 2: Krylov FSP$(A, p_0, t_f, \epsilon, X_0, m, tol)$

$t_0 := 0$; $\hat{X}_0 := X_0$;

$A_0 := submatrix(X_0)$; $\hat{A}_0 := A_0$;

expandFSP := FALSE ;

for $k := 0, 1, 2, \ldots$ until $t_k = t_f$ do

  $[V_{m+1}, \overline{H}_{m+1}] := Arnoldi\ (\hat{A}_k, p_k, m)$;

  repeat { enforce numerical accuracy.................................. }

    $\tau_k := $ step-size ;

    $p_{k+1} := \beta V_{m+1} \exp(\tau_k \overline{H}_{m+1}) e_1$;

    err := numerical-error-estimate ;

  until err $\leq 1.2\ tol$;

  $\Gamma_{k+1} := \mathbb{1}^T p_{k+1}$;

  while $\Gamma_{k+1} < 1 - \epsilon \frac{t_k + \tau_k}{t_f}$ do { enforce FSP criteria............. }

    expandFSP := TRUE ;

    $\tau_k := \frac{1}{2}\tau_k$;

    $p_{k+1} := \beta V_{m+1} \exp(\tau_k \overline{H}_{m+1}) e_1$;

    $\Gamma_{k+1} := \mathbb{1}^T p_{k+1}$;

  endwhile

  if expandFSP

    $X_{k+1} := expand(X_k)$; $\hat{X}_{k+1} := select(X_{k+1})$;

    $A_{k+1} := submatrix(X_{k+1})$; $\hat{A}_{k+1} := columns(\hat{X}_{k+1})$;

    expandFSP := FALSE ;

  endif

  $t_{k+1} := t_k + \tau_k$;

endfor

return $p(t_k)$

Often information about the system will be required over a prolonged time period. This is a strength of the new algorithm since in a single sweep the information at various time points is generated anyway and so may be recorded at virtually no extra cost.

The original FSP algorithm always expands the projection and gives equal importance to all states, even though some regions may no longer contribute much to the total probability of lying inside the projection. This leads to the next improvement.

4.3. Inexact matrix-vector product. In some ways, an implementation of the original FSP algorithm that used a Krylov subspace projection for the computation of the exponential could be thought of as using an inexact matrix-vector product [4, 18], since a truncated approximation A_k is used instead of the full operator A. In biological applications however, the norm of the difference between the approximation and the true operator may be arbitrarily large, depending on the model. Despite this the FSP approximation performs quite well, as certified by the FSP theorems and experimental evidence. The reason why it works so well is because it captures the support of the distribution. Taking this idea one step further it is natural to track the support even more closely, forming a nested projection at each stage. This nested projection is used to define another submatrix $\hat{A}_k$ derived from A_k (the matrix that would normally be used at the kth step) by zeroing the columns of $\boldsymbol{A}_k$ that correspond to states *not* contained in the nested projection. This second-level submatrix $\hat{\boldsymbol{A}}_k$ is then used to approximate the matrix-vector products required by the Arnoldi process to compute (4.3). That is, the matrix-vector multiplication is effectively approximated as

$$\boldsymbol{A}_k\boldsymbol{v} = \begin{bmatrix} \boldsymbol{a}_1^{(k)} & \boldsymbol{a}_2^{(k)} & \cdots \end{bmatrix} \begin{bmatrix} v_1 \\ v_2 \\ \vdots \end{bmatrix} \approx \hat{\boldsymbol{A}}_k\boldsymbol{v} = \sum_{i \in \hat{\boldsymbol{X}}_k} v_i \boldsymbol{a}_i^{(k)}.$$

Since the support is captured, the inexact matrix-vector product may be quite a good approximation, at least for the first few applications of the operator (i.e., at the beginning of the Arnoldi process). It is seen in [4] that good results may be obtained from Krylov methods using inexact matrix-vector products, especially if the first few applications of the operator were relatively faithful. Further investigation is needed to analyse the type of inexact Krylov method used here and draw connections with the theory in [18].

Overall therefore, our proposed Krylov FSP algorithm simultaneously keeps track of a pair of nested projections, $\boldsymbol{X}_k$ and $\hat{\boldsymbol{X}}_k$, that are updated at the end of each time-step. The larger projection $\boldsymbol{X}_k$ grows monotonically in size and represents the original FSP projection that would be required if the current time were to be the final time. Inside this is a second projection, $\hat{\boldsymbol{X}}_k$, essentially capturing a suitably large proportion of the support of the probability distribution at that instant. This latter projection need not grow monotonically, it may also shrink and indeed the algorithm's striking performance in the second model system in section 5.2 is attributable to this. This can be seen in Figure 3. The support shifts over time but remains confined to a small region. This is the region tracked by $\hat{\boldsymbol{X}}_k$. The smaller this region, the smaller the corresponding submatrix $\hat{\boldsymbol{A}}_k$, making our algorithm highly efficient in such circumstances.

The nested projection is obtained by sorting the solution at the present step in decreasing order and selecting the states of highest probabilities. The underlying idea is that states whose probabilities quickly sum to one are the most relevant. A partial, heuristic sort is acceptable for the purposes here, so the sort routine is not nearly as expensive as, say, a full sort implemented via quick-sort or other well-known methods. We select the first components $\hat{\boldsymbol{p}}_k(t_k)$ that satisfy

$$\hat{\Gamma}_k = \mathbb{1}^T \hat{\boldsymbol{p}}_k(t_k) \geq 1 - \epsilon\sqrt{\frac{t_k}{t_f}}.$$

Owing to (4.4), the above criteria ensures that $\Gamma_k \geq \hat{\Gamma}_k$. All states will otherwise be retained if $\hat{\Gamma}_k$ is not chosen less than Γ_k. In general, the selection at each time-step can be guided by the distribution at the previous time-step as well as the natural geometry and reachability of the model.

It should be noted that there are alternative truncation strategies that preserve the Markov property. One way of doing this is to perturb the diagonal entries of the truncated matrix so as to recover the zero column sum condition. However, attempting to incorporate these strategies in the FSP suffers from a major drawback because they enforce the conservation of probability $\Gamma_k = 1$, and so it is no longer possible to use the error bound (3.3) derived for the FSP as a stopping criterion.

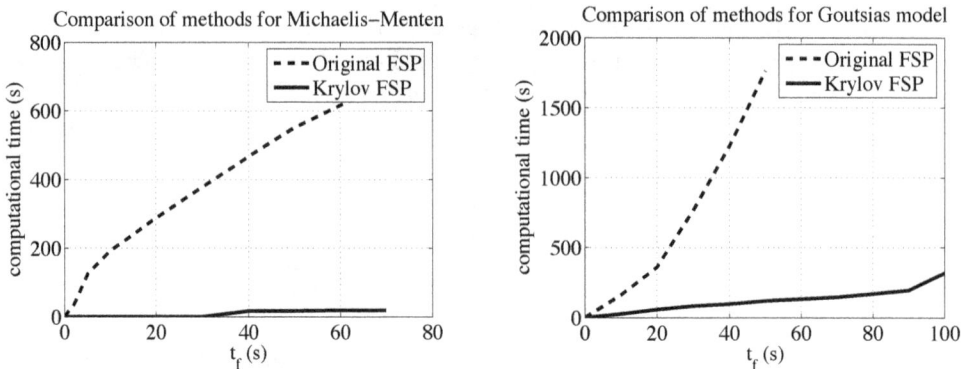

FIG. 1. *Comparison of both methods by computational time required to compute solution to the CME for various final times t_f. Both methods were applied to the Michaelis-Menten model (left) and the Goutsias model (right), described in sections 5.2 and 5.3. The superiority of the Krylov FSP over the original FSP is evident.*

5. Numerical Experiments.
Experiments were conducted on a HPC system hosted at the University of Queensland and running the Linux operating system. The system is a multiprocessor SGI Altix machine known as "Gust", with 16 Intel Itanium 2 CPUs and 30 GB of memory. However, only a single processor execution is used in our experiments, and moreover, the experiments have been attempted on a SUN desktop and completed as well, albeit with longer execution time.

The Krylov FSP algorithm was implemented in FORTRAN and was applied to three model systems and the performance compared with the original FSP, which we also re-implemented in FORTRAN using Expokit for the matrix exponential evaluation. Both codes uses $tol = 1.5 \times 10^{-8}$ for the numerical accuracy tolerance in the matrix exponential approximation and $\epsilon = 10^{-5}$ for the FSP sum criteria. However, given that the Krylov FSP uses inexact matrix-vector products, its numerical results may be slightly less accurate than the original FSP. Note that the largest example dealt with in Munsky and Khammash [13], the Pap-pili model, is small compared to the last example here. Applying this technique to the model of Michaelis-Menten enzyme kinetics (described below) showed that it was possible to compute the solutions much more quickly, see Figure 1.

Note that the results of these computations are probability densities for discrete state processes and so would naturally be displayed as something like a histogram. The figures here are all smooth curves however (think of the tops of the histograms as having been connected) to emphasise the implications of the diffusive nature of the peak for computation. Also the propensity functions defined in the tables here, use the notation [X] to denote the number of copies of chemical species X, so it is a nonnegative integer and *not* a concentration.

5.1. Pap-pili epigenetic switch.
A simplified model of a component of the Pap-pili bacterial gene network provided the original example that the FSP algorithm was applied to. The model consists of ten reactions and six species. The reactions are listed here in Table 1, but see [13, Example 2] for more details. The state of the system includes the number of PapI proteins and the configuration of the genetic element, of which there are only four possibilities. Thus the model is essentially one dimensional. In principle it is countably infinite, as there is no bound on the potential number of

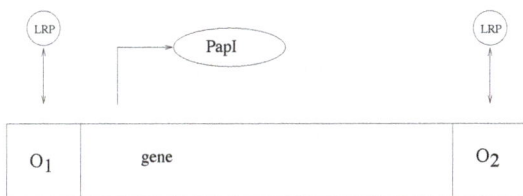

FIG. 2. *Schematic of gene regulatory network modeled in original FSP work. It is intended to represent an extremely simplified component of the Pap-pili epigenetic switch. The Leucine-Responsive Protein (LRP) may bind or unbind at either of the regulatory sites O_1 or O_2, giving a total of four configurations for the genetic element. The gene produces the protein PapI, only in the configurations in which operator O_1 is bound and operator O_2 is empty. (Adapted from Munsky and Khammash [13]).*

PapI proteins. In practice however, the FSP algorithm truncated the corresponding matrix at a side length of only about 124, which indicates that the size of the problem is quite small. Both the original and new methods solve this system quite quickly, in about a second, because it is so small and the final time at which the solution was computed is only $t_f = 10s$.

	reaction	propensity function (s^{-1})
1.	LRP + DNA $\longrightarrow$ LRP.DNA	$\alpha_1 = 1$
2.	LRP.DNA $\longrightarrow$ LRP + DNA	$\alpha_2 = 2.50 - 2.25(\frac{r}{1+r})$
3.	LRP + DNA $\longrightarrow$ DNA.LRP	$\alpha_3 = 1$
4.	DNA.LRP $\longrightarrow$ DNA + LRP	$\alpha_4 = 1.20 - 0.20(\frac{r}{1+r})$
5.	LRP.DNA $\longrightarrow$ LRP.DNA.LRP	$\alpha_5 = 0.01$
6.	LRP.DNA.LRP $\longrightarrow$ LRP.DNA + LRP	$\alpha_6 = 1.20 - 0.20(\frac{r}{1+r})$
7.	DNA.LRP $\longrightarrow$ LRP.DNA.LRP	$\alpha_7 = 0.01$
8.	LRP.DNA.LRP $\longrightarrow$ DNA.LRP + LRP	$\alpha_8 = 2.50 - 2.25(\frac{r}{1+r})$
9.	LRP.DNA $\longrightarrow$ LRP.DNA + r	$\alpha_9 = 10$
10.	r $\longrightarrow \emptyset$	$\alpha_{10} = r$

TABLE 1

Model of the Pap-pili epigenetic switch used in the original FSP. Here r denotes the number of PapI proteins, DNA denotes the DNA template free of LRP, DNA.LRP denotes the genetic element with operator O_1 bound, LRP.DNA denotes the genetic element with operator O_2 bound and LRP.DNA.LRP denotes the configuration with both operators bound. Reaction 1 to 9 can occur only when the genetic element is in the configuration appearing on the left hand side of the reaction. Propensities $2, 4, 6, 8, 10$ depend on the state only through the number of PapI proteins.

5.2. Michaelis-Menten enzyme kinetics. Michaelis-Menten enzyme kinetics involve four species and three chemical reactions. The species are: substrates (S), enzymes (E), enzyme-substrate complexes (ES) and products (P). These interact according to the chemical reactions listed in Table 2. We use reaction rate constants that match those in [10].

Note from reaction 3 in Table 2 that forming a product from the enzyme-substrate complex is an irreversible reaction, so the part of the state-space that is reachable shrinks with time, and it contains an absorbing state corresponding to the situation where all substrates have been converted to products. The state-space of the Markov chain resembles a triangle, as depicted in Figure 4. The examples in [10] have a maximum of 100 substrates, which keeps the state-space quite small. The resulting

reaction	propensity	rate constant (s^{-1})
1. $S + E \xrightarrow{\kappa_1} ES$	$\alpha_1 = \kappa_1 \ [S] \ [E]$	$\kappa_1 = 1.0$
2. $ES \xrightarrow{\kappa_2} E + S$	$\alpha_2 = \kappa_2 \ [ES]$	$\kappa_2 = 1.0$
3. $ES \xrightarrow{\kappa_3} P + E$	$\alpha_3 = \kappa_3 \ [ES]$	$\kappa_3 = 0.1$

TABLE 2

Michaelis-Menten model reactions and propensities. ([X] is the number of copies of the species X.)

matrix is of order 5151 and we take $t_f = 70$. From the numerical results in Figure 1 (left), we see that our Krylov approach is at least an order of magnitude faster than the original FSP implementation.

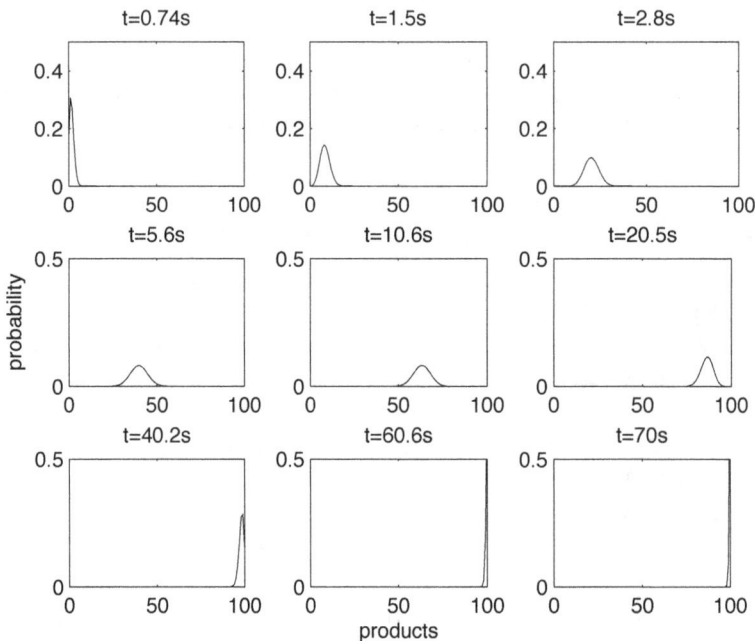

FIG. 3. *Time progression of probability distribution for Michaelis-Menten enzyme kinetics over a period of 70s. Compare to Figure 1(b) in Goutsias [10]. Initially the system is started with 100 substrates and 1000 enzymes and gradually all of the substrate is converted to product. The marginal probability distributions (solutions to CME, summed along the products) at 0.74s, 1.5s, 2.8s, 5.6s, 10.6s, 20.5s, 40.2s, 60.6s and 70s can be seen to move to the right with time. Using the Krylov FSP this took about 20s to compute on a single processor on UQ's Gust machine. The example shows that the support is dynamic and at various times may shrink in size or stay the same size but shift, demonstrating the importance of tracking the projection closely.*

5.3. Goutsias. Goutsias [10] presents a simplified model of transcription regulation, which is about the smallest non-trivial model one can imagine and thus provides a testing ground for CME solvers. It is a simplified model of a subsystem of the λ-phage, comprised of ten reactions, representing auto-regulation through dimerization and repression. The numbers of one kind of mRNA, its protein product as a monomer and also as a dimer are monitored, so the state-space may be thought of

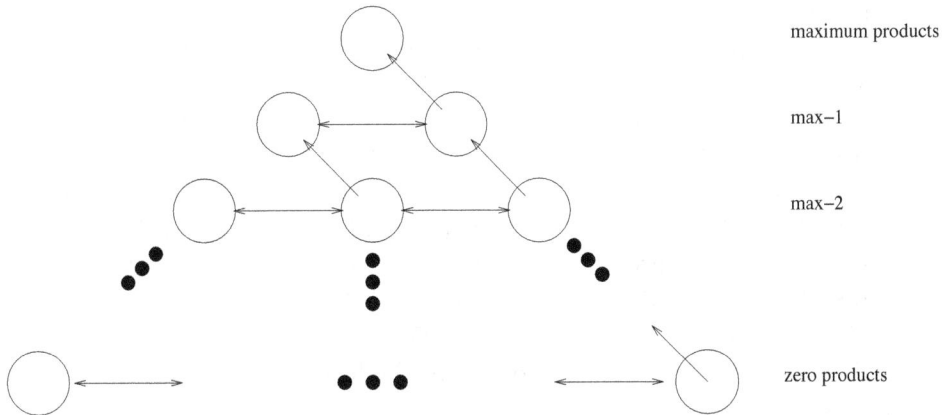

FIG. 4. *Schematic representation of the 'top' portion of the state-space for the Michaelis-Menten enzyme kinetics model. Note that the conversion of a substrate to a product is an irreversible reaction, so the triangle is gradually climbed until the absorbing state at the top is reached. Each layer of the triangle represents the possible combinations of enzymes and enyzme-substrate complexes, for a fixed number of products. This representation assumes that the substrates are outnumbered by the enzymes as in example (b) from Goutsias but note that this situation is always eventually reached.*

as only about 3D. The model also tracks the configuration of the operators but given the initial copy numbers of the DNA there is only a small number of possible configurations for these, e.g. 6 in the example Goutsias gives. Here the propensities (and so the matrix being exponentiated) are fixed, which represents a slight simplification of the model presented by Goutsias where they allow certain propensities to change over time with the cell cycle. Even so the number of configurations of the system explodes and a matrix side length in the millions would be needed to capture a sufficiently large portion of the total probability. The original FSP implementation would be unlikely to handle this in a reasonable time.

Goutsias studies this example by simulation, and here that approach is complemented by solving the associated chemical master equation and so obtaining probability densities. Goutsias uses the model to demonstrate the efficiency of a new simulation method based on quasi-steady state approximations, and it is remarked here that an analogous approximation can be made to reduce the dimension of the problem when solving the CME, however this will be pursued in a later work.

Snapshots of the solution as computed by the Krylov FSP approach are given in Figure 6. The distribution begins with most of the mass close to zero. Then over time Figure 6 shows that the peak appears to be shifting to the right as well as flattening in height. This is a well known phenomena and such a spread of the support is an example of a worse case scenerio for the Krylov FSP approach. Indeed, in this model, the size of the nested projection grows much more quickly than linearly with time. As an indication, beginning with a size of one, after ten seconds it has grown to only about a dozen, but by 100s it has grown to 9 000 states, by 200s it has grown to 40 000 states and by 300s it has grown to 130 000 states. This is of course related to the diffusion characterizing function associated with the Markov model. Although the complexity of the matrix vector product is linear in the size of this projection, the Arnoldi sweeps require this to be repeated many times, depending on the dimension

	reaction	propensity	rate constant (s^{-1})
1.	RNA $\xrightarrow{\kappa_1}$ RNA+M	$\alpha_1 = \kappa_1$ [RNA]	$\kappa_1 = 0.043$
2.	M $\xrightarrow{\kappa_2} \emptyset$	$\alpha_2 = \kappa_2$ [M]	$\kappa_2 = 0.0007$
3.	DNA.D $\xrightarrow{\kappa_3}$ RNA +DNA.D	$\alpha_3 = \kappa_3$ [DNA.D]	$\kappa_3 = 0.078$
4.	RNA $\xrightarrow{\kappa_4} \emptyset$	$\alpha_4 = \kappa_4$ [RNA]	$\kappa_4 = 0.0039$
5.	DNA+D $\xrightarrow{\kappa_5}$ DNA.D	$\alpha_5 = \kappa_5$ [DNA][D]	$\kappa_5 = \frac{0.012 \times 10^9}{AV}$
6.	DNA.D $\xrightarrow{\kappa_6}$ DNA +D	$\alpha_6 = \kappa_6$ [DNA.D]	$\kappa_6 = 0.4791$
7.	DNA.D + D $\xrightarrow{\kappa_7}$ DNA.2D	$\alpha_7 = \kappa_7$ [DNA.D] [D]	$\kappa_7 = \frac{0.00012 \times 10^9}{AV}$
8.	DNA.2D $\xrightarrow{\kappa_8}$ DNA.D +D	$\alpha_8 = \kappa_8$ [DNA.2D]	$\kappa_8 = 0.8765 \times 10^{-11}$
9.	M+M $\xrightarrow{\kappa_9}$ D	$\alpha_9 = \kappa_9$ [M]([M]-1)	$\kappa_9 = \frac{0.05 \times 10^9}{AV}$
10.	D $\xrightarrow{\kappa_{10}}$ M+M	$\alpha_{10} = \kappa_{10}$ [D]([D]-1)	$\kappa_{10} = 0.5$

TABLE 3

Goutsias model of regulated transcription. Avogadro's number is $A = 6.0221415 \times 10^{23}$ and V is the volume of the cell, about $10^{-15} L$.

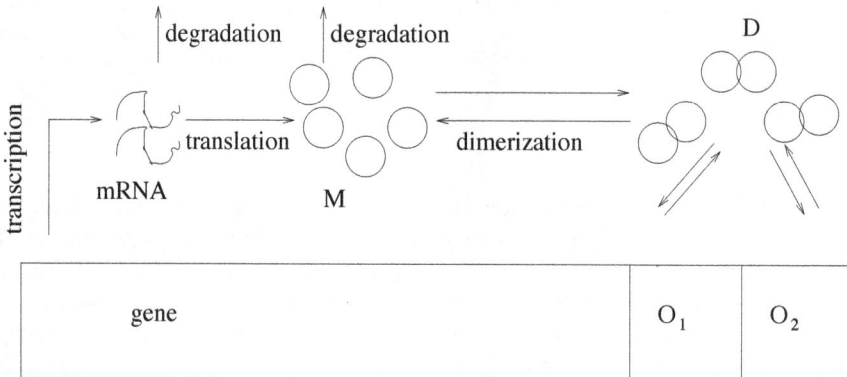

FIG. 5. *Schematic of gene regulatory network modeled in Goutsias. It is intended to represent an extremely simplified component of the λ-phage switch. The monomers may dimerize in a reversible reaction, and the dimers may sequentially bind to the operator sites O_1 and then O_2 in a reversible reaction. The gene produces the mRNA via transcription only when the operator site O_1 is bound and the remaining operator site is empty. Transcription is an irreversible reaction, as is translation and the degradation of either the transcript or the monomer. (Adapted from Goutsias [10])*

of the Krylov subspace and this slows down each time-step by a multiplicative factor, of say, around thirty to forty.

From the numerical results in Figure 1 (right), it is seen that the size of the system grows to unmanageable proportions for the original FSP after 50s. The dimension attains several millions. Our Krylov FSP approach remains an order of magnitude smaller and copes much better with respect to memory issues.

6. Discussion. The Krylov FSP algorithm, although sometimes less accurate, performs much faster than the original FSP and is much more effective in terms of memory management, which is an important step towards larger and more realistic models. Extensive numerical simulations for the Michaelis-Mentem model, in which both accuracy and efficiency considerations were monitored, suggest that our approach is especially well-suited to problems that have one or both of the following properties.

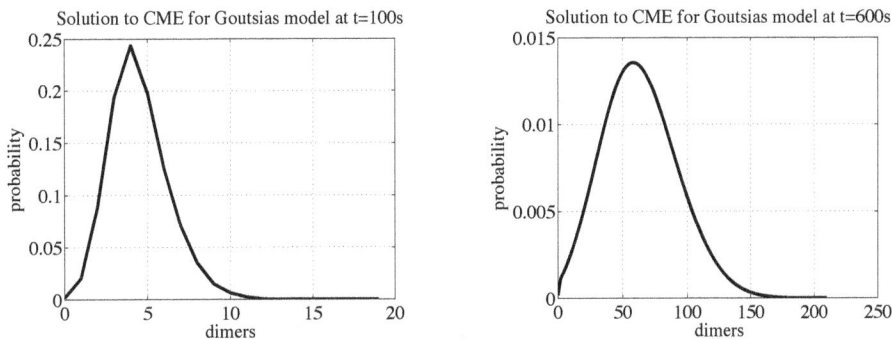

FIG. 6. *Solution to the chemical master equation for the Goutsias model of regulated transcription after a relatively short time of $t = 100s$ (left) and a much larger time of $t = 600s$ (right). The solution depicted here represents the marginal distribution for the number of dimers. Note that the scales differ by about an order of magnitude. The solution at $t = 100s$ was computed using the Krylov FSP algorithm and took only about 5 minutes on a single processor on UQ's Gust machine. Note that the distribution is relatively peaked, and so the nested projection that tracks the elevated part of the support is still relatively small, including only around 9 000 states, making the inexact Krylov method relatively fast. The solution for the larger final time of ten minutes has a distribution spread across many states, slowing down the Krylov method, which took 8.5 hours.*

Firstly, the support spreads only relatively slowly, and secondly, the support is dynamic so that some states originally included in the projection are eventually dropped once their contribution becomes too small. Additionally it computes solutions at intermediate times at no extra cost, which is advantageous compared to the original. It has been applied to two new models, the Goutsias model of transcription regulation as well as Michaelis-Menten enzyme kinetics, one of which is substantially larger than that on which the original FSP was demonstrated.

A simplifying assumption of both methods is that the propensity matrix is constant, giving a temporally homogeneous Markov chain. More realistic models would allow propensities to vary with time (to take account of cell growth or temperature for example) but implementing them would be considerably more difficult. In contrast this can be taken into account via a simulation algorithm such as SSA relatively easily.

Computing solutions for large final times presents a challenge for both methods, although the Krylov FSP algorithm goes further. In the example on which the original FSP algorithm is computed, the solution is obtained at a final time of only $t_f = 10s$. However for many applications the solution may be required at much larger times. For an *E.coli* cell, for example, the cell cycle takes about thirty five minutes, and so solutions at times in the order of $t_f = 2\,100s$ are likely to be of interest. A solution at around the third of the cell cycle time has been presented here. With the Krylov method presented here the computational complexity increases linearly (through the matrix-vector product) with the size of the projection. The numerical experiments presented here however have shown that the required projection size tends to increase much more quickly than linearly with time.

Overall more theoretical and computational considerations need to be addressed before any of the algorithms for the solution of the chemical master equation become truly scalable. While approaches that enforce the Markov property when truncating large state spaces are mathematically attractive, they may not be so compelling from a computational perspective. However this issue needs further consideration. On the other hand, progress in the computation of the matrix exponential would be very helpful, since the methods are so closely related to this area. Models arising from biological applications often exhibit extra structure, and exploiting this seems a natural next step. Finding new approximations suited to distributions spread across huge numbers of states would also seem a necessary step towards tackling larger models. Future work will involve addressing a number of these issues and additionally applying these inexact Krylov techniques to other applications.

Acknowledgements. The first named author would like to thank the Australian Research Council (ARC) for his funding of a Federation Fellowship. We would also like to thank the Queensland Parallel Supercomputing Foundation (QPSF) for the use of its HPC computational infrastructure. The support of the Australian Centre of Bioinformatics is also acknowledged.

REFERENCES

[1] A. Arkin and H. McAdams. It's a noisy business! *Trends Genet.*, 15(2):65–69, 1999.

[2] A. Arkin, J. Ross, and H. McAdams. Stochastic kinetic analysis of developmental pathway bifurcation in phage lambda-infected *Escherichia coli* cells. *Genetics*, 149:1633–1648, 1998.

[3] W. J. Blake, M. Kærn, C. R. Cantor, and J. J. Collins. Noise in eukaryotic gene expression. *Nature*, 422:633–7, April 2003.

[4] A. Bouras and V. Frayssé. Inexact matrix-vector products in Krylov methods for solving linear systems: A relaxation strategy. *SIAM J. Mat. Anal. Appl.*, 26(3):660–678, 2005.

[5] J. Butcher. *Numerical Methods for Ordinary Differential Equations.* Wiley, 2003.

[6] B. D. Davis, R. Dulbecco, H. N. Eisen, and H. S. Ginsberg. *Bacterial Physiology: Microbiology, Second Edition.* Maryland: Harper and Row, 1973.

[7] D. T. Gillespie. Exact stochastic simulation of coupled chemical reactions. *J. Phys. Chem.*, 81:2340–2361, 1977.

[8] D. T. Gillespie. *Markov Processes: An Introduction for Physical Scientists.* Academic Press, Harcourt Brace Jovanovich, 1992.

[9] D. T. Gillespie. Approximate accelerated stochastic simulation of chemically reacting systems. *J. Chem. Phys.*, 115:1716–1733, July 2001.

[10] J. Goutsias. Quasiequilibrium approximation of fast reaction kinetics in stochastic biochemical systems. *J. Chem. Phys.*, 122(18):184102, May 2005.

[11] G. Latouche and V. Ramaswami. *Introduction to Matrix Analytic Methods in Stochastic Modeling.* ASA-SIAM Series on Statistics and Applied Probability. SIAM, Philadelphia, 1999.

[12] B. Lewin. *Genes V.* Oxford University Press, 1994.

[13] B. Munsky and M. Khammash. The finite state projection algorithm for the solution of the chemical master equation. *J. Chem. Phys.*, 124:044104, 2006.

[14] M. Ptashne. *A Genetic Switch.* Blackwell Science and Cell Press, 1992.

[15] R. B. Sidje. Expokit software. *http://www.expokit.org*.

[16] R. B. Sidje. Expokit: A software package for computing matrix exponentials. *ACM Transactions on Mathematical Software*, 24(1):130–156, 1998.

[17] R. B. Sidje and W. J. Stewart. A numerical study of large sparse matrix exponentials arising in Markov chains. *Comput. Statist. Data Anal.*, 29:345–368, 1999.

[18] V. Simoncini and D. B. Szyld. Theory of inexact Krylov subspaces methods an applications to scientific computing. *SIAM J. Sci. Comput.*, 25(2):454–476, 2003.

[19] W. J. Stewart. *Introduction to the Numerical Solution of Markov Chains.* Princeton University Press, 1994.

[20] T. Tian and K. Burrage. Binomial leap methods for simulating stochastic chemical kinetics. *J. Chem. Phys.*, 121:10356, 2004.

[21] T. Tian and K. Burrage. Bistability and switching in the lysis/lysogeny genetic regulatory network of bacteriophage lambda. *J. Theor. Biol.*, 227(2):229–237, March 2004.

[22] N. G. van Kampen. *Stochastic Processes in Physics and Chemistry.* Elsevier Science, 2001.

FIVE PERFORMABILITY ALGORITHMS: A COMPARISON

LUCIA CLOTH AND BOUDEWIJN R. HAVERKORT*

Abstract. Since the introduction by John F. Meyer in 1980 [21], various algorithms have been proposed to evaluate the performability distribution. In this paper we describe and compare five algorithms that have been proposed recently to evaluate this distribution: Picard's method, a uniformisation-based method, a path-exploration method, a discretisation approach and a fully Markovian approximation.

As a result of our study, we recommend Picard's method not to be used (due to numerical stability problems). Furthermore, the path exploration method turns out to be heavily dependent on the branching structure of the Markov-reward model under study. For small models, the uniformisation method is preferable; however, its complexity is such that it is impractical for larger models. The discretisation method performs well, also for larger models; however, it does not easily apply in all cases. The recently proposed Markovian approximation works best, even for large models; however, error bounds cannot be given for it.

Key words. Performability evaluation, Markov-reward models, computational techniques

AMS subject classifications. 60J22, 60J27, 65C20, 90B25

1. Introduction. Over the last 25 years, many algorithms for the computation of the performability distribution, that is, the distribution of accumulated reward up to some time t in a Markov reward model (MRM), have been proposed; for overviews we refer to [16, 17, 26, 35]. Early work was restricted to *acyclic* MRMs [2, 9–11]. Some algorithms for possibly cyclic MRMs are based on Laplace transforms and are therefore only suitable for MRMs with relatively small state spaces [19, 32]. In this context, however, the application of newer algorithms for Laplace transform inversion (cf. Dingle et al. [7]) has not been investigated. Other authors have considered only availability models with two different reward classes [5, 29, 30].

The contribution of this paper is that we compare five algorithms that have been proposed recently to evaluate the performability distribution in general Markov reward models. As far as we can see, this is the first time that these five algorithms are described and compared, using the same notation and using the same cases. Thus, we are able to make comparative statements about complexity (space, time) and accuracy. The five considered algorithms are: Picard's method [24], Sericola's uniformisation-based method [31], the path exploration method [4, 26], a discretisation method [34], and a fully Markovian approximation. This fifth algorithm has been developed by us (and has been described concisely in [14, 15]).

As a result of the comparison, we conclude that for small models (less than a few dozens of states) the uniformisation-based method of Sericola is the best choice. For larger models, that may even include inhomogeneities, the Markovian approximation is preferred; for some models also the discretisation algorithm appears to be a good choice. Disadvantage of both these latter methods, though, is the lack of a clear accuracy statement. Picard's method should not be used as it neither provides accurate results nor is reasonably fast. Furthermore, the performance of the path exploration method heavily depends on the model under study, and therefore cannot always be recommended.

This paper is further organised as follows. In Section 2 we recapitulate the definitions of Markov reward models and the accumulated reward distribution. Section 3

---

*Design and Analysis of Communication Systems, P.O. Box 217, University of Twente, 7500 AE Enschede, The Netherlands, (`[lucia,brh]@cs.utwente.nl`).

describes the five algorithms for the computation of the performability distribution. Section 4 then compares the algorithms with respect to accuracy and scalability by means of a small example MRM, as well as a larger MRM representing a multiprocessor system (taken from the literature). We conclude the paper in Section 5.

2. Markov Reward Models. An MRM $\mathcal{M}$ consists of a continuous-time Markov chain (CTMC) $(X_t, t \geq 0)$ and a reward structure ρ. The CTMC is defined by its state space S and the generator matrix $\mathbf{Q} = (Q_{ss'})_{s,s' \in S}$. For a state $s \in S$, the value of ρ_s indicates the rate at which reward is accumulated in state s. We only consider the case where all reward rates are nonnegative. The initial distribution of the CTMC is given by the discrete probability distribution α over all states in S. While X_t is the random variable that describes the state of the CTMC at time t, Y_t is the accumulated reward up to time t. It depends on the reward rates of the states the CTMC has visited in its evolution until time t:

$$(2.1) \qquad Y_t = \int_0^t \rho_{X_u} du.$$

Y_t is a random variable, and, hence, we are interested in the distribution $F_Y(t, y) = \Pr\{Y_t \leq y\}$. Meyer called this the performability distribution [21, 22]. The algorithms presented in Section 3 compute the joint distribution $\Upsilon_{ss'}(t, y)$ of the state of the CTMC X_t and the accumulated reward Y_t, given the starting state X_0, that is,

$$\Upsilon_{ss'}(t, y) = \Pr\{X_t = s', Y_t \leq y \mid X_0 = s\}$$

The distribution of the accumulated reward at time t is then given as

$$(2.2) \qquad \Pr\{Y_t \leq y\} = \sum_{s \in S} \alpha_s \cdot \sum_{s' \in S} \Upsilon_{ss'}(t, y).$$

The joint distributions for state pairs $s, s' \in S$, $\Upsilon_{ss'}(t, y)$, are characterised by a set of partial differential equations [24, 31]

$$(2.3) \qquad \frac{\partial \Upsilon_{ss'}(t, y)}{\partial t} + \rho_s \cdot \frac{\partial \Upsilon_{ss'}(t, y)}{\partial y} = \sum_{z \in S} Q_{sz} \cdot \Upsilon_{zs'}(t, y),$$

with initial values

$$(2.4) \qquad \Upsilon_{ss'}(0, y) = \begin{cases} 1, & s = s' \text{ and } y \geq 0, \\ 0, & \text{otherwise.} \end{cases}$$

Applying the method of characteristics [24] leads to the set of integral equations

$$(2.5) \quad \Upsilon_{ss'}(t, y) = e^{Q_{ss}t} \Upsilon_{ss'}(0, y - \rho_s t) + \int_0^t \sum_{z \neq s} e^{Q_{ss}x} Q_{sz} \Upsilon_{zs'}(t - x, y - \rho_s x) dx$$

with the same initial values.

3. Five algorithms. In this section we discuss five algorithms for the computation of $\Upsilon_{ss'}(t, y)$. We start with a straightforward solution of the integral equation (2.5) using successive approximations (Picard's method), as proposed by Pattipati et al. [24]. The second algorithm is based on uniformisation and was developed by Sericola [31]. The third algorithm explicitly explores the possible realisations (paths) of the MRM. It was first presented by Qureshi and Sanders [25, 26] and later also used in the CSRL model checking context [4, 18]. We then describe the discretisation algorithm by Tijms and Veldman [34]. The last algorithm presented is the Markovian approximation first presented in [14, 15].

3.1. Picard's method. The set of integral equations (2.5) has to be evaluated by a fixed point computation because terms involving $\Upsilon_{ss'}(t, y)$ appear on both sides. One numerical algorithm for fixed point computations is known as "Picard's method:" it generates a sequence of approximations $\Upsilon_{ss'}^{(n)}(t, y)$ that converges to the correct solution, that is,

$$\lim_{n \to \infty} \Upsilon_{ss'}^{(n)}(t, y) = \Upsilon_{ss'}(t, y).$$

The first approximation is given by

$$\Upsilon_{ss'}^{(0)}(t, y) = e^{Q_{ss}t} \Upsilon_{ss'}(0, y - \rho_s t),$$

where $\Upsilon_{ss'}(0, y - \rho_s t)$ is known from (2.4). The subsequent approximations are computed using the integral equation:

$$\Upsilon_{ss'}^{(n+1)}(t, y) = e^{Q_{ss}\tau} \Upsilon_{ss'}^{(n)}(0, y - \rho_s t) + \int_0^t \sum_{z \neq s} e^{Q_{ss}x} Q_{sz} \Upsilon_{zs'}^{(n)}(t - x, y - \rho_s x) dx.$$

The iteration is terminated if the absolute value of the difference between two subsequent iterations drops below a given accuracy threshold.

Each iteration step involves the evaluation of $|S| - 1$ integrals over the previous approximation of the joint distribution. The integration can only be performed numerically. Different integration schemes are possible; for the sake of simplicity we restrict the description to the trapezoidal rule. The integration interval $[0, t]$ is divided into subintervals of size Δt. The integrals are then approximated as follows:

$$\int_0^t e^{Q_{ss}x} Q_{sz} \Upsilon_{zs'}^{(n)}(t - x, y - \rho_s x) dx \approx$$

$$Q_{sz} \cdot \left(\frac{1}{2} \Upsilon_{zs'}^{(n)}(t, y) + \sum_{i=1}^{\frac{t}{\Delta t} - 1} e^{Q_{ss}i\Delta t} \Upsilon_{zs'}^{(n)}(t - i\Delta t, y - \rho_s i\Delta t) + \frac{1}{2} e^{Q_{ss}t} \Upsilon_{zs'}^{(n)}(0, y - \rho_s t) \right).$$

The integration scheme shows that it does not just suffice to compute $\Upsilon_{ss'}(t, y)$ in each iteration step. Instead we need sample points $\Upsilon_{ss'}(i\Delta t, j\Delta t)$, for all $i = 0, \cdots, \frac{t}{\Delta t}$, and $j = 0, \cdots, \frac{y}{\Delta t}$. Actually, we might also need sample points for negative values of j, but then the distribution is zero anyway and it is not necessary to compute/store these values.

The multitude of numerical integrations plus the approximate nature of the outer iteration make it impossible to indicate an estimate of the resulting numerical error for this method.

3.2. Analytical solution using uniformisation. Sericola [31] derives a uniformisation-based [12,13] solution for the system of partial differential equations that describes the complementary joint distribution of state and accumulated reward $\overline{\Upsilon}_{ss'}(t, y) = \Pr\{X_t = s', Y_t > y \mid X_0 = s\}$ for an MRM. The joint distribution $\overline{\Upsilon}_{ss'}(t, y)$ is conditioned on the number of steps n in the uniformised MRM and on the number k of transitions that happen before a certain threshold y_h (see below) and the reward bound y, as follows:

$$(3.1) \qquad \overline{\Upsilon}_{ss'}(t, y) = \sum_{n=0}^{\infty} PP(\lambda t, n) \sum_{k=0}^{n} \binom{n}{k} y_h^k (1 - y_h)^{n-k} C_{ss'}^{(h)}(n, k),$$

where $y_h = \frac{y - r_{h-1}t}{r_h t - r_{h-1}t}$, for $y \in [r_{h-1}t, r_h t)$, and $\binom{n}{k} y_h^k (1 - y_h)^{n-k}$ is the probability that exactly k of the n transitions have happened by time $\frac{y - r_{h-1}t}{r_h}$.

The value of $C_{ss'}^{(h)}(n, k)$ is then the complementary distribution $\overline{\Upsilon}_{ss'}(t, y)$ conditioned on n and k. The $C_{ss'}^{(h)}(n, k)$-values are computed recursively. For details of this recursion we refer to [3, 31].

Using uniformisation, it is not possible to evaluate the complete infinite sum (3.1). It has to be truncated at some $N \in \mathbb{N}$. The error induced by this truncation is bounded by $\varepsilon = 1 - \sum_{n=0}^{N} PP(\lambda t, n)$, exactly as it is the case with traditional uniformisation for the computation of $\Pi_{ss'}(t) = \Pr\{X_t = s' \mid X_0 = s\}$. From the complementary probability $\overline{\Upsilon}_{ss'}(t, y)$ we compute $\Upsilon_{ss'}(t, y)$ by

$$\Upsilon_{ss'}(t, y) = \Pi_{ss'}(t) - \overline{\Upsilon}_{ss'}(t, y).$$

3.3. Path exploration. We now present a uniformisation-based method, where the $\Upsilon_{ss'}(t, y)$ are not conditioned on the number of steps taken until time t, but on the precise path taken up to this time. Let $\sigma = (s_0, \cdots, s_n) \in S^n$ be a so-called uniformised path of length $|\sigma| = n$. Its probability of occurrence is $P(\sigma) = U_{s_0 s_1} \cdot \cdots \cdot U_{s_{n-1} s_n}$, where $\mathbf{U} = (U_{ss'})_{s, s' \in S}$ is the generator matrix uniformised with uniformisation parameter λ, that is, $\mathbf{U} = \mathbf{I} + \mathbf{Q}/\lambda$. The set of all uniformised paths is denoted $uPath$, $first(\sigma)$ and $last(\sigma)$ denote the first and last state of a uniformised path σ. Then $\Upsilon_{ss'}(t, y)$ conditioned on uniformised paths is given by

$$(3.2) \qquad \Upsilon_{ss'}(t, y) = \sum_{n=0}^{\infty} PP(\lambda t, n) \cdot \sum_{\substack{\sigma \in uPath \\ |\sigma| = n \\ first(\sigma) = s \\ last(\sigma) = s'}} P(\sigma) \cdot \Pr\{Y_t \leq y \mid \sigma\}.$$

Following this expression, we consider the uniformised paths that start in s and end in s', calculate the reward distribution conditioned on each of these paths and compute the weighted sum of all conditioned probabilities. Two questions arise with this approach:

i) How do we calculate the conditional reward distribution?
ii) Is it possible to consider all relevant uniformised paths?

These two issues are addressed in the following.

The conditional reward distribution. The state space of the MRM $\mathcal{M}$ can be divided into $K + 1$ distinct reward classes. States with identical reward rate constitute one such reward class. Without loss of generality, the reward classes are ordered such that

$$r_0 > r_2 > \ldots > r_K \geq 0.$$

For the computation of $\Pr\{Y_t \leq y \mid \sigma\}$ it is not necessary to consider the complete information contained in the path σ but one only has to know how many epochs of the uniformised path have been spent in each of the reward classes. A vector $\mathbf{k} = (k_1, \ldots, k_K)$ recording these visit counts is called a *colouring*. The value k_i indicates the number of epochs the MRM has spent in states of reward class i. The term colouring stems from the idea of assigning the same colour to states with identical reward rate [6]. The computation of the distribution of Y_t given a colouring $\mathbf{k}$ boils down to the computation of the distribution of a linear combination of uniform order statistics [31]. We are aware of 3 methods for the calculation of this type of distribution. The approaches of Weisberg [36] and Matsunawa [20] use involved computations and tend to be numerically unstable. The recently proposed method of Diniz et al. [8] is based on a very simple recursion scheme and is numerically stable. It is therefore the one we use in this context.

Which paths to explore? The set of all uniformised paths starting in s and ending in s' in (3.2) is partitioned according to the number of steps (the length) within a path. For a fixed starting state s there is exactly one uniformised path with 0 steps, namely s itself. The starting state has up to $|S|$ successor states, so there are $\mathcal{O}(|S|)$ uniformised paths of length 1. Repeating this argument, there are $\mathcal{O}(|S|^n)$ uniformised paths of length n. The total number of paths is of course infinite, but even if we only take into account paths up to a given length N, the number grows exponentially with N. Hence, the consideration of all paths in (3.2) is infeasible.

The probability $P(\sigma)$ of a uniformised path is used in (3.2) as a weight for the conditional reward distribution. Qureshi and Sanders [26] introduce a threshold $w \in (0,1)$ for $P(\sigma)$: only if $P(\sigma) > w$, the path σ is included in the summation. Additionally, a maximum length N is fixed for the uniformised paths. Define the set of uniformised paths of length n that are actually considered for the computation as

$$Considered(s, s', w, n) = \{\sigma \in uPath \mid \text{first}(\sigma) = s, \text{last}(\sigma) = s', P(\sigma) > w \text{ and } |\sigma| = n\}.$$

This leads to the following approximation for the reward distribution:

$$(3.3) \qquad \Upsilon_{ss'}(t, y) \approx \sum_{n=0}^{N} PP(\lambda t, n) \cdot \sum_{\sigma \in Considered(s, s', w, n)} P(\sigma) \cdot \Pr\left\{Y_t \leq y \mid \mathbf{k}(\sigma)\right\},$$

where $\mathbf{k}(\sigma)$ is the colouring arising from path σ. For the calculation of (3.3), all paths contained in one of the sets $Considered(s, s', w, n)$ for $n = 0, \ldots, N$, have to be generated one by one. An error bound for the approximation can be determined in the course of the path exploration. For details on the exploration algorithm and the error bound we refer to [3, 4, 26].

3.4. Discretisation. A wide variety of general purpose numerical solution methods for ODEs and PDEs are based on the idea of *discretising* the continuous parameters. Tijms and Veldman published an approximate discretisation algorithm for the computation of $\Upsilon_{ss'}(t, y)$ that uses the same step size Δ for both time and accumulated reward [34].

Like any distribution, $\Upsilon_{ss'}(t, y)$ is a definite integral over the corresponding density:

$$(3.4) \qquad \Upsilon_{ss'}(t, y) = \int_0^y v_{ss'}(t, x) dx.$$

For fixed $t > 0$ and step size Δ we can use the rectangular approximation:

$$(3.5) \qquad \Upsilon_{ss'}(t, y) \approx \sum_{j=1}^{\frac{y}{\Delta}} v_{ss'}(t, j \cdot \Delta) \Delta.$$

For $\Delta \to 0$ we obtain again (3.4). Other approximation schemes, e.g., trapezoid, are possible.

We discretise the time up to t and the accumulated reward up to y in steps of size Δ and consider the density at times $0, \Delta, \cdots, \frac{t}{\Delta}\Delta$, and for accumulated rewards $0, \Delta, \cdots, \frac{y}{\Delta}\Delta$. The densities $v_{ss'}(\tau, x)$ are also not determined exactly but approximated by $v_{ss'}^{\Delta}(\tau, x)$ assuming that at most one transition has occurred in a time interval of length Δ. The possibility that two or more transitions occur is neglected. This is a reasonable assumption if Δ is small. The initial values for $\tau = 0$ are given by

$$v_{ss'}^{\Delta}(0, x) = \begin{cases} \frac{1}{\Delta}, & s = s' \text{ and } x = 0, \\ 0, & \text{otherwise.} \end{cases}$$

Only if $s = s'$ and $x = 0$ the approximate density can be positive at time 0. Since it is a derivative we have to fit it to the step size Δ.

By assuming that either no transition or exactly one transition has occurred in the time interval $[\tau, \tau + \Delta)$, the quantity $v_{ss'}^{\Delta}(\tau + \Delta, x)$ can recursively be calculated as follows:

$$v_{ss'}^{\Delta}(\tau + \Delta, x) = v_{ss'}^{\Delta}(\tau, x - \rho_{s'}\Delta) \cdot (1 + Q_{s's'} \cdot \Delta)$$
$$(3.6) \qquad + \sum_{z \neq s'} v_{sz}^{\Delta}(\tau, x - \rho_{s'}\Delta) \cdot Q_{zs'} \cdot \Delta.$$

The above recursion only operates correctly if $(1 + Q_{ss}\Delta)$ and $(Q_{sz}\Delta)$ are indeed probabilities, that is, if Δ is small enough. This is the case for any state s' if $\Delta \leq -\frac{1}{Q_{ss}}$ for all $s \in S$ and all τ and x [33]. No error bound is known for this method.

3.5. Markovian approximation. The last algorithm we present is an approximation for the joint distribution of state and accumulated reward that is based on the transient solution of a derived CTMC, that is, no rewards are involved. It was described in [14, 15].

The joint distribution of state and accumulated reward, can be rewritten by summing over evenly-sized subintervals of the reward interval $[0, y]$:

$$\Upsilon_{ss'}(t, y) = \Pr\{X_t = s', Y_t \in [0, \Delta y] \mid X_0 = s\}$$
$$+ \sum_{j=1}^{\frac{y}{\Delta y}-1} \Pr\{X_t = s', Y_t \in (j\Delta y, (j+1)\Delta y] \mid X_0 = s\}.$$

We want to approximate the terms $\Pr\{X_t = s', Y_t \in (j\Delta y, (j+1)\Delta y] \mid X_0 = s\}$ in such a way that the computation is done for a pure CTMC (without rewards). An MRM $\mathcal{M}$ can be seen as having an infinite and uncountable state space $S \times \mathbb{R}_{\geq 0}$. A joint state (s, y) indicates that CTMC underlying the MRM is in state s and that the accumulated reward is y. For our approximation, we break down the uncountable state space to an infinite but countable one. Define a CTMC C^{∞} with infinite state space $S \times \mathbb{N}$. The probability of being in state (s', j) is then the desired approximation:

$$\Pr\{X_t = s', Y_t \in (j\Delta y, (j+1)\Delta y] \mid X_0 = s\} \approx \Pi_{(s,0)(s',j)}^{C^{\infty}},$$

where $\Pi_{(s,0)(s',j)}^{C^{\infty}}$ is the transient probability of CTMC C^{∞} to reside in state (s', j) at time t, having started in state $(s, 0)$. The generator matrix $\mathbf{Q}^{\infty}$ must contain transitions from one reward level $j\Delta y$ to the next reward level $(j+1)\Delta y$. The rate at which reward is accumulated in a state s in the original MRM is ρ_s. The natural choice for the rate of accumulating Δy reward, that is, reaching the next reward level, is $\frac{\rho_s}{\Delta y}$. Transitions between states at the same reward levels are transfered from the original MRM, leading to the following generator matrix:

$$
Q^\infty_{(s,i)(s',j)} = \begin{cases}
Q_{ss'}, & s \neq s' \text{ and } i = j, \\
Q_{ss'} - \frac{\rho_s}{\Delta y}, & s = s' \text{ and } i = j, \\
\frac{\rho_s}{\Delta y}, & s = s' \text{ and } j = i+1, \\
0, & \text{otherwise.}
\end{cases}
$$

The generator matrix $\mathbf{Q}^\infty$ has a block diagonal structure, with the original generator matrix $\mathbf{Q}$ (with adapted diagonal entries) appearing on the diagonal and the matrix $\mathbf{D}/\Delta y$ (where $\mathbf{D}$ is the diagonal matrix arising from the reward structure ρ) appearing as upper off-diagonal. The infinite CTMC $\mathcal{C}^\infty$ is a quasi-birth process (a subclass of quasi-birth-death processes [23]). Transitions are only possible within a level (matrix $\mathbf{Q}$) or to the next higher level (matrix $\mathbf{D}/\Delta y$).

The transient probabilities $\Pi^{\mathcal{C}^\infty}_{(s,0)(s',j)}(t)$ needed for the approximation can efficiently be computed using uniformisation even though the CTMC $\mathcal{C}^\infty$ has infinite state space [27, 28].

We are not able to indicate the error introduced by this approximation. Of course, the approximation will get more accurate for smaller Δy. The accuracy is also influenced by the error bound used for computing the transient probabilities using uniformisation.

3.6. Complexity of the algorithms. Table 3.1 indicates the space and time complexity of Picard's method, uniformisation, discretisation and the Markovian approximation. For all four algorithms the complexity of computing $\Upsilon_{ss'}(t,y)$ for *all* pairs $s, s' \in S$ is given. For the path exploration algorithm it is not reasonable to indicate the complexity. The number of paths explored is potentially exponential, but the exploration is restricted by the weight w and the maximal path length N.

	space	time
Picard's method	$\mathcal{O}\left(\dfrac{\|S\|^2 ty}{(\Delta t)^2}\right)$	$\mathcal{O}\left(\dfrac{\|S\|^3 N t^2 y}{(\Delta t)^3}\right)$
Uniformisation	$\mathcal{O}(\|S\|^2 \lambda t K)$	$\mathcal{O}(\|S\|^3 (\lambda t)^2)$
Discretisation	$\mathcal{O}\left(\dfrac{\|S\|^2 y}{\Delta}\right)$	$\mathcal{O}\left(\dfrac{\|S\|^3 ty}{\Delta^2}\right)$
Markovian approx.	$\mathcal{O}\left(\dfrac{\|S\|^2 y}{\Delta y}\right)$	$\mathcal{O}\left(\dfrac{\|S\|^3 ty}{(\Delta y)^2}\right)$

TABLE 3.1
Time and space complexity

The four algorithms included in the table have a complexity that is cubic in the number of states. If only a single starting state s is considered, the complexity w.r.t. the number of states is only quadratic for Picard's method, Discretisation and the Markovian approximation. If one uses a sparse representation of the generator matrix $\mathbf{Q}$, one factor $|S|$ can be replaced by ν, the average number of transitions originating from a state.

The uniformisation approach suffers from stiffness problems as normal uniformisation does, if the rates in the generator differ several orders of magnitude. In the discretisation approach, the step size Δ has to be smaller than $1/Q_{ss}$ for any state s, which may lead to an excessively high number of steps. The Markovian approximation relies again on transient solution via uniformisation and therefore also exhibits bad performance when dealing with stiff models.

4. Comparison. In this section we compare and discuss the five algorithms described in the previous section. For the comparison of the accuracy and execution times we use the small four-state illustrating example of [3]. The behaviour of the algorithms for larger state spaces is evaluated using a scalable model of a multiprocessor computer system [32]. On the basis of the numerical results we make a recommendation for the use of the different algorithms.

Experiments in the CSRL setting [3] have shown that Picard's method performs poorly with respect to both accuracy and execution time. Additionally, we have frequently encountered numerical problems with this algorithm. This algorithm is therefore not further considered.

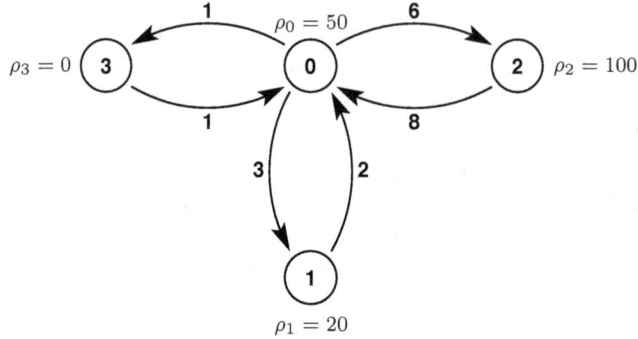

FIG. 4.1. *Four-state MRM*

4.1. Accuracy and execution times. Consider the MRM of Figure 4.1. Its four states have reward rates $\rho_0 = 50$, $\rho_1 = 20$, $\rho_2 = 100$ and $\rho_3 = 0$. In Figure 4.2 the performability distribution for this model is depicted for $t \in [0, 5]$ and $y \in [0, 200]$ and the initial distribution $\alpha = (1, 0, 0, 0)$. For $t = 0$, the performability measure is one, since $Y_0 = 0 \leq y$ for any y. With growing t and constant y, the performability measure decreases because more reward is accumulated. For constant t and growing y the performability increases because it is more likely to stay below the reward bound.

In the following we apply the algorithms to compute the performability measure for $t = 0.2$ and $y = 5$, that is, $\Pr\{Y_{0.2} \leq 5\}$.

Uniformisation. We start the discussion with the uniformisation algorithm because it is the only one that provides us with an *a priori* error bound for the results. The results are then used to compute the relative error for the numerical results of the other algorithms. Table 4.1 shows the numbers of steps, the resulting probability and the run time (user time) in seconds. The number of considered steps increases with the required accuracy, and so does the required time. Further experiments (see also [3]) have shown that also the time t has influence on the number of steps, as expected for a uniformisation-based algorithm. In contrast, the value of y has no impact on the execution time. The execution time for this small example is reasonable, staying far below one second.

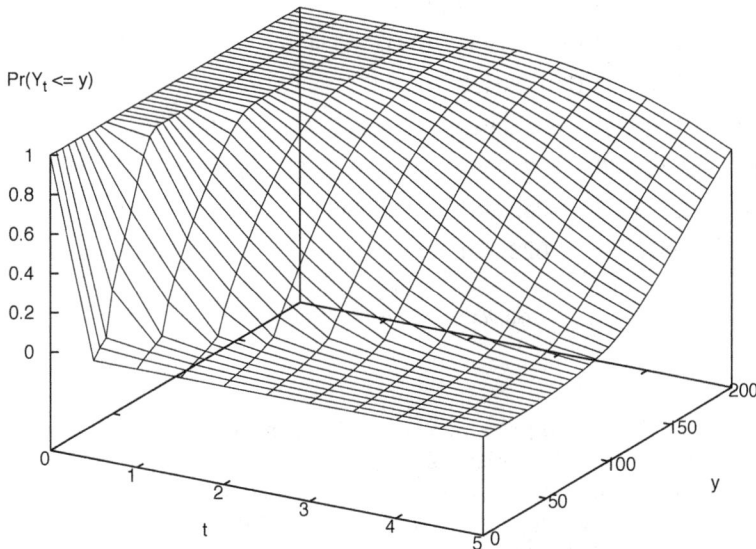

Pr(Y_t <= y)

FIG. 4.2. *Performability distribution for the four-state MRM*

ε	N	$\Pr\{Y_{0.2} \le 5\}$	time
10^{-2}	6	0.1319195485	0.005
10^{-4}	9	0.1330107485	0.010
10^{-8}	14	0.1330224790	0.022
10^{-16}	22	0.1330224800	0.051

TABLE 4.1
Results for the uniformisation algorithm

Path exploration. With path exploration, an error bound can only be determined a posteriori. In Table 4.2 we list the results of this algorithm. Each row shows the weight for the paths, the number c of colourings considered for this weight, the probability, the a posteriori error bound, the relative error with respect to the results of the uniformisation algorithm and the execution time in seconds. We have chosen the maximum number of steps such that it does not influence the path generation; hence, paths are only discarded because of their weight. Clearly, with decreasing path weight w, the error bounds decrease and the execution time increases. Other experiments [3] have shown that the error bound and the execution time increase with growing t. As it was the case for the uniformisation algorithm, y does not have any influence on the execution time, nor does it affect the error bound obtained. The execution times are reasonable for this small example but for comparable accuracies, uniformisation is faster. Uniformisation also has the advantage of providing an a priori error bound.

w	c	$\Pr\{Y_{0.2} \le 5\}$	absolute error	relative error	time
10^{-1}	24	0.0844766620	0.2845779895	0.3649444668	0.003
10^{-2}	173	0.1286911957	0.0483092418	0.0325605437	0.020
10^{-3}	597	0.1327120417	0.0037866433	0.0023337283	0.086
10^{-4}	1709	0.1330023912	0.0002745972	0.0001510179	0.421
10^{-5}	3357	0.1330217567	0.0000122965	0.0000054377	2.000
10^{-6}	4758	0.1330224461	0.0000005541	0.0000002551	9.758
10^{-7}	5870	0.1330224791	0.0000000176	0.0000000063	43.69

TABLE 4.2
Results for the path exploration

Discretisation. Table 4.3 presents the resulting probability, the relative error and the execution time of the discretisation algorithm subject to the step size Δ. With decreasing Δ, the relative error decreases and the execution time increases. Further experiments [3] have shown that both t and y have an influence on the execution time. Dividing the step size Δ by 10 increases the execution time by a factor 100 which confirms that the algorithm has a time complexity in Δ^{-2}. The values show that the discretisation algorithm provides usable results in tolerable time but is much slower than uniformisation and path exploration.

Δ	$\Pr\{Y_{0.2} \le 5\}$	relative error	time
10^{-2}	0.0773048006	0.4188591236	0.006
10^{-3}	0.1270651788	0.0447841688	0.543
10^{-4}	0.1324221136	0.0045132699	57.18

TABLE 4.3
Results for discretisation

Markovian approximation. The last algorithm to evaluate is the Markovian approximation. Table 4.4 shows the probability, the relative error and the execution time depending on the step size Δy for the accumulated reward. For the calculation of the transient probabilities via (ordinary) uniformisation we have chosen the error bound $\varepsilon = 10^{-16}$. The execution time grows with Δy^{-2}: if we divide Δy by 10, the run time is multiplied with 100. Further experiments [3] have shown that the execution time also depends linearly on t and y. The Markovian approximation is quite fast while at the same time providing adequately accurate results.

Δy	$\Pr\{Y_{0.2} \le 5\}$	relative error	time
10^{-1}	0.1294067747	0.0271811597	0.003
10^{-2}	0.1324884190	0.0040148179	0.232
10^{-3}	0.1329690459	0.0004016921	20.56

TABLE 4.4
Results for the Markovian approximation

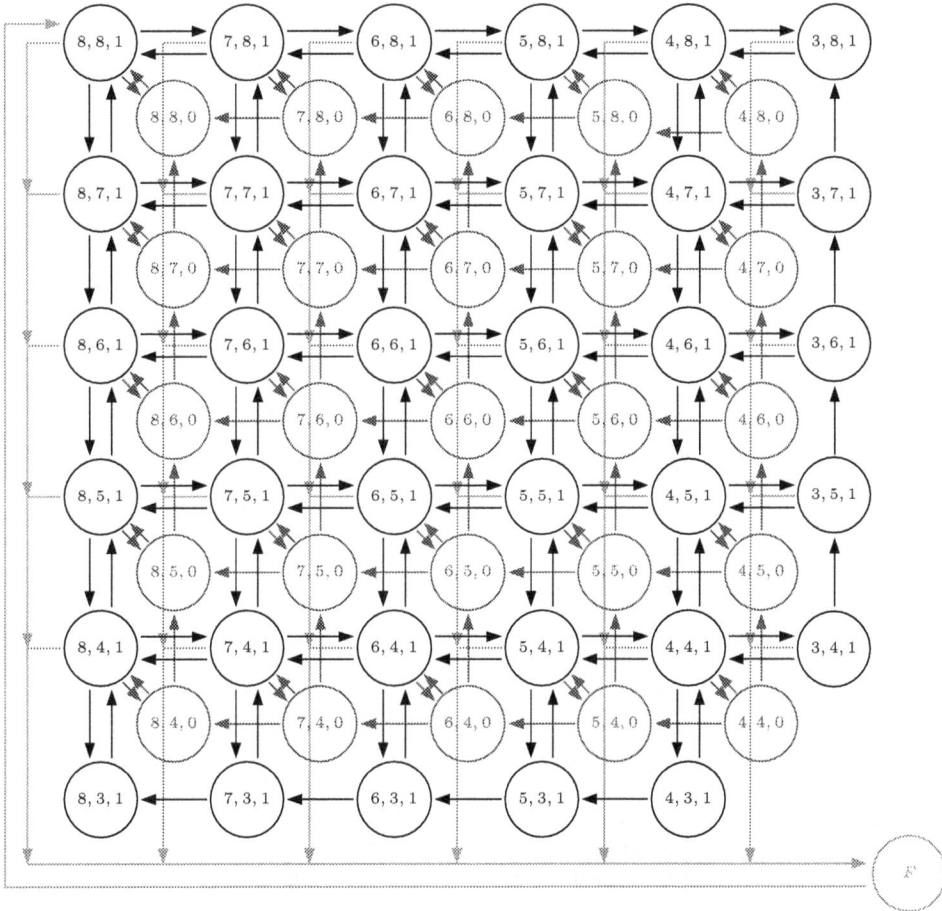

FIG. 4.3. *State-transition diagram for the multiprocessor model with $M = 8$*

4.2. Scaling the state space. The results shown in the previous section suggest that the uniformisation algorithm is fast and the most accurate of the four algorithms, However, we have not shown the behaviour of the algorithms for different sizes of the state space. In this section we compare the execution times for growing state spaces. For this purpose we adapt the multiprocessor model from [32, Section III].

The multiprocessor model. We address a multiprocessor system with M processors and M memory modules and an interconnecting network (crossbar switch). The states of the MRM are triples (i, j, k), where i is the number of operating processors, j is the number of operating memories, and $k = 1$ if the network is operational and $k = 0$ if the network is broken. Each of the components can fail. After the failure of a processor or a memory a reconfiguration takes place leading to a state where one of the component indices is decreased. The probability $c = 0.9$ for this reconfiguration to be successful is known as coverage. The component is then locally repaired by a single repair unit specific to each component type. In case the reconfiguration does not complete successfully, a global failure state F is entered. A global repair takes longer than a local component repair and always restores the system in fully operational state.

To be operational, a minimum of four operating processors and four operating memories is required, and the switch has to be operational as well. We assume that there are no further failures once the system is non-operational. With this restriction, the MRM for a system with M processors and memories has $(M - 2)^2 + (M - 3)^2$ states. The state-transition diagram for this model with $M = 8$ can be found in Figure 4.3.

We consider a single initial state (instead of an initial distribution), namely the state where all components are non-failed. The parameters of the system are given in Table 4.5; the repair rates are taken from the original model in [32]. The original failure rates are scaled up to get more meaningful results for the numerical comparison.

transition		rate
single processor failure rate	λ	0.0689
single memory failure rate	γ	0.2241
switch failure rate	δ	0.2024
processor repair rate	ν	2.0
memory repair rate	η	1.0
switch repair rate	ϵ	0.5
global repair rate	μ	0.2

TABLE 4.5
Rates for the multiprocessor model

The reward rate of an operational state is defined to be the average number of busy memories. Following [1], this equals

$$\rho_{(i,j,1)} = m \left(1 - \left(1 - \frac{1}{m} \right)^l \right),$$

where $l = \min(i, j)$ and $m = \max(i, j)$. The reward rate of non-operational states is zero. The accumulated reward Y_t can then be interpreted as the amount of work that has been completed up to time t.

By varying the number of processors and memory modules M between 4 and 240 we obtain MRMs with 5 to 112813 states. For these models the performability measure $\Pr\{Y_{10} \leq 10\}$, that is, the probability that the amount of performed work at time instant 10 is at most 10, should be computed. As basis for a comparison we take the smallest MRM with 5 states and compute the performability measure using the uniformisation algorithm with error bound $\varepsilon = 10^{-5}$. For the other algorithms we choose the parameters in such a way that the relative error is below 3% and keep these algorithm settings as such.

Figure 4.4 shows the execution time in seconds (y-axis) for the computation of $\Pr\{Y_{10} \leq 10\}$ in the size-varying MRMs (x-axis in logarithmic scale) for the different algorithms, as long as they stay below 5 minutes. For path exploration, discretisation and the Markovian approximation there are two implementations each, one using a dense and one using a sparse representation of the generator matrix. Uniformisation (which has cubic time complexity in the number of states) is only possible for small state spaces; already for $M = 8$ (61 states), the execution time exceeds 5 minutes.

Discretisation is only directly suited for MRMs with integer reward rates. The reward rates of the multiprocessor model are non-integers. To overcome this problem we can scale the reward rates *and* the reward bound y by a factor that makes the rates integers. Unfortunately, this factor also impacts the execution time of the algorithm, since it increases the number of reward steps to be taken, cf. Section 3.4. We choose to scale the rates and y by a factor 10 only and take the integer part of that. In order to have an relative error w.r.t. uniformisation for $M = 4$ below 3%, we set $\Delta = 0.02$. As can be seen from the figure, even with this small factor, discretisation with a dense representation of $\mathbf{Q}$ is slower than uniformisation, while the sparse version performs slightly better.

For the Markovian approximation, the allowed relative error of 3% leads to a step size $\Delta y = 1$. The dense implementation of the algorithm manages up to 1741 states ($M = 32$) in less than 5 minutes. The quadratic dependency on the number of states in clearly visible. The implementation using a sparse representation has an execution time of about 5 minutes for a model with more than 112000 states ($M = 240$).

Finally, both implementations of the path exploration algorithm are not able to deliver a result for $M = 4$ with a relative error below 3% in less than 5 minutes. This algorithm is therefore not included in Figure 4.4.

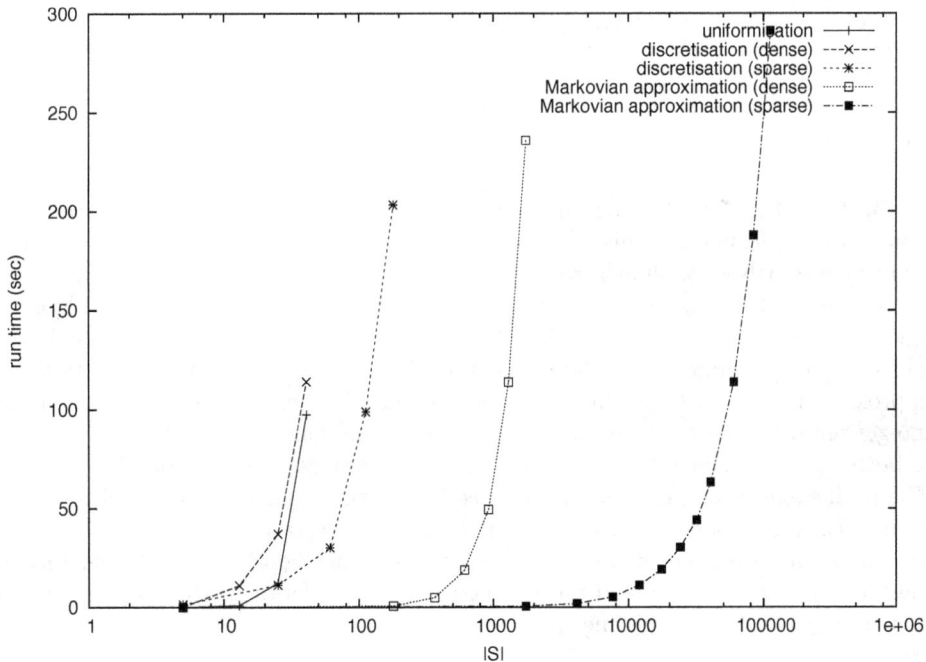

FIG. 4.4. *Execution times for uniformisation, discretisation and Markovian approximation when scaling the size of the state space*

Table 4.6 shows the values for accomplishing an amount of work less than 10 by time 10, i.e., $\Pr\{Y_{10} \le 10\}$, computed by uniformisation, the Markovian approximation and discretisation. This probability is quite high for the model with only four processors and memory modules (43.4%). When increasing the number of processors and memories M, the performability measure decreases, as the maximum bandwidth increases (from less than 3 for $M = 4$ to approximately 20 for $M = 32$) and it is less likely to reside in the failure state, having started in the state where all components are operational. The relative error for the Markovian approximation and the discretisation behaves in a non-predictable way.

	uniformisation	Markovian approximation		discretisation	
M	$\Pr\{Y_{10} \leq 10\}$	$\Pr\{Y_{10} \leq 10\}$	rel. error	$\Pr\{Y_{10} \leq 10\}$	rel. error
4	0.43406876	0.42224274	0.02724458	0.42504043	0.02079930
5	0.21462346	0.22717613	0.05848695	0.20817395	0.03005031
6	0.16067183	0.17011596	0.05877896	0.15614911	0.02814883
7	0.13737533	0.14419586	0.04964886	0.12886252	0.06196754
8	-	0.12984323	-	0.11656597	-
10	-	0.11441596	-	0.10151163	-
12	-	0.10601227	-	0.09367540	-
32	-	0.08532634	-	-	-
64	-	0.08038803	-	-	-
128	-	0.07809976	-	-	-
240	-	0.07706989	-	-	-

TABLE 4.6
Performability measure $\Pr\{Y_{10} \leq 10\}$ for the multiprocessor model

5. Conclusions. In this paper we discussed five algorithms for the computation of the performability distribution of MRMs with rate rewards. All of them are applicable to MRMs with arbitrary structure.

Based on the numerical results, we recommend the uniformisation algorithm whenever the model is small (a few dozens of states). It is the only algorithm for which an error bound can be determined a priori. For larger models, the Markovian approximation seems to be the only applicable algorithm. For a larger model with integer reward rates, the discretisation algorithm could also be employed. The results of both algorithms have to be handled with care because no error bound is provided. The path exploration algorithm performed badly when applied to the multiprocessor model. For other models, where less paths through the state space are possible or the probability mass is concentrated on a few path only, it might be competitive. Picard's method has only been studied for completeness, it is inferior to the other algorithms in accuracy and execution time.

REFERENCES

[1] D. BHANDARKAR, *Analysis of memory interference in multiprocessors*, IEEE Transactions on Computers, C-24 (1975), pp. 897–908.
[2] B. CICIANI AND V. GRASSI, *Performability evaluation of fault-tolerant satellite systems*, IEEE Transactions on Communications, 35 (1987), pp. 403–409.
[3] L. CLOTH, *Model Checking Algorithms for Markov Reward Models*, PhD thesis, University of Twente, 2006.
[4] L. CLOTH, J.-P. KATOEN, M. KHATTRI, AND R. PULUNGAN, *Model checking Markov reward models with impulse rewards*, in International Conference on Dependable Systems and Networks (DSN'05), IEEE Press, 2005, pp. 722–731.

[5] E. DE SOUZA E SILVA AND H. R. GAIL, *Calculating cumulative operational time distributions of repairable computer systems*, IEEE Transactions on Computers, C-35 (1986), pp. 322–332.

[6] E. DE SOUZA E SILVA, H. R. GAIL, AND R. VALLEJOS CAMPOS, *Calculating transient distributions of cumulative reward*, ACM SIGMETRICS Performance Evaluation Review, 23 (1995), pp. 231–240.

[7] N. J. DINGLE, P. G. HARRISON, AND W. J. KNOTTENBELT, *Response time densities in generalised stochastic Petri net models*, in Proceedings of the 3rd international workshop on Software and performance (WOSP'02), ACM Press, 2002, pp. 46–54.

[8] M. C. DINIZ, E. DE SOUZA E SILVA, AND H. R. GAIL, *Calculating the distribution of a linear combination of uniform order statistics*, INFORMS Journal on Computing, 14 (2002), pp. 124–131.

[9] L. DONATIELLO AND B. R. IYER, *Analysis of a composite performance reliability measure for fault-tolerant systems*, Journal of the ACM, 34 (1987), pp. 179–199.

[10] D. FURCHTGOTT AND J. F. MEYER, *Performability solution method for degradable nonrepairable systems*, IEEE Transactions on Computers, 33 (1984), pp. 550–553.

[11] A. GOYAL AND A. TANTAWI, *Evaluation of performability for degradable computer systems*, IEEE Transactions on Computers, 36 (1987), pp. 738–744.

[12] W. K. GRASSMANN, *Finding transient solutions in Markovian event systems through randomization*, in Numerical Solution of Markov Chains, W. Stewart, ed., Marcel Dekker Inc., 1991, pp. 357–371.

[13] D. GROSS AND D. R. MILLER, *The randomization technique as a modeling tool and solution procedure for transient Markov processes*, Operations Research, 32 (1984), pp. 343–361.

[14] B. R. HAVERKORT, L. CLOTH, H. HERMANNS, J.-P. KATOEN, AND C. BAIER, *Model checking performability properties*, in Proceedings of the International Conference on Dependable Systems and Networks (DSN'02), IEEE Press, 2002, pp. 102–112.

[15] B. R. HAVERKORT, H. HERMANNS, J.-P. KATOEN, AND C. BAIER, *Model checking CSRL-specified performability properties*, in Proceedings of the 5th International Workshop on Performability Modeling of Computer and Communications Systems (PMCCS'01), 2001, pp. 105–109.

[16] B. R. HAVERKORT, R. MARIE, G. RUBINO, AND K. S. TRIVEDI, eds., *Performability Modelling*, John Wiley & Sons, 2001.

[17] B. R. HAVERKORT AND I. G. NIEMEGEERS, *Performability modelling tools and techniques*, Performance Evaluation, 25 (1996), pp. 17–40.

[18] M. KHATTRI AND R. M. PULUNGAN, *Model checking Markov reward models with impulse rewards*, Master's thesis, University of Twente, Department of Computer Science, 2004.

[19] V. G. KULKARNI, V. F. NICOLA, R. M. SMITH, AND K. S. TRIVEDI, *Numerical evaluation of performability and job completion time in repairable fault-tolerant systems*, in Proceedings of the 16th International Symposium on Fault-Tolerant Computing (FTCS'85), 1985, pp. 252–257.

[20] T. MATSUNAWA, *The exact and approximate distributions of linear combinations of selected order statistics from uniform distributions*, The Annals of the Institute of Statistical Mathematics, 37 (1985), pp. 1–16.

[21] J. F. MEYER, *On evaluating the performability of degradable computing systems*, IEEE Transactions on Computers, 29 (1980), pp. 720–731.

[22] ——, *Performability: a retrospective and some pointers to the future*, Performance Evaluation, 14 (1992), pp. 139–156.

[23] M. F. NEUTS, *Matrix-Geometric Solutions in Stochastic Models. An Algorithmic Approach*, Dover Publications, 1994.

[24] K. R. PATTIPATI, R. MALLUBHATLA, V. GOPALAKRISHNA, AND N. VISWANATHAM, *Markov-reward models and hyperbolic systems*, in Performability Modelling, John Wiley & Sons, 2001, pp. 83–106.

[25] M. A. QURESHI, *Reward model solution methods with impulse and rate rewards: An algorithm and numerical results*, Master's thesis, University of Arizona, 1992.

[26] M. A. QURESHI AND W. H. SANDERS, *Reward model solution methods with impulse and rate rewards: an algorithm and numerical results*, Performance Evaluation, 20 (1994), pp. 413–436.

[27] A. REMKE, L. CLOTH, AND B. R. HAVERKORT, *Uniformization with representatives: Transient analysis of infinite-state Quasi-Birth-Death processes. Submitted for publication*, 2006.

[28] A. REMKE, B. R. HAVERKORT, AND L. CLOTH, *Model checking infinite-state Markov chains*, in Proceedings of the 11th International Conference on Tools and Algorithms for the Construction and Analysis of Systems (TACAS'05), Lecture Notes in Computer Science 3440, Springer-Verlag, 2005, pp. 237–252.

[29] G. RUBINO AND B. SERICOLA, *Interval availability distribution computation*, in Proceedings of the 23rd IEEE International Symposium on Fault Tolerant Computing (FTCS'93), IEEE Press, 1993, pp. 48–55.

[30] B. SERICOLA, *Closed-form solution for the distribution of the total time spent in a subset of states of a homogeneous Markov process during a finite observation period*, Journal of Applied Probability, 27 (1990), pp. 713–719.

[31] ———, *Occupation times in Markov processes*, Communications in Statistics — Stochastic Models, 16 (2000), pp. 479–510.

[32] R. SMITH, K. S. TRIVEDI, AND A. RAMESH, *Performability analysis: Measures, an algorithm and a case study*, IEEE Transactions on Computers, 37 (1988), pp. 406–417.

[33] W. J. STEWART, *Introduction to the Numerical Solution of Markov Chains*, Princeton University Press, 1994.

[34] H. TIJMS AND R. VELDMAN, *A fast algorithm for the transient reward distribution in continuous-time Markov chains*, Operations Research Letters, 26 (2000), pp. 155–158.

[35] K. S. TRIVEDI, J. K. MUPPALA, S. WOOLET, AND B. R. HAVERKORT, *Composite performance and dependability analysis*, Performance Evaluation, 14 (1992), pp. 197–215.

[36] H. WEISBERG, *The distribution of linear combinations of order statistics from the uniform distribution*, Annals of the Institute of Statistics, 42 (1971), pp. 704–709.

MODELING THROUGH MARKOV CHAINS: WHERE IS THE RISK ?

RAYMOND A. MARIE *

Abstract. The interests of Markovian models are well known from the specialists of modeling. Nevertheless, a common criticism is that, very often, the random variables that represent durations of some activities inside the Markovian model are exponentially distributed while this is not the case in the reality. In general people are often arguing just for a few percents or a few ten percents on a relative error. The purpose of this presentation aims to show that larger errors may be made just because of a slight misunderstanding of the considered application under the modeling stage. The example is taken in the area of dependability in a situation where a spare part shortage may occur very rarely. For this real application, a classical heuristic is used by the specialists from the field. In this presentation, we show that, while this heuristic may give "exact" results in some simple cases, this approach may underestimate some criterion such as the unavailability by a factor of hundreds. We propose a new approach and give the formal expressions of some dependability metrics in the case of a classical redundant architecture. For larger applications, the use of a numerical solution stays quite tractable.

Key words. Markov chains, Modeling Process, Dependability, Causality factors

AMS subject classifications. 60J27, 90B25

1. Introduction. The interests of Markovian models are well known from the specialists of modeling. Nevertheless, a common criticism (specially from the non specialists) is that, very often, the random variables that represent durations of some activities inside the Markovian model are exponentially distributed while this is not the case in the reality. The answer of the specialists to that criticism is that it is possible to fit other probability distributions by increasing the state space. However, a trade off may have to be done between tractability and accuracy because the criterion of faithfulness of the model may demand a huge number of states. This is why the specialists of modeling also may use simulations to test their Markovian models with respect to a more realistic model with the scope of convincing the people of the application of the validity of their model. In fact people are often arguing just for a few percents on a relative error. While much larger errors may be made just because of a misunderstanding of the considered application under the modeling stage. The purpose of this presentation aims to give an example of such a situation where a small inattention of a specialist may result in a very significant error.

The example is taken in the area of dependability. When we consider complex systems, a high reliability of the system is partly achieved thanks to redundancy. Breakdowns of elements may occur frequently but in general the global system conserves its operational status. However, due to redundancy, the utilization rate of spare parts may be consequent and it is part of the job of the engineer to predict the initial quantities of spares in order to keep the system at a level of high availability. Nevertheless, a spare part shortage may occur, hopefully with a low probability. Because of the size of the model to be considered for such a complex system, practitioners try to take this possibility (of spare part shortage) into consideration but in a way to keep the model tractable. From a dependability point of view, a complex system can be seen as a reliability diagram composed of sub diagrams. A classical case is the one where the global diagram corresponds to a series of sub diagrams, each sub diagram representing a redundant substructure. In order to emphasize the key point, we will just concentrate our discussion on a unique redundant substructure.

In section 2, we will present the classical approach for a non redundant structure and show that the use of such a heuristic is quite reasonable. In section 3, the case of a redundant substructure is considered, first following the classical heuristic and then we will investigate

---

* University of Rennes 1, IRISA, Campus de Beaulieu, 35042 Rennes Cedex, France. (marie@irisa.fr).

a more rigorous approach and present the new expressions of dependability metrics. Finally, numerical examples will convince the reader that very significant errors can be done quite easily.

2. The Simplest Case with a non Redundant Element. In order to describe a possible approach, let us consider the simplest model with just one element (cf. figure 2.1). The element breaks down according to an exponential distribution with rate λ. Given that a spare part is available, the exchange time is also exponential distributed, with rate μ. Therefore the model is a continuous time Markov chain (CTMC), its transition graph being represented on figure 2.1, where state U (resp. D) corresponds to the up state (resp. down state).

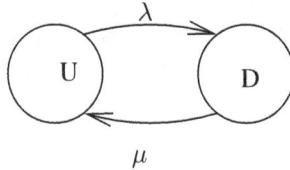

FIGURE 2.1. *Markovian transition graph - Non redundant case - Exponential distribution of exchange times.*

It is a very simple exercise to check that the steady state unavailability $\overline{A}$ can be expressed as $\frac{\rho}{1 + \rho}$, if ρ denotes the ratio $\frac{\lambda}{\mu}$.

Let us consider now that a spare part shortage may happen. Let us assume that new spare parts arrive to the exchange place according to a Poisson process with rate γ. We also assume that there is a low probability α that a spare part shortage occurs (assuming that this value α has been evaluated from the field). Thus, there exists a new mean repair time (denoted by $1/\hat{\mu}$) obtained as follows (using conditional expectation) :

$$\frac{1}{\hat{\mu}} = (1 - \alpha)\frac{1}{\mu} + \alpha(\frac{1}{\gamma} + \frac{1}{\mu}) = \frac{1}{\mu}\left(1 + \frac{\alpha}{\phi}\right),$$

where $\phi = \frac{\gamma}{\mu}$. So the term $\frac{\alpha}{\phi}$ represents the relative increase of the mean repair time when the exchange operation may have to be delayed due to a spare part shortage.

Then, the reaction of the practitioner will be to consider the Markovian model of figure 2.1 with a new mean repair rate $\hat{\mu}$.

The corresponding steady state unavailability (that we denote $\overline{A}$) becomes

$$\overline{A} = \frac{\rho\left(1 + \dfrac{\alpha}{\phi}\right)}{1 + \rho\left(1 + \dfrac{\alpha}{\phi}\right)}$$

Given that ρ is much smaller than one, the new unavailability will be, as a first order approximation, multiplied by the factor $(1 + \frac{\alpha}{\phi})$. For example, if $\alpha = \phi = 10^{-2}$, then the new unavailability will be, as a first order approximation, multiplied by 2.

Let us note that under the assumptions made above on the real behavior of the system, the CTMC corresponding faithfully to the given descriptions has the transition graph represented on figure 2.2, where G denotes the good element, E denotes the element being exchanged and A denotes the impossibility of exchanging the broken element due to the spare part shortage.

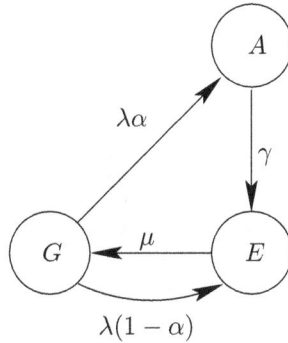

FIGURE 2.2. *Markovian transition graph - Non redundant case - With spare part shortage.*

When looking for the steady state distribution of this CTMC, we find, when π_e denotes the steady state probability that the chain is in state e :

$$\pi_A = \frac{\alpha\rho}{\alpha\rho + \phi\rho + \phi} \qquad \pi_E = \frac{\phi\rho}{\alpha\rho + \phi\rho + \phi} \qquad \pi_G = \frac{\phi}{\alpha\rho + \phi\rho + \phi} ,$$

From these probabilities we get the steady state unavailability :

$$\overline{A} = \pi_A + \pi_E = \frac{\rho\left(1 + \dfrac{\alpha}{\phi}\right)}{1 + \rho\left(1 + \dfrac{\alpha}{\phi}\right)}$$

which is the same expression as in the simplified model. The steady state availability being obviously equal to π_G.

In addition, let us consider the determination of the mean up time (MUT) and of the mean down time (MDT). On the one hand the steady state availability A can be expressed as follows :

$$(2.1) \qquad\qquad A = \frac{MUT}{MUT + MDT}$$

On the other hand, the frequency of the transitions from state G toward the subset $\{E, A\}$ which is here just equal to $\lambda\pi_G$ has to be equal to the inverse of the mean time between failures (MTBF) with MTBF = MUT + MDT :

$$(2.2) \qquad\qquad \lambda\pi_G = \frac{1}{MUT + MDT}$$

From these two equations, we verify that the term λMUT is still equal to one. Now, using this last result, together with equation (2.2), we also obtain :

$$(2.3) \qquad\qquad MDT = \frac{1 - \pi_G}{\lambda\pi_G} = \frac{1}{\mu}\left(1 + \frac{\alpha}{\phi}\right) = \frac{1}{\hat{\mu}}$$

And we conclude that the approach consisting in using the simple two-states model of figure 2.1 with a new rate $\hat{\mu}$ gives the correct answers in terms of the dependability metrics we are concerned with, i.e., the steady state unavailability $\overline{A}$, the mean time between failures MTBF, the mean up time MUT and the mean down time MDT. At this point, we may think that the heuristic used by the practitioners is quite interesting.

3. The Case with a Redundant Substructure.

We now consider a redundant substructure composed of two elements working in parallel where the word "parallel" is taken in the meaning of the reliability. If one element breaks down, it can be exchanged without stopping the system (the service remains available). The lifetime of each element is exponentially distributed with rate λ. All the random variables corresponding to the different lifetimes are independent and identically distributed (*iid*). We also assume that the exchange times are all *iid* according to an exponential distribution with rate μ. If one element breaks down while the other element is already down, the two exchanges are done by two distinct repairmen.

In a first step, we assume that spares are always available. We consider the CTMC corresponding to the description of the behavior of this simple redundant substructure. The CTMC is said to be in state i when i elements are in good state, $i = 0, 1, 2$. Its transition graph is represented on figure 3.1.

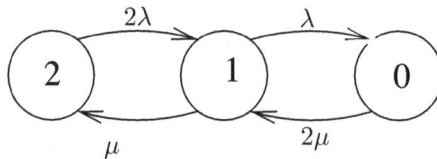

FIGURE 3.1. *Markovian transition graph - Redundant case - Exponential distribution of exchange times.*

Using again the notation $\rho = \lambda/\mu$, it is easy to check that the probability π_0 that the system is unavailable is equal to :

$$\pi_0 = \frac{1}{1 + \dfrac{1}{2\rho} + \dfrac{1}{\rho^2}} = \frac{\rho^2}{1 + 2\rho + \rho^2},$$

This probability π_0 corresponds to the steady state unavailability $\overline{A}$. Let us remarks that the behaviors of the two elements are independent (they works in parallel and there is no competition for a unique repairman), so the system unavailability is the product of the two element unavailabilities :

$$\overline{A} = \left(\frac{\rho}{1 + \rho} \right)^2$$

(3.1)

For example, if $\rho = 10^{-3}$, $\overline{A}$ is slightly lower than 10^{-6} when a second element is added for redundancy.

Because of the memoryless property of the exponential distribution, the sojourn of the MDT in state 0 is exponentially distributed with rate 2μ. This give us the expectation of a down time of the system[1] : MDT $= 1/2\mu$.

With this substructure, the mean up time MUT is the expected sojourn time in the subset $\{1, 2\}$ of the state space. Observing the transition graph of the CMTC, we conclude that these

---

[1] Since in this paper we only consider a substructure, we also use this term to refer to it.

sojourn times in the subset $\{1, 2\}$ are *iid* and also that π_0 can be seen as the time proportion that the CMTC spends in state 0. This allows us to write :

$$\pi_0 = \frac{MDT}{MUT + MDT} \,,$$

and from that

$$
\begin{aligned}
MUT &= MDT\frac{(1 - \pi_0)}{\pi_0} \\
&= \frac{1}{2\mu}\frac{(1 + 2\rho)}{\rho^2} \,,
\end{aligned}
$$

we finally get :

$$\lambda MUT = \frac{(1 + 2\rho)}{2\rho} = 1 + \frac{1}{2\rho} \,.$$

Again, by adding a second element for redundancy purpose, we see that if, for example, $\rho = 10^{-3}$, then λMUT is multiplied by 501. And because the new MDT is divided by 2 due to the existence of a second repairman, we may check that the steady state unavailability $\overline{A}$ is roughly divided by 1000. Which agrees with what was said above thanks to equation (3.1).

The steady state availability A is obtained directly from this equation (3.1) :

(3.2)
$$A = 1 - \overline{A} = \frac{1 + 2\rho}{1 + 2\rho + \rho^2} \,.$$

In a second step, we consider again that a spare part shortage may happen. Let us assume again that new spares arrive to the exchange place according to a Poisson process with rate γ and that there is a low probability α that the spare part shortage occurs. Let us first examine the results that the classical heuristic gives when we replace the rate μ by the rate $\hat{\mu}$. We get :

$$\lambda MUT = 1 + \frac{1}{2\rho\left(1 + \dfrac{\alpha}{\phi}\right)} \,,$$

$$MDT = \frac{1}{2\hat{\mu}} = \frac{1 + \alpha/\phi}{2\mu} \,,$$

(3.3)   and $$\overline{A} = \frac{\rho^2\left(1 + \dfrac{\alpha}{\phi}\right)^2}{1 + 2\rho\left(1 + \dfrac{\alpha}{\phi}\right) + \rho^2\left(1 + \dfrac{\alpha}{\phi}\right)^2} \,.$$

With respect to the situation where the spare part shortage did not happen, we observe a decrease of the mean up time MUT. For example, if $\alpha = \phi = 10^{-2}$, then the quantity $(\lambda MUT - 1)$ is divided by 2. Given that ρ is much smaller than one, this classical heuristic will produce a new unavailability that will be, as a first order approximation, multiplied by the factor $\left(1 + \dfrac{\alpha}{\phi}\right)^2$. With $\alpha = \phi = 10^{-2}$, then the new unavailability will be slightly multiplied by 4.

Let us now consider a more elaborated model where we try to capture the real behavior of the system in case of spare part shortage. This new model[2] corresponds to the new CMTC with the transition graph on figure 3.2. On that figure, a state denoted by xy means that one element of the structure is in state x while the other is in state y; where, again G denotes a good element, E denotes an element being exchanged and A denotes the impossibility of exchanging a broken element due to a spare part shortage. For example, in state GA, one element is up while the other is down and cannot be exchanged due to a spare part shortage. If the up element also breaks down, obviously no spare part is available for this second element and the CMTC moves to state AA.

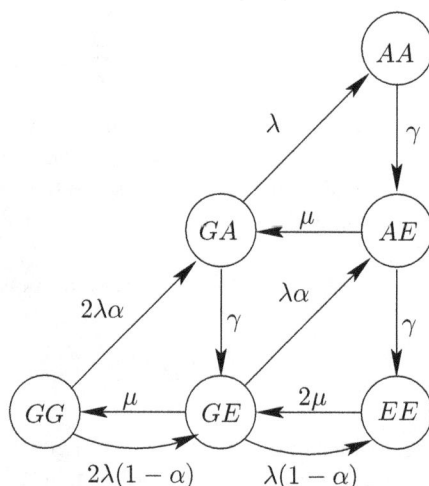

FIGURE 3.2. *Markovian transition graph - Redundant case - With spare part shortage*

In order to simplify the notations, let us rename the states of the CMTC by using numbers in the way of the figure 3.3.

Let us write the Chapman and Kolmogorov steady state equations for states $0, 1, 2, 3$ and 5 :

$$\gamma \pi_0 = \lambda \pi_3$$
$$(\mu + \gamma)\pi_1 = \gamma \pi_0 + \lambda \alpha \pi_4$$
$$2\mu \pi_2 = \gamma \pi_1 + \lambda(1 - \alpha)\pi_4$$
$$(\lambda + \gamma)\pi_3 = 2\lambda \alpha \pi_5 + \mu \pi_1$$
$$2\lambda \pi_5 = \mu \pi_4$$

Using the definitions $\rho \stackrel{\triangle}{=} \lambda/\mu$ and $\phi \stackrel{\triangle}{=} \gamma/\mu$, we can rewrite these equations as follows :

---

[2]We could build a larger model including the logistic loop of the non-functioning and of the repaired spares, but we don't need that to capture the phenomena we want to exhibit and also we would certainly not get the formal expressions obtained in this paper.

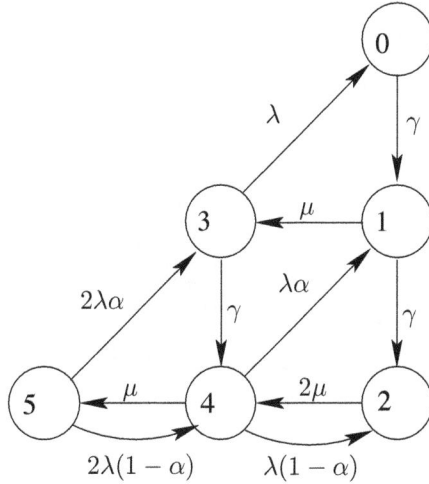

FIGURE 3.3. *Second Markovian transition graph - With numbered states*

$$(3.4) \qquad \phi\pi_0 = \rho\pi_3$$

$$(3.5) \qquad (1 + \phi)\pi_1 = \phi\pi_0 + \rho\alpha\pi_4$$

$$(3.6) \qquad 2\pi_2 = \phi\pi_1 + \rho(1 - \alpha)\pi_4$$

$$(3.7) \qquad (\rho + \phi)\pi_3 = 2\rho\alpha\pi_5 + \pi_1$$

$$(3.8) \qquad 2\rho\pi_5 = \pi_4$$

Thanks to equation (3.8), equation (3.7) is rewrite as

$$(3.9) \qquad (\rho + \phi)\pi_3 = \alpha\pi_4 + \pi_1$$

that allows us, with the help of equation 3.4, to rewrite equation 3.5 as :

$$(1 + \phi)\pi_1 = \phi\pi_0 + \rho[(\rho + \phi)\pi_3 - \pi_1]$$
$$= \phi\pi_0 + \rho[(\rho + \phi)\frac{\phi}{\rho}\pi_0 - \pi_1]$$

This gives us the following equation :

$$(3.10) \qquad \pi_1 = \phi\pi_0$$

Thanks to equations (3.4) and (3.10), the equation (3.9) allows us to obtain the probability π_4 as a function of π_0 :

$$(3.11) \qquad \pi_4 = \frac{\phi^2}{\alpha\rho}\pi_0$$

Combining this last equation with equation (3.8), we also obtain the probability π_5 as a function of π_0 :

$$(3.12) \qquad \pi_5 = \frac{\phi^2}{2\alpha\rho^2}\pi_0$$

Finally, equations (3.6), (3.10) and (2.1) allows us to obtain the probability π_2 as a function of π_0 :

$$(3.13) \qquad \pi_2 = \frac{\phi^2}{2\alpha}\pi_0$$

Let $D \stackrel{\triangle}{=} (2\alpha\rho^2(1+\phi) + \rho^2\phi^2 + 2\alpha\rho\phi + 2\rho\phi^2 + \phi^2)$. The normalizing equation gives us :

$$(3.14) \qquad \pi_0 = \frac{2\alpha\rho^2}{D}$$

The steady state unavailability $\overline{A}$ of this redundant structure is now equal to the sum of probabilities $(\pi_0 + \pi_1 + \pi_2)$, while the steady state availability A is now equal to the sum of probabilities $(\pi_3 + \pi_4 + \pi_5)$.

In order to exhibit the new expression of λMUT, let us consider the partition $\{\{3, 4, 5\}, \{0, 1, 2\}\}$ of the state space of the CMTC. Because this partition has just two elements, the frequency of departure from one subset toward the other has to be equal to the inverse of the sum of the two mean sojourn times in the subsets. With the help of figure 3.3, we see that the frequency of transitions from the subspace $\{3, 4, 5\}$ toward the subspace $\{0, 1, 2\}$ is equal to $\lambda(\pi_3 + \pi_4)$. This frequency is equal to the inverse of the sum (MUT +MDT) ; therefore :

$$(3.15) \qquad \lambda(MUT + MDT) = \frac{1}{\pi_3 + \pi_4}$$

But the steady state availability can also be written as :

$$A = \frac{\lambda MUT}{\lambda(MUT + MDT)},$$

and from these two points, we may write

$$\lambda MUT = \frac{\pi_3 + \pi_4 + \pi_5}{\pi_3 + \pi_4} = 1 + \frac{\pi_5}{\pi_3 + \pi_4} \ .$$

Then, using the expressions of the probabilities previously obtained, we get after some simplifications

$$\lambda MUT = 1 + \frac{1}{2\rho}\frac{\phi}{\alpha + \phi} = 1 + \frac{1}{2\rho} \times \frac{1}{1 + \alpha/\phi}$$

With respect to the classical heuristic, we get exactly the same expression. At this point, we may think that the heuristic used by the practitioners is quite smart!

Let us now look for the new expression of $2\mu MDT$ which was previously equal to one in the case of no spare part shortage and to $(1 + \alpha/\phi)$ according to the classical heuristic in the case of spare part shortage.

Let us start by getting the expression of λMDT using relation (3.15):

$$
\begin{aligned}
\lambda MDT &= \frac{1}{\pi_3 + \pi_4} - \lambda MUT \\
&= \frac{1}{\pi_3 + \pi_4} - \frac{\pi_3 + \pi_4 + \pi_5}{\pi_3 + \pi_4} \\
&= \frac{\pi_0 + \pi_1 + \pi_2}{\pi_3 + \pi_4} \\
&= \frac{\pi_0(1 + \phi + \phi^2/2\alpha)}{\pi_0(\phi/\rho + \phi^2/\alpha\rho)} \\
&= \frac{\rho[2\alpha(1 + \phi) + \phi^2]}{2\phi(\alpha + \phi)} \\
&= \frac{\rho}{2}\left[1 + \frac{\alpha(2 + \phi)}{\phi(\alpha + \phi)}\right]
\end{aligned}
$$

This gives us immediately the result:

$$
2\mu MDT = 1 + \frac{\alpha}{\phi} \times \left(\frac{1 + 2/\phi}{1 + \alpha/\phi}\right)
$$

This result is different from the one obtained by the classical heuristic.

Here, for example, if $\alpha = \phi = 10^{-2}$, then the quantity $2\mu MDT$ is multiplied by 101,5....while, with the classical heuristic, this quantity is just multiplied by 2. and if ρ is still equal to 10^{-3}, because of the spare part shortage, the steady state unavailability $\overline{A}$ of this redundant structure is multiplied by 202.5... while, using the classical heuristic, the steady state unavailability $\overline{A}$ would be just multiplied by 3.99 !

So, with respect to the heuristic consisting in using a new repair rate $\hat{\mu}$ as a way to take the spare part shortage into account, we see that, in case of a redundant substructure, this heuristic is not really worthwhile since it does not take the causality factor into consideration (given that an exchange operation is blocked due to a spare part shortage, if the working element breaks down, then the new exchange operation is blocked with probability 1 (due to the spare part shortage).

4. Numerical Examples. On figure 4.1 we represent the unavailabilities as a function of the probability α, obtained both with the classical heuristic and with the new approach ; when $\phi = 10^{-2}$ and $\rho = 10^{-3}$. We observe first the high influence of the shortage probability on the unavailability that increase from 10^{-6} for $\alpha = 0.0$ to 2×10^{-3} for $\alpha = 0.1$. We also note that the gap between the two answers is quite significant. This denies the interest of the classical heuristic in such a situation where the substructure is redundant and for these values of the parameters.

On figures 4.2a and 4.2b we plot the same functions but for different values of the parameter ϕ. On figure 4.2a, $\phi = 10^{-1}$ and the variation of unavailability is less affected than on figure 4.1, however the relative gap between the two answers stays important. On figure 4.2b, where $\phi = 10^{-3}$, the variation of unavailability is so important that with the new approach the unavailability is greater than 0.1. And the gap between the two answers stays quite significant.

On figure 4.3 we represent the ratio of the unavailability obtained by the new approach over the unavailability given by the classical heuristic as a function of the probability α, when $\phi = 10^{-2}$ and $\rho = 10^{-3}$. Of course, for $\alpha = 0$, this ratio equals 1, but increases up to 50

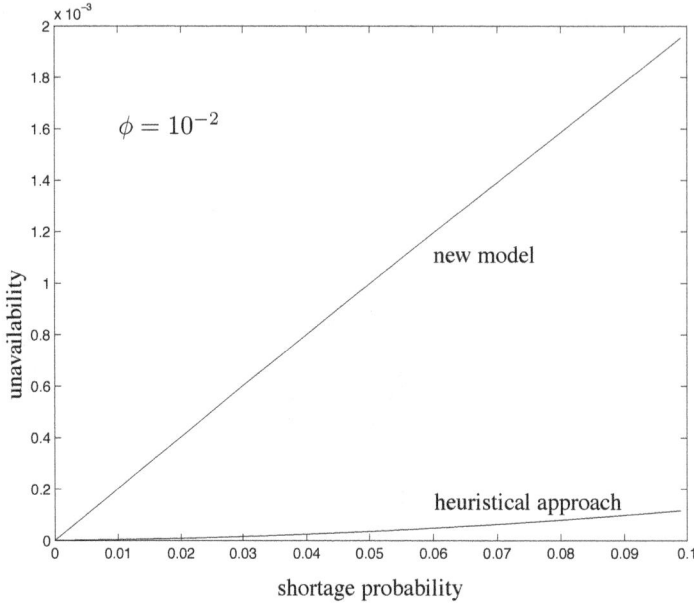

FIGURE 4.1. *Comparaison of the UN availabilities given by the two approaches ; $\rho = 10^{-3}$.*

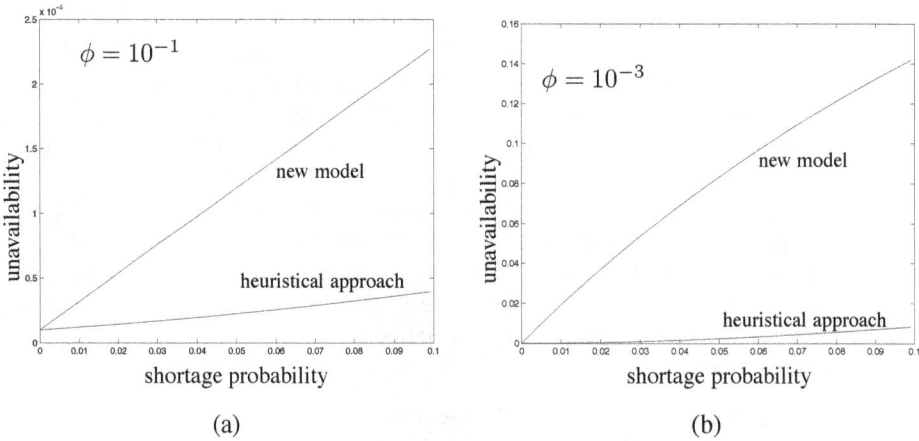

(a)   (b)

FIGURE 4.2. *Comparaison of the UN availabilities given by the two approaches ; $\rho = 10^{-3}$.*

when α approaches 10^{-2} and then decreases slowly when α keeps increasing. Note that this ratio close to 50 matches with the one we already obtained at the end of the previous section.

When $\phi = 10^{-1}$ (figure 4.4a) the ratio curve is again unimodal and reaches a maximum value close to 5.5. With $\phi = 10^{-3}$ (figure 4.4b) the maximum is close to 500. Although we believe that such a value of ϕ has to be exceptional, we may say that this result is astonishing. It is also interesting to remark that for the three cases, the maxima of the ratios are obtained when α is almost equal to ϕ.

Let us recall that the expressions of the λMUT are identical for the two approaches. This is easy to check that its derivative with respect to ϕ is positive while its derivative with respect to α is negative. On figure 4.5 we represent the expression of λMUT as a function of the probability α, when $\phi = 10^{-2}$ and $\rho = 10^{-3}$. For $\alpha = 0$, λMUT is equal to 500 (for any

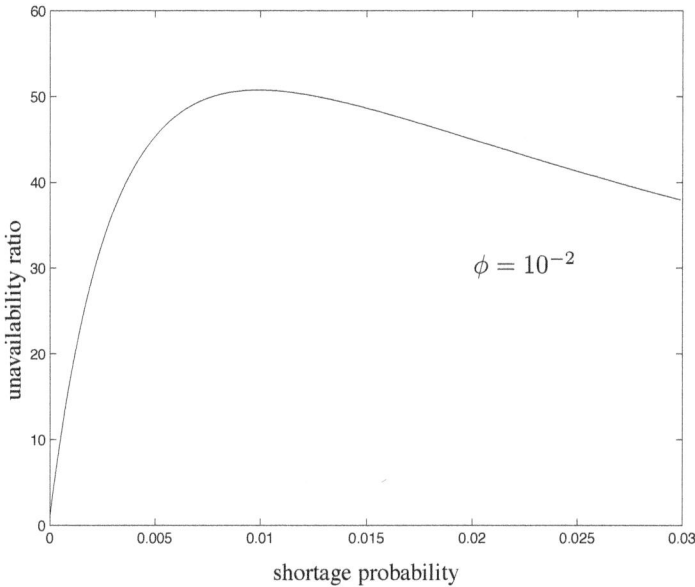

FIGURE 4.3. *Unavailability ratio ;* $\rho = 10^{-3}$.

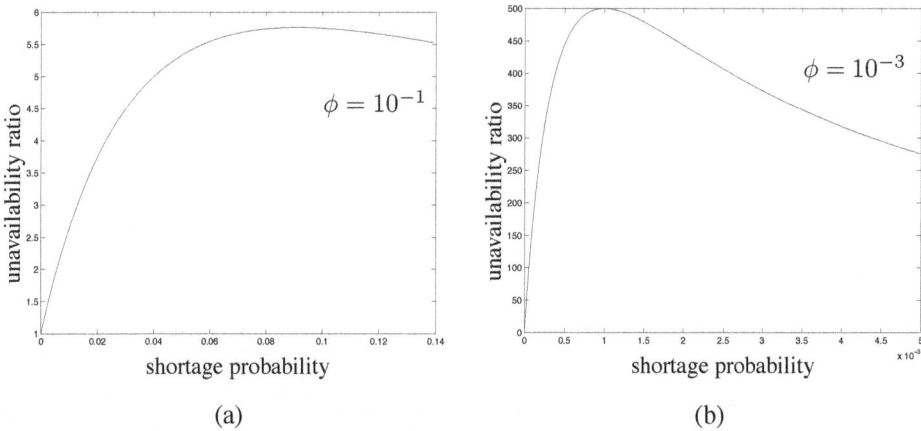

(a)   (b)

FIGURE 4.4. *Unavailability ratio ;* $\rho = 10^{-3}$.

value of ϕ).

On figures 4.6a and 4.6b we plot the same function λMUT but for different values of the parameter ϕ. When $\phi = 10^{-1}$ (figure 4.6a) the slope of the curve is less negative than on figure 4.5 since when $\alpha = 0.1$, the value of the function is about 250. On the opposite, with $\phi = 10^{-3}$ (figure 4.6b) the slope of the curve is very negative and the value of the function is about 50 when α is only 0.01.

On figure 4.7 we represent the ratio of the MDT obtained by the new approach over the unavailability given by the classical heuristic as a function of the probability α, when $\phi = 10^{-2}$ and $\rho = 10^{-3}$. Of course, for $\alpha = 0$, this ratio equals 1, but increases up to 50 when α approaches 10^{-2} and then decreases slowly when α keeps increasing. With these parameters, the unavailability is still relatively low (lower than 10^{-3}) and this ratio of the MDT behaves similarly to the ratio of the unavailability (cf. equation (2.1)).

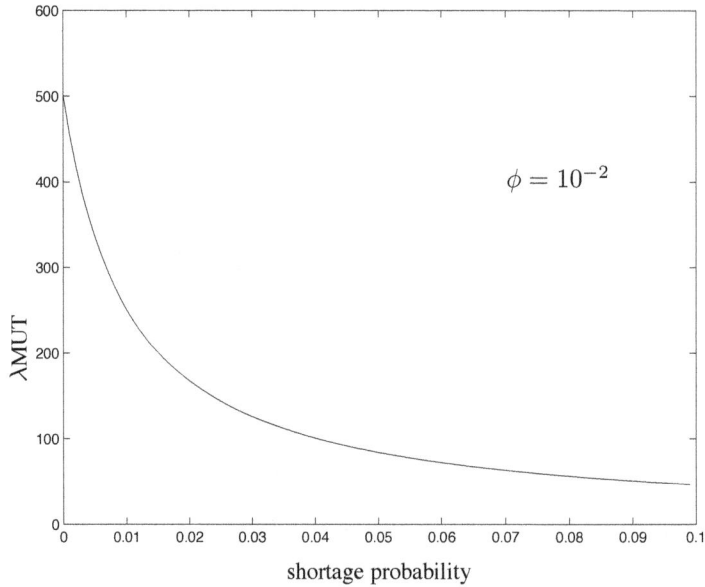

FIGURE 4.5. λMUT ; $\rho = 10^{-3}$.

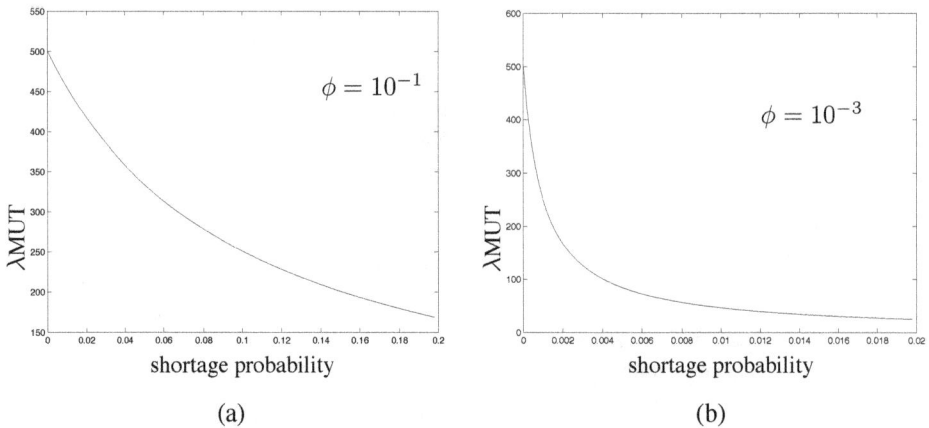

(a)   (b)

FIGURE 4.6. λMUT ; $\rho = 10^{-3}$.

When $\phi = 10^{-1}$ (figure 4.8a) the ratio curve is again unimodal and reaches a maximum value close to 5.5. With $\phi = 10^{-3}$ (figure4.8b) the maximum is close to 500. Theses figures help us to see that the gap between the two models is particularly large when the ratio α/ϕ is around one. When α is larger than ϕ, then the ratio decreases, quicker if ϕ is smaller.

To conclude this section, let us assume that a global structure consists of a series of 100 such substructures, with $\rho = 10^{-3}$. If no spare part shortage occurs, then the value of the unavailability of the global structure is close to 10^{-4}. Suppose now that spare part shortage can occur and that $\alpha = \phi = 10^{-2}$, then, the answer obtained with the classical heuristic will be close to 4×10^{-4}, while with the new approach (that we believe to be the right way) the answer will be close to 2×10^{-2}.

5. Conclusions. In this study, we have emphasized the point that modeling through Markov chain may be a question of expertise. We used an example taken from the area of

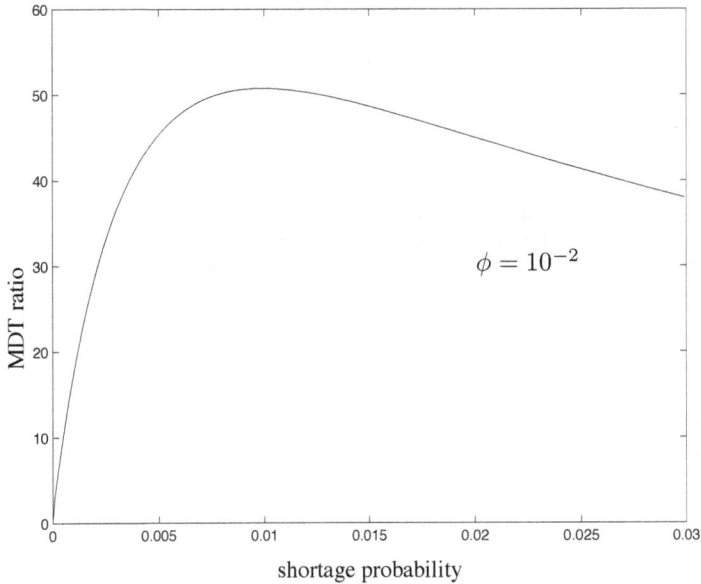

FIGURE 4.7. *MDT ratio* ; $\rho = 10^{-3}$.

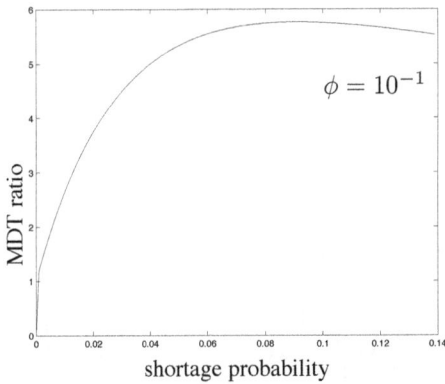

(a)

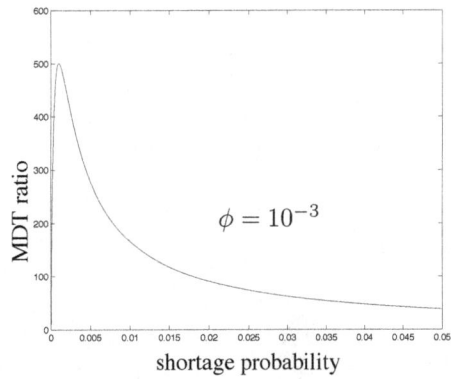

(b)

FIGURE 4.8. *MDT ratio* ; $\rho = 10^{-3}$.

dependability, in a situation where a spare part shortage may occur very rarely. For this real application, a classical heuristic is used by the specialists from the field. In this presentation, we show that, while a classical heuristic may give "exact" results in some simple cases, this approach may underestimate some criterion such as the unavailability by a factor of hundreds. We proposed a new approach and give the formal expressions of several dependability metrics in the case of a simple but classical redundant architecture. For larger applications, it would be with no doubts straitforward to get the steady state distribution of the Markov chain directly through a numerical solver.

As we shown it in the previous section, dramatic gaps may exist between a (bad) prediction and the (not yet known) real answer. While, very often, approximations on the real probability distribution of the duration of a given activity have little effects on the results of the prediction.

REFERENCES

[1] R.E. BARLOW AND F. PROSCHAN, *Statistical Theory of Reliability and Life Testing*, Holt, Rinehart and Winston, NY, 1975.

[2] E. CINLAR, *Introduction to stochastic Processes*, Prentice Hall, New-Jersey, 1975.

[3] W. FELLER, *An Introduction to Probability Theory and its Applications*, Volumes I and II, Third ed., John Wiley and Sons, 1968.

[4] D. P. HEYMAN AND M. J. SOBEL, *Stochastic models in operations research : vol.I : stochastic processes and operating characteristics*, McGraw-Hill, 1982.

[5] A. PAPOULIS, *Probability, Random Variables and Stochastic Processes*, Third ed., McGraw-Hill, 1991.

[6] R. A. SAHNER, K. S. TRIVEDI AND A. PULIAFITO, *Performance and Reliability Analysis of Computer Systems: An Example-Based Approach Using the SHARPE Software Package*, Kluwer Academic Publishers, 1996.

STRUCTURED MARKOV CHAINS IN APPLIED PROBABILITY AND NUMERICAL ANALYSIS.

GUY LATOUCHE *

Abstract.
About thirty years ago, Quasi-Birth-and-Death processes and Skip-Free Markov chains came to the attention of applied probabilists. One of their prominent features is that their analysis requires the resolution of nonlinear equations, involving matrix-polynomial or matrix power series. At first, these were tackled 'in-house' and very soon several algorithms appeared which had their justification grounded, to a large extent, in probabilistic thinking.

Soon, these equations caught the attention of numerical analysts who brought to bear their own special way of thinking about such problems and, not surprisingly, obtained improved algorithms in terms of convergence speed or numerical accuracy.

The interaction between the two lines of approach are very exciting and this is an attempt to illustrate how the one meshes into the other.

Key words. Structured Markov chains, Quasi-Birth-and-Death processes, matrix polynomial equations

AMS subject classifications. 60K15, 60K37

1. Introduction. *Structured Markov chains* is a keyword for a family of Markov processes with a particular repetitive structure which may be described as follows: the state space forms a strip in the plane, along the horizontal axis, say, and the transition probability from (x, y) to (x', y') depends individually on y and y' but only on $x' - x$ if $x > 0$, not on x and x' individually; the states of the form $(0, y)$ are called *boundary* states and have a special status.

Because of this, the transition matrix has a block-Toeplitz structure, except, again, for the rows and columns corresponding to the boundary states; this exception reflects the special status of these states. Of course, we write "transition matrix" only if the state space is discrete, as when x takes values in the non-negative integers and y in an interval $\{1, 2, \ldots, m\}$ for some finite m.

The theory is not restricted to discrete state spaces but the algorithmic developments concern that case, for the most part, and we shall deal in the first part of this paper with discrete state space Markov chains.

One simple way of thinking about these Markov processes is that the two dimensions in the state space correspond to different systems: x is the number of customers in a queue, say, or the number of items in a buffer, and it is called the *level*. That number cannot be negative but it may be arbitrarily large, assuming that the buffer has an infinite capacity. The level is driven by a random environment, and y is the state of that environment, also called the *phase*. When the phase is y, the rules of evolution of the level are not the same as when the phase is y', because, for instance, y represents a condition when the arrival process to the queue has a high intensity while the converse is true when the phase is y'. It must be emphasized that applications can be much more involved than in this example.

In order to compute the stationary distribution, one must solve a nonlinear matrix equation, which may be done in a variety of ways. Some of the algorithms which have appeared may easily be justified by thinking about the dynamic behavior of

---

*Université Libre de Bruxelles, Département d'Informatique, CP 212, Boulevard du Triomphe, 1050 Bruxelles, Belgium.

the stochastic processes; for others, the justification is mostly analytical. Two early references are Evans [8] and Wallace [30] but it is really with the work of Neuts [21, 22] that the field took off in the applied probability community; it caught the attention of numerical analysts a few years later, in Bini and Meini [6] and Stewart [27].

The paper is organized as follows. We start with Quasi-Birth-and-Death (QBD) processes, for which we need to deal with *quadratic* equations only. We present some of the simplest resolution algorithms in Section 2 and the fastest ones in Section 3. Section 4 is devoted to the general case of skip-free processes and polynomial equations. We systematically assume in these sections that we deal with discrete-time, irreducible, positive recurrent Markov chains. In the last section we present a class of continuous-time processes, called *Fluid Queues,* for which the level takes arbitrary nonnegative *real* values; here the equations to be solved are algebraic Riccati equations and one observes, again, the interplay between Probability Theory and Numerical Analysis.

Full details about the cited theorems may be found in the classical books of Neuts [22, 23], in Latouche and Ramaswami [17] and in Bini, Latouche and Meini [5]; the last reference follows the approach from Numerical Analysis mostly, while the other three heavily rely on Probability Theory. References about fluid queues are given in the last section. The books of Asmussen [1], Nelson [20], Stewart [28] and Tran-Gia [29] also have chapters on these matters.

2. QBDs — Quadratic Equations. Discrete-time QBDs $\{(X_n, \varphi_n) : n \geq 0\}$ are processes on the two-dimensional state space $\{0, 1, 2, \ldots\} \times \{0, 1, 2, \ldots, m\}$ such that the only feasible transitions from the state (i, j) is to one of the states $(i+1, j')$, (i, j') or $(i-1, j')$, for some j', provided that $i-1$ does not become negative. With X_n being the level at time n and φ_n the phase, this means that the process may move from one level to the next but may not jump several levels at a time, while there is no a priori restriction on the change of phase. If one thinks of the queueing system example, this means that customers arrive or depart one at a time.

The matrix of transition probabilities[1] is given as

$$
(2.1) \qquad P = \begin{bmatrix} B_0 & A_1 & & & & \\ A_{-1} & A_0 & A_1 & & & \\ & A_{-1} & A_0 & A_1 & & \\ & & A_{-1} & A_0 & \ddots & \\ & & & & \ddots & \ddots \end{bmatrix},
$$

where B_0, A_{-1}, A_0 and A_1 are nonnegative matrices of order m; the component $(A_1)_{j,j'}$, for instance, is the transition probability from (n, j) to $(n+1, j')$, for any n, and the elements of the matrices A_0, A_{-1} and B_0 have similar interpretations.

We denote by $\boldsymbol{\pi}$ the stationary probability vector, that is, the unique solution of the system $\boldsymbol{\pi}^{\mathrm{T}} = \boldsymbol{\pi}^{\mathrm{T}} P$, $\boldsymbol{\pi}^{\mathrm{T}}\mathbf{1} = 1$, and we partition it as $\boldsymbol{\pi} = (\boldsymbol{\pi}_n : n = 0, 1, \ldots)$, with $\pi_{n,j}$ being the stationary probability of the state (n, j).

THEOREM 2.1. *If the QBD is positive recurrent, then*

$$
\boldsymbol{\pi}_n^{\mathrm{T}} = \boldsymbol{\pi}_0^{\mathrm{T}} R^n, \qquad for \ n \geq 0
$$

where R is the minimal nonnegative solution of the matrix-quadratic equation

$$
(2.2) \qquad X = X^2 A_{-1} + X A_0 + A_1
$$

---

[1]Notations for the elements of that matrix vary; we follow those in [5].

and $\boldsymbol{\pi}_0^{\mathrm{T}}$ is the unique solution of the system

$$\boldsymbol{\pi}_0^{\mathrm{T}}(B_0 + A_1 G) = \boldsymbol{\pi}_0^{\mathrm{T}}$$
$$\boldsymbol{\pi}_0^{\mathrm{T}}(I - R)^{-1}\mathbf{1} = 1$$

where G is the minimal nonnegative solution of

$$(2.3) \qquad X = A_{-1} + A_0 X + A_1 X^2.$$

Furthermore,

$$(2.4) \qquad R = A_1(I - A_0 - A_1 G)^{-1}.$$

This is known as the *matrix-geometric* distribution; the matrices R and G each have a probabilistic interpretation which is very useful in the development of the theory:

- the component $R_{i,j}$ is the expected number of visits that the QBD will make to the state $(1,j)$ in level one before it returns to level zero for the first time, if we assume that the QBD starts in state $(0,i)$ in level zero at time zero;
- the component $G_{j,k}$ is the probability that at some time in the future, the QBD will be in level zero and that the first state visited in level zero is $(0,k)$, if we assume that the QBD starts in state $(1,j)$ in level one at time zero.

The equation $G = A_{-1} + A_0 G + A_1 G^2$ for G is actually very simple to understand: starting from $(1,j)$, the QBD may move down to level zero at once — that is the first term in the right-hand side, or it may stay in level one from where it still has to go down to level zero — that is the second term, or it may move up to level two, from where it will have to go down to level one, and then it will still have to go down to level zero — that is the third term. The proof that R is the minimal solution of (2.2) is slightly more involved.

The reason why the inverse matrix in (2.4) exists is that its components count the expected number of visits to the states in level one, assuming that the process starts from *level one*, while avoiding the level zero: it is well known that for irreducible processes, such expected number of visits to some states, while avoiding some other subset of states, are always finite and that, in short, is the reason why *all* the inverse matrices which will appear in our equations do exist.

The matrix G is stochastic (it is such that $G\mathbf{1} = \mathbf{1}$) and that makes it a slightly better matrix to work with than the matrix R. This is the reason why the work on algorithms focused on the equation (2.3) for G, with the matrix R being eventually determined through (2.4).

We obtain two other equations for G by simple formal manipulation of (2.3):

$$(2.5) \qquad X = (I - A_0)^{-1}(A_{-1} + A_1 X^2)$$
$$(2.6) \qquad X = (I - A_0 - A_1 X)^{-1}A_{-1}.$$

Each of these fixed point equations may be solved by functional iteration, (2.6) being particularly interesting with that respect, as stated in the following two theorems.

THEOREM 2.2. *The sequence* $\{X_n\}_{n\geq 0}$*, with* $X_0 = 0$ *and*

$$(2.7) \qquad X_{n+1} = (I - A_0 - A_1 X_n)^{-1}A_{-1}$$

monotonically converges to the matrix G, that is, $X_{n+1} - X_n \geq 0$ for all n and $\lim_n X_n = G$.

To prove this property is quite straightforward as soon as one realizes that the successive iterations may be interpreted in terms of the dynamics of the stochastic process: assume that the QBD starts in state $(1, j)$ in level one at time zero; the component $(X_n)_{j,k}$ is the probability that at some time in the future, the QBD will be in level zero and that the first state visited in level zero is $(0, k)$ *and* in the interval, the process has not moved beyond level n.

One may also define convergent sequences based on (2.3, 2.5) but the sequence of Theorem 2.2 is the more interesting, for two reasons. The first is that it is the sequence which converges fastest among the three: if we respectively denote by $\{X_n^{(N)}\}_{n \geq 0}$ and $\{X_n^{(T)}\}_{n \geq 0}$ the sequences obtained from (2.3) and (2.5), starting with $X_0^{(N)} = X_0^{(T)} = 0$, then $X_n^{(N)} \leq X_n^{(T)} \leq X_n$ for any $n \geq 0$. The second reason is that it is the simplest to interpret (see Lucantoni [18] for the interpretation of $X_n^{(N)}$).

Another way to show that $\{X_n\}$ is the best of the three sequences, is through a comparison of their convergence rates

$$r = \lim_n \|G - X_n\|^{\frac{1}{n}}, \qquad r_N = \lim_n \|G - X_n^{(N)}\|^{\frac{1}{n}}, \qquad r_T = \lim_n \|G - X_n^{(T)}\|^{\frac{1}{n}}$$

to G.

THEOREM 2.3. *The convergence rates of the three sequences* $\{X_n\}_{n \geq 0}$, $\{X_n^{(N)}\}_{n \geq 0}$ *and* $\{X_n^{(T)}\}_{n \geq 0}$ *are given by*[2]

$$r = \rho(R), \qquad r_N = \rho(R_N), \qquad r_T = \rho(R_T),$$

where R is given by (2.4) and

$$R_N = A_0 + A_1 + A_1 G$$
$$R_T = (A_1 + A_1 G)(I - A_0)^{-1}.$$

Moreover, if the QBD is positive recurrent, then $\rho(R) \leq \rho(R_N) \leq \rho(R_T) < 1$ and if it is null recurrent, then $\rho(R) = \rho(R_N) = \rho(R_T) = 1$.

The proof is based on the fact that the matrices R, R_N and R_T are obtained from three different regular splittings of the nonsingular M-matrix $I - A_0 - A_1 - A_1 G$.

A direct consequence of this theorem is that, if the QBD process is positive recurrent, then the convergence of the three sequences is linear, with the sequence $\{X_n\}$ converging faster than the other two; if the QBD is null recurrent, then the convergence of all three sequences is sub-linear. Now, a simple way to accelerate the convergence is to start with a matrix X_0 which is already stochastic. Take the sequence

$$Y_0 = I, \qquad Y_{n+1} = (I - A_0 - A_1 Y_n)^{-1} A_{-1}$$

for instance. The sequence $\{Y_n\}$ is well defined and converges faster than the sequence $\{X_n\}$ which starts from the null matrix; moreover, each matrix Y_n is stochastic. One might start with Y_0 being some other stochastic matrix but one needs to exercise some caution, in order to ensure that the matrix $I - A_0 - A_1 Y_n$ remains nonsingular at each step. Details are given in [17, Section 8.3].

---

[2]We denote by $\rho(M)$ the spectral radius of the matrix M

Since both G and the Y_n's are stochastic matrices, the differences $G - Y_n$ are no longer nonnegative and the error analysis is more involved. In the case where the matrix R is irreducible, the result is simple to express, and we give it below; we refer to [5, Section 6.2.3] for the general case.

THEOREM 2.4. *If the QBD is positive recurrent and if R is irreducible, then the convergence rate of the sequence $\{Y_n\}$ is strictly less than $\rho(G^T \otimes R)$, which is itself equal to $\rho(R)$.*

An important observation which we make here is that the sequences defined so far are self-correcting: errors introduced at any step by floating point arithmetics are corrected during subsequent iterations. This is not true for cyclic reduction, a much faster procedure which we describe below.

3. Quadratic Convergence. Let us define $\mathcal{F}(X) = X - (I - A_0 - A_1 X)^{-1} A_{-1}$ so that G is a solution of $\mathcal{F}(X) = 0$ by (2.6). Newton's method may be applied to that equation and it has been shown that the iteration is well defined and that the convergence is monotonic and quadratic under very weak conditions.

THEOREM 3.1. *If the QBD is positive recurrent and if the matrix $A_{-1} + A_0 + A_1$ is irreducible, if Z_0 is chosen such that $0 \le Z_0 \le G$, and $\mathcal{F}(Z_0) \le 0$, then Newton's iterates*

$$Z_{n+1} = Z_n - [\mathcal{F}'_{Z_n}]^{-1}(\mathcal{F}(Z_n)), \qquad \text{for } n \ge 0,$$

where $\mathcal{F}'_X$ is the Gateaux derivative of $\mathcal{F}$ evaluated at X, are well defined, form a nondecreasing sequence and converge to G as n tends to infinity. There exists a positive constant c such that $\|Z_{n+1} - G\| \le c\|Z_n - G\|^2$, for all $n \ge 0$.

In order to implement this procedure, we need to evaluate expressions of the form $[\mathcal{F}'_X]^{-1}(\mathcal{F}(X))$. That means that for a given matrix X, we need to find the unique matrix W such that $[\mathcal{F}'_X](W) = \mathcal{F}(X)$ or

$$(3.1) \qquad W - H(X)A_1 W H(X)A_{-1} = H(X)A_{-1},$$

with $H(X) = (I - A_0 - A_1 X)^{-1}$.

Another quadratically convergent procedure is the cyclic reduction algorithm of Bini and Meini [6], designed specifically to exploit the Toeplitz-like structure of the transition matrix P in (2.1). It is similar to the logarithmic reduction algorithm of Latouche and Ramaswami [16] and slightly more efficient, and for that reason we only describe the former.

The equation $G = A_{-1} + A_0 G + A_1 G^2$ may be transformed into the linear system

$$\begin{bmatrix} I - A_0 & -A_1 & & \\ -A_{-1} & I - A_0 & -A_1 & \\ & -A_{-1} & I - A_0 & \ddots \\ & & \ddots & \ddots \end{bmatrix} \begin{bmatrix} G \\ G^2 \\ G^3 \\ \vdots \end{bmatrix} = \begin{bmatrix} A_{-1} \\ 0 \\ 0 \\ \vdots \end{bmatrix}.$$

If we apply an even–odd permutation to the blocks of rows and columns, followed by one step of Gaussian elimination to remove the even-numbered blocks, we obtain

$$\begin{bmatrix} I - \widehat{A}_0^{(1)} & -A_1^{(1)} & & \\ -A_{-1}^{(1)} & I - A_0^{(1)} & -A_1^{(1)} & \\ & -A_{-1}^{(1)} & I - A_0^{(1)} & \ddots \\ & & \ddots & \ddots \end{bmatrix} \begin{bmatrix} G \\ G^3 \\ G^5 \\ \vdots \end{bmatrix} = \begin{bmatrix} A_{-1} \\ 0 \\ 0 \\ \vdots \end{bmatrix}.$$

which has nearly the same structure. Repeating this procedure, we have after n steps the system

(3.2)
$$\begin{bmatrix} I - \widehat{A}_0^{(n)} & -A_1^{(n)} & & \quad 0 \\ -A_{-1}^{(n)} & I - A_0^{(n)} & -A_1^{(n)} & \\ & -A_{-1}^{(n)} & I - A_0^{(n)} & \ddots \\ 0 & & \ddots & \ddots \end{bmatrix} \begin{bmatrix} G \\ G^{2^n+1} \\ G^{2\cdot 2^n+1} \\ G^{3\cdot 2^n+1} \\ \vdots \end{bmatrix} = \begin{bmatrix} A_{-1} \\ 0 \\ 0 \\ \vdots \end{bmatrix}$$

where

(3.3)
$$A_{-1}^{(n+1)} = A_{-1}^{(n)}(I - A_0^{(n)})^{-1}A_{-1}^{(n)},$$
$$A_0^{(n+1)} = A_0^{(n)} + A_{-1}^{(n)}(I - A_0^{(n)})^{-1}A_1^{(n)} + A_1^{(n)}(I - A_0^{(n)})^{-1}A_{-1}^{(n)},$$
$$A_1^{(n+1)} = A_1^{(n)}(I - A_0^{(n)})^{-1}A_1^{(n)},$$
$$\widehat{A}_0^{(n+1)} = \widehat{A}_0^{(n)} + A_1^{(n)}(I - A_0^{(n)})^{-1}A_{-1}^{(n)},$$

for $n = 0, 1, \ldots$, with $\widehat{A}_0^{(0)} = A_0$, $A_i^{(0)} = A_i$, $i = -1, 0, 1$.

THEOREM 3.2. *The sequence $\{X_n^{(C)}\}$ with $X_n^{(C)} = (I - \widehat{A}_0^{(n)})^{-1}A_{-1}$ monotonically converges to G. If the QBD is positive recurrent, then there exists some $\gamma > 0$ such that $\|G - X_n^{(C)}\| \le \gamma r^{2^n}$ for all n.*

The whole procedure is well-defined because the first equation in (3.2), which we write as

$$G = A_{-1} + \widehat{A}_0^{(n)}G + A_1^{(n)}G^{2^n+1},$$

is interpreted as follow: to go down from level one to level zero, the process may do so at time one (first term) *or* at some unspecified time later, before going up to level 2^n (second term) *or* after going up to level 2^n (third term). This implies also that $I - \widehat{A}_0^{(n)}$ is non singular for any n and that $\lim_n A_1^{(n)} = 0$.

Each step of cyclic reduction costs one matrix inversion and six matrix multiplications. This is more than each step in the iterations of Theorem 2.2, but this is usually amply compensated by the much smaller number of iterations which are required to reach a given precision. However, we have mentioned earlier that the algorithms based on functional iteration are self-correcting. This is not true of cyclic reduction, which may be a disadvantage in some cases.

A comparison between Newton's scheme and cyclic reduction needs to be made. Equation (3.1) is a Sylvester equation, for which Gardiner *et al.* [9] provide a $O(m^3)$ resolution algorithm. In Latouche [15], the equation is solved through the traditional algorithm (Lancaster and Tismenetsky [13]), which has a complexity of $O(m^6)$ and made Newton's method a terribly inefficient procedure, when compared to cyclic reduction.

As a matter of fact, both may be interpreted in terms of some limited set of trajectories which the QBD is allowed to follow at step n, with the restrictions being weakened as n goes to infinity. Under normal circumstances, the trajectories of cyclic reduction are more restricted than those of the Newton scheme, so that, in general, Newton's scheme requires fewer iterations than cyclic reduction, although each iteration takes a little more time. Nevertheless, there are examples where Newton's scheme requires more iterations.

Guo [10] has shown that the convergence of cyclic reduction is linear when the QBD is *null recurrent*, with rate $1/2$. This entails that the convergence of cyclic reduction may be slow in some cases, even with the QBD being positive recurrent.

Actually, the convergence rate of cyclic reduction is the product of the spectral radius r of R and the spectral radius g of G, so that $\|G - X_n^{(C)}\| \le \gamma(gr)^{2^n}$ for some γ. The matrix G being stochastic, its spectral radius is 1 and this explains why we gave the bound γr^{2^n} in Theorem 3.2. In order to accelerate the convergence, one might shift the eigenvalue one of G to zero, solve the problem and shift the eigenvalue back to 1. This is accomplished as follows: take any vector $\boldsymbol{u} > \boldsymbol{0}$, such that $\boldsymbol{u}^T \boldsymbol{1} = 1$, define $Q = \boldsymbol{1}\boldsymbol{u}^T$ and

$$\widetilde{A}_{-1} = A_{-1} - A_{-1}Q, \qquad \widetilde{A}_0 = A_0 - A_1 Q, \qquad \widetilde{A}_1 = A_1.$$

If $\widetilde{X}$ is the minimal nonnegative solution of the equation $X = \widetilde{A}_{-1} + \widetilde{A}_0 X + \widetilde{A}_1 X^2$, then $G = \widetilde{X} + Q$. Details are to be found in He, Meini and Rhee [12] and in [5, Section 8.2.1].

4. Skip-free Markov chains — Polynomial Equations. The techniques described in the sections above may be extended to the case of skip-free Markov chains, for which the transition matrices have the M/G/1-type structure

(4.1)
$$P = \begin{bmatrix} B_0 & B_1 & B_2 & B_3 & \cdots \\ A_{-1} & A_0 & A_1 & A_2 & \cdots \\ & A_{-1} & A_0 & A_1 & \ddots \\ & & A_{-1} & A_0 & \ddots \\ \mathbf{0} & & & \ddots & \ddots \end{bmatrix}$$

or the GI/M/1-type structure

(4.2)
$$P = \begin{bmatrix} \widehat{B}_0 & \widehat{A}_1 & & & \mathbf{0} \\ \widehat{B}_{-1} & \widehat{A}_0 & \widehat{A}_1 & & \\ \widehat{B}_{-2} & \widehat{A}_{-1} & \widehat{A}_0 & \widehat{A}_1 & \\ \widehat{B}_{-3} & \widehat{A}_{-2} & \widehat{A}_{-1} & \widehat{A}_0 & \ddots \\ \vdots & \vdots & \ddots & \ddots & \ddots \end{bmatrix}.$$

Thinking in terms of queues, (4.1) corresponds to situations where customers arrive in groups and departs singly, while (4.2) corresponds to single arrivals and group departures.

The key to the evaluation of the stationary distribution is the minimal nonnegative solution G to the equation

(4.3)
$$X = A_{-1} + A_0 X + A_1 X^2 + A_2 X^3 + \cdots$$

in the first case, and the minimal nonnegative solution R of

(4.4)
$$X = \widehat{A}_1 + X\widehat{A}_0 + X^2 \widehat{A}_{-1} + X^3 \widehat{A}_{-2} + \cdots$$

in the second case.

Again, we concentrate on algorithms to determine G since they may directly be transformed into algorithms for R through the duality principle in Ramaswami [24]. The best method based on functional iteration is a direct analogue of (2.7). The sequence

$$X_{n+1} = (I - A_0 - A_1 X_n - A_2 X_n^2 - \cdots)^{-1} A_{-1}$$

with $X_0 = O$ monotonically and linearly converges to G. If $X_0 = I$, convergence is also guaranteed and the successive approximations are all stochastic matrices.

Newton's method is applicable to $\mathcal{F}(X) = X - (I - A_0 - A_1 X - A_2 X^2 - \cdots)^{-1} A_{-1}$ and produces quadratically convergent sequences. However, each iteration requires that one solves a linear equation of the form

$$W - H_n \sum_{i \geq 1} A_i \sum_{0 \leq j \leq i-1} X_n^j W X_n^{i-1-j} H_n A_{-1} = H_n A_{-1}$$

with $H_n = (I - A_0 - A_1 X_n - A_2 X_n^2 - \cdots)^{-1}$, for which there is no simple efficient procedure.

Cyclic reduction *is* a viable, quadratically convergent alternative. The principle remains the same: separate the odd- from the even-numbered levels, perform one step of Gaussian elimination to remove the even-numbered levels, and repeat as needed. The proof of convergence remains the same also. The implementation, however, is more delicate since, in addition to the errors introduced by floating point arithmetics, one has to deal with the unavoidable approximations of the power series $A_0 + A_1 X + A_2 X^2 + \cdots$).

Algorithm 7.5 in [5] has complexity $O(m^3 d + m^2 d \log d)$, where d is the maximum index of the blocks A_i which are significantly different from zero, it relies on the Toeplitz structure of the transition matrix and on FFT for matrix polynomial products.

5. Fluid Queues — Riccati Equations. Fluid queues are like QBDs in continuous time, except that the level is not restricted to being integer-valued. Formally, we are dealing with processes $\{(X(t), \varphi(t)) : t \geq 0\}$ on the state space $\mathbb{R} \times \{1, 2, \ldots, m\}$, with $\{\varphi(t)\}$ itself being an irreducible Markov process, and such that over intervals of time when $\varphi(t)$ remains equal to i, say, $X(t)$ increases or decreases linearly at a constant rate r_i which may be positive or negative. When the rate is negative and $X(t)$ becomes equal to zero, it remains equal to zero until the rate becomes positive again.

The connection with structured Markov chains has been made in Ramaswami [25], some important results are presented in Asmussen [2] and Rogers [26].

In order to simplify the presentation, we assume that $r_i \neq 0$ for all i: the rates are either strictly positive or strictly negative. We further assume that the phases are numbered so as to put together those for which the rate is positive and those for which it is negative. The transition matrix T of the process $\{\varphi(t)\}$ may then be partitioned as follows:

$$(5.1) \qquad\qquad T = \begin{bmatrix} T_{++} & T_{+-} \\ T_{-+} & T_{--} \end{bmatrix}$$

where the subscripts $+$ and $-$ respectively refer to indices i for which $r_i > 0$ and $r_i < 0$.

THEOREM 5.1. *If the process (X, φ) is positive recurrent, then the stationary density for $x > 0$ is given by*

$$\pi(x) = \boldsymbol{p}_- T_{-+} e^{Kx} [I \ \Psi]$$

where $K = T_{++} + \Psi T_{-+}$ and Ψ is the minimal nonnegative solution of the equation

$$(5.2) \qquad T_{+-} + T_{++}\Psi + \Psi T_{--} + \Psi T_{-+}\Psi = 0.$$

The vector $\boldsymbol{p}_-$ is the unique solution of the system $\boldsymbol{p}_- U = \boldsymbol{0}$, $\boldsymbol{p}_-(\boldsymbol{1} - 2T_{-+}K^{-1}\boldsymbol{1}) = 1$, where $U = T_{--} + T_{-+}\Psi$.

This theorem indicates that the whole resolution depends on finding the minimal nonnegative solution Ψ of the algebraic Riccati equation (5.2).

Ramaswami [25] has shown that in order to evaluate that matrix Ψ, it suffices to solve a matrix-quadratic equation of the form (2.3). He uniformizes the phase process and defines $P = I + 1/\mu T$, where $\mu \geq \max_{i \in \mathcal{S}} |T_{ii}|$; he decomposes the matrix P in a manner conformant to the partition of T and defines

$$(5.3) \qquad A_1 = \begin{bmatrix} \frac{1}{2}I & 0 \\ 0 & 0 \end{bmatrix} \qquad A_0 = \begin{bmatrix} \frac{1}{2}P_{++} & 0 \\ P_{-+} & 0 \end{bmatrix} \quad \text{and} \quad A_{-1} = \begin{bmatrix} 0 & \frac{1}{2}P_{+-} \\ 0 & P_{--} \end{bmatrix}.$$

The minimal nonnegative solution of (2.3) is

$$G = \begin{bmatrix} 0 & \Psi \\ 0 & P_{--} + P_{-+}\Psi \end{bmatrix}.$$

Ramaswami's proof is by probabilistic argument but once the equations have been found, it is easy to check the property by direct substitution in the matrix-quadratic equation.

These results have prompted a new line of enquiry in the use of cyclic reduction or logarithmic reduction to solve algebraic Riccati equations of the form

$$XCX - AX - XD + B = 0$$

when the matrix

$$M = \begin{bmatrix} D & -C \\ -B & A \end{bmatrix}$$

is an M-matrix, either nonsingular or singular and irreducible, of which the matrix T in (5.1) is a special case: see Bini *et al.* [3] and Guo and Higham [11].

6. Conclusion. Other stochastic processes lead to more complex functional equations: Markov chains as in Section 3 but without constraints on the size of the displacement and processes with a tree-like structure are two examples which are covered in [5]. More cases have started to appear in the fields of Risk Theory and Markov Reward Processes. One may be confident that other interesting properties await us.

REFERENCES

[1] S. Asmussen. *Applied Probability and Queues.* John Wiley, New York, 1987.
[2] S. Asmussen. Stationary distributions for fluid flow models with or without Brownian noise. *Comm. Statist. Stochastic Models*, 11:21–49, 1995.

[3] D. A. Bini, B. Iannazzo, G. Latouche, and B. Meini. On the solution of Riccati equations arising in fluid queues. *Linear Algebra Appl.*, 413:474–494, 2006.

[4] D. A. Bini, G. Latouche, and B. Meini. Solving matrix polynomial equations arising in queueing problems. *Linear Algebra Appl.*, 340:225–244, 2002.

[5] D. A. Bini, G. Latouche, and B. Meini. *Numerical Methods for Structured Markov Chains.* Numerical Mathematics and Scientific Computation. Oxford University Press, Oxford, 2005.

[6] D. A. Bini and B. Meini. On cyclic reduction applied to a class of Toeplitz-like matrices arising in queueing problems. In W. J. Stewart, editor, *Computations With Markov Chains*, pages 21–38. Kluwer Academic Publishers, Boston, MA, 1995.

[7] A. da Silva Soares and G. Latouche. A matrix-analytic approach to fluid queues with feedback control. *I. J. of Simulation*, 6:4–12, 2005.

[8] R. V. Evans. Geometric distribution in some two-dimensional queueing systems. *Oper. Res.*, 15:830–846, 1967.

[9] J. D. Gardiner, A. J. Laub, J. J. Amato, and C. B. Moler. Solution of the Sylvester matrix equation $AXB^t + CXD^t = E$. *ACM Trans. Math. Softw.*, 18:223–231, 1992.

[10] C.-H. Guo. Convergence analysis of the Latouche-Ramaswami algorithm for null recurrent Quasi-Birth-and-Death processes. *SIAM J. Matrix Anal. Appl.*, 23:744–760, 2001/02.

[11] C.-H. Guo and N. J. Higham. Iterative solution of a nonsymmetric algebraic Riccati equation. Technical Report MIMS EPrint 2005.48, The University of Manchester, 2005.

[12] C. He, B. Meini, and N. H. Rhee. A shifted cyclic reduction algorithm for Quasi-Birth-Death problems. *SIAM J. Matrix Anal. Appl.*, 23:673–691, 2001/02.

[13] P. Lancaster and M. Tismenetsky. *The Theory of Matrices*. Academic Press, New York, 1985.

[14] G. Latouche. Algorithms for infinite Markov chains with repeating columns. In C. D. Meyer and R. J. Plemmons, editors, *Linear Algebra, Markov Chains and Queueing Models*, pages 231–265. Springer-Verlag, New York, 1993.

[15] G. Latouche. Newton's iteration for nonlinear equations in Markov chains. *IMA J. Numer. Anal.*, 14:583–598, 1994.

[16] G. Latouche and V. Ramaswami. A logarithmic reduction algorithm for quasi-birth-and-death processes. *J. Appl. Probab.*, 30:650–674, 1993.

[17] G. Latouche and V. Ramaswami. *Introduction to Matrix Analytic Methods in Stochastic Modeling.* ASA-SIAM Series on Statistics and Applied Probability. SIAM, Philadelphia PA, 1999.

[18] D. M. Lucantoni. *An Algorithmic Analysis of a Communication Model With Retransmission of Flawed Messages.* Res. Notes in Math. 81. Pitman Books Limited, London, 1983.

[19] B. Meini. An improved FFT-based version of Ramaswami's formula. *Comm. Statist. Stochastic Models*, 13:223–238, 1997.

[20] R. D. Nelson. *Probability, Stochastic Processes, and Queueing Theory. The Mathematics of Computer Performance Modeling.* Springer-Verlag, New York, 1995.

[21] M. F. Neuts. Markov chains with applications in queueing theory, which have a matrix-geometric invariant vector. *Adv. in Appl. Probab.*, 10:185–212, 1978.

[22] M. F. Neuts. *Matrix-Geometric Solutions in Stochastic Models: An Algorithmic Approach.* The Johns Hopkins University Press, Baltimore, MD, 1981.

[23] M. F. Neuts. *Structured Stochastic Matrices of M/G/1 Type and Their Applications.* Marcel Dekker, New York, 1989.

[24] V. Ramaswami. A duality theorem for the matrix paradigms in queueing theory. *Comm. Statist. Stochastic Models*, 6:151–161, 1990.

[25] V. Ramaswami. Matrix analytic methods for stochastic fluid flows. In D. Smith and P. Hey, editors, *Teletraffic Engineering in a Competitive World (Proceedings of the 16th International Teletraffic Congress)*, pages 1019–1030. Elsevier Science B.V., Edinburgh, UK, 1999.

[26] L. C. G. Rogers. Fluid models in queueing theory and Wiener-Hopf factorization of Markov chains. *Ann. Appl. Probab.*, 4:390–413, 1994.

[27] G. W. Stewart. On the solution of block Hessenberg systems. *Numerical Linear Algebra with Applications*, 2(3):287–296, 1995.

[28] W. J. Stewart. *Introduction to the Numerical Solution of Markov Chains.* Princeton University Press, Princeton, NJ, 1994.

[29] P. Tran-Gia. *Analytische Leistungsbewertung verteilter Systeme - Eine Einfhrung.* Springer, Berlin, Heidelberg, NewYork, 1996.

[30] V. Wallace. *The Solution of Quasi Birth and Death Processes Arising from Multiple Access Computer Systems.* PhD thesis, Systems Engineering Laboratory, University of Michigan, 1969.

ON THE GENERALITY OF BINARY TREE-LIKE MARKOV CHAINS

K. SPAEY , B. VAN HOUDT , AND C. BLONDIA *

Abstract. In this paper we show that an arbitrary tree-like Markov chain can be embedded in a binary tree-like Markov chain with a special structure. When combined with [7], this implies that any tree structured QBD Markov chain can be reduced to a binary tree-like process. Furthermore, a simple relationship between the V, R_s and G_s matrices of the original and the binary tree-like Markov chain is established. We also explore the effectiveness of computing the steady state probabilities from the reduced binary chain.

Key words. matrix analytic methods, (binary) tree-like Markov chains, embedded Markov chains

1. Introduction. Tree structured Quasi-Birth-Death (QBD) Markov chains were first introduced in 1995 by Takine et al [5] and later, in 1999, by Yeung et al [10]. More recently, Bini et al [1] have defined the class of tree-like processes as a specific sub-class of the tree structured QBD Markov chains. In [7] it was shown that any tree structured QBD Markov chain can be embedded in a tree-like process. Moreover, the natural fixed point iteration (FPI) to the nonlinear matrix equation $V = B + \sum_{s=1}^{d} U_s(I - V)^{-1}D_s$ that solves the tree-like process, was proven to be equivalent to the more complicated iterative algorithm presented by Yeung and Alfa [10]. In this paper, we demonstrate that any tree-like process can be reduced to a binary tree-like process (i.e., a tree-like process with $d = 2$). Thus, combined with [7], this implies that any tree structured QBD Markov chain can be embedded in a binary tree-like process. We also clarify the relationship between the V, R_s and G_s matrices of the original tree-like process and the binary one. The contribution made by this paper is mostly of theoretical interest, because a careful study on the effectiveness of computing the steady state probabilities from the reduced binary chain, seems to indicate that the reduction technique does not give rise to a speed-up of the iterative algorithms involved.

Typical applications of tree-like processes include preemptive and non-preemptive single server queues with a LCFS service discipline that serves customers of different types, where each type has a different service requirement [5, 10, 3, 2, 11]. Tree structured QBD Markov chains have also been used to evaluate conflict resolution algorithms of the Capetanakis-Tsybakov-Mikhailov-Vvedenskaya (CTMV) type [6, 9]. Some recent work also indicates that tree-like processes can be used to study FCFS priority queues with three service classes [8].

2. Tree-like quasi-birth-death processes - a review. The set of tree-like processes [1] was first introduced as a subclass of the set of tree structured Quasi-Birth-Death Markov chains and afterward shown to be equivalent [7]. This section provides some background information on this type of discrete time Markov chains (MCs). Consider a discrete time bivariate MC $\{(X_t, N_t), t \geq 0\}$ in which the values of X_t are represented by nodes of a d-ary tree, for $d \geq 2$, and where N_t takes integer values between 1 and m. We will refer to X_t as the *node* and to N_t as the *auxiliary* variable of the MC at time t. With some abuse of notation, we shall refer to this MC as (X_t, N_t). The root node of the d-ary tree is denoted as $\emptyset$ and the remaining

* University of Antwerp - IBBT, Department of Mathematics and Computer Science, Performance Analysis of Telecommunication Systems Research Group, Middelheimlaan, 1, B-2020 Antwerp - Belgium, {kathleen.spaey,benny.vanhoudt,chris.blondia}@ua.ac.be.

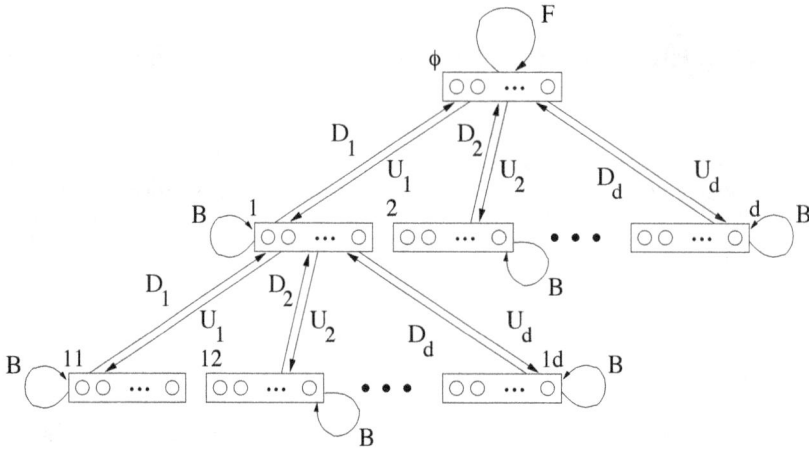

FIG. 2.1. *The structure of a tree-like Markov chain and the matrices characterizing its transitions.*

nodes are denoted as strings of integers, where each integer takes a value between 1 and d. For instance, the k-th child of the root node is represented by k, the l-th child of the node k by kl, and so on. Throughout this paper, we use the '+' to denote the concatenation on the right and '−' to represent the deletion from the right. For example, if $J = k_1 k_2 \ldots k_n$, then $J + k = k_1 k_2 \ldots k_n k$. Let $f(J, k)$, for $J \neq \emptyset$, denote the k rightmost elements of the string J, then $J - f(J, 1)$ represents the parent node of J.

The following restrictions need to apply for an MC (X_t, N_t) to be a tree-like process. At each step the chain can only make a transition to its parent (i.e., $X_{t+1} = X_t - f(X_t, 1)$, for $X_t \neq \emptyset$), to itself ($X_{t+1} = X_t$), or to one of its children ($X_{t+1} = X_t + s$ for some $1 \leq s \leq d$). Moreover, the state of the chain at time $t+1$ is determined as follows:

$$P[(X_{t+1}, N_{t+1}) = (J', j) | (X_t, N_t) = (J, i)] =$$
$$\begin{cases} f^{i,j} & J' = J = \emptyset, \\ b^{i,j} & J' = J \neq \emptyset, \\ d_k^{i,j} & J \neq \emptyset, f(J, 1) = k, J' = J - f(J, 1), \\ u_s^{i,j} & J' = J + s, s = 1, \ldots, d, \\ 0 & otherwise. \end{cases}$$

Notice, the transition probabilities between two nodes depend only on the spacial relationship between the two nodes and not on their specific values.

We can now define the $m \times m$ matrices D_k, B, F and U_s with respective $(i, j)^{th}$ elements given by $d_k^{i,j}$, $b^{i,j}$, $f^{i,j}$ and $u_s^{i,j}$. This completes the description of the tree-like process. Notice, a tree-like process is fully characterized by the matrices D_k, B, U_s and F (see Figure 2.1).

Next, we introduce a number of matrices that play a crucial role when studying the stability and stationary behavior of a tree-like process. The fundamental period of a tree-like process starting in state $(J + k, i)$ is defined as the first passage time from the state $(J + k, i)$ to one of the states (J, j), for $j = 1, \ldots, m$. Let G_k, for $1 \leq k \leq d$, denote the matrix whose $(i, v)^{th}$ element is the probability that the MC is in state (J, v) at the end of a fundamental period which started in state $(J + k, i)$. Let the $(i, v)^{th}$ element of the matrix R_k, for $1 \leq k \leq d$, denote the expected number of visits to state $(J + k, v)$ before visiting node J again, given that $(X_0, N_0) = (J, i)$. Finally, let V denote the matrix whose $(i, v)^{th}$ element is the taboo probability that starting from state $(J + k, i)$, the process eventually returns to node $J + k$ by visiting $(J + k, v)$, under the taboo of the node J. Notice, due to the restrictions on the transition probabilities, the matrix V does not depend on k. Yeung and Alfa [11] were able to show that the following expressions hold for these matrices:

$$G_k = (I - V)^{-1} D_k,$$

$$R_k = U_k (I - V)^{-1},$$

$$V = B + \sum_{s=1}^{d} U_s G_s.$$

Combining these equations, we have the following relation:

$$V = B + \sum_{s=1}^{d} U_s (I - V)^{-1} D_s.$$

Provided that the tree-like process $\{(X_t, N_t), t \geq 0\}$ is ergodic (which is the case if and only if all the G_k matrices are stochastic or likewise if and only if the spectral radius of $R = R_1 + \ldots + R_d$ is less than one [10]), define its steady state probabilities as

$$\pi_i(J) = \lim_{t \to \infty} P[X_t = J, N_t = i],$$
$$\pi(J) = (\pi_1(J), \pi_2(J), \ldots, \pi_m(J)).$$

The vectors $\pi(J)$ can be computed from $\pi(\emptyset)$ using the relation $\pi(J + k) = \pi(J) R_k$. $\pi(\emptyset)$ is found by solving the boundary condition $\pi(\emptyset) = \pi(\emptyset)(\sum_k R_k D_k + F)$ with the normalizing restriction that $\pi(\emptyset)(I - R)^{-1} e = 1$ (where e is a column vector with all its entries equal to one).

In this paper we will consider a tree-like process with a somewhat more general boundary condition as a starting point. We extend the state space of (X_t, N_t) with a single node $\emptyset_s$ consisting of m_s states. This node acts as a *super* root and transitions from and to this node can only occur via the node $\emptyset$. Transitions from, to and within node $\emptyset_s$ are characterized by the $m_s \times m$, $m \times m_s$ and $m_s \times m_s$ matrices $F_{s \to}, F_{\to s}$ and F_s, respectively. Apart from $\pi(\emptyset)$ and $\pi(\emptyset_s)$, the computation of the vectors $\pi(J)$ is not affected by adding the node $\emptyset_s$. We expand the MC with the node $\emptyset_s$, because the reduced tree-like process in [7] contains such a node, as do some applications (e.g., [8]).

3. Constructing the binary tree-like processes. The idea behind the construction used to reduce a tree-like process to a binary one exists in representing each integer part of a string J as a star followed by a series of zeros. For instance, $X_t = J = j_1 \ldots j_n$ is represented in its binary form as

$$\psi(J) \stackrel{def.}{=} * \overbrace{0 \ldots 0}^{j_1 - 1} * \overbrace{0 \ldots 0}^{j_2 - 1} * \ldots * \overbrace{0 \ldots 0}^{j_n - 1}.$$

We will also denote $\psi(\emptyset) = \emptyset$ and $\psi(\emptyset_s) = \emptyset_s$. Obviously, simply representing all strings J by a binary string does not make $(\psi(X_t), N_t)$ a binary tree-like process, as a single transition may add/remove a series of zeros preceded by a single star.

To reduce $(\psi(X_t), N_t)$ into a binary tree-like process we construct an expanded MC $(\mathcal{X}_n, \mathcal{N}_n = (\mathcal{Q}_n, \mathcal{M}_n))$. The technique used to set-up this expanded MC has some similarity with Ramaswami's [4] to reduce a classic M/G/1-type MC to a QBD MC or to the approach taken in Van Houdt et al [9] to construct a tree structured QBD.

The MC $(\mathcal{X}_n, \mathcal{N}_n)$ is defined on the state space $\Omega = \{(\emptyset_s, (0, i)) | i = 1, \ldots, m_s\} \cup \{(\emptyset, (0, i)) | i = 1, \ldots, m\} \cup \{(J, (a, i)) | J = *j_1 j_2 \ldots j_n; j_k = 0 \text{ or } *; k = 1, \ldots, n; n \geq 0; a = -(d-1), \ldots, d-1; i = 1, \ldots, m\}$[1]. We will establish a one-to-one correspondence between the state (J, i) of the original chain and the state $(\psi(J), (0, i))$ of the expanded chain (for all J and i). The key idea behind establishing this association is that whenever a transition occurs that adds a series of $k-1$ zeros preceded by a star to the node variable $\psi(X_t) = J$, we split this transition into k transitions that each add one symbol at a time. Similarly, the removal of a star followed by $k-1$ zeros from $\psi(X_t)$ will be split into k transitions that each remove one symbol. The role of the random variable $\mathcal{Q}_n$ is as follows. Having $\mathcal{Q}_n = a < 0$, implies that a series of zeros is being removed and so far $-a$ zeros have been removed. While $\mathcal{Q}_n = a > 0$ indicates that a more zeros need to be added to the string $\mathcal{X}_n$.

More formally, consider a realization $(X_t(w), N_t(w))$ of the Markov chain (X_t, N_t). The corresponding realization of the expanded chain $(\mathcal{X}_n, \mathcal{N}_n = (\mathcal{Q}_n, \mathcal{M}_n))$ is defined as follows.

Initial state: If $(X_0(w), N_0(w)) = (J, i)$, then set $(\mathcal{X}_0(w), \mathcal{N}_0(w)) = (\psi(J), (0, i))$. Also, set $t = 0$ and $n = 0$; t represents the steps of the original chain (X_t, N_t) and n represents the steps of the expanded chain.

Transition Rules: We distinguish between three possible cases: $\mathcal{Q}_n(w) = 0$, $\mathcal{Q}_n(w) > 0$ and $\mathcal{Q}_n(w) < 0$.

1. $\mathcal{Q}_n(w) = 0$, consider $(X_t(w), N_t(w))$, and do one of the following:
   a. Suppose $X_{t+1}(w) = X_t(w) + k = J + k$, for some $1 \leq k \leq d$ and string $J \neq \emptyset_s$. Let $\mathcal{X}_{n+1}(w) = \mathcal{X}_n(w) + * = \psi(J) + *$ and $\mathcal{N}_{n+1}(w) = (k-1, N_{t+1}(w))$.
   b. Given $X_{t+1}(w) = X_t(w) - k = J$, for some $1 \leq k \leq d$ and string $J \neq \emptyset_s$. Notice, if $k > 1$, then $\psi(J+k)$ ends on a zero (while for $k = 1$, it ends on a star) and we can define $\mathcal{X}_{n+1}(w) = \psi(J+k) - 0$ and $\mathcal{N}_{n+1}(w) = (-1, N_t(w))$ (notice, $\mathcal{M}_{n+1}(w) = N_t(w)$ and not $N_{t+1}(w)$). For $k = 1$, set $\mathcal{X}_{n+1}(w) = \psi(J+k) - * = \psi(J)$ and $\mathcal{N}_{n+1}(w) = (0, N_{t+1}(w))$.
   c. In all other cases (with $\mathcal{Q}_n(w) = 0$), set $\mathcal{X}_{n+1}(w) = \psi(X_{t+1}(w))$ and $\mathcal{N}_{n+1}(w) = (0, N_{t+1}(w))$. These cases include the transitions to and from $\emptyset_s$ and those for which $X_{t+1}(w) = X_t(w)$.
   Next, both n and t are incremented by one.
2. $\mathcal{Q}_n(w) > 0$, define $\mathcal{X}_{n+1}(w) = \mathcal{X}_n(w) + 0$ and $\mathcal{N}_{n+1}(w) = (\mathcal{Q}_n(w) - 1, \mathcal{M}_n(w))$. Remark, if $\mathcal{Q}_{n+1}(w)$ becomes zero, then $\mathcal{X}_{n+1}(w) = \psi(X_t(w))$. Next, increase n by one and do not alter the value of t.
3. $\mathcal{Q}_n(w) < 0$, consider $\mathcal{X}_n(w)$ and distinguish between the following two cases:
   a. Assume $f(\mathcal{X}_n(w), 1) = *$. Let $\mathcal{X}_{n+1}(w) = \mathcal{X}_n(w) - * = \psi(X_t(w))$ and $\mathcal{N}_{n+1}(w) = (0, N_t(w))$.

---

[1] Many of these states will be transient. However, their corresponding entries in the steady state probability vectors will automatically become zero, so there is no need to remove them.

b. Assume $f(\mathcal{X}_n(w), 1) = 0$. Set $\mathcal{X}_{n+1}(w) = \mathcal{X}_n(w) - 0$ and $\mathcal{N}_{n+1}(w) = (\mathcal{Q}_n(w) - 1, \mathcal{M}_n(w))$.

Next, increase n by one and do not alter the value of t.

Next, we show that $(\mathcal{X}_n, \mathcal{N}_n)$ is a (binary) tree-like process with a generalized boundary condition. Indeed, if we remove the nodes $\emptyset$ and $\emptyset_s$ from the state space Ω we end up with a state space of a standard tree-like process, where the star node figures as the root node. Moreover, the string $\mathcal{X}_n$ never grows/shrinks by more than one symbol at a time and it can be readily seen from its construction that the transition between different nodes only depends upon their spacial relationship as required. When describing the transition matrices that characterize $(\mathcal{X}_n, \mathcal{N}_n)$, we will add a hat to all matrices involved (if a conflict arises with earlier notations). Furthermore, the transition matrices Z between two nodes J and J' that both start with a star, are partitioned into nine submatrices as follows:

$$Z = \begin{bmatrix} Z_{-,-} & Z_{-,0} & Z_{-,+} \\ Z_{0,-} & Z_{0,0} & Z_{0,+} \\ Z_{+,-} & Z_{+,0} & Z_{+,+} \end{bmatrix},$$

where $Z_{x,y}$ with $x, y \neq 0$ are square matrices of dimension $(d-1)m$, while $Z_{0,0}$ is a square matrix of size m and all other matrices have an appropriate dimension such that Z is square. The subscripts of these matrices refer to the signs of $\mathcal{Q}_n$ and $\mathcal{Q}_{n+1}$. Let I_x denote the unit matrix of size x. In case $x = m$ (with m the number of values N_t and $\mathcal{M}_n$ can take), we drop the subscript.

A star is only added to $\mathcal{X}_n$ in case 1a, while the addition of a zero only occurs in case 2. Hence,

$$\begin{bmatrix} (U_*)_{0,0} & (U_*)_{0,+} \end{bmatrix} = \begin{bmatrix} U_1 & U_2 & U_3 & \dots & U_d \end{bmatrix},$$

$$\begin{bmatrix} (U_0)_{+,0} & (U_0)_{+,+} \end{bmatrix} = \begin{bmatrix} I & 0 & 0 & \dots & 0 \\ 0 & I & 0 & \ddots & 0 \\ \vdots & \ddots & \ddots & \ddots & \vdots \\ 0 & \dots & 0 & I & 0 \end{bmatrix},$$

where $U_1, \dots, U_d$ are the $m \times m$ matrices belonging to the MC (X_t, N_t). All other blocks of the matrices U_0 and U_* are zero. According to cases 1b, 3a and 3b, the non-zero blocks of the matrices D_0 and D_* equal

$$\begin{bmatrix} (D_*)_{-,0} \\ (D_*)_{0,0} \end{bmatrix} = \begin{bmatrix} D_2^T \Delta^{-1} & D_3^T \Delta^{-1} & \dots & D_d^T \Delta^{-1} & D_1^T \end{bmatrix}^T,$$

$$\begin{bmatrix} (D_0)_{-,-} & (D_0)_{-,0} \\ (D_0)_{0,-} & (D_0)_{0,0} \end{bmatrix} = \begin{bmatrix} 0 & I & 0 & \dots & 0 \\ 0 & 0 & I & \ddots & 0 \\ \vdots & \vdots & \ddots & \ddots & \vdots \\ 0 & 0 & \dots & 0 & I \\ \Delta & 0 & \dots & 0 & 0 \end{bmatrix},$$

where Δ is a diagonal matrix with entries[2] $D_k e$ and ^{T} denotes the transpose of a matrix. In practice (e.g., [6, 9]) Δ need not be invertible as some of its diagonal

---

[2] Remark, the vectors $D_k e$, $1 \leq k \leq d$, are all identical.

entries might be zero, say those appearing on rows $\mathcal{S} \subset \{1, \ldots, m\}$. In this case all the states of the form $(J, (a, s))$ are transient for $a < 0$ and $s \in \mathcal{S}$ and can be removed at once. Hence, it suffices that Δ can be inverted after removing the rows and columns corresponding to these states. For ease of notation, we assume that $\mathcal{S}$ is empty. The identity matrices in D_0 are a consequence of case 3b. The appearance of the Δ matrix is caused by case 1b for $k > 1$, as we remove a zero and keep $\mathcal{M}_{n+1}(w)$ equal to $\mathcal{M}_n(w)$ irrespective of the value of k. The matrices of the form $\Delta^{-1}D_k$ make sure that $\mathcal{M}_{n+1}(w) = N_t(w)$ in case 3a. This construction is needed as we cannot determine the correct value of k until we have removed all the necessary zeros (which are counted by $\mathcal{Q}_n(w)$). Transitions of the MC $(\mathcal{X}_t, \mathcal{N}_t)$ from a node J, which differs from $\emptyset_s$ and $\emptyset$, to itself are captured by the matrix $\hat{B}$; where $\hat{B}_{0,0} = B$, while all other blocks of $\hat{B}$ are identical to zero. Finally, the transitions among the nodes $\emptyset$ and $\emptyset_s$ are still characterized by the matrices F, F_s, $F_{\to s}$ and $F_{s \to}$, while the transition matrix from node $\emptyset$ to node $*$ is identical to $[(U_*)_{0,-} \; (U_*)_{0,0} \; (U_*)_{0,+}]$ and from node $*$ to node $\emptyset$ is given by $[(D_*)_{-,0}^T \; (D_*)_{0,0}^T \; (D_*)_{+,0}^T]^T$.

4. Structural properties of the $\hat{V}$, R and G matrices.

In this section, the structural properties of the matrices $\hat{V}$, R_0, R_*, G_0 and G_* of the MC $(\mathcal{X}_n, \mathcal{N}_n)$ will be discussed and their relationship with the matrices V, G_k and R_k, for $k = 1, \ldots, d$, of the original MC (X_t, N_t) will be identified. $A \otimes B$ denotes the Kronecker product of the matrices A and B.

Consider the matrix $\hat{V}$ whose elements labeled $((a, i), (a', i'))$ are the the taboo probabilities that starting from a state $(J + k, (a, i))$, for $k = 0$ or $*$, the process $(\mathcal{X}_n, \mathcal{N}_n)$ eventually returns to the node $J+k$ by visiting the state $(J+k, (a', i'))$, under the taboo of node J. By construction of $(\mathcal{X}_n, \mathcal{N}_n)$, every sample path in $(\mathcal{X}_n, \mathcal{N}_n)$ that starts and ends in a state with $\mathcal{Q}_n = 0$, and that does not visit any other such state, corresponds to a single transition in (X_t, N_t). So with every path in $(\mathcal{X}_n, \mathcal{N}_n)$ that starts in the state $(J + k, (0, i))$ and that eventually returns to the node $J + k$ under the taboo of node J by visiting the state $(J + k, (0, i'))$, there corresponds exactly one path in (X_t, N_t), namely a path starting in state $(\psi^{-1}(J + k), i)$ that eventually returns to node $\psi^{-1}(J + k)$ by visiting the state $(\psi^{-1}(J + k), i')$. By construction of $(\mathcal{X}_n, \mathcal{N}_n)$, both these sample paths occur with the same probability. As a consequence, $\hat{V}_{0,0} = V$. Any sample path of $(\mathcal{X}_n, \mathcal{N}_n)$ starting from a state $(J + k, (0, i))$ to the same node $J + k$, under taboo of its parent node J, is either of length one, or starts by adding a star to the string $J + k$. Due to the structure of $\hat{B}$ and D_*, this implies that $\hat{V}_{0,+} = 0$ and $\hat{V}_{0,-} = 0$.

If a path in $(\mathcal{X}_n, \mathcal{N}_n)$ starts in a node $(J+k, (a, i))$ with $a < 0$, this means that the process is in the course of removing symbols from the string $J + k$, so only transitions to node J are possible from such a state. So by definition of $\hat{V}$, $\hat{V}_{-,-} = \hat{V}_{-,+} = 0$ and $\hat{V}_{-,0} = 0$. In case a path in $(\mathcal{X}_n, \mathcal{N}_n)$ starts in a node $(J + k, (a, i))$ with $a > 0$, the process will, with probability one, reach the state $(J', (0, i))$ after making a transitions, where $J' = J + k + s_1 + \ldots + s_a$, with all $s_i = 0$, for $i = 1, \ldots, a$. Having reached $(J', (0, i))$, the process will follow a path starting from this state that passes l times through the node J' again, before eventually reaching node $J' - f(J', 1)$ by visiting some state $(J' - f(J', 1), (-1, i'))$. The probability of all these paths is given by the (i, i')-th element of the matrix $\left(\sum_{l=0}^{\infty} \hat{V}_{0,0}^l \Delta \right) = \left(\sum_{l=0}^{\infty} V^l \right) \Delta = (I - V)^{-1} \Delta$. In case $a = 1$, the process has now reached the node $J + k$ again, otherwise it will still make $a - 1$ transitions with probability one, after which it reaches the node $J + k$ again via the state $(J + k, (-a, i'))$. So $\hat{V}_{+,-} = I_{d-1} \otimes (I - V)^{-1} \Delta$, $\hat{V}_{+,0} = 0$, and $\hat{V}_{+,+} = 0$.

Due to the structure of $\hat{V}$, the only non-zero block of $\hat{V}^k$, $k \geq 2$, is $(\hat{V}^k)_{0,0} = V^k$. Hence,

$$(I - \hat{V})^{-1} = \sum_{k=0}^{\infty} \hat{V}^k = \begin{bmatrix} I_{d-1} \otimes I & 0 & 0 \\ 0 & (I - V)^{-1} & 0 \\ I_{d-1} \otimes (I - V)^{-1} \Delta & 0 & I_{d-1} \otimes I \end{bmatrix}.$$

Using the expressions $R_0 = U_0(I_{(2d-1)m} - \hat{V})^{-1}$ and $R_* = U_*(I_{(2d-1)m} - \hat{V})^{-1}$, we find that the non-zero entries of R_0 and R_* are equal to

$$\begin{bmatrix} (R_0)_{+,-} & (R_0)_{+,0} & (R_0)_{+,+} \end{bmatrix} = \\ \begin{bmatrix} 0 & 0 & (I - V)^{-1} & 0 & 0 \\ I_{d-2} \otimes (I - V)^{-1} \Delta & 0 & 0 & I_{d-2} \otimes I & 0 \end{bmatrix},$$

and

$$\begin{bmatrix} (R_*)_{0,-} & (R_*)_{0,0} & (R_*)_{0,+} \end{bmatrix} = \begin{bmatrix} R_2\Delta & \cdots & R_d\Delta & R_1 & U_2 & \cdots & U_d \end{bmatrix}.$$

Remark, $(R_0)^d = 0$, which is as expected as there can be at most $d - 1$ consecutive zeros in a binary representation $\psi(J)$ of any string J. Analogously, the non-zero components of $G_0 = (I_{(2d-1)m} - \hat{V})^{-1} D_0$ and $G_* = (I_{(2d-1)m} - \hat{V})^{-1} D_*$ can be written as

$$\begin{bmatrix} (G_0)_{-,-} & (G_0)_{-,0} \\ (G_0)_{0,-} & (G_0)_{0,0} \\ (G_0)_{+,-} & (G_0)_{+,0} \end{bmatrix} = \begin{bmatrix} 0 & I_{d-1} \otimes I \\ (I - V)^{-1} \Delta & 0 \\ 0 & I_{d-1} \otimes (I - V)^{-1} \Delta \end{bmatrix},$$

and

$$\begin{bmatrix} (G_*)_{-,0}^T & (G_*)_{0,0}^T & (G_*)_{+,0}^T \end{bmatrix}^T = \begin{bmatrix} D_2^T \Delta^{-1} & \cdots & D_d^T \Delta^{-1} & G_1^T & \cdots & G_d^T \end{bmatrix}^T.$$

Remark that $\Delta e = D_k e$, meaning $\Delta^{-1} D_k e = e$, for $k \in \{1, \ldots, d\}$. As a consequence, G_0 and G_* are stochastic if and only if all the matrices G_k, $k \in \{1, \ldots, d\}$, of (X_t, N_t) are stochastic (as $(I - V)^{-1} \Delta e = (I - V)^{-1} D_k e = G_k e$). This means that the binary tree-like process $(\mathcal{X}_n, \mathcal{N}_n)$ is ergodic if and only if the tree-like process (X_t, N_t) is ergodic.

5. Computing steady state probabilities. From the previous section, it is clear that any algorithm that computes $\hat{V}$ produces the matrix V as a by-product and vice versa. The steady state probabilities of $(\mathcal{X}_n, \mathcal{N}_n)$ and (X_t, N_t), provided that the MC is stationary, can be easily computed from $\hat{V}$ and V, respectively, as explained in Section 2. In this section we will demonstrate that the use of some algorithms to compute $\hat{V}$, is equivalent to the computation of V via the same algorithm, provided that we make use of the structural properties of the matrices involved.

5.1. Fixed point iteration (FPI) [10, 1]. This algorithm computes $\hat{V}$ as follows. Set $\hat{V}[0] = \hat{B}$ and compute $\hat{V}[N + 1]$ as

$$(5.1) \quad \hat{V}[N + 1] = \hat{B} + U_*(I_{(2d-1)m} - \hat{V}[N])^{-1} D_* + U_0(I_{(2d-1)m} - \hat{V}[N])^{-1} D_0.$$

In this case $\hat{V}[N]$ monotonically converges to $\hat{V}$. It is easily seen that more iterations are required to compute $\hat{V}$, when compared to computing V via the FPI algorithm. We can improve the convergence by taking the specific structure of the $\hat{V}$ matrix into account. That is, it suffices to compute the component $\hat{V}_{0,0}[N + 1]$ via (5.1) and to update the other entries such that $\hat{V}[N + 1]$ has the same form as $\hat{V}$. Using the expressions for U_*, U_0, D_* and D_0 we have

$$\hat{V}_{0,0}[N + 1] = B + \sum_{i=1}^{d} U_i(I - \hat{V}_{0,0}[N])^{-1}D_i,$$

which is identical to applying the FPI algorithm to V.

5.2. Reduction to quadratic equations (RQE) [1]. This algorithm allows us to compute G_0 and G_* iteratively, from which we can derive the matrix $\hat{V}$. We start with $G_0[0] = G_*[0] = 0$ and solve the following two quadratic equations to obtain $G_0[N + 1]$ and $G_*[N + 1]$ from $G_0[N]$ and $G_*[N]$:

$$(5.2) \quad 0 = D_0 + \left(\hat{B} - I_{(2d-1)m} + U_*G_*[N]\right)G_0[N + 1] + U_0G_0^2[N + 1],$$

$$(5.3) \quad 0 = D_* + \left(\hat{B} - I_{(2d-1)m} + U_0G_0[N + 1]\right)G_*[N + 1] + U_*G_*^2[N + 1],$$

where $G_0[N]$ and $G_*[N]$ converge to G_0 and G_*, respectively. This iterative procedure converges more slowly when compared to applying the RQE algorithm to the MC (X_t, N_t). This can be seen by realizing that the $G_*[N]$ and $G_0[N]$ matrices hold the first-passage probabilities from the node $**$ and $*0$ to the node $*$ in the trees $\mathcal{T}_{N,*}$ and $\mathcal{T}_{N,0}$, respectively, with the tree $\mathcal{T}_{N,*} = \{\emptyset_s, \emptyset, *\} \cup \{*(*^{s_1}0^{s_2})^N i | i = 0$ or $*; s_1, s_2 \geq 0\}$ and $\mathcal{T}_{N,0} = \{\emptyset_s, \emptyset, *\} \cup \{*0^{s_1}(*^{s_2}0^{s_3})^{N-1}i | i = 0$ or $*; s_1, s_2, s_3 \geq 0\}$, for $N \geq 1$. Applying the RQE algorithm to the MC (X_t, N_t), however, generates a sequence of matrices $G_k[N]$, for $k = 1, \ldots, d$, where $G_k[N]$ holds the first-passage probabilities from the node k to the node $\emptyset$ in the tree $\mathcal{T}_{N,k} = \{\emptyset_s, \emptyset\} \cup \{(k)^{s_1} \ldots 1^{s_k}(d^{s_{k+1}} \ldots 1^{s_{d+k}})^{N-1}i | i = 1, \ldots, d; s_j \geq 0; j = 1, \ldots, d+k\}$. Thus, a sample path that visits a node J containing a series of identical integers $k > 1$ is more rapidly taken into account in the latter case.

We can improve upon (5.2)-(5.3) by taking the structure of the matrices involved into account. More specifically, the matrix equation $D_0 + (\hat{B} - I_{(2d-1)m} + U_*G_*)G_0 + U_0G_0^2 = 0$ can be simplified to

$$0 = \Delta + \left(B - I + \sum_{\substack{1 \leq j \leq d \\ j \neq k}} U_jG_j\right)(I - V)^{-1}\Delta + U_kG_k(I - V)^{-1}\Delta.$$

If we post-multiply this equation by $\Delta^{-1}D_k$ for $k = 1, \ldots, d$, we end up with the d quadratic equations used by the RQE algorithm when applied to the MC (X_t, N_t).

5.3. Newton's iteration (NI) [1]. The NI algorithm can be used to compute the $m \times m$ matrices G_k of a tree-like process in a quadratically converging manner. However, each step requires the solution of a (large) linear system of equations of the form: $\sum_{k=1}^{d} H_kXK_k = X + L$ for some square matrices H_k, K_k and L of dimension m. Thus, after applying the reduction technique presented in this paper, it suffices to develop an efficient algorithm to solve a system of the form $H_1XK_1 + H_2XK_2 = X + L$, which is identical to a generalized Sylvester matrix equation, except for the X appearing on the right-hand side. Currently, it is unclear whether such a simplification can result in a computational gain.

Acknowledgements. The second author is a postdoctoral Fellow of the FWO-Flanders. This work was partly funded by the IWT project CHAMP "Cross-layer planning of Home and Access networks for Multiple Play".

REFERENCES

[1] D.A. BINI, G. LATOUCHE, AND B. MEINI, *Solving nonlinear matrix equations arising in tree-like stochastic processes*, Linear Algebra Appl., 366 (2003), pp. 39–64.

[2] Q. HE AND A.S. ALFA, *The MMAP[K]/PH[K]/1 queues with a last-come-first-serve preemptive service discipline*, Queueing Systems, 28 (1998), pp. 269–291.

[3] ———, *The discrete time MMAP[K]/PH[K]/1/LCFS-GPR queue and its variants*, in Proc. of the 3rd Int. Conf. on Matrix Analytic Methods, Leuven (Belgium), 2000, pp. 167–190.

[4] V. RAMASWAMI, *The generality of QBD processes*, in Advances in Matrix Analytic Methods for Stochastic Models, Notable Publications Inc., Neshanic Station, NJ, 1998, pp. 93–113.

[5] T. TAKINE, B. SENGUPTA, AND R.W. YEUNG, *A generalization of the matrix M/G/1 paradigm for Markov chains with a tree structure*, Stochastic Models, 11 (1995), pp. 411–421.

[6] B. VAN HOUDT AND C. BLONDIA, *Stability and performance of stack algorithms for random access communication modeled as a tree structured QBD Markov chain*, Stochastic Models, 17 (2001), pp. 247–270.

[7] ———, *Tree structured QBD Markov chains and tree-like QBD processes*, Stochastic Models, 19 (2003), pp. 467–482.

[8] ———, *Analyzing priority queues with 3 classes using tree-like processes*, Submitted for publication, (2005).

[9] ———, *Throughput of Q-ary splitting algorithms for contention resolution in communication networks*, Communications in Information and Systems, 4 (2005), pp. 135–164.

[10] R.W. YEUNG AND A.S. ALFA, *The quasi-birth-death type Markov chain with a tree structure*, Stochastic Models, 15 (1999), pp. 639–659.

[11] R.W. YEUNG AND B. SENGUPTA, *Matrix product-form solutions for Markov chains with a tree structure*, Adv. Appl. Prob., 26 (1994), pp. 965–987.

PERFORMANCE ANALYSIS OF ASSEMBLY SYSTEMS[*]

MARCEL VAN VUUREN[†] AND IVO J. B. F. ADAN[‡]

Abstract. In this paper we present an approximation for the performance analysis of assembly systems with finite buffers and generally distributed service times. The approximation is based on decomposition of the assembly system in subsystems. Each subsystem can be described by a finite-state quasi-birth-and-death process, the parameters of which are determined by an iterative algorithm. Numerical results show that the approximation accurately predicts performance characteristics such as throughput and mean sojourn time.

Key words. queueing system, approximation, decomposition, finite buffer, matrix analytic method

AMS subject classifications. 60K25, 68M20

1. Introduction. Queueing networks consisting of single-server or multi-server nodes with finite buffers have been studied extensively in the literature; see, e.g., [3] and [11]. These models have many applications in manufacturing, communication and computer systems. In manufacturing systems, it often occurs that different parts arrive at a node (machine), where they are assembled into one product. The performance analysis of assembly nodes is much more complicated, and did not receive much attention in the literature. In this paper we study an assembly node in isolation, with general service times, finite buffers and blocking after service (BAS), and we propose a method for the approximative performance analysis. We are interested in the steady-state queue-length distribution of each buffer; these distributions may be used to determine performance characteristics, such as the throughput and mean sojourn time.

We consider a queueing system (denoted by L; see Fig. 1.1) assembling n parts into one product. The parts are labeled $1, \ldots, n$. The arrival processes of parts are modeled as follows. Type i parts are generated by a so-called arrival server, denoted by M_i, $i = 1, \ldots, n$. For example, in manufacturing systems, arrival server M_i may typically represent the upstream production line producing type i parts. Arrival server M_i serves one part at a time and is never starved (i.e., there is always a new part available). The generic random variable S_i denotes the service (or inter-arrival) time of server M_i; S_i is generally distributed with rate μ_i and coefficient of variation c_i. After service completion at M_i, type i parts are put in buffer B_i, where they wait for assembly. The size of buffer B_i is b_i. Server M_i operates according to the BAS blocking protocol: if upon service completion, buffer B_i is full, then server M_i becomes blocked and the finished part waits until space becomes available in buffer B_i. The parts in the buffers $B_1, \ldots, B_n$ are assembled into one product by (assembly) server M_a. The assembly can start as soon as a part of each type is available. If some are not available yet, the other ones can wait in the assembly server (i.e., they are removed from the buffer). The generic random variable S_a denotes the assembly time of server M_a; S_a is generally distributed with rate μ_a and coefficient of variation c_a.

The method, proposed in this paper, to approximate the steady-state queue-length distribution of the buffers is based on decomposition of the assembly system into subsystems. Each buffer is considered in isolation, and the interaction with other buffers is incorporated in the service time: it consists of a so-called wait-to-assembly time and the actual assembly

---

[*]This research is supported by the Technology Foundation STW, applied science division of NWO and the technology programme of the Dutch Ministry of Economic Affairs.

[†]Department of Mathematics and Computer Science, University of Technology Eindhoven, P.O. Box 513, 5600 MB, Eindhoven, The Netherlands (`m.v.vuuren@tue.nl`).

[‡]Department of Mathematics and Computer Science, University of Technology Eindhoven, P.O. Box 513, 5600 MB, Eindhoven, The Netherlands (`i.j.b.f.adan@tue.nl`).

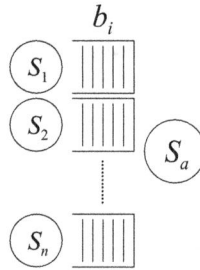

FIG. 1.1. *A schematic representation of an assembly system.*

time. The wait-to-assembly time reflects that a part may have to wait for other parts to arrive, and the parameters of the wait-to-assembly time (such as the first two moments) are determined by an iterative algorithm. In this algorithm, the inter-arrival times and service times are approximated by fitting simple phase type distribution on the first two moments; then each buffer can be described by a finite-state quasi-birth-and-death process (QBD), the steady-state distribution of which can be efficiently determined by matrix-analytic techniques.

Assembly queueing systems have been studied by several authors. Hemachandra and Eedupuganti [6] look at a fork-join queue in an open system. Rao and Suri [2] and Krishnamurti et al. [8] also treat a fork-join queue, but then as a closed system. These references develop approximations. An exact analysis of an assembly system is presented by Gold [5]. None of these references, however, consider general inter-arrival and assembly times, and some of them only look at assembly systems for two parts. This paper considers general service times and any number of parts and is therefore a valuable contribution to the existing literature.

The paper is organized as follows. In Section 2 we explain the decomposition of the assembly system in subsystems. In the section thereafter we take a close look at the subsystems. Section 4 describes the iterative algorithm. Numerical results are presented in Section 5. Finally, Section 6 contains some concluding remarks.

2. Decomposition of the assembly system. We decompose the original assembly system L into n subsystems $L_1, L_2, \ldots, L_n$. Subsystem L_i describes the processing of type i parts in isolation; it consists of a finite buffer of size b_i, arrival-server M_i in front of the buffer, and a so-called departure-server D_i behind the buffer. In Figure 2.1 we show the decomposition of assembly system L in Figure 1.1.

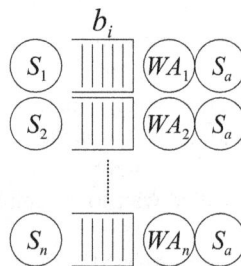

FIG. 2.1. *Decomposition of the assembly system in Figure 1.1.*

The service time of departure-server D_i consists of two components. The random variable WA_i denotes the wait-to-assembly time in subsystem L_i, $i = 1, \ldots, n$. This random variable WA_i represents the time that elapses from the departure of an assembled product until the moment that all parts $j \neq i$ are available for assembly; part i is excluded, because its presence is explicitly modeled by the subsystem. Note that the clock for WA_i starts to tick immediately after a departure, irrespective of whether there is a part in buffer B_i or not; also, WA_i maybe equal to zero, namely when the buffers B_j, $j \neq i$, are nonempty just after departure. Thus, if it takes A time units for the next part i to become available, then the next assembly can start after $\max\{A, WA_i\}$ time units. An important (approximation) assumption is that the successive wait-to-assembly times in subsystem L_i are *independent and identically distributed*. The second part of the service time of departure server D_i is the assembly time S_a itself.

In the next section we elaborate further on the subsystems.

3. The subsystems. In this section we describe how the wait-to-assembly times of subsystem L_i are determined, and subsequently, how the steady-state queue length distribution of subsystem L_i can be found by employing matrix-analytic techniques. Crucial in the analysis is that the distributions of the random variables involved are represented by simple phase-type distributions matching the first two moments. Below we first explain which phase-type distributions will be used.

3.1. Two moment fit. To obtain an approximating distribution of a positive random variable X, one may fit a phase-type distribution on the mean $\mathbb{E}[X])$ and the coefficient of variation c_X by using the following approach [13]. First of all, a random variable X is defined to have to a Coxian distribution of order k if it has to go through up to at most k exponential phases, where phase n has rate μ_n, $n = 1, 2, \ldots, k$. It starts in phase 1 and after phase n, $n = 1, 2, \ldots, k - 1$, it ends with probability $1 - p_n$, whereas it enters phase $n + 1$ with probability p_n. Finally, by definition p_k is equal to zero.

Now, the distribution of X is approximated as follows. If $c_X^2 > 1$, then the rate and coefficient of variation of the Coxian$_2$ distribution matches with $\mathbb{E}[X]$ and c_X, provided the parameters are chosen as (cf., [10]):

$$\mu_1 = 2/\mathbb{E}[X], \quad p_1 = \frac{1}{2c_X^2}, \quad \text{and} \quad \mu_2 = p_1\mu_1.$$

If $1/k \leq c_X^2 \leq 1/(k - 1)$ for some $k \geq 2$, then the rate and coefficient of variation of the Erlang$_{k-1,k}$ distribution, which is a special case of a Coxian distribution of order k, matches with $\mathbb{E}[X]$ and c_X, provided the parameters are chosen as (cf. [13]):

$$\begin{aligned} p_n &= 1, & n = 1, 2, \ldots, k - 2, \\ p_{k-1} &= 1 - \frac{kc_X^2 - \sqrt{k(1 + c_X^2) - k^2 c_X^2}}{1 + c_X^2}, \\ \mu_1 &= \mu_2 = \ldots = \mu_k = (k - p)/\mathbb{E}[X]. \end{aligned}$$

Of course, also other phase-type distributions may be fitted on the mean and the coefficient of variation, but numerical experiments suggest that other distributions only have a minor effect on the results, as shown in [7].

3.2. The wait-to-assembly time. As said before, the wait-to-assembly time at subsystem L_i is the time that elapses from the departure of an assembled product in subsystem L_i until the moment that all parts $j \neq i$ are available at the assembly server. So, WA_i is the maximum of the residual inter-arrival times RA_j of the parts $j \neq i$; note that the residual inter-arrival time RA_j is equal to 0 when buffer B_j is nonempty just after the departure of the assembled product (i.e., the next part j is immediately available). So, we have that

$$WA_i = \max_{j \neq i} RA_j.$$

Below we determine the first two moments of WA_i and its probability mass at 0. Denote by $p_{e,j}$ the probability that buffer B_j is empty just after the departure of an assembled product; so RA_j is positive with probability $p_{e,j}$ (and zero otherwise). In Section 3.4 we determine, for each subsystem L_j, the probability $p_{e,j}$ and the first two moments of the *conditional* RA_j, i.e., RA_j given that it is positive. Then, by adopting the approximation assumption that the random variables RA_j are *independent*, we have the ingredients to recursively compute the first two moments of WA_i and its mass at 0. The computation is based on the following relation for the maximum of the first k residual inter-arrival times:

$$(3.1) \qquad \max_{1 \leq j \leq k} RA_j = \max\{RA_k, \max_{1 \leq j \leq k-1} RA_j\}.$$

Hence, once the first two moments of the first $k-1$ residual inter-arrival times are known, we condition on whether the two random variables RA_k and $\max_{1 \leq j \leq k-1} RA_j$ are positive or not; note that, by the independence assumption,

$$P(\max_{1 \leq j \leq k-1} RA_j = 0) = \prod_{1 \leq j \leq k-1} (1 - p_{e,j}).$$

Then we fit phase-type distributions on the first two moments of the conditional random variables (according to the recipe of Section 3.1) to compute the first two moments of their maximum. The exact computation of the maximum of two independent phase-type distributed random variables is presented in the next section.

Thus, by repeated application of (3.1), we can compute the first two moments of WA_i, and its probability mass at 0, denoted by $p_{ne,i}$, which immediately follows from the assumption that the random variables RA_j are independent, yielding

$$p_{ne,i} = \prod_{j \neq i} (1 - p_{e,j}).$$

A representation of WA_i is shown in Figure 3.1, where WAC_i is the conditional wait-to-assembly time. The distribution of WAC_i is approximated by a phase-type distribution, matching its first two moments,

$$(3.2) \qquad \mathbb{E}(WAC_i) = \frac{\mathbb{E}(WA_i)}{1 - p_{ne,i}},$$

$$(3.3) \qquad \mathbb{E}(WAC_i^2) = \frac{\mathbb{E}(WA_i^2)}{1 - p_{ne,i}}.$$

3.3. The maximum of two phase-type random variables. In this section we calculate the first two moments of the maximum of two independent Erlang distributed random variables. Let E_i denote an Erlang$_{k_i}$ distributed random variable with scale parameter μ_i, $i = 1, 2$, and assume that E_1 and E_2 are independent. The maximum of E_1 and E_2 is phase-type distributed, where the first (random number of) exponential phases have rate $\mu_1 + \mu_2$. These phases are followed by a (random) number of exponential phases with rate μ_1 or rate

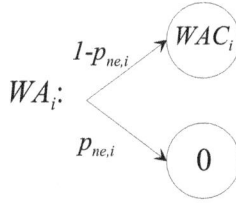

FIG. 3.1. *The wait-to-assembly time of subsystem L_i.*

μ_2, depending on which of the random variables E_1 and E_2 finishes first. Let $q_{1,j}$ with $0 \leq j \leq k_2 - 1$ be the probability that E_2 has completed j phases when E_1 completes its final phase, and similarly, let $q_{2,i}$ with $0 \leq i \leq k_1 - 1$ be the probability that E_1 has completed i phases when E_2 completes its final phase. It is easily verified that $q_{1,j}$ and $q_{2,i}$ both follow a Negative Binomial distribution, i.e.,

$$q_{1,j} = \binom{k_1 - 1 + j}{k_1 - 1} \left(\frac{\mu_2}{\mu_1 + \mu_2} \right)^j \left(\frac{\mu_1}{\mu_1 + \mu_2} \right)^{k_1}, \quad 0 \leq j \leq k_2 - 1,$$

$$q_{2,i} = \binom{k_2 - 1 + i}{k_2 - 1} \left(\frac{\mu_1}{\mu_1 + \mu_2} \right)^i \left(\frac{\mu_2}{\mu_1 + \mu_2} \right)^{k_2}, \quad 0 \leq i \leq k_1 - 1.$$

Conditioned on the event that E_1 finishes first and E_2 has then completed j phases, the maximum of E_1 and E_2 is Erlang distributed with $k_1 + k_2$ phases, the first $k_1 + j$ of which have rate $\mu_1 + \mu_2$ and the last $k_2 - j$ have rate μ_2. Let $M_{1,j}$ denote this conditional maximum, then

$$\mathbb{E}M_{1,j} = \frac{k_1 + j}{\mu_1 + \mu_2} + \frac{k_2 - j}{\mu_2},$$

$$\mathbb{E}M_{1,j}^2 = \frac{(k_1 + j)(k_1 + j + 1)}{(\mu_1 + \mu_2)^2} + \frac{(k_1 + j)(k_2 - j)}{(\mu_1 + \mu_2)\mu_2} + \frac{(k_2 - j)(k_2 - j + 1)}{\mu_2^2}.$$

Similarly, let $M_{2,i}$ denote the maximum of E_1 and E_2, conditioned on the event that E_2 finishes first and E_1 has then completed i phases. For the first two moments of $M_{2,i}$ we have

$$\mathbb{E}M_{2,i} = \frac{k_2 + i}{\mu_1 + \mu_2} + \frac{k_1 - i}{\mu_1},$$

$$\mathbb{E}M_{2,i}^2 = \frac{(k_1 + j)(k_1 + j + 1)}{(\mu_1 + \mu_2)^2} + \frac{(k_1 + j)(k_2 - j)}{(\mu_1 + \mu_2)\mu_2} + \frac{(k_2 - j)(k_2 - j + 1)}{\mu_2^2}.$$

Now, the first two moments of the maximum of E_1 and E_2 can easily be computed by conditioning on the above events, yielding

$$\mathbb{E}(\max\{E_1, E_2\}) = \sum_{j=0}^{k_2-1} q_{1,j}\mathbb{E}M_{1,j} + \sum_{i=0}^{k_1-1} q_{2,i}\mathbb{E}M_{2,i},$$

$$\mathbb{E}(\max\{E_1, E_2\}^2) = \sum_{j=0}^{k_2-1} q_{1,j}\mathbb{E}M_{1,j}^2 + \sum_{i=0}^{k_1-1} q_{2,i}\mathbb{E}M_{2,i}^2.$$

Note that, if E_1 and E_2 are both probabilistic mixtures of Erlang random variables, then the first two moments of the maximum of E_1 and E_2 can be computed from the above equations by conditioning on the number of phases of E_1 and E_2.

3.4. Subsystem analysis. In this subsection we analyze substem L_j (and in the remainder of this section we drop the subscript j). The conditional wait-to-assembly time WAC in the subsystem is represented by a phase-type random variable, the first two moments of which match (3.2) and (3.3). Further, we also fit simple phase-type distributions (according to the recipe in Section 3.1) on the first two moments of the service time S of the arrival server, and the assembly time S_a. In doing so, the subsystem can be described by a finite-state Markov process, with states (i, j, k). The state variable i denotes the total number of parts in the subsystem (at the assembly server or waiting in the buffer). Thus, i is at least 0 and at most $b + 1$. The state variable j indicates the phase of the service time S of the arrival server, and k indicates the phase of the residence time $WA + S_a$ at the assembly server.

To define the generator of the Markov process we use the Kronecker product: If A is an $n_1 \times n_2$ matrix and B is an $n_3 \times n_4$ matrix, the Kronecker product $A \otimes B$ is defined by

$$
A \otimes B = \begin{pmatrix} A(1,1)B & \cdots & A(1,n_2)B \\ \vdots & & \vdots \\ A(n_1,1)B & \cdots & A(n_1,n_2)B \end{pmatrix}.
$$

By ordering the states lexicographically and partitioning the state space into levels $0, 1, \ldots, b + 1$, where level i is the set of all states with i parts in the system, it is immediately seen that the Markov process is a QBD, the generator $\mathbf{Q}$ of which has the following form:

$$
\mathbf{Q} = \begin{pmatrix} B_{00} & B_{01} & & & & \\ B_{10} & A_1 & A_0 & & & \\ & A_2 & \ddots & \ddots & & \\ & & \ddots & \ddots & A_0 & \\ & & & A_2 & A_1 & C_{10} \\ & & & & C_{01} & C_{00} \end{pmatrix}
$$

Below we specify the submatrices in $\mathbf{Q}$. To describe the service processes of the arrival server and assembly server we use the concept of a Markovian Arrival Process (MAP); see [1]. In general, a MAP is defined in terms of a continuous-time Markov process with finite state space $\{0, \cdots, m-1\}$ and generator $G_0 + G_1$. The element $G_{1,(i,j)}$ denotes the intensity of transitions from i to j accompanied by an arrival, whereas for $i \neq j$ element $G_{0,(i,j)}$ denotes the intensity of the remaining transitions from i to j and the diagonal elements $G_{0,(i,i)}$ are negative and chosen such that the row sums of $G_0 + G_1$ are zero.

The service process of the arrival server can be straightforwardly represented by a MAP, the states of which correspond the phases of the service time S. Its generator can be expressed as $AR_0 + AR_1$, where the transition rates in AR_1 are the ones that do correspond to a service completion, i.e., an arrival in the buffer. Hence, in $\mathbf{Q}$, the transitions in AR_1 lead to a transition from level i to $i + 1$, whereas the ones in AR_0 correspond to transitions within level i.

The MAP for the service process of the assembly server will be described in more detail. Let us assume that the distribution of the conditional wait-to-assembly time WAC can be represented by a phase-type distribution with n_{wac} phases, numbered $1, \ldots, n_{wac}$; the rate of phase i is ν_i and p_i is the probability to proceed to phase $i+1$, and $1-p_i$ is the probability that the wait-to-assembly time is finished. Similarly, the distribution of the assembly time S_a can be represented by a phase-type distribution with n_{s_a} phases, with rates μ_i and transition probabilities q_i, $i=1,\ldots,n_{s_a}$. Now the states of the MAP are numbered $1,\ldots,n_{wac}+n_{s_a}$, and its generator can be expressed as DE_0+DE_1, where the transition rates in DE_1 are the ones corresponding to a service completion, i.e., a departure from the system. So, in $\mathbf{Q}$, the transitions in DE_1 lead to a transition from level i to $i-1$, whereas the ones in DE_0 correspond to transitions within level i. The non-zero elements of DE_0 and DE_1 are specified below.

$$
\begin{aligned}
DE_0(i,i) &= -\nu_i, & i &= 1,\ldots,n_{wac}, \\
DE_0(i,i+1) &= p_i\nu_i, & i &= 1,\ldots,n_{wac}-1, \\
DE_0(i,n_{wac}+1) &= (1-p_i)\nu_i, & i &= 1,\ldots,n_{wac}, \\
DE_0(i,i) &= -\mu_i, & i &= n_{wac}+1,\ldots,n_{wac}+n_{s_a}, \\
DE_0(i,i+1) &= q_i\mu_i, & i &= n_{wac}+1,\ldots,n_{wac}+n_{s_a}-1, \\
DE_1(i,1) &= (1-p_{ne})(1-q_i)\mu_i, & i &= n_{wac}+1,\ldots,n_{wac}+n_{s_a}, \\
DE_1(i,n_{wac}+1) &= p_{ne}(1-q_i)\mu_i, & i &= n_{wac}+1,\ldots,n_{wac}+n_{s_a}.
\end{aligned}
$$

Now we can describe the submatrices in $\mathbf{Q}$. The transition rates from levels $1 \le i \le b$ are given by

$$
\begin{aligned}
A_0 &= AR_1 \otimes I_{n_{wac}+n_{s_a}}, \\
A_1 &= AR_0 \otimes I_{n_{wac}+n_{s_a}} + I_{n_a} \otimes DE_0, \\
A_2 &= I_{n_a} \otimes DE_1,
\end{aligned}
$$

where I_n is the identity matrix of size n.

If the subsystem is empty and the wait-to-assembly time elapsed, the assembly can not start yet (but has to wait for a part to arrive). This implies that the transition rates from level 0 are slightly different from the ones at higher levels. Therefore, we introduce the square matrix $\widetilde{DE}_0$ of size $n_{wac}+n_{s_a}$. The transitions from states $1,\ldots,n_{wac}$ remain the same, but now n_{wac+1} is an absorbing state, indicating that the wait-to-assembly time has been finished. The non-zero elements of $\widetilde{DE}_0$ are

$$
\begin{aligned}
\widetilde{DE}_0(i,i) &= -\nu_i, & i &= 1,\ldots,n_{wac}, \\
\widetilde{DE}_0(i,i+1) &= p_i\nu_i, & i &= 1,\ldots,n_{wac}-1, \\
\widetilde{DE}_0(i,n_{wac}+1) &= (1-p_i)\nu_i, & i &= 1,\ldots,n_{wac}.
\end{aligned}
$$

Hence, for the transition rates at level 0 we have

$$
\begin{aligned}
B_{01} &= AR_1 \otimes I_{n_{wac}+n_{s_a}}, \\
B_{00} &= AR_0 \otimes I_{n_{wac}+n_{s_a}} + I_{n_a} \otimes \widetilde{DE}_0, \\
B_{10} &= I_{n_a} \otimes DE_1,
\end{aligned}
$$

Finally, we have

$$C_{01} = AR_1 \otimes I_{n_{wac}+n_{sa}},$$
$$C_{00} = I_{n_a} \otimes DE_0,$$
$$C_{10} = I_{n_a} \otimes DE_1,$$

This completes the description of the QBD. The steady-state distribution can be determined by the matrix geometric method. More specifically, we use the efficient techniques developed by Latouche and Ramaswami [9], and Naoumov et al. [12]. If we denote the equilibrium probability vector of level i by π_i, then π_i has the matrix-geometric form

(3.4) $$\pi_i = x_1 R^{i-1} + x_b \hat{R}^{b-i}, \qquad i = 1, \ldots, b.$$

Here, R is the minimal nonnegative solution of matrix-quadratic equation

$$A_0 + RA_1 + R^2 A_2 = 0,$$

and $\hat{R}$ is the minimal nonnegative solution of equation

$$A_2 + \hat{R}A_1 + \hat{R}^2 A_0 = 0.$$

The matrices R and $\hat{R}$ are determined by using an iterative algorithm developed by Naoumov et al. [12]. The algorithm for R is listed in Figure 3.2; $\hat{R}$ can be determined by interchanging A_0 and A_2 in the algorithm.

```
N  := A₁
L  := A₀
M  := A₂
W  := A₁
dif := 1

while dif > ε
{
    X  := -N⁻¹L
    Y  := -N⁻¹M
    Z  := LY
    dif := ‖Z‖
    W  := W + Z
    N  := N + Z + MX
    Z  := LX
    L  := Z
    Z  := MY
    M  := Z
}
R  := -A₀W⁻¹
```

FIG. 3.2. *Algorithm of Naoumov et al. [12] for finding the rate matrix R, where $\|.\|$ denotes a matrix-norm and ϵ some positive number.*

The final step is to determine x_1, x_b, π_0, and π_{b+1}. The balance equations at the boundary levels $0, 1, b$ and $b + 1$ are given by

$$0 = \pi_0 B_{00} + \pi_1 B_{10},$$
$$0 = \pi_0 B_{01} + \pi_1 A_1 + \pi_2 A_2,$$
$$0 = \pi_{b-1} A_0 + \pi_b A_1 + \pi_{b+1} C_{01},$$
$$0 = \pi_b C_{10} + \pi_{b+1} C_{00}.$$

Eliminating π_0 and π_{b+1} from the equations above, and then substituting the form (3.4) for π_1 and π_b yields

$$0 = x_1(A_1 + RA_2 - B_{10}B_{00}^{-1}B_{01}) + x_b(\hat{R}^{b-1}A_1 + \hat{R}^{b-2}A_2 - \hat{R}^{b-1}B_{10}B_{00}^{-1}B_{01}),$$

$$0 = x_1(R^{b-2}A_0 + R^{b-1}A_1 - R^{b-1}C_{10}C_{00}^{-1}C_{01}) + x_b(\hat{R}A_0 + A_1 - C_{10}C_{00}^{-1}C_{01}).$$

These equations have, together with the normalization equation, a unique solution x_1 and x_b.

From the queue-length distribution we can readily derive performance measures, such as throughput, mean buffer content and mean sojourn time (where the sojourn time is the time that elapses from arrival in the buffer until service completion at the assembly server). Also, the probability p_e that the buffer is empty just after a departure and the distribution of the conditional residual inter-arrival time $RA|RA > 0$ can be obtained; namely

$$p_e = \frac{\pi_1 B_{10}e}{T},$$

where e is a vector of ones and T is the throughput. The probability vector α, given by

$$\frac{\pi_1 B_{10}}{\pi_1 B_{10}e},$$

is the distribution of the phase of the inter-arrival time S just after a departure leaving behind an empty buffer, and thus α can be used to determine the distribution, and the first two moments in particular, of the conditional residual inter-arrival time.

4. The iterative algorithm. We now describe the iterative algorithm for approximating the characteristics of the assembly system L. The algorithm is based on the decomposition of L in n subsystems $L_1, L_2, \ldots, L_n$. Before going into detail in Section 4.2, we present the outline of the algorithm in Section 4.1.

4.1. Outline of the algorithm.
- Step 0: For each subsystem $L_i, i = 1, \ldots, n$: choose initial characteristics of the wait-to-assembly time WA_i.
- Step 1: For each subsystem $L_i, i = 1, \ldots, n$: Determine $p_{ne,i}$ and the first two moments of WAC_i.
- Step 2: For each subsystem $L_i, i = 1, \ldots, n$: Determine the queue-length distribution.
- Repeat Step 1 and 2 until the characteristics of the wait-to-assembly times have converged.

4.2. Details of the algorithm.

Step 0: Initialization
The first step of the algorithm is to initially assume that the wait-to-assembly times are zero. This means that the probabilities $p_{e,j}$ are set to 0. More sophisticated initializations, allowing faster convergence, are probably possible, but the present initialization already works well.

Step 1: The wait-to-assembly times

By using the probabilities $p_{e,j}$ that buffer j is empty just after a departure and the first two moments of the conditional residual inter-arrival times (obtained from the initialization or the previous iteration), we determine for subsystem L_i the (new) first two moments of the wait-to-assembly time and its mass at 0 (as described in Section 3.2). Step 1 is performed for each subsystem L_i, $i = 1, \ldots, n$.

Step 2: Analysis of subsystem L_i

Based on the (new) estimates for the first two moments and the mass at 0 of the wait-to-assembly time, we determine the steady-state queue length distribution of subsystem L_i, as described in Section 3.4. Then, by using the steady-state queue length distribution, we calculate the probability $p_{e,i}$ that buffer B_i is empty just after a departure and the conditional residual inter-arrival time, as well as performance characteristics such as throughput and mean sojourn time. Step 2 is performed for each subsystem L_i, $i = 1, \ldots, n$.

After completion of Step 1 and 2 we check whether the iterative algorithm has converged or not. This can be done by comparing the new estimates for the probabilities $p_{e,i}$ with the ones from the previous iteration. If the sum of the absolute values of the differences between these estimates is less than ε, the algorithm stops; otherwise Step 1 and 2 are repeated.

Of course, other stop-criteria may be used as well; for example, we may consider the throughput instead of the probabilities $p_{e,i}$. Bottom line is that we go on until 'nothing' changes anymore.

Remark: In all experiments we observed that the throughput of each of the subsystem converged, and that all throughputs converged to exactly the same value. However, we have not been able to rigorously prove that all throughputs converge to the same value.

4.3. Complexity analysis. In the iterative algorithm, the solution of a subsystem consumes most of the time. In one iteration step, n subsystems are solved. The number of iterations needed is difficult to predict, but in practice this number is about 10 to 15 iterations. The time consuming part of the solution of a subsystem is the calculation of the R-matrix. This can be done in $O(k_i^3)$ time, where k_i is the size of the R matrix of subsystem i. Then, the time complexity of one iteration becomes $O(n \max_i(k_i^3))$. This means that the time complexity is polynomial in the number of parts and the number of phases for each process.

5. Numerical Results. To investigate the quality of the proposed approximation we compare, for a large number of cases, the estimates for the mean sojourn time of each part and the throughput with the ones produced by discrete-event simulation. We are especially interested in investigating under which circumstances the approximation method gives satisfying results. Each simulation run is sufficiently long such that the widths of the 95% confidence intervals of the mean sojourn time and the throughput are smaller than 1%. Overall, we can remark that a simulation run takes on average about 100 times more time compared to the calculation of the approximation.

We use a broad set of parameters for the tests. The number of parts in the assembly system is varied between 2, 4 and 8. All buffers have the same size, which is varied between 0, 2, 4, and 8. The average assembly time of the assembly server is varied between 0.75 and 1, and the squared coefficient of variation (SCV) of the assembly time is varied between 0.5 and 1. We consider balanced and imbalanced systems. In the balanced cases we set the service rates of the arrival servers all to 1. Also the SCV of the service time of each arrival server is the same and is varied between 0.2, 0.5, 1 and 2. We further investigate two kinds of imbalance. We test imbalance in the average service times of the arrival servers by making the first arrival server $1/3$ faster then the last one, and by letting the service rates of the arrival servers in between change linearly (such that the overall service rate is maintained at 1). For example, in case of 4 arrival servers we get service rates $(1.143, 1.048, 0.952, 0.857)$. Imbalance in the SCV of the service times of the arrival servers is tested in the same way, but now the SCV of the service time of the last server is three times the SCV of the first server and the SCVs of the service times of the arrival servers in between change linearly (such that the average SCV over the arrival servers is equal to one of SCVs mentioned above for the balanced cases). This leads to a total of $4^2 2^4 3 = 768$ test cases. The results for each category are summarized in Table 5.1. Each (sub)table lists the average error in the throughput and the mean sojourn times compared with simulation results. Table 5.2 summarizes all cases with the average errors and for 3 error-ranges the percentage of the cases which fall in that range.

Average error in	throughput	mean sojourn time
Imb. SCV		
no	1.46 %	2.68 %
yes	1.49 %	2.85 %
Imb. mean		
no	1.74 %	3.01 %
yes	1.21 %	2.52 %
No. parts		
2	0.61 %	1.53 %
4	1.37 %	2.72 %
8	2.44 %	4.04 %
Buffers		
0	2.35 %	3.29 %
2	1.68 %	2.58 %
4	1.05 %	2.09 %
8	0.82 %	3.10 %
Occ. rate		
0.75	1.34 %	2.97 %
1	1.61 %	2.56 %
SCV as. ser.		
0.5	1.24 %	2.47 %
1	1.71 %	3.06 %
SCV ar. ser.		
0.2	1.99 %	4.33 %
0.5	1.52 %	2.86 %
1	1.30 %	2.22 %
2	1.09 %	1.65 %

TABLE 5.1

Results for the assembly system with one parameter fixed.

Perf. char.	Avg.	0-5 %	5-10 %	> 10 %
Throughput	1.5 %	97.4 %	2.6 %	0.0 %
Mean sojourn time	2.8 %	84.9 %	13.4 %	1.7 %

TABLE 5.2

Overall results for the assembly system.

Overall we can conclude from the above results that the approximation method works very well. The average error in the throughput is around 1.5 % and the average error in the mean sojourn time is around 2.8 %.

Now let us take a look at the results in more detail. If we look at Table 5.1, we see that the quality of the results for the throughput and mean sojourn times are nearly insensitive to both types of imbalance, the occupation rate and the SCV of the assembly system, because the differences in average errors are at most 0.5%. We see that, as expected, the errors get larger when the number of parts increases. Also, the approximation for the throughput becomes more accurate for large buffers. Finally, the quality of the approximation is slightly better for high SCVs of the service times of the arrival servers. Note that all these results are highly acceptable. We expect that the accuracy of the approximation will deteriorate when the number of parts significantly increases. However, a setting with many (more than 8) parts seems not common in practice.

In Table 5.2 we see that the error in the throughput is almost always within 5%, which is very reliable and robust. For the mean sojourn the times the approximation gives slightly higher errors, but almost all cases are within 10%.

6. Concluding remarks. In this paper we developed an algorithm for approximating an assembly queueing system with general arrival and service times. We used a decomposition approach and developed an iterative algorithm to approximate the performance characteristics of the assembly queue. Therefore, we accurately determine the characteristics of the wait-to-assembly time at a queue. The queue-length distributions of the subsystems are determined by using a matrix geometric method.

We tested the algorithm by comparing it with a discrete-event simulation and the results are very promising. After testing many cases, we concluded that the average errors in the throughput are around 1.5% and the errors in the mean sojourn time are around 3%. The next step is to incorporate this algorithm in a network setting.

Acknowledgements. The authors would like to thank Marco Vijfvinkel for his support with the numerical experiments.

REFERENCES

[1] S. Asmussen and G. Koole (1993) Marked Point Processes as Limits of Markovian Arrival Streams. *Journal of Applied Probability* 30, 365-372.

[2] P.C. Rao and R. Suri (1994) Approximate Queueing Network Models for Closed Fabrication/Assembly Systems Part 1: Single Level Systems. *Production and Operations Management* 3(4), 244-275.

[3] Y. Dallery and S.B. Gershwin (1992) Manufacturing flow line systems: a review of models and analytical results. *Queueing Systems* 12, 3-94.

[4] S.B. Gershwin and M.H. Burman (2000) A Decomposition Method for Analyzing Inhomogeneous Assembly/Disassembly Systems. *Annals of Operation Research* 93, 91-115.

[5] H. Gold (1998) A Markovian Single Server with Upstream Job and Downstream Demand Arrival Stream. *Queueing Systems* 30, 435-455.

[6] N. Hemachandra and S.K. Eedupuganti (2003) Peformance Analysis and Buffer Allocations in some Open Assembly Systems. *Computers and Operations Research* 30, 695-704.

[7] M.A. Johnson (1993) An Empirical Study of Queueing Approximations based on Phase-Type Distributions. *Stochastic Models* 9(4), 531-561.

[8] A. Krishnamurthy, R. Suri and M. Vernon (2004) Analysis of a Fork/Join Synchronization Station with Inputs from Coxian Servers in a Closed Queueing Network. *Annals of Operations Research* 125, 69-94.

[9] G. Latouche and V. Ramaswami (1999) *Introduction to Matrix Analytic Methods in Stochastic Modeling.* ASA-SIAM Series on Statistics and Applied Probability 5.

[10] R.A. Marie (1980) Calculating equilibrium probabilities for $\lambda(n)/C_k/1/N$ queue. *Proceedings Performance '80, Toronto*, 117-125.

[11] H.G. Perros (1989) A Bibliography of Papers on Queueing Networks with Finite Capacity Queues. *Perf. Eval.* 10, 255-260.

[12] V. Naoumov, U.R. Krieger and D. Wagner (1997) Analysis of a Multiserver Delay-Loss System with a General Markovian Arrival Process. *Matrix-Analytic Methods in Stochastic Models (A.S.Alfa and S.R.Chakravarthy eds), Lecture Notes in Pure and Applied Mathematics*, 183, Marcel Dekker, New York, 1996.

[13] H.C. Tijms (1994) *Stochastic models: an algorithmic approach.* John Wiley & Sons, Chichester.

THE DYNAMIC ANALYSIS AND DESIGN OF A COMMUNICATION LINK WITH STATIONARY AND NONSTATIONARY ARRIVALS

WENHONG TIAN, HARRY PERROS
DEPARTMENT OF COMPUTER SCIENCE
NORTH CAROLINA STATE UNIVERSITY
RALEIGH, NC 27965
WTIAN@UNITY.NCSU.EDU, HP@CSC.NCSU.EDU

Abstract. Most research in queueing theory is typically based on the steady-state analysis. In today's dynamically changing traffic environment, the steady-state analysis may not provide enough information to operators regarding Quality of Service (QoS) requirements and dynamic design. In addition, the steady state analysis is not practical for nonstationary arrivals. In this paper, we consider the time-dependent behavior of a communication link depicted by an Erlang loss queue with stationary and nonstationary arrival rates. The time-dependent analysis for stationary arrival rates captures the dynamic nature of the system during its transient phase. The time-dependent analysis for nonstationary arrivals is of great interest since the arrival rate in most communication systems varies over time.

In this paper, we review and compare various methods that have been proposed in the literature for the time-dependent analysis of the nonstationary Erlang loss system for both stationary arrivals and nonstationary arrivals. We try to classify five practical methods into two categories: (1) closed-form exact solution and closed-form approximation; (2) numerical exact solution and numerical approximation. Our work tries to compare their computation complexity and accuracy. We apply some of these techniques to dimensioning dynamically a single communication system.

Key words. Markov chain, Stationary arrivals, Nonstationary arrivals, Blocking probability, Dynamic dimensioning

Introduction. In this paper, we consider the time-dependent behavior and design of a communication link with stationary and also with nonstationary arrival rates. Time-dependent (transient) analysis is motivated by the dynamic nature of the traffic. Riordan [14] introduced different methods for the transient analysis of a single service center. The necessary and sufficient conditions for a queueing network to have a transient product-form solution are provided by Taylor and Boucherie [15]. A new dimensioning approach for optical networks under nonstationary arrival rates was introduced by Nayak and Sivarajan [13].

Queueing models with nonstationary arrival rates have been studied extensively by Abdalla and Boucherie [1], Alnowibet [2], Alnowibet and Perros [3], Jagerman [6], Karagiannis et al. [7], Massey and Whitt [10], Massey [11], and Nayak and Sivarajan [13]. The nonstationary blocking probability for an Erlang loss queue was first obtained by Jagerman [6] using the modified offered load (MOL) approach. Massey and Whitt [10] developed analytical bounds on the error between the MOL approximation and the exact solution for an Erlang loss queue with a nonstationary arrival rate. A modified offered load approximate product-form solution was introduced by Abdalla and Boucherie [1] for mobile networks. A survey for the nonstationary analysis of the Erlang loss queue can be found in Alnowibet and Perros [3]. Mandjes and Ridder [9] proposed large deviation solutions for the transient analysis of the Erlang loss model with a stationary arrival rate. Massey [11] analyzed different queues with time-varying arrival rate for telecommunication models. Nayak and Sivarajan [13] introduced a dynamic dimensioning approach for optical networks under nonstationary arrival rates. Karagiannis et al. [7] showed that the traffic of the internet backbone network can be characterized by a nonstationary Poisson process.

In this paper, we review and compare various techniques that have been reported in the literature for the calculation of transient blocking probabilities of an Erlang loss queue assuming a stationary and nonstationary arrival rate. We also dimension a communication link, modelled by an Erlang loss queue for both stationary and nonstationary arrivals.

The paper is organized as follows. In section 1 we describe the behavior of an Erlang loss queue as a function of time assuming that the arrival rate is either constant or nonstationary, i.e. a function of time t. We also show how an Erlang loss queue can be dimensioned using time-dependent blocking probabilities. In section 2, we review closed-form solutions of the transient behavior of an Erlang loss queue assuming constant and nonstationary arrival rates. Section 3 reviews an approximation method based on a property of truncated Markov processes, and section 4 describes a numerical procedure known as the fixed point approximation (FPA). An alternative approach to dimensioning a communication link, based on the method of large deviations, is presented in section 5. Numerical results are given in section 6. Finally the conclusions are summarized in section 7.

1. The Nonstationary Erlang Loss System. An Erlang loss queue is a system consisting of s servers and no waiting room. A customer is lost if it arrives at a time when all servers are busy. The loss queue is commonly used to model the telephone network. It has been extensively studied in the stationary case, i.e., assuming that the arrival process is a homogeneous Poisson process, or more generally, an Interrupted Poisson processes, and the service rate is exponentially distributed. (It has been shown that in a loss system, the blocking probability is insensitive to the service distribution but it only depends on its mean). The nonstationary loss queue, where the arrival rate is time-dependent is also of great interest.

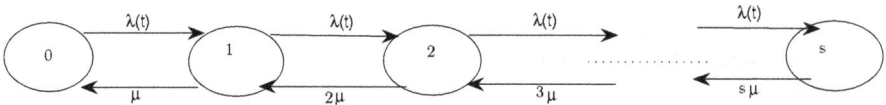

FIG. 1.1. *The Markov chain of Erlang loss model for a single service center*

The rate diagram of a loss queue with nonstationary arrivals $(M_t/M/s/s)$ is shown in FIG.1.1, where s is the total number of servers, $\lambda(t)$ is the time-dependent arrival rate and μ is the service rate. We say that the arrival process is stationary if it is time-independent, i.e., $\lambda(t) = \lambda$, and nonstationary if it is time-dependent (or time-varying). In this case $\lambda(t)$ is a single continuous or discrete function of time. We discuss these two cases in the following two subsections.

1.1. Stationary Arrivals. Let us consider a loss queue M/M/s/s with a time-independent Poisson arrival rate λ. Each arrival requests a service that requires an exponential amount of time with mean $1/\mu$, and it is performed by a single server. The queue has s identical servers and there is no waiting room. The probability that there are n, n=0, 1, ..., s, customers in the queue at time t, $P_n(t)$, is given by the following set of forward differential equations:

(1.1)
$$P_0'(t) = \mu P_1(t) - \lambda P_0(t)$$

(1.2)
$$P_n'(t) = \lambda P_{n-1}(t) + (n+1)\mu P_{n+1}(t) - (\lambda + n\mu)P_n(t),$$

(1.3)
$$P_s'(t) = \lambda P_{s-1}(t) - s\mu P_s(t)$$

where $P_0(t)+P_1(t)+P_2(t)+...+P_s(t)$=1, and $0 \le P_n(t) \le 1$, for $t \ge 0$ and n=0, 1, 2, ..., s, with initial conditions: $P_0(0)$=1 and $P_n(0)$=0, n=1, 2, 3, ..., s.
A numerical example of the time-dependent blocking probability is shown in FIG.1.2. These probabilities were obtained by solving equations (1.1)-(1.3) using an ordinary differential equation (ODE) solver. We note that the blocking probability reaches steady-state when

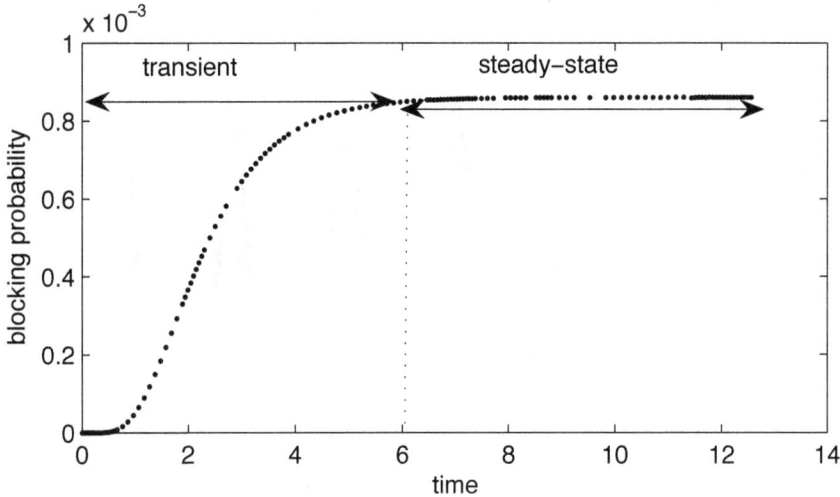

FIG. 1.2. *The time-dependent blocking probabilities of the stationary arrival for $M/M/8/8$, with offered load* $\rho = 2$

$t = 6$. We also note that the steady-state probability can be very different from that during the transient state. Therefore, dimensioning the network based on the steady-state may result in overprovisioning during the transient period. This may have little impact if the duration of the transient period is very short. However, if the transient period is long, for example, a few months for optical networks, then it may be advantageous to dimension the network using the time-dependent blocking probability instead of the steady-state probability.

1.2. Time-varying Arrivals. Let us consider a loss queue M(t)/M/s/s with a time-dependent arrival rate $\lambda(t)$. Each arrival requests a service that requires an exponential amount of time with mean $1/\mu$. The probability that there are n, n=0, 1, ... , s, customers in the queue at time t, $P_n(t)$, is represented by the following forward differential equations:

$$(1.4) \qquad P_0'(t) = \mu P_1(t) - \lambda(t) P_0(t)$$

$$(1.5) \qquad P_n'(t) = \lambda(t) P_{n-1}(t) + (n+1)\mu P_{n+1}(t) - (\lambda(t) + n\mu) P_n(t),$$

$$(1.6) \qquad P_s'(t) = \lambda(t) P_{s-1}(t) - s\mu P_s(t)$$

where $P_0(t) + P_1(t) + P_2(t) + ... + P_s(t)$=1, $t \geq 0$, and $0 \leq P_n(t) \leq 1$, for $t \geq 0$ and n=0, 1, 2, ..., s, with initial conditions: $P_0(0)$=1 and $P_n(0)$=0, n=1, 2, ..., s.

In FIG.1.3, we show a numerical example of the time-dependent blocking probability for a single link obtained assuming a periodic arrival rate function $\lambda(t) = 180 + 50\sin(2(t+20))$. These probabilities were calculated numerically by solving equations (1.4)-(1.6) using an ODE solver. We note that the blocking probability in this case also has a transient period followed by repeating periods. The periodic behavior looks like the steady-state behavior of the stationary arrival case but the blocking probabilities follow a repeating pattern.

1.3. Dimensioning A Single Link . A link can be dimensioned using the time-dependent blocking probability for both stationary and nonstationary arrivals. This is done by calculating the number of servers s so that the blocking probability is under a given threshold at any time t.

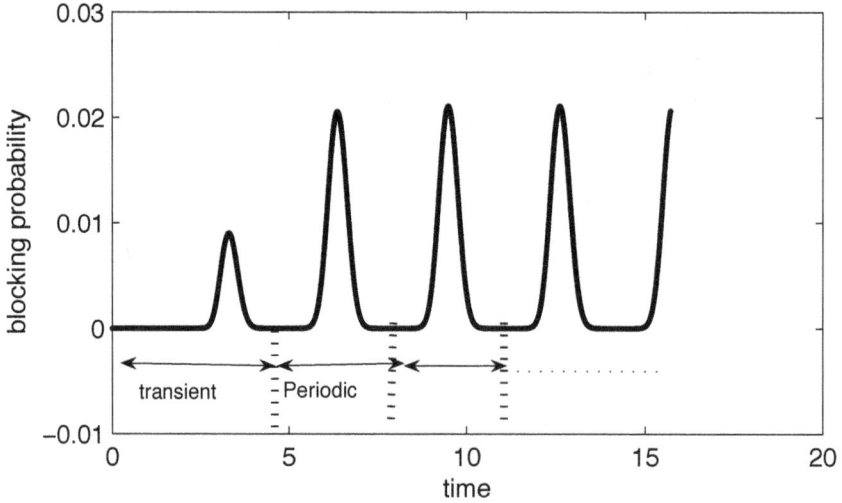

Fig. 1.3. *The time-dependent blocking probabilities of the nonstationary arrival where* $\lambda(t) = 180 + 50sin(2(t+20))$ *and s=220*

1.3.1. Stationary Arrivals. Let us consider an Erlang loss queue with $\lambda = 10$. The number of servers required so that the blocking probability is less than 0.01 is given in FIG.1.4. The solid line labelled 'steady-state dimensioning' gives the optimum number of servers calculated using the steady-state blocking probability of an Erlang loss model with $\lambda = 10$. The dotted line labelled 'time-dependent dimensioning' gives the result using the time-dependent blocking probability of the arrival rate, obtained using differential equations (1.1) to (1.3). We calculate the number of servers iteratively until the blocking probability is less than 0.01. Note that these two curves are the same after the transient phase is over. As we can see, the dimensioning results are quite different for these two scenarios with the time-dependent dimensioning requiring fewer servers.

1.3.2. Time-varying arrivals. We consider an example where the arrival rate varies as shown in FIG.1.5. We assume that time is divided into 12 periods, where each period for instance can be a month. During each period i, the arrival rate is constant. In FIG.1.5, the 12 arrival rates are: $\lambda(t)$=[8, 1, 3, 6, 2, 5, 12, 9, 11, 4, 7, 10]. We dimension the link so that at any time, the blocking probability is less than 0.01. The dimensioning results are also shown in FIG.1.5, and they were obtained assuming that all servers are free at time $t = 0$. These results were obtained as follows: we first calculate the number of servers for the first period using equations (1.4)-(1.6), assuming an empty system at time $t = 0$ and service rate $\mu = 1$, so that the nonstationary blocking probability is less than 0.01. This is done as before, in an iterative manner. We repeat this process for the second period assuming that at the beginning of the period the number of customers in it is equal to the average number of customers in the system.

This process is repeated until all 12 periods have been analyzed. We note that we solve this problem by breaking it into periods and analyzing each period separately. Alternatively, we could solve equations (1.4)-(1.6) for the entire 12 period arrival process. In this case, we will not have to approximate the initial condition for each period. However, it is difficult to define $\lambda(t)$ as a single function over the entire 12 periods.

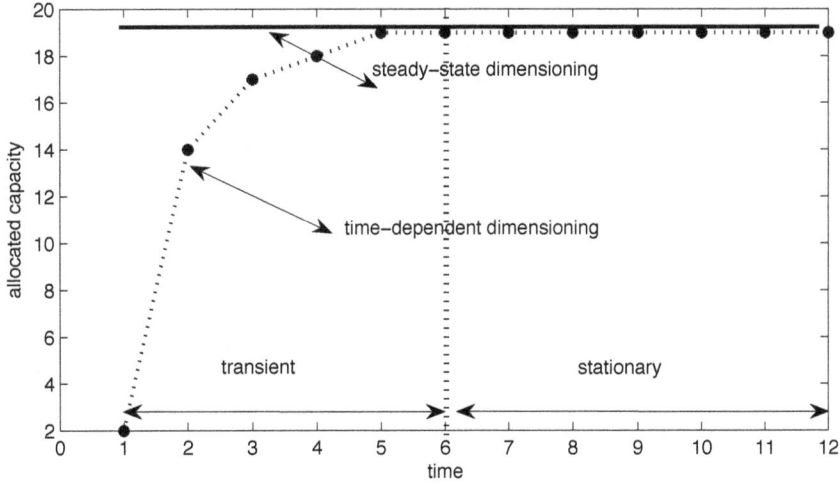

FIG. 1.4. *A dimensioning example of a single link with stationary arrival rates*

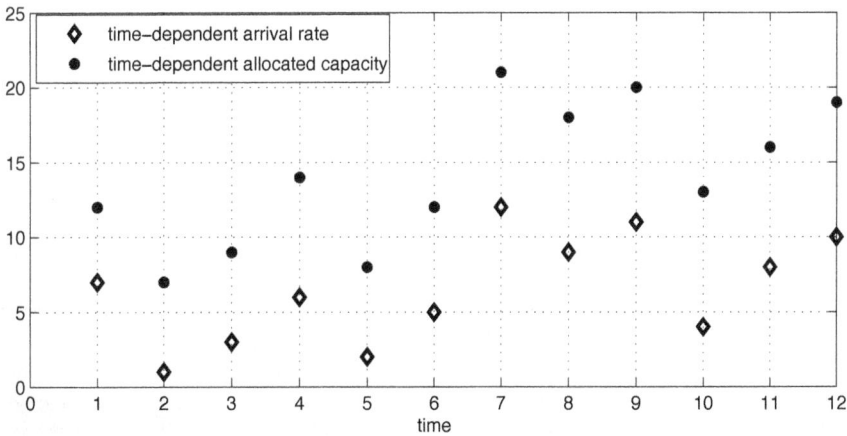

FIG. 1.5. *A dimensioning example of a single link with nonstationary arrival rates*

We note that the dimensioning results are very sensitive to the initial conditions and the arrival rates, and the required number of servers follows the arrival process.

2. Exact Closed-form Solutions for the Time-dependent Blocking Probability. In this section, we review closed-form solutions for the transient analysis of the Erlang loss queue under a constant and nonstationary arrivals.

2.1. Stationary Arrivals. In this case, the time-dependent blocking probability can be obtained using equations (1.1)-(1.3). We have $P_n'(t) \to 0$ for all n as $t \to \infty$ in the differential equations (1.1)-(1.3). The differential equations (1.1)-(1.3) reduce to a set of linear equations from which we can obtain the closed-form solution for the probability P_n that there are n customers in the system.

$$(2.1) \qquad P_n = \lim_{t \to \infty} P\{q(t) = n\} = \frac{\rho^n/n!}{\sum_{i=0}^{s} \rho^i/i!}, n = 0, 1, ..., s$$

where $\rho = \lambda/\mu$, and $q(t)$ is the number of customers in the system at time t. The probability of blocking B_p is:

$$(2.2) \qquad B_p = \lim_{t \to \infty} P\{q(t) = s\} = \frac{\rho^s/s!}{\sum_{i=0}^{s} \rho^i/i!}$$

This is the well-known Erlang B formula. The average number of customers in the system (i.e. the average number of busy servers) is: $\lim_{t \to \infty} E[q(t)] = E[q] = (1 - B_p)\rho$.

The time-dependent blocking probability function at time t, can be obtained from the differential equations (1.1)-(1.3). We have:

$$(2.3) \qquad P_s(t) = \beta e^{(Q^T)t} \alpha$$

where α is the initial state probability vector , β is an all-zero row vector except that the last entry is 1, and Q is the infinitesimal generator matrix of the underlying Markov chain, defined as:

$$(2.4) \qquad \begin{bmatrix} -\lambda & \lambda & 0 & \cdots & 0 \\ \mu & -\mu - \lambda & \lambda & \cdots & 0 \\ 0 & 2\mu & -2\mu - \lambda & \cdots & 0 \\ \vdots & \vdots & \vdots & \vdots & \vdots \\ 0 & 0 & \cdots & \lambda & 0 \\ 0 & \cdots & (s-1)\mu & -(s-1)\mu - \lambda & \lambda \\ 0 & \cdots & 0 & s\mu & -s\mu \end{bmatrix}$$

The computation complexity of equation (2.3) is $O(s^3)$ using a scaling and squaring algorithm with a Pade approximation [12]. The solution to the differential equations (1.1)-(1.3) can be found using different methods. An elegant closed-form solution can be obtained using Sylvester matrix theorem. Let us assume that we start from an empty system and let Q^T be the transpose of Q, $P(t) = [P_0(t), ..., P_s(t)]^T$, and $P(0) = [1, 0, ..., 0]^T$. Now the system of differential equations (1.1)-(1.3) can be written in the following form:

$$(2.5) \qquad P'(t) = Q^T P(t),$$

Solving this equation by applying the property of exponential function, we get

$$(2.6) \qquad P(t) = e^{(Q^T)t} P(0),$$

Let $D = e^{Q^T t} = [d_0, d_1, ..., d_s]$ where d_i is the column vector of D. Then we have $P(t) = d_0$ and $P_i(t) = d_{0,i}$.

Theorem 1 : All the $(s+1)$ eigenvalues of Q are real and distinct and one eigenvalue of Q is zero .

Proof: We first change Q to a symmetric tridiagonal matrix A with positive subdiagonal elements by the similarity transform (this does not change the eigenvalues of Q). We have: A=$D^{-1}QD$, where D is a diagonal matrix. Let us set A^* as the conjugate transpose of matrix

A. Then we know that A^*=A.

We now show that all eigenvalues of A are real as follow: let (λ, x) be the (eigenvalue, eigenvector) pair of matrix A.

$$(2.7) \qquad\qquad\qquad\qquad Ax = \lambda x,$$

Multiplying both sides of equation (2.7) by x^*, we have

$$(2.8) \qquad\qquad\qquad\qquad x^* A x = \lambda x^* x,$$

Taking the transpose of equation (2.8) and noticing that $A^* = A$, we get:

$$(2.9) \qquad\qquad\qquad\qquad x^* A x = \lambda^* x^* x,$$

where λ^* is the conjugate transpose of λ. Comparing equation (2.8) and (2.9), we see that $\lambda = \lambda^*$, which means that all eigenvalues of A are real.

A is a tridiagonal matrix, so we can compute the characteristic polynomial with a three-term recurrence (just do a determinant expansion) to construct a Sturm sequence. Since off-diagonal elements are positive, the matrix can only have simple eigenvalues (by the property of Sturm sequence). Since det(Q)=0, then det(Q-0*I)=0, which means that zero is one of the eigenvalues of matrix Q. This can also be seen by the fact that row sum of matrix Q is zero.

Theorem 2: Sylvester's matrix theorem for distinct roots (eigenvalues) (see Frazer [5]): If $G(U)$ is any polynomial of the square matrix U, and if x_i represents one of the n distinct eigenvalues of U, then

$$(2.10) \qquad\qquad G(U) = \sum_{i=1}^{n} \frac{G(x_i) Adj(x_i I - U)}{\prod_{j \neq i}(x_j - x_i)}$$

An important application of Sylverster's theorem is in finding a closed-form solution for the matrix exponential. The following explains how to use this theorem to get the closed-form solution to our $M/M/s/s$ transient analysis problem. Let $x_0, x_1, ..., x_s$ be the $(s + 1)$ eigenvalues of Q. From Sylvester's theorem we have

$$(2.11) \qquad\qquad D = e^{Q^T t} = \sum_{r=0}^{s} e^{x_r t} \frac{Adj(x_r I - Q^T)}{\prod_{i \neq r}(x_r - x_i)}$$

where Adj(U) is the Adjoint matrix of U. Especially, the probability that all servers are busy at time t can be further simplified as:

$$(2.12) \qquad\qquad P_s(t) = \sum_{r=0}^{s} e^{x_r t} \frac{(-1)^{s+1} det(M)}{\prod_{i \neq r}(x_r - x_i)}$$

where $det(M) = \lambda^s$, λ is the average arrival rate and M is the submatrix of Q^T with size $s \times s$. We can obtain all the eigenvalues of matrix Q using the fast algorithm with complexity $O(s^2)$), reported in [4]. So the computation complexity of equation (2.12) is roughly $O(s^3)$. From $P_s(t)$ we can also know the steady-state probability. The steady-state probability is just the constant part (which corresponds to the zero eigenvalue of matrix Q) of $P_s(t)$. Other quantities of interest such as $P_n(t)$ and average number of busy servers at time t can also be calculated.

2.2. Nonstationary Arrivals. The closed-form solution to differential equations (1.4)-(1.6) is complex even for fairly small systems with special arrival rate function $\lambda(t)$, see Alnowibet [2]. An explicit solution is provided in Jagerman [6] by using the probability generating functions of the state probabilities and the corresponding binomial moments where the arrival rate function $\lambda(t)$ is considered to be continuous. Following Jagerman [6], we have that the probability of j calls arriving in the time interval (0, t) is given by

$$(2.13) \qquad \frac{[\int_0^t a(u)du]^j}{j!} exp(-\int_0^t a(u)du)$$

where a(t) is Poisson-offered load given by $a(t)=\lambda(t)/\mu$. We normalize the service rate $\mu= 1$, so that $a(t)$ is measured in Erlangs. Let us define the Volterra operator K_r

$$(2.14) \qquad K_r f = \int_0^t K_r(t,\tau)f(\tau)d\tau, r = 0, 1, ..., N.$$

The time-dependent blocking probability that all servers are busy at time t is

$$(2.15) \qquad P_s(t) = \frac{\gamma(t,0)}{s!} - K_s(t,\tau)P_s(t)$$

hence the explicit form of the solution is :

$$(2.16) \qquad P_s(t) = \frac{\Lambda(t)^s}{s!} - K_s \frac{\Lambda(t)^s}{s!} + K_s^2 \frac{\Lambda(t)^s}{s!} - ...,$$

where

$$(2.17) \qquad \Lambda(t) = e^{-t} \int_0^t e^u a(u)du$$

and K_s is a Volterra operator defined by the kernel

$$(2.18) \qquad K_s(t,\tau) = \sum_{j=0}^{s-1} \frac{\gamma(t,0)^j}{j!} e^{-(s-j)(t-\tau)} \binom{N}{s-j-1} a(\tau)$$

where

$$(2.19) \qquad \gamma(t,\tau) = e^{-t} \int_\tau^t e^u a(u)du$$

Note that $\Lambda(t) = \gamma(t,0)$. We see that the above explicit solution is quite complicated for an arbitrary arrival rate $\lambda(t)$. The computation complexity of equation (2.16) is approximately $O(s^3)$ depending on how many terms used in the series. In view of this, it is not useful in practice.

3. The Truncated Markov Process Approximation. The following Corollary holds for truncated reversible Markov process (see Kelly [8]).
Corollary 1: If a reversible Markov process X_t with state space S and equilibrium distribution $\Pi(j), j \in S$, is truncated to the set of $S_1 \subset S$, then the resulting Markov process Y_t is reversible in equilibrium and has the equilibrium distribution:

$$(3.1) \qquad \Pi_1(j) = \frac{\Pi(j)}{\sum_{k\in S_1} \Pi(k)}, j \in S_1.$$

It is interesting to note that the equilibrium distribution of the truncated process is just the conditional (renormalized within the truncated state space) probability of the original process. An efficient way to obtain the stationary distribution of the M/M/s/s queue is to use the fact that the M/M/s/s queue is a truncated process of an $M/M/\infty$, which is a reversible Markov process. Therefore:

$$(3.2) \qquad P\{q = n\} = P\{q_\infty = n | q_\infty < s\} = \frac{\rho^n/n!}{\sum_{i=0}^{s} \rho^i/i!}$$

For the time-dependent analysis of M/M/s/s queue, we also can apply this truncation property approximately. First let us consider the transient behavior of the $M/M/\infty$ queue Riordan [14]. We have the following differential equations:

$$(3.3) \qquad P_n'(t) = -(\lambda + n\mu)P_n(t) + \lambda P_{n-1}(t) + (n+1)\mu P_{n+1}(t),$$

$$(3.4) \qquad P_0'(t) = -\lambda P_0(t) + \mu P_1(t),$$

where

$$(3.5) \qquad \sum_{i=0}^{\infty} P_i(t) = 1, t \geq 0$$

and

$$(3.6) \qquad 0 \leq P_n(t) \leq 1, for\ t \geq 0\ and\ n = 0, 1, 2, ..., \infty,$$

with initial conditions: $P_0(0)=1$ and $P_n(0)=0$, n=1, 2, 3, ..., ∞. Applying the z-transform approach, we can obtain the transient distribution of $M/M/\infty$ Riordan [14] :

$$(3.7) \qquad P_n^\infty(t) = \frac{m^n e^{-m}}{n!}, m \equiv m(t) = \rho(1 - e^{-\mu t}).$$

For the time-dependent analysis of an M/M/s/s queue with constant arrival rate, we can apply the truncation property approximately as follows:

$$(3.8) \qquad P_n^s(t) \approx \frac{P_n^\infty(t)}{\sum_k P_k^\infty(t)}, k \in \{0, 1, ...s\}$$

Notice that this equation will be close to the exact solution as time increases for small blocking probabilities. Also it is a good approximation for low blocking probabilities. Despite its appealing closed-form solution, equation (3.8) is non-trivial to compute for large value of s since it involves factorial terms which may cause numerical instability problems such as overflows. So, we can adapt the well-known recursive formula from the steady-state as follows:

$$(3.9) \qquad B(k + 1, t) = \frac{m(t)B(k, t)}{k + 1 + m(t)B(k, t)}$$

where $B(s, t)$ is the probability that all s servers are busy at time t and $B(0, t) = 1$.

The M(t)/M/s/s can also be approximated by truncating the $M(t)/M/\infty$ queue. This method is referred to as the modified offered load (MOL) method and it was first proposed by

Jagerman [6]. For the $M(t)/M/\infty$ queue, the time-dependent blocking probability is given by:

$$(3.10) \qquad\qquad P_n^\infty(t) = \frac{\rho(t)^n e^{-\rho(t)}}{n!}.$$

where $\rho(t) = e^{-t} \int_0^t \lambda(u) e^u du$. For the M(t)/M/s/s queue, the probability $P_n(t)$ that there are n customers in the system is:

$$(3.11) \qquad\qquad P_n^\infty(t) \approx P\{q_\infty(t) = n | q_\infty(t) < s\} = \frac{\rho(t)^n/n!}{\sum_{i=0}^s \rho(t)^i/i!}$$

In this case, the following recursive formula can be used:

$$(3.12) \qquad\qquad B(k+1, t) = \frac{\rho(t) B(k, t)}{k + 1 + \rho(t) B(k, t)}$$

with $B(0, t)=1$ as the initial condition. $B(s, t)$ is the probability that all s servers are busy at time t, $P_s^s(t)$. The truncated $M/M/\infty$ provides an exact solution to an M/M/s/s queue in the steady-state due to the reversibility property. However, this property is lost in the nonstationary case [2]. Hence the truncated $M(t)/M/\infty$ provides an approximate solution to the M(t)/M/s/s. Massy and Whitt [10] developed analytical bounds on the error between the MOL approximation and the exact solution of the M(t)/M/s/s queue. The computation complexity of equation (3.10) and (3.12) is $O(s)$.

Experiments showed that the actual blocking probability of the M(t)/M/s/s queue should be less than 0.1 in order for the MOL to provide a good approximation, see Alnowibet and Perros [3]. As expected, the MOL underestimates the blocking probability of a loss queue with high load, i.e. when the exact blocking probability is high.

4. Numerical Solutions.

4.1. Differential Equations Solver. Equations (1.1)-(1.3) and (1.4)-(1.6) can be solved numerically using an ODE (ordinary differential equation) solver. The numerical results in section 2 were obtained using the ODE solver of Matlab 6.5, which can solve efficiently an Erlang loss queue with a few hundreds servers. However, as reported in Moler et al. [12], ODE solver may be very expensive. We also found that ODE solver is not suitable for dimensioning.

4.2. The Fixed Point Approximation (FPA) Method . The fixed point approximation (FPA) method was proposed by Alnowibet and Perros [3]. This method calculates numerically the time-dependent mean number of customers and blocking probability functions in a nonstationary loss queue. The FPA method was also extended to nonstationary queueing networks of multi-rate loss queues and nonstationary queueing networks with population constraints, see Alnowibet and Perros [3]. The main idea of the FPA method is as follows:

Given a loss queue M(t)/M/s/s with time-dependent rate $\lambda(t)$, the time-dependent average number of customers $E[Q(t)]$ can be expressed as the difference between the effective arrival rate and the departure rate at time t. That is:

$$(4.1) \qquad\qquad E'[Q(t)] = \lambda(t)(1 - B_p(t)) - \mu E[Q(t)]$$

We note that the time-dependent mean number of customers is given by the expression: $E[Q(t)]=\rho(t)(1-B_p(t))$, from which we can develop the following expression for the offered load $\rho(t)$:

$$(4.2) \qquad\qquad \rho(t) = E[Q(t)]/(1 - B_p(t)).$$

where

(4.3)
$$B_p(t) = \frac{\rho(t)^s/s!}{\sum_{i=0}^{s} \rho(t)^i/i!}$$

Using equations (4.1)–(4.3), we can calculate the blocking probability iteratively as follows.

    1). Choose an appropriate Δt, final time T_f and tolerance ϵ.

    2). Choose initial conditions for E[Q(t)]. Set E[Q(0)]=0.

    3). Evaluate $\lambda(t)$ at $t=0, \Delta t, 2\Delta t, ..., T_f$.

    4). Start with an initial blocking probability $B_p^0(t)=0$, t=0,Δt,2Δt,...,T_f.

    5). Set the iteration counter $k=0$.

    6). Solve numerically for $E[Q^k(t)]$ using the following equation:
$E[Q^k(t + \Delta t)]=E[Q^k(t)]+\lambda(t)(1 - B_p^k(t))\Delta t - \mu E[Q^k(t)]\Delta t$.

    7). Calculate $\rho^k(t)=E[Q^k(t)]/(1 - B_p^k(t))$,t=0,2$\Delta$ t,2Δ t,...,T_f.

    8). Calculate the blocking probability $B_p^{k+1}(t)=[\rho^k(t)]^s/s!/(\sum_{i=0}^{s}[\rho^k(t)]^i/i!)$,t=0,$\Delta$ t,2Δ t,...,T_f.

    9). If $\|(B_p^k(t) - B_p^{k+1}(t))\| < \epsilon$, then $B_p^k(t)$ has converged and the algorithm stops. Else, set $k = k + 1$, and go to step 6).

The FPA algorithm does not require a closed-form expression for the arrival rate function. It only requires that the arrival rate function be defined at time points equally spaced by Δt. In view of this, any arrival rate function can be used despite whether we know its closed-form or not. Since this algorithm discretizes the arrival rate function, the continuity and differentiability properties of the arrival rate function are not necessary. The computation complexity of this algorithm to find blocking probability is $O(sT_f/\Delta t)$. In all the experiments the FPA results were very close to the exact numerical results or within the simulation confidence intervals. This leads to the conjecture that the blocking probability obtained by FPA is exact (see, Alnowibet and Perros [2]). For dimensioning purpose, we need to slightly modify the algorithm in order to obtain the capacity for any time t given a blocking probability threshold. This can be done by adding an iterative procedure into the main algorithm.

5. The Large Deviation Approach. In this section, we obtain an expression for dimensioning the Erlang loss queue using the large deviation method. For stationary arrivals, Mandjes and Ridder [9] have obtained approximate expressions for $P_n(t)$, the probability of having n customers at time t. This expression is extended in the case of nonstationary arrivals. The large deviation theory is similar to the Central Limit theory (CLT). The CLT governs random fluctuations only near the mean, which are of the order of $\delta/\sqrt{n}$, where δ is the standard deviation. Fluctuations which are of the order of δ are, relative to typical fluctuations, much bigger: they are large deviations from the mean. They happen only rarely, and so the large deviation theory is often described as the theory of rare events, that is, events which take place away from the mean, out in the tails of the distribution. The main idea of the large deviation approach for the nonstationary Erlang loss queue is as follows. An asymptotic regime is obtained by scaling the arrival process. This is done by replacing $\lambda(t)$ with $n\lambda(t)$. The number of sources active at time t are partitioned into the sources that were active at time 0 and are still active at time t, and the sources that became active in (0,t) and are still active at time t. We then can apply Cramer's theorem and Chernoff's formula to obtain the result.

5.1. Stationary Arrivals. Assuming exponential service time distribution, Mandjes and Ridder [9] obtained the following expression:

(5.1)
$$P_s(t) \approx e^{s(ln(\gamma(t))-\gamma(t)+1)}, \gamma(t) = \lambda_1/\mu(1 - e^{-t})$$

where $\lambda = s\lambda_1$ and s is the total number of servers.

$P_s(t)$ can be better approximated using Bahadur-Rao approximation [9]. We have:

$$(5.2) \qquad P_s(t) \approx \frac{1}{\sqrt{2\pi s\delta\theta}} e^{s(log(\gamma(t))-\gamma(t)+1)}, \gamma(t) = \lambda_1/\mu(1 - e^{-t}),$$

where $\theta = -log(\gamma(t))$, $\delta^2 = \frac{M''(\theta)}{M(\theta)}$ and $M(\theta) = e^{\gamma(t)(e^\theta - 1)}$.

Mandjes and Ridder [9], pointed out that the Bahadur-Rao approximation (5.2) is more accurate than (5.1). We note that the large deviation theory yields simple expressions of the time-dependent blocking probability for stationary arrivals.

5.2. Nonstationary Arrivals. Let us assume that the service time is exponentially distributed with unit mean for nonstationary arrivals. Then expression (5.1) can be extended as follows:

$$(5.3) \qquad\qquad P_s(t) \approx e^{s(log(\gamma(t))-\gamma(t)+1)}$$

where $\gamma(t) = e^{-t}\int_0^t \lambda_1(u)e^u du, \lambda(t) = s\lambda_1(t)$.

The Bahadur-Rao approximation given by (5.2) can be extended as follows:

$$(5.4) \qquad\qquad P_s(t) \approx \frac{1}{\sqrt{2\pi s\delta\theta}} e^{s(log(\gamma(t))-\gamma(t)+1)},$$

where $\gamma(t) = e^{-t}\int_0^t \lambda_1(u)e^u du$, $\theta = -log(\gamma(t))$, $\delta^2 = \frac{M''(\theta)}{M(\theta)}$ and $M(\theta) = e^{\gamma(t)(e^\theta - 1)}$. The computation complexity of equation (5.2) and (5.4) is $O(1)$.

For both stationary and nonstationary cases, the number of servers C for which the blocking probability is ϵ can be obtained using an iterative procedure: starting by the candidate allocation $C = n_0$, the candidate allocation is increased by one until the blocking probability is below the threshold ϵ.

6. Numerical Results. In FIG.6.1, we give the blocking probability calculated at a specific time $t = 4.8$ using four different methods for $\lambda_1(t) = 0.7 + 0.2sin(2\pi t)$. The time $t=4.8$ was chosen because by that time the system is out of the transient state for the given periodic arrival rate function. The initial condition was set to an empty system. The blocking probability is plotted in the logarithmic scale against the total number of servers s. The graph labelled "LD" shows the results obtained from the large deviation theory using equation (5.3), the graph labelled "BR" gives the results obtained using the Bahadur-Rao equation (5.4), and, the graph labelled "TR" gives the results obtained using equation (3.8) from the truncated Markov process approximation. The exact solution is obtained using the fixed point approximation (FPA).

Running more examples by varying the arrival rate with (high load, medium load, low load), we note that the truncated Markov process approximation provides a very good approximation to the exact solution but underestimates the blocking probabilities. The large deviations approximation differs considerably from the exact blocking probability and is less accurate than the Bahadur-Rao approximation. Because of page limit, we do not provide all the examples here.

In FIG.6.2, we show a dimensioning example for a communication link over 20 observation periods. We assume that the arrival rate is constant during each period. The values of the arrival rates are given in FIG.6.2. We calculated the capacity, i.e., the number of servers, iteratively so that at any time the blocking probability is less than 0.005. The dimensioning results are shown in FIG.6.2, where we assume that all servers are free at time t=0. 'BR' represents the results obtained using the Bahadur-Rao equation (5.4), and 'FPA' gives the results

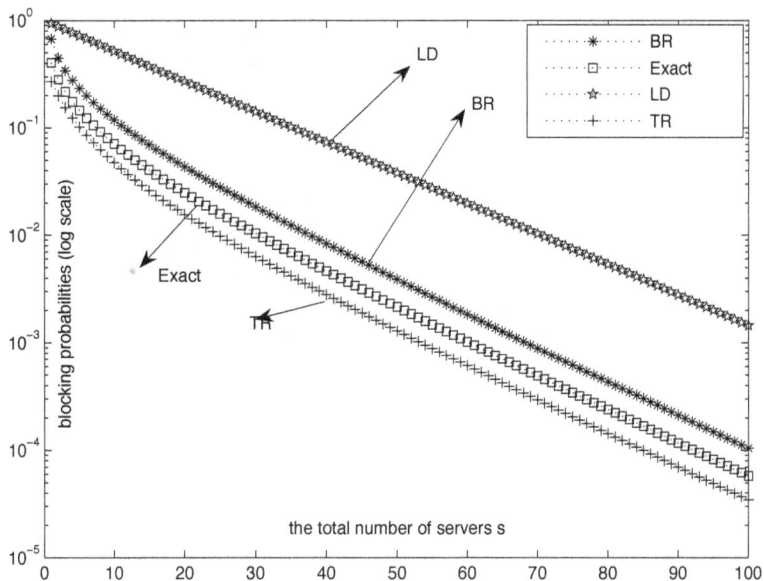

FIG. 6.1. *A comparison among the exact solution (Exact), the truncated Markov process approximation (TR), the Bahadur-Rao approximations(BR) and the large deviations approximations (LD)*

obtained using the fixed point approximation (see, Alnowibet and Perros [3]). The results obtained using the other two methods, i.e. the truncated Markov process approximation and equation (5.3) from the large deviation theory, are not shown because they have a large error. We also carried out a variety of numerical results under different loads, and here we only show a representative sample of these results, since they are all similar. We observed that Bahadur-Rao (BR) approximation is very close to the exact results but it overestimates the capacity. We note that both BR and exact allocated capacity follow the pattern of the arrival rates.

7. Conclusions. In this work, we reviewed and compared various time-dependent analysis techniques of a single loss queue with stationary and nonstationary arrivals. The aim of time-dependent analysis is to dimensioning a link in a more efficient way.

It is difficult to answer the question "Which method is the best ? ". One method maybe preferable over the others when considering computation complexity and accuracy of the results. We have the following observations:

1) For stationary arrivals, the exact closed-form solution and the truncated Markov process approximation are CPU efficient and easy to implement for medium size systems whereas the large deviation approach is preferred if the system is large.

2)For the nonstationary arrivals:

If the arrival rate is a single continuous function, then the truncated Markov process approximation (MOL) and FPA method will be a good choice for medium size systems, and the large deviation approach is a better choice if the system is large.

If the arrival rate is not a continuous function, the FPA method is a better choice.

The FPA method can work for both stationary arrivals and nonstationary arrivals and it can be used for medium size systems. For large systems, we may consider using the Bahadur-

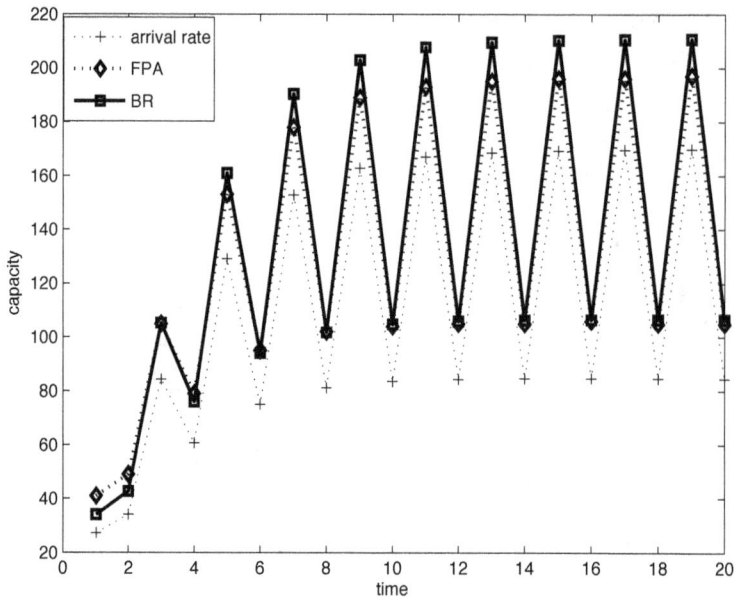

FIG. 6.2. *A dimensioning example of a single link with nonstationary arrival rates*

Rao approximation for dimensioning purposes for both stationary and nonstationary cases since it is the fastest and provides a tight upper bound.

The size of a medium and a large system is relative to the computer used. Our results were obtained on a Pentum(R)4 PC with a 3GHz CPU and a RAM of 504MB. In this context, a medium size system has less than two hundred servers, and a large size system has more than two hundred servers.

For future work we will develop efficient methods for the dynamic dimensioning of an entire network.

Acknowledgments. The authors would like to thank the three anonymous reviewers for their useful comments and suggestions.

REFERENCES

[1] A. Abdalla and R. J. Boucherie, *Blocking probabilities in mobile communications networks with time-varying rates and redialing subscribers* , Annals of Operations Research 112 (2002), pp. 15-34.

[2] K. Alnowibet, *Nonstationary Erlang Loss Queues and Networks*, Ph.D Dissertaion, North Carolina State University, 2004.

[3] K. Alnowibet and H. Perros, *The nostationary loss queue: A survey*, in: Modelling to Computer Systems and Networks, Ed: J. Barria, Imperial College Press, 2005.

[4] I. S. Dhillon, *A New $O(n^2)$ Algorithm for the Symmetric Tridiagonal Eigenvalue/Eigenvector Problem.*, Ph.D. thesis, Computer Science Division (EECS), University of California at Berkeley, 1997.

[5] R. A. Frazer et al. , *Elementary matrices*, Oxford, Cambridge Univ. Press, 1938, UK.

[6] D. L. Jagerman, *Nonstationary Blocking in Telephone Traffic*, The Bell System Technical Journal, 54 (1975), pp. 626-661.

[7] T. Karagiannis, M. Molle , M. Faloutsos and A. Broido, *A nonstationary Poisson view of Internet traffic*, In the Proceedings of INFOCOM 2004, Vol. 3, pp. 1558-1569.

[8] F. Kelly, *Markov Processes and Reversibility*,Wiley Chichester, 1979.

[9] M. Mandjes and Ad. Ridder, *A large deviations approach to the transient of the Erlang loss model*, Performance Evaluation, Vol. 43 (2001), pp. 181-198.

[10] W. A. Massey and W. Whitt, *An Analysis of the Modified Offered-load Approximation for the Nonstationary Loss Model*, Annals of Applied Probability, 4 (1994), pp. 1145-1160.

[11] W. A. Massey, *The Analysis of Queues with Time-Varying Rates for Telecommunication Models*, Telecommunication Systems, Vol. 21:2-4 (2002), pp. 173-204.

[12] C. Moler and C. Van Loan, *Nineteen dubious ways to compute the exponential of a matrix, twenty-five years later.* , SIAM Rev., 45(1): pp. 3-49 (electronic), 2003.

[13] T. K. Nayak and K. N. Sivarajan, *Dimensioning Optical Networks Under Traffic Growth Models*, IEEE/ACM Transactions on Networking, Vol. 11, No. 6, December 2003.

[14] J. Riordan, *Stochastic Service Systems*, John Wiley & Sons, Inc., 1962.

[15] P. G. Taylor and R. Boucherie, *Transient product form distributions in queueing networks*, Discrete Event Dynamic Systems, Vol. 3 (1993), pp. 375-395.

BOUNDS FOR THE COUPLING TIME IN QUEUEING NETWORKS PERFECT SIMULATION*

JANTIEN G. DOPPER, BRUNO GAUJAL AND JEAN-MARC VINCENT [†]

Abstract. In this paper, the duration of perfect simulations for Markovian finite capacity queuing networks is studied. This corresponds to hitting time (or coupling time) problems in a Markov chain over the Cartesian product of the state space of each queue. We establish an analytical formula for the expected simulation time in the one queue case which provides simple bounds for acyclic networks of queues with losses. These bounds correspond to sums on the coupling time for each queue and are either almost linear in the queue capacities under light or heavy traffic assumptions or quadratic, when service and arrival rates are similar.

Key words. Perfect simulation, Markov chain, Hitting time

AMS subject classifications. 60J10, 60J22, 65C40, 65C20, 68U20

1. Introduction. Markov chains are an important tool in modelling systems. Amongst others, Markov chains are being used in the theory of queueing systems, which itself is used in a variety of applications as performance evaluation of computer systems and communication networks. In modelling any queueing system, one of the main points of interest is sampling the behavior of the system in the long run. For an irreducible, ergodic (i.e. aperiodic and positive-recurrent) Markov chain with probability matrix P, this long run behavior follows the stationary distribution of the chain given by the unique vector π which satisfies the linear system $\pi = \pi P$. However, it may be hard to compute this stationary distribution, especially when the finite state space is huge which is frequent in queuing models. In that case, several approaches have been proposed to get samples of the long run behavior of the system.

The most classical methods are indirect. They consists in first computing an estimation of π and then sample according to this distribution (by classical methods such as p.d.f. inverse, rejection or aliasing).

Estimating π can be done through efficient *numerical iterative methods* solving the linear system $\pi = \pi P$. [10]. Even if they converge fast, they do not scale when the state space (and thus P) grows.

Another approach to estimate π is to use a *regenerative simulation* [4, 7] based on the fact that on a trajectory of a Markov chain returning to its original state, the frequency of the visits to each state is steady state distributed. This technique does not suffer from statistical biais but is very sensitive to the return time to the regenerative state. This means that the choice of the initial state is crucial but also that regenerative simulation complexity increases fast with the state space which is exponential in the number of queues. This can be partially overcome by using importance sampling [3] or semi-regenerative simulation [2]. However, the simulation times still typically exhibit a multiplicative behavior with the number of queues.

There also exist direct techniques to sample states of Markov chain according to its stationary distribution. The classical method has been *Monte Carlo simulation* for many years. This method is based on the fact that for an irreducible aperiodic

---

*This work was partially supported by the French ACI SurePath project and SMS ANR

[†]Jantien Dopper, Mathematical institute, Leiden University, NL (jgdopper@math.leidenuniv.nl) (this author was partially supported by a grant from INRIA), Bruno Gaujal and Jean-Marc Vincent, Laboratoire ID-IMAG, Mescal Project INRIA-UJF-CNRS-INPG, 51, avenue Jean Kuntzmann, F-38330 Montbonnot, France, e-mail : ({Bruno.Gaujal,Jean-Marc.Vincent}@imag.fr)

finite Markov chain with initial distribution $\pi^{(0)}$, the distribution $\pi^{(n)}$ of the chain at time n converges to π as n gets very large:

$$\lim_{n \to \infty} \pi^{(n)} = \lim_{n \to \infty} \pi^{(0)} P^n = \pi.$$

After running the Markov chain long enough, the state of the chain will not depend on the initial state anymore. However, the question is how long is long enough? That is, when is n sufficiently large so that $|\pi^{(n)} - \pi| \leqslant \epsilon$ for a certain $\epsilon > 0$? Moreover, the samples generated by this method will always be biased.

In 1996, Propp and Wilson[8] solved these problems for Markov chain simulation by proposing an algorithm which returns exact samples of the stationary distribution very fast. The striking difference between Monte Carlo simulation and this new algorithm is that Propp and Wilson do not simulate into the future, but go backwards in time. The main idea is, while going backwards in time, to run several simulations, starting with all $s \in S$ until the state at $t = 0$ is the same for all of them. If the output is the same for all runs, then the chain has coupled. Because of this coupling property and going backwards, this algorithm has been called Coupling From The Past (from now on: CFTP). A more detailed description of this algorithm will be presented in section 2.

When the coupling from the past technique is applicable, one gets in a finite time one state with steady-state distribution. Then one can use either a one long-run simulation from this state avoiding the estimation of the initial transient problem or replicate independently the CFTP algorithm to get a sample of independent steady-state distributed variables. The analysis of the choice could be done exactly as in [1]. The replication technique has been applied successfully in finite capacity queueing networks with blocking and rejection (very large state-space) [12]. The efficiency of the simulation allows also the estimation of rare events (blocking probability, rejection rate) is done in [11].

The aim of this paper is to study the simulation time needed to get a stationary sample for finite capacity networks. We show that for *monotone* systems CFTP scales very well with the state space explosion accompanying the increase in the number of queues.

More precisely, we study the coupling time τ of a CFTP algorithm (*i.e.* the number of steps needed to provide one sample). Our main interest is setting bounds on the expected coupling time. We first obtain exact analytical values for the expected simulation time for one M/M/1/C queue which serves as a building block for the following).

As for networks of queues, we show how upper bounds on the mean simulation time can be obtained as sums of the coupling times for all queues. One of the main result of this paper is to show that for acyclic networks with rejection in case of overflow,

$$\mathbb{E}\tau \leqslant \sum_i \frac{\Lambda}{\Lambda_i} \frac{C_i + C_i^2}{2},$$

where Λ is the global event rate in the network, Λ_i is the rate of events affecting Queue i and C_i is the capacity of Queue i. This result can be refined under light or heavy traffic assumptions to almost linear bounds in the capacities. All these bounds scale very well with the number of queues. This explains why perfect simulation of monotone queueing networks is so fast, especially when dealing with large scale networks as in [11] where systems with up to 32 queues of capacity 30 (the state space is of size $31^{32} \approx 10^{47}$) are sampled over a classical desktop computer is less than 20 milli-seconds. This is good enough to estimate rare event probabilities.

The paper is organized as follows. We first introduce the coupling from the past algorithm in Section 2. Then we show general properties of the coupling time for open Markovian queueing networks in Section 3. We will investigate the M/M/1/c queue in Section 4 providing exact computation for the expected coupling time and the case of acyclic networks in Section 5 where bounds are derived, together with several experimental tests assessing their quality.

2. Coupling from the Past. Let $\{X_n\}_{n \in \mathbb{N}}$ be an irreducible and aperiodic discrete time Markov chain with a finite state space S and a transition matrix $P = (p_{i,j})$. Let

$$\phi : S \times \mathcal{E} \to S,$$

encode the chain, which means that it verifies the property $\mathbb{P}(\phi(i, e) = j) = p_{i,j}$ for every pair of states $(i, j) \in S$ and for any e, a random variable distributed on $\mathcal{E}$. The function ϕ could be considered as a construction algorithm and e is the *innovation* for the chain. In the context of discrete event systems, e is an *event* and ϕ is the *transition function*. Now, the evolution of the Markov chain is described as a stochastic recursive sequence

$$(2.1) \qquad X_{n+1} = \phi(X_n, e_{n+1}),$$

with X_n the state of the chain at time n and $\{e_n\}_{n \in \mathbb{N}}$ an independent and identically distributed sequence of random variables.

Let $\phi^{(n)} : S \times \mathcal{E}^n \to S$ denote the function whose output is the state of the chain after n iterations and starting in state $s \in S$. That is,

$$(2.2) \qquad \phi^{(n)}(s, e_{1 \to n}) = \phi(\dots \phi(\phi(s, e_1), e_2), \dots, e_n).$$

This notation can be extended to set of states. So for a set of states $A \subset S$ we note

$$\phi^{(n)}(A, e_{1 \to n}) = \left\{ \phi^{(n)}(s, e_{1 \to n}), s \in A \right\}.$$

In the following, $|X|$ denotes the size of set X.

THEOREM 2.1 ([8]). *Let ϕ be a transition function on $S \times \mathcal{E}$. There exists an integer l^* such that*

$$\lim_{n \to +\infty} \left| \phi^{(n)}(S, e_{1 \to n}) \right| = \ell^* \text{ almost surely.}$$

The system *couples* if $\ell^* = 1$. Then the *forward coupling time* τ^f defined by

$$\tau^f = \min\{n \in \mathbb{N}; \text{ such that } \left| \phi^{(n)}(S, e_{1 \to n}) \right| = 1\},$$

is almost surely finite. The coupling property of a system ϕ depends only on the structure of ϕ. The probability measure on $\mathcal{E}$ does not affect the coupling property, provided that all events in $\mathcal{E}$ have a positive probability. Moreover, the existence of some pattern $e^*_{1 \to n^*_0}$ that ensures coupling, guarantees that τ^f is stochastically upper bounded by a geometric distribution

$$(2.3) \qquad \mathbb{P}(\tau^f \geqslant k.n^*_0) \leqslant \left(1 - p(e^*_1).p(e^*_2) \dots p(e^*_{n_0})\right)^k;$$

where $p(e) > 0$ is the probability of event e.

At time τ^f, all trajectories issued from all initial states at time 0 have collapsed in only one trajectory. Unfortunately, the distribution of X_{τ^f} is not stationary. In [6] an example is given that illustrates why it is not possible to consider that this process has the stationary regime.

In fact, this iteration scheme could be reversed in time as it is usually done in the analysis of stochastic point processes. For that, one needs to extend the sequence of events to negative indexes and build the reversed scheme on sets by

$$A_n = \phi^{(n)}\left(\mathcal{S}, e_{-n+1 \to 0}\right).$$

It is clear that the sequence of sets A_n is non-decreasing $(A_{n+1} \subset A_n)$. Consequently, the system couples if the sequence A_n converges almost surely to a set with only one element. Almost surely, there exists a finite time τ^b, the *backward coupling time*, defined by

$$\tau^b = \min\{n \in \mathbb{N}; \text{ such that } \left|\phi^{(n)}\left(\mathcal{S}, e_{-n+1 \to 0}\right)\right| = 1\}.$$

PROPOSITION 2.2 ([13]). *The backward coupling time τ^b and the forward coupling time τ^f have the same probability distribution.*

The main result of the backward scheme is the following theorem.

THEOREM 2.3 ([8]). *Provided that the system couples, the state when coupling occurs for the backward scheme, is steady state distributed.*

From this fact, a general algorithm (1) sampling the steady state can be constructed.

Algorithm 1 Backward-coupling simulation (general version)

---

for all $s \in \mathcal{S}$ **do**
    $y(s) \leftarrow s$ {choice of the initial value of the vector y, $n = 0$}
end for
repeat
    e $\leftarrow$ Random_event; {generation of e_{-n+1}}
    for all $s \in \mathcal{S}$ **do**
        $y(s) \leftarrow y(\phi(s, e))$;
        {$y(s)$ state at time 0 of the trajectory issued from s at time $-n + 1$}
    end for
until All $y(x)$ are equal
return $y(x)$

The mean complexity c_ϕ of this algorithm is $c_\phi = O(\mathbb{E}[\tau^b]|\mathcal{S}|)$. The coupling time τ^b is of fundamental importance for the efficiency of the sampling algorithm. To improve its complexity, we could reduce the factor $|\mathcal{S}|$ and reduce the coupling time. When the state space is partially ordered by a partial order $\prec$ and the transition function is monotone for each event e, it is sufficient to simulate trajectories starting from the maximal and minimal states [8]. Denote by MAX and MIN the set of maximal, respectively minimal elements of $\mathcal{S}$ for the partial order $\prec$. The monotone version of algorithm (1) is given by algorithm (2). In this case, we need to store the sequence of events in order to preserve the coherence between the trajectories driven from $MAX \cup MIN$.

Algorithm 2 Backward-coupling simulation (monotone version)

---

n=1;

R[n]=Random_event;{array R stores the backward sequence of events }

repeat

  n=2.n;

  for all $s \in MAX \cup MIIN$ **do**

    $y(s) \leftarrow s$ {Initialize all trajectories at time $-n$}

  end for

  for i=n downto n/2+1 **do**

    R[i]=Random_event; {generates all events from time $-n + 1$ to $\frac{n}{2} + 1$}

  end for

  for i=n downto 1 **do**

    for all $s \in MAX \cup MIN$ **do**

      $y(s) \leftarrow \phi(y(s), R[i])$

      {$y(s)$ is the state at time $-i$ of the trajectory starting in s at time $-n$}

    end for

  end for

until All $y(s)$ are equal

return $y(s)$

The doubling scheme (first step in the loop) leads to a mean complexity

$$(2.4) \qquad c_\phi = O(\mathbb{E}[\tau^b](|MAX| + |MIN|)).$$

3. Open Markovian queueing networks. Consider an open network Q consisting of K queues $Q_1, \ldots, Q_K$. Each queue Q_i has a finite capacity, denoted by C_i, $i = 1, \ldots K$. Thus the state space of a single queue Q_i is $\mathcal{S}_i = \{0, \ldots C_i\}$. Hence, the state space $\mathcal{S}$ of the network is $\mathcal{S} = \mathcal{S}_1 \times \ldots \times \mathcal{S}_K$. The state of the system is described by a vector $s = (s_1, \ldots, s_K)$ with s_i the number of customers in queue Q_i. The state space is partially ordered by the component-wise ordering and there are a maximal state MAX when all queues are full and a minimal state when all queues are empty.

The network evolves in time due to exogenous customer arrivals from outside of the network and to service completions of customers. After finishing his service at a server, a customer is either directed to another queue by a certain routing policy or leaves the network. A routing policy determines to which queue a customer will go, taking into account the global state of the system. Moreover, the routing policy also decides what happens with a customer if he is directed to a queue with a buffer filled with C_i customers.

An event in this network is characterized by the movements of some clients between queues, modeling the routing strategy and the Poisson process defines the occurrence rate of the events. For example, consider the acyclic queueing network of figure 3.1, made of 4 queues and 6 events.

Since the network is open, clients are able to enter and leave the network. We assume that customers who enter from outside the network to a given queue arrive according to a Poisson process. Furthermore, suppose that the service times at server i are independent and exponentially distributed with parameter μ_i.

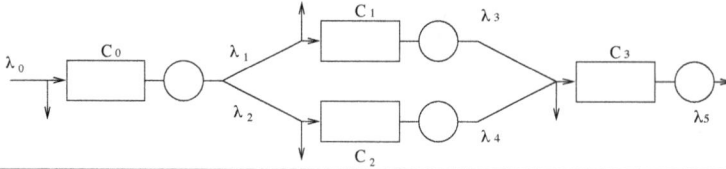

	rate	origin	destination	enabling condition	routing policy
e^0	λ_0	Q_{-1}	Q_0	none	rejection if Q_0 is full
e^1	λ_1	Q_0	Q_1	$s_0 > 0$	rejection if Q_1 is full
e^2	λ_2	Q_0	Q_2	$s_0 > 0$	rejection if Q_2 is full
e^3	λ_3	Q_1	Q_3	$s_1 > 0$	rejection if Q_3 is full
e^4	λ_4	Q_2	Q_3	$s_2 > 0$	rejection if Q_3 is full
e^5	λ_5	Q_3	Q_{-1}	$s_3 > 0$	none

FIG. 3.1. *Network with rejection*

For example, for event e^1 (fig 3.1) we get

$$\phi(., e^1) : (s_0, s_1, s_2, s_3) \longmapsto \begin{cases} (s_0 - 1, s_1 + 1, s_2, s_3) & \text{if } s_0 \geqslant 1 \text{ and } s_1 < C_1; \\ (s_0 - 1, s_1, s_2, s_3) & \text{if } s_0 \geqslant 1 \text{ and } s_1 = C_1 (Q_1 \text{ full}); \\ (s_0, s_1, s_2, s_3) & \text{if } s_0 = 0 (Q_0 \text{ empty}). \end{cases}$$

DEFINITION 3.1. *An event e is monotone if $\phi(x, e) \leqslant \phi(y, e)$ for every x, y in $\mathcal{S}$ with $x \leqslant y$.*

It should be clear that event e^1 is monotone. Moreover usual events such as routing with overflow and rejection, routing with blocking and restart, routing with a index policy rule (eg Join the shortest queue) are monotone events [5, 12].

Denote by $\mathcal{E} = \{e_1, \ldots, e_M\}$ the finite collection of events of the network. With each event e_i is associated a Poisson process with parameter λ_i. If an event occurs which does not satisfy the enabling condition the state of the system is unchanged.

To complete the construction of the discrete-time Markov chain, the system is uniformized by a Poisson process with rate $\Lambda = \sum_{i=1}^{M} \lambda_i$. Hence, one can see this Poisson process as a clock which determines when an event transition takes place. To choose which specific transition actually takes place, the collection $\mathcal{E}$ of events of the network is randomly sampled with

$$p_i = \mathbb{P}\left(\text{event } e_i \text{ occurs}\right) = \frac{\lambda_i}{\Lambda}.$$

By construction, the following proposition should be clear.

PROPOSITION 3.2. *The uniformized Markov chain has the same stationary distribution as the queueing network, and so does the embedded discrete time Markov chain.*

Provided that events are monotone, the CFTP algorithm can be applied on queueing networks to build steady-state sampling of the network.

In our example of Figure 3.1 we ran the CFTP algorithm and produced samples of coupling time. The parameters used for the simulation are the following. Queues capacity : for all $i = 1, \ldots, 4$, $C_i = 10$. Event rates: $\lambda_1 = 1.4$, $\lambda_2 = 0.6$, $\lambda_3 = 0.8$, $\lambda_4 = 0.5$ and $\lambda_5 = 0.4$. The global input rate λ_0 is varying. The number of samples used to estimate the mean coupling time is 10000. The result is displayed in Figure 3.2.

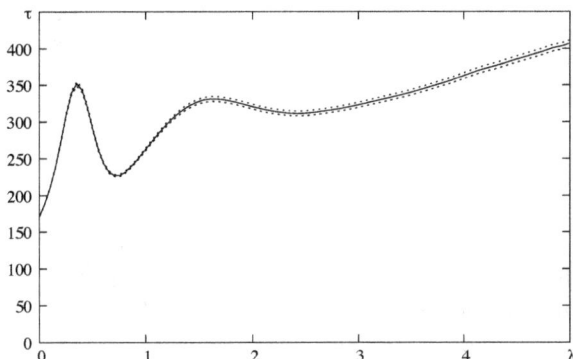

FIG. 3.2. *The mean coupling time for the acyclic network of Figure 3.1 varies from 160 to 400 events when the input rate ranges from 0 to 4, with 95% confidence intervals.*

This type of curve is of fundamental importance because the coupling time corresponds to the simulation duration and is involved in the simulation strategy (long run versus replication). These first results can be surprising because they exhibit a strong dependence on parameters values. The aim of this paper is now to understand more deeply what are the critical values for the network and to build bounds on the coupling time that are non-trivial.

Let N_i be the projection function from $\mathcal{S}$ to $\mathcal{S}_i$ with $N_i(s_1,\ldots,s_K) = s_i$. So N_i is the number of customers in queue Q_i. As in section 2, τ^b refers to the *backward coupling time* of the Markov chain, describing the queueing network.

DEFINITION 3.3. *Let τ_i^b denote the backward coupling time on coordinate i of the state space. Thus τ_i^b is the smallest n for which*

$$\left| \left\{ N_i \left(\phi^{(n)} \left(\mathcal{S}, e_{-n+1}, \ldots, e_0 \right) \right) \right\} \right| = 1.$$

Because coordinate s_i refers to queue Q_i, the random variable τ_i^b represents the coupling time from the past of queue Q_i. Once all queues in the network have coupled, the CFTP algorithm returns one value and hence the chain has coupled. Thus

$$(3.1) \qquad \tau^b = \max_{1 \leqslant i \leqslant K} \{\tau_i^b\} \leqslant_{st} \sum_{i=1}^{K} \tau_i^b.$$

By taking expectation and interchanging sum and expectation we get:

$$(3.2) \qquad \mathbb{E}\left[\tau^b\right] = \mathbb{E}\left[\max_{1 \leqslant i \leqslant K} \{\tau_i^b\}\right] \leqslant \mathbb{E}\left[\sum_{i=1}^{K} \tau_i^b\right] = \sum_{i=1}^{K} \mathbb{E}\left[\tau_i^b\right]$$

It follows from Proposition 2.2 that τ^b and τ^f have the same distribution. The same holds for τ_i^f and τ_i^b. Hence $\mathbb{E}\left[\tau_i^b\right] = \mathbb{E}\left[\tau_i^f\right]$ and

$$(3.3) \qquad \mathbb{E}[\tau^b] \leqslant \sum_{i=1}^{K} \mathbb{E}\left[\tau_i^f\right].$$

The bound given in Equation 3.3 is interesting because $\mathbb{E}\left[\tau_i^f\right]$ is sometimes amenable to explicit computations, as shown in following sections. In order to derive those bounds, one may provide yet other bounds, by making the coupling state explicit.

DEFINITION 3.4. *The hitting time $h_{j \to k}$ in a Markov chain X_n is defined as*

$$h_{j \to k} = \inf\{n \in \mathbb{N} \ s.t. \ X_n = k | X_0 = j\} \ with \ j, k \in \mathcal{S}.$$

In the queueing framework, $h_{0 \to C_i}$ represents the number of steps it takes for queue Q_i to go from state 0 to state C_i. Now we consider queue Q_i out of the network and examine it independently. Suppose that $h_{0 \to C_i} = n$ for the sequence of events $e_1, \ldots e_n$. Because of monotonicity of ϕ we have

$$\phi^{(n)}(0, e_1, \ldots, e_n) \leqslant \phi^{(n)}(s, e_1, \ldots, e_n) \leqslant \phi^{(n)}(C_i, e_1, \ldots, e_n) = 0,$$

with $s \in \mathcal{S}_i$. Hence, coupling has occurred. So $h_{0 \to C_i}$ is an upper bound on the forward coupling of queue Q_i. The same argumentation holds for $h_{C_i \to 0}$. Thus

$$(3.4) \qquad \mathbb{E}\left[\tau_i^f\right] \leqslant \mathbb{E}\left[\min\{h_{0 \to C_i}, h_{C_i \to 0}\}\right].$$

Hence,

$$(3.5) \mathbb{E}[\tau^b] \leqslant \sum_{i=1}^{K} \mathbb{E}\left[\tau_i^f\right] \leqslant \sum_{i=1}^{K} \mathbb{E}\left[\min\{h_{0 \to C_i}, h_{C_i \to 0}\}\right] \leqslant \sum_{i=1}^{K} \min(\mathbb{E}h_{0 \to C_i}, \mathbb{E}h_{C_i \to 0}),$$

by Jensen's Inequality.

4. Coupling time in a M/M/1/C queue. The M/M/1/C queue is well known and there is no need to run simulations to get the distribution of its stationary distribution. However, the computation of hitting times provided here is new and will serve as a building block for the following section on networks.

In a M/M/1/C model, we have a single queue with one server. Customers arrive at the queue according to a Poisson process with rate λ and the service time is distributed according to an exponential distribution with parameter μ. In the queue there is only place for C customers. So the state space $\mathcal{S} = \{0, \ldots, C\}$. If a customer arrives when there are already C customers in the queue, he immediately leaves without entering the queue. After uniformization, we get a discrete time Markov chain which is governed by the events e_a with probability $p = \frac{\lambda}{\lambda + \mu}$ and e_d with probability $q = 1 - p$. Event e_a represents an arrival and event e_d represents an end of service with departure of the customer.

In order to estimate the expectation of the coupling time from the past $\mathbb{E}[\tau^b]$ we use inequality 3.5. Since there is only one queue, the first two inequalities in 3.5 become equalities. Indeed, when applying forward simulation, the chain only can couple in state 0 or state C. This follows since for $r, s \in \mathcal{S}$ with $0 < r < s < C$ we have $\phi(r, e_a) = r + 1 < s + 1 = \phi(s, e_a)$ and $\phi(r, e_d) = r - 1 < s - 1 = \phi(s, e_d)$ So the chain cannot couple in a state s with $0 < s < C$. Furthermore we have $\phi(C, e_a) = C = \phi(C - 1, e_a)$ and $\phi(0, e_d) = 0 = \phi(1, e_d)$. Hence, forward coupling can only occur in 0 or C:

$$(4.1) \qquad \mathbb{E}\left[\tau^b\right] = \mathbb{E}\left[\min\{h_{0 \to C}, h_{C \to 0}\}\right].$$

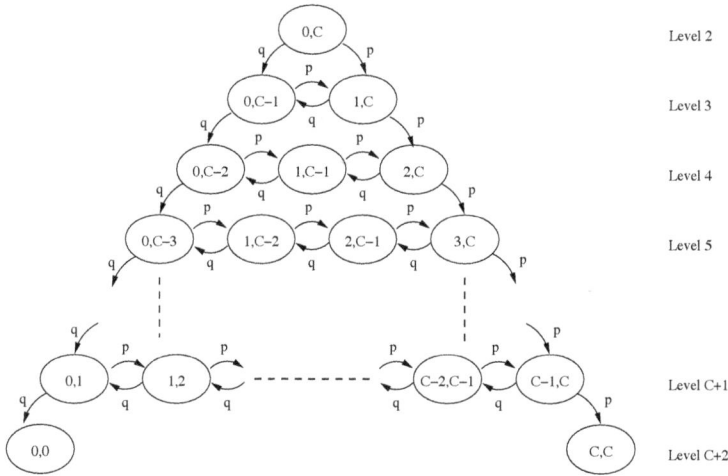

FIG. 4.1. *Markov chain $X(q)$ corresponding to $H_{i,j}$*

4.1. Explicit calculation of $\mathbb{E}\left[\tau^b\right]$. In order to compute $\min\{h_{0\to C}, h_{C\to 0}\}$ we have to run two copies of the Markov chain for a M/M/1/C queue simultaneously .(whose states are x and y respectively). One copy starts in state 0 and the other one starts in state C. We stop when either the chain starting in 0 reaches state C or when the copy starting in state C reaches state 0.

Therefore, let us rather consider a product Markov chain $X(q)$ with state space $\mathcal{S} \times \mathcal{S} = \{(x,y), x = 0, \ldots, C, y = 0, \ldots, C\}$. The Markov chain $X(q)$ is also governed by the two events e_a and e_d and the function ϕ is:

$$\psi\left((x,y), e_a\right) = ((x+1) \wedge C, (y+1) \wedge C)$$
$$\psi\left((x,y), e_d\right) = ((x-1) \vee 0, (y-1) \vee 0).$$

Without any loss of generality, we may assume that $x \leqslant y$. This system corresponds with the *pyramid Markov chain $X(q)$* displayed in Figure 4.1. The rest of this section is devoted to establishing a formula for the expected exit time of the pyramid. Although the technique used here (one step analysis) is rather classical, it is interesting to notice how this is related to random walks on the line (this also explains the shifted indexes associated to the levels of the pyramid).

Since we can only couple in 0 or C, this coupling occurs as soon as the chain $X(q)$ reaches states $(0,0)$ or (C,C). Define

$H_{i,j} :=$ number of steps to reach state $(0,0)$ or (C,C) starting from state (i,j)

with $(i,j) \in \mathcal{S} \times \mathcal{S}$. By definition, $\min\{h_{0\to C}, h_{C\to 0}\} = H_{0,C}$. Now $H_{i,j}$ represents the hitting time of the coupling states $(0,0)$ and (C,C) (also called absorption time) in a product Markov chain. Using a one step analysis, we get the following system of equations for $\mathbb{E}[H_{i,j}]$:

$$(4.2) \quad \begin{cases} \mathbb{E}[H_{i,j}] = 1 + p\mathbb{E}[H_{(i+1)\wedge C, (j+1)\wedge C}] + q\mathbb{E}[H_{(i-1)\vee 0, (j-1)\vee 0}], & i \neq j, \\ \mathbb{E}[H_{i,j}] = 0, & i = j \end{cases}$$

Two states (i,j) and (i',j') are said to be at the same level if $|j - i| = |j' - i'|$. In Figure 4.1 we can distinguish $C + 1$ levels. Because of monotonicity of ϕ, $|j - i|$

cannot increase. Hence, starting at a level with $|j - i|$, the chain will gradually pass all intermediate levels to reach finally the level with $|j - i| = 0$ in state $(0,0)$ or (C,C). Thus, starting in state $(0,C)$, the chain will run through all levels to end up at the level with $|j - i| = 0$. So, $H_{0,C} = \min\{h_{0 \to C}, h_{C \to 0}\}$. To determine $\mathbb{E}[H_{0,C}]$ we determine the mean time spent on each level and sum up over all levels.

A state (i,j) belongs to level m if $|j - i| = C + 2 - m$. Then state $(0,C)$ belongs to level 2 and the states $(0,0)$ and C,C belong to level $C + 2$. To get from $(0,C)$ into either $(0,0)$ or (C,C), the chain $X(q)$ needs to cross all levels between the levels 2 and $C + 2$. Let T_m denote time it takes to reach level $m + 1$, starting in level m. Then

$$(4.3) \qquad\qquad H_{0,C} = \sum_{m=2}^{C+1} T_m.$$

In order to determine T_m, we associate to each level m a random walk R_m on $0, \ldots, m$ with absorbing barriers at state 0 and state C. In the random walk, the probability of going up is p and of going down is $q = 1 - p$. We have the following correspondence between the states of the random walk R_m and the states of $X(q)$ (see Figure 4.2).

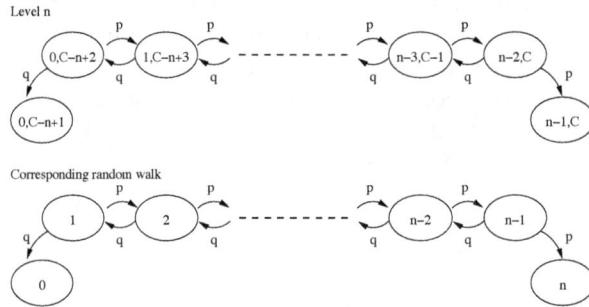

FIG. 4.2. *Relationship between level m and random walk R_m.*

State 0 of R_m corresponds with state $(0, C - m + 1)$ of $X(q)$,
State i of R_m corresponds with state $(i - 1, C - m + 1 + i)$ of the $X(q)$,
$$1 \leqslant i \leqslant m - 1,$$
State m of R_m corresponds with state $(m - 1, C)$ of $X(q)$.

Now the time spent on level m in $X(q)$ is the same as the time spent in a random walk R_m before absorption. Therefore, on can use the two following results on random walks in determining T_m, which are known as ruin problems (see for example [9]).

Let $\alpha_{i \to 0}^m$ denote the probability of absorption in state 0 of the random walk R_m starting in i. Then:

$$(4.4) \qquad\qquad \alpha_{i \to 0}^m = \begin{cases} \frac{a^m - a^i}{a^m - 1}, & p \neq \frac{1}{2}, \\[2mm] \frac{m - i}{m}, & p = \frac{1}{2}, \end{cases}$$

where $a = q/p$.
Now, absorption occurs in R_m once the state 0 or C has been achieved.

LEMMA 4.1. *Let $\widetilde{T}_i^m$ denote the mean absorption time of a random walk R_m starting in i. Then:*

(4.5)
$$\mathbb{E}[\widetilde{T}_i^m] = \begin{cases} \dfrac{i - m(1 - \alpha_{i\to 0}^m)}{q - p}, & p \neq \tfrac{1}{2}, \\[2mm] i(m - i), & p = \tfrac{1}{2}. \end{cases}$$

Now, let β_0^m (resp. β_m^m) denote the probability that absorption occurs in 0 (resp. m). Then

(4.6)
$$\beta_0^m = \sum_{i=0}^{m} \alpha_{i\to 0}^m \mathbb{P}\left(R_m \text{ starts in state } i\right),$$

and $\beta_m^m = 1 - \beta_0^m$. From the structure of the Markov chain $X(q)$ and the correspondence between $X(q)$ and the random walks, we have that (see Figure 4.2):

$$\mathbb{P}\left(\text{enter level } m+1 \text{ at } (0, C - m + 1)\right) = \mathbb{P}\left(\text{absorption in 0 in } R_m\right) = \beta_0^m.$$

Now one has:

(4.7)
$$\begin{aligned} \mathbb{E}\left[T_m\right] &= \mathbb{E}[\widetilde{T}_1^m]\beta_0^{m-1} + \mathbb{E}[\widetilde{T}_{m-1}^m]\beta_{m-1}^{m-1} \\ &= \mathbb{E}[\widetilde{T}_{m-1}^m] + \left(\mathbb{E}[\widetilde{T}_1^m] - \mathbb{E}[\widetilde{T}_{m-1}^m]\right)\beta_0^{m-1}. \end{aligned}$$

4.1.1. Case $p = 1/2$. $\mathbb{E}[T_m]$ can be calculated explicitly for $p = \tfrac{1}{2}$. Since the random walk is symmetric, we have $\beta_0^m = \beta_n^m = \tfrac{1}{2}$. So:

(4.8)
$$\mathbb{E}\left[T_m\right] = \mathbb{E}[\widetilde{T}_1^m]\beta_0^{m-1} + \mathbb{E}[\widetilde{T}_{m-1}^m]\beta_{m-1}^{m-1} = m - 1.$$

Hence,

$$\mathbb{E}\left[H_{0,C}\right] = \sum_{m=2}^{C+1} \mathbb{E}\left[T_m\right] = \sum_{m=2}^{C+1} m - 1 = \frac{C^2 + C}{2}.$$

LEMMA 4.2. *For a $M/M/1/C$ with $\lambda = \mu$, $\mathbb{E}\tau^b = \frac{C^2+C}{2}$.*

4.1.2. Case $p \neq 1/2$. Entering the random walk R_m corresponds to entering level m in $X(q)$. Since we can only enter level m in the state $(0, C - m + 2)$ or $(m - 2, C)$ this means we can only start the random walk in state 1 or $m-1$. Therefore (4.6) becomes:

$$\begin{aligned} \beta_0^m &= \sum_{i=0}^{m} \alpha_{i\to 0}^m \mathbb{P}\left(R_m \text{ starts in state } i\right) \\ &= \alpha_{1\to 0}^m \mathbb{P}\left(R_m \text{ starts in 1}\right) + \alpha_{m-1\to 0}^m \mathbb{P}\left(R_m \text{ starts in } m-1\right) \\ &= \frac{a^m - a^{m-1}}{a^m - 1} + \frac{a^{m-1} - a}{a^m - 1}\beta_0^{m-1}. \end{aligned}$$

This gives the recurrence:

(4.9)
$$\begin{cases} \beta_0^m = \dfrac{a^m - a^{m-1}}{a^m - 1} + \dfrac{a^{m-1} - a}{a^m - 1}\beta_0^{m-1} & m > 2; \\[2mm] \beta_0^2 = 2. \end{cases}$$

Thus we obtain,

PROPOSITION 4.3. *For a M/M/1/C queue, using the foregoing notations,*

$$(4.10) \qquad \mathbb{E}\tau^b = \mathbb{E}\left[H_{0,C}\right] = \sum_{m=2}^{C+1} \mathbb{E}[\tilde{T}_{m-1}^m] + \left(\mathbb{E}[\tilde{T}_1^m] - \mathbb{E}[\tilde{T}_{m-1}^m]\right)\beta_0^{m-1},$$

with β_0^m defined by (4.9)and $\mathbb{E}[\tilde{T}_{m-1}^m]$ and $\mathbb{E}[\tilde{T}_1^m]$ defined by (4.5).

4.1.3. Comparison between the cases $p = 1/2$ and $p \neq 1/2$.

PROPOSITION 4.4. *The coupling time in a M/M/1/C queue is maximal when the input rate λ and the service rate μ are equal.*

Proof. By definition, $\lambda = \mu$ corresponds to $p = q = 1/2$. The proof holds by induction on C. The result is obviously true when $C = 0$, because whatever q, $\mathbb{E}\left[H_{0,C}\right] = 0$.

For $C + 1$, let q be an arbitrary probability with $q > 1/2$ (the case $q < 1/2$ is symmetric). We will compare the expected time for absorption of three Markov chains. The first one is the Markov chain $X := X(1/2)$ displayed in Figure 4.1, with $q = p = 1/2$. The second one is the Markov chain $X' = X(q)$ displayed in Figure 4.1 and the last one X'' is a mixture between the two previous chains: The first C levels are the same as in X while the last level $(C + 1)$ is the same as in X'.

The expected absorption time for the first C levels is the same for X and for X'': $\sum_{m=2}^{C} \mathbb{E}T_m = \sum_{m=2}^{C} \mathbb{E}T_m''$. By induction, this is larger than for X': we have $\sum_{m=2}^{C} \mathbb{E}T_m = \sum_{m=2}^{C} \mathbb{E}T_m'' \geqslant \sum_{m=2}^{C} \mathbb{E}T_m'$. Therefore, we just need to compare the expected exit times out of the last level, namely $\mathbb{E}T_{C+1}, \mathbb{E}T_{C+1}'$ and $\mathbb{E}T_{C+1}''$, to finish the proof.

Let us first compare $\mathbb{E}T_{C+1}$ and $\mathbb{E}T_{C+1}''$. In both cases, the Markov chain enters level $C + 1$ in state $(0, 1)$ with probability $1/2$.

Equation (4.8) says that $\mathbb{E}T_{C+1} = C$ and Equation (4.5) gives after straightforward computations, $\mathbb{E}T_{C+1}'' = 1/2 \frac{1-C(1-\alpha_{1\to0}^C)}{q-p} + 1/2 \frac{C-1-C(1-\alpha_{C-1\to0}^C)}{q-p} = \frac{C}{2q}\frac{a^C-a}{a^C-1} \leqslant C/(2q) < C = \mathbb{E}T_{C+1}$.

In order to compare $\mathbb{E}T_{C+1}'$ and $\mathbb{E}T_{C+1}''$, let us first show that β_0^m is larger than $1/2$, for all $m \geqslant 2$. This is done by an immediate induction on Equation (4.9). If $\beta_0^{m-1} \geqslant 1/2$, then $\beta_0^m \geqslant \frac{2a^m - a^{m-1} - a}{2(a^m-1)}$ Now, $\frac{2a^m - a^{m-1} - a}{2(a^m-1)} \geqslant 1/2$ if $2a^m - a^{m-1} - a \geqslant a^m - 1$, *i.e.* after recombining the terms, $(a-1)(a^{m-1}-1) \geqslant 0$. This is true as soon as $q \geqslant 1/2$.

To end the proof, it is enough to notice that for the chain X', time to absorption starting in 1, $\mathbb{E}\tilde{T}_1^{m'}$ is smaller that time to absorption starting in $m-1$, $\mathbb{E}\tilde{T}_{m-1}^{m'}$ for all m. The difference $\mathbb{E}\tilde{T}_{m-1}^{m'} - \mathbb{E}\tilde{T}_1^{m'}$ is

$$\frac{ma^m - ma^{m-1} + ma - m - 2a^m + 2}{p(a^m-1)(a-1)} = \frac{m(a-1)\left(\frac{a^{m-1}+1}{2} - \frac{1+a+\cdots+a^{m-1}}{m}\right)}{p(a^m-1)(a-1)} \geqslant 0,$$

by convexity of $x \mapsto a^x$. Finally,

$$\mathbb{E}T_{C+1}' = \beta_0^{C+1}\mathbb{E}\tilde{T}_1^{C+1'} + (1-\beta_0^{C+1})\mathbb{E}\tilde{T}_C^{C+1'} \leqslant \frac{1}{2}\mathbb{E}\tilde{T}_1^{C+1'} + \frac{1}{2}\mathbb{E}\tilde{T}_C^{C+1'} = \mathbb{E}T_{C+1}''.$$

$\square$

4.2. Explicit Bounds. Equation (4.10) provides a quick way to compute $\mathbb{E}[H_{0,C}]$ using recurrence equation (4.9). However, it may also be interesting to get a simple closed form for an upper bound for $\mathbb{E}[H_{0,C}]$. This can be done using the last inequality in Equation (3.5) that gives an upper bound for $\mathbb{E}[H_{0,C}]$ amenable to direct computations.

$$(4.11) \qquad \mathbb{E}[H_{0,C}] = \mathbb{E}[\min\{h_{0 \to C}, h_{C \to 0}\}] \leqslant \min\{\mathbb{E}[h_{0 \to C}], \mathbb{E}[h_{C \to 0}]\}.$$

The exact calculation of $\mathbb{E}[h_{0 \to C}]$ can be done using a one step analysis. Let F_i be the time to go from state i to 0. Then, $h_{0 \to C} = F_C$ and for all $i > 0$,

$$(4.12) \qquad \mathbb{E}[F_i] = 1 + p\mathbb{E}[F_{(i+1) \wedge C}] + q\mathbb{E}[F_{i-1}].$$

With an approach derived from [9] one can condition on the next event. Let T_i denote the time to go from state i to $i+1$. Then

$$(4.13) \qquad \mathbb{E}[h_{0 \to C}] = \sum_{i=0}^{C-1} \mathbb{E}[T_i].$$

To get an expression for T_i, with $0 < i \leqslant C$, we condition on the first event. Therefore let $\mathbb{E}[T_i | e]$ denote the conditional expectation of T_i knowing that the next event is e. Since $\mathbb{E}[T_i \mid e_a] = 1$ and $\mathbb{E}[T_i \mid e_d] = 1 + \mathbb{E}[T_{i-1}] + \mathbb{E}[T_i]$, conditioning delivers the following recurrent expression for the $\mathbb{E}[T_i]$:

$$(4.14) \qquad \mathbb{E}[T_i] = \left\{ \begin{array}{ll} \frac{1}{p} + \frac{q}{p}\mathbb{E}[T_{i-1}], & 0 < i < C \\ \frac{1}{p}, & i = 0. \end{array} \right.$$

By induction one can show that $\mathbb{E}[T_i] = \frac{1}{p}\sum_{k=0}^{i}\left(\frac{q}{p}\right)^k$. Hence, $\mathbb{E}[T_i] = \frac{1-\left(\frac{q}{p}\right)^{i+1}}{p-q}$ and from (4.13) it follows that

$$(4.15) \qquad \mathbb{E}[h_{0 \to C}] = \sum_{i=0}^{C-1} \frac{1 - \left(\frac{q}{p}\right)^{i+1}}{p-q} = \frac{C}{p-q} - \frac{q(1 - \left(\frac{q}{p}\right)^C)}{(p-q)^2}.$$

By reasons of symmetry, we have

$$(4.16) \qquad \mathbb{E}[h_{C \to 0}] = \frac{C}{q-p} - \frac{p(1 - \left(\frac{p}{q}\right)^C)}{(q-p)^2}$$

The curves of $\mathbb{E}[h_{0 \to C}]$ and $\mathbb{E}[h_{C \to 0}]$ intersect in $C^2 + C$ when $p = q$. If $p > q$ then $\mathbb{E}[h_{0 \to C}] < \mathbb{E}[h_{C \to 0}]$ and because of symmetry, if $p < q$ then $\mathbb{E}[h_{0 \to C}] > \mathbb{E}[h_{C \to 0}]$. Since also $\frac{C^2+C}{2}$ is an upper bound corresponding to the critical case $p = q$ on the mean coupling time $\mathbb{E}\tau^b$, as shown in Proposition 4.4, one can state:

PROPOSITION 4.5. *The mean coupling time $\mathbb{E}\tau^b$ of a $M/M/1/C$ queue with arrival rate λ and service rate μ is bounded using $p = \lambda/(\lambda + u)$ and $q = 1 - p$.*

Critical bound:	$\forall p \in [0, 1],$	$\mathbb{E}[\tau^b] \leqslant \frac{C^2+C}{2}.$
Heavy traffic Bound:	if $p > \frac{1}{2},$	$\mathbb{E}[\tau^b] \leqslant \frac{C}{p-q} - \frac{q(1-\left(\frac{q}{p}\right)^C)}{(p-q)^2}.$
Light traffic bound:	if $p < \frac{1}{2},$	$\mathbb{E}[\tau^b] \leqslant \frac{C}{q-p} - \frac{p(1-\left(\frac{p}{q}\right)^C)}{(q-p)^2}.$

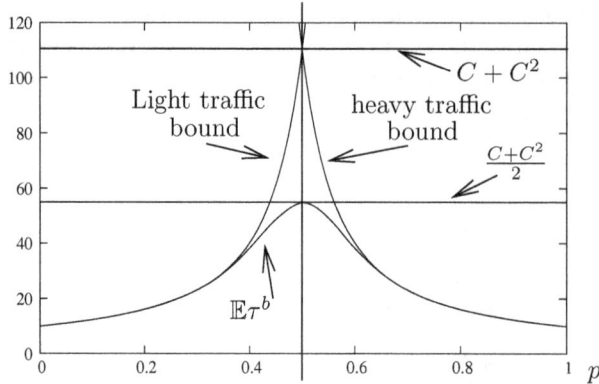

FIG. 4.3. *Expected coupling time in an M/M/1/10 queue when q varies from 0 to 1 and the three explicit bounds given in Proposition 4.5*

Figure 4.3 displays both the exact expected coupling time for a queue with capacity 10 as given by Equation (4.10) as well as the three explicit bounds given in Proposition 4.5. Note that the bounds are very accurate under light or heavy traffic ($q \leqslant 0.4$ and $q > 0.6$). In any case, the ratio is never larger than 1.2.

5. Coupling in acyclic queueing networks. This section is dedicated to the effective computation of a bound of the coupling time in acyclic networks. In the acyclic network given in Figure 3.1, the coupling time has a peak at $\lambda_0 = 0.4$, as one can see in Figure 3.2. This corresponds to the case when the input rate and service rate in Queue 3 are equal. This should not be surprising regarding the result for a single queue, which says that the coupling time is maximal when the rates are equal. Then a second peak occurs around $\lambda_0 = 1.4$ when coupling in Queue 0 is maximal. The rest of the curve shows a linear increase of the coupling time which may suggest an asymptotic linear dependence in λ_0. In this part, an explicit bound on the coupling time which exhibits these two features will be derived.

The first result concerns an extension of inequality (3.5) to distributions. The second part shows how the results for a single M/M/1/C queue can be used to get an effective computation of bounds for acyclic networks on queues.

In the following, the queues $Q_0, \ldots Q_K$ are numbered according to the topological order of the network. Thus, no event occurring in queue Q_i has any influence on the state of queue Q_j as long as $i > j$.

5.1. Computation of an upper bound on the coupling time. Here, an acyclic network of ./M/1/C queues with an arbitrary topology and Bernoulli routings is considered. The events here are of only two types: exogenous arrivals (Poisson with rate γ_i in queue i) and routing of one customer from queue i to queue j after service completion in queue i (with rate μ_{ij}). Queue $K + 1$ is a dummy queue representing exits: routing a customer to queue $K + 1$ means that the customer exits the network forever. In case of overflow, the new customer trying to enter the full queue is lost. The service rate at queue i is also denoted $\mu_i = \sum_{i=0}^{K+1} \mu_{ij}$.

Let us introduce new random variables. $\tau^b(s_j = x)$ is the backward coupling time of the network, over the set of all intial states with the j-th coordinate equal to x.

Namely,

$$\tau^b(s_j = x) = \min\left\{n \text{ s.t. } \left|\phi^{(n)}\left(\mathcal{S} \cap \{s_j = x\}, e_{-n+1}, \ldots, e_0\right)\right| = 1\right\}.$$

$\tau_i^b(s_j = x)$ is the backward coupling time on coordinate i given $s_j = x$:

$$\tau_i^b(s_j = x) = \min\left\{n \text{ s.t. } \left|N_i\left(\phi^{(n)}\left(\mathcal{S} \cap \{s_j = x\}, e_{-n+1}, \ldots, e_0\right)\right)\right| = 1\right\}.$$

It should be obvious that $\tau^b(s_j = x) \leqslant_{st} \tau^b$ and for all i, $\tau_i^b(s_j = x) \leqslant_{st} \tau_i^b$. We also have the same notions for forward coupling times:

$$\tau^f(s_j = x) = \min\left\{n \in \mathbb{N}; \text{ s.t. } \left|\phi^{(n)}\left(\mathcal{S} \cap \{s_j = x\}, e_{1 \to n}\right)\right| = 1\right\},$$

$\tau_i^f(s_j = x)$ being defined in the same manner, and for hitting times:

$$h_{C_i \to 0}(s_j = x) = \min\{n \in \mathbb{N}; \text{ s. t. } \phi^{(n)}\left(\mathcal{S} \cap \{s_i = C_i, s_j = x\}, e_{1 \to n}\right) \in \mathcal{S} \cap \{s_i = 0\}\}.$$

Now, sweeping the list of queues in the topological order, one can construct a sequence of backward simulations in the following way.

First simulate the queueing system from the past up to coupling of queue 0. The number of steps is by definition τ_0^b. Queue Q_0 has coupled in a random state X_0. Then, run a second backward simulation up to coupling of Queue Q_1 given $s_0 = X_0^0$. This simulation takes $\tau_i^b(s_0 = X_0^0)$ steps and the state at time $t = 0$ is X_1^1 for Q_1 and X_0^1 for Q_0.

This construction goes on up to the backward simulation up to coupling of Queue Q_K given $s_0 = X_0^{K-1}, s_1 = X_1^{K-1}, \ldots, s_{K-1} = X_{K-1}^{K-1}$. The last simulation takes $\tau_i^b(s_0 = X_0^{K-1}, s_1 = X_1^{K-1}, \ldots, s_{K-1} = X_{K-1}^{K-1})$ steps and the coupling state of Q_K is X_K^K.

LEMMA 5.1. *One has* $\tau^b \leqslant_{st} \sum_{i=0}^{K} \tau_i^b(s_0 = X_0^{i-1}, \ldots, s_{i-1} = X_{i-1}^{i-1})$, *and for all i, $(X_0^i, \ldots, X_i^i)$ is steady state distributed for $Q_0, \ldots, Q_i$. Furthermore, for all i,*

$$(5.1) \qquad \tau^b \leqslant_{st} \sum_{i=0}^{K} h_{C_i \to 0}(s_0 = X_0^{i-1}, \ldots, s_{i-1} = X_{i-1}^{i-1}).$$

Proof. From the previous sequence of backward simulations one can construct a single one by appending them in the reverse order: the backward simulation for Queue Q_K preceded by the simulation of Q_{K-1}, and so forth up to the simulation of Q_0. This is a backward simulation of the system (the last state is $(X_0^K, \ldots, X_K^K)$). This construction is illustrated in the case of two queues in tandem in Figure 5.1.

A straightforward consequence, using acyclicity, is that $(X_0^i, \ldots, X_i^i)$ is steady state distributed for $Q_0, \ldots, Q_i$ for all i.

Furthermore, one gets in distribution

$$\tau^b \leqslant_{st} \sum_{i=0}^{K} \tau_i^b(s_0 = X_0^{i-1}, \ldots, s_{i-1} = X_{i-1}^{i-1}) = \sum_{i=0}^{K} \tau_i^f(s_0 = X_0^{i-1}, \ldots, s_{i-1} = X_{i-1}^{i-1})$$

$$\leqslant_{st} \sum_{i=0}^{K} h_{C_i \to 0}(s_0 = X_0^{i-1}, \ldots, s_{i-1} = X_{i-1}^{i-1}),$$

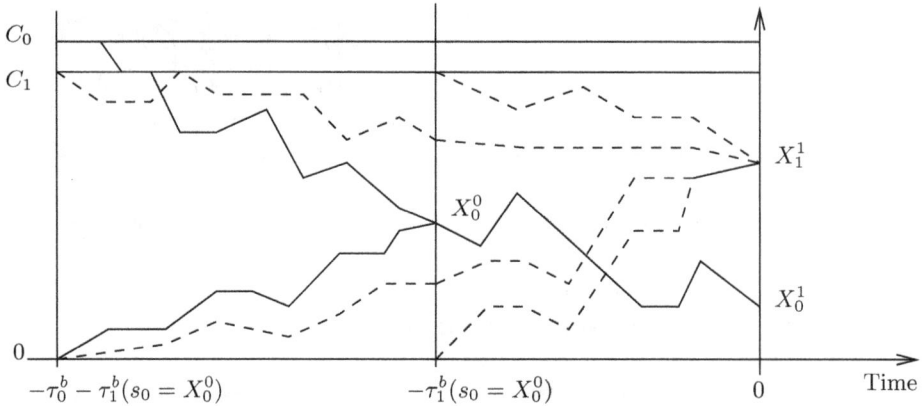

FIG. 5.1. *The trajectories of the state in Q_0 are displayed with solid lines while the trajectories for Q_1 are dashed. Starting at time $-\tau_0^b - \tau_1^b(s_0 = X_0^0)$, the state of Q_0 has coupled in X_0^0 at time $-\tau_1^b(s_0 = X_0^0)$. From then on, Q_0 stays coupled and Q_1 couples at some time before 0.*

by independence of the variables, given the initial states X^{i-1}.

□

Let us now consider a new circuit with one difference from the original one: all queues are replaced by infinite queues, except for queue Q_i which stays the same. In the following, all the notations related to this new network will be expressed by appending the ∞ symbol to all variables corresponding to this new circuit.

The new circuit up to Queue i is product form and using Burke's Theorem, the input stream in Queue i is Poisson. The rate of the input stream in queue i is given by ℓ_i, the solution of the flow equations:

$$\ell_i = \sum_{j<i} \ell_j \frac{u_{ji}}{\mu_j} + \gamma_i.$$

The network is said to be *stable* for Queue i as soon as $\ell_i < \mu_i$. We assume stability for all i in the following.

One can construct a sequence of backward simulations for the new network in the same way as for the original network. This provides the quantities $^\infty X_j^{i-1}$, $^\infty \tau_i^b(s_0 = {}^\infty X_0^{i-1}, \ldots, s_{i-1} = {}^\infty X_{i-1}^{i-1})$, $^\infty \tau_i^f(s_0 = {}^\infty X_0^{i-1}, \ldots, s_{i-1} = {}^\infty X_{i-1}^{i-1})$, and $^\infty h_{C_i \to 0}(s_0 = {}^\infty X_0^{i-1}, \ldots, s_{i-1} = {}^\infty X_{i-1}^{i-1})$.

The monotony property given above implies that $X_i^j \leqslant_{st} {}^\infty X_i^j$ and

$$h_{C_i \to 0}(s_0 = X_0^{i-1}, \ldots, s_{i-1} = X_{i-1}^{i-1}) \leqslant_{st} {}^\infty h_{C_i \to 0}(s_0 = {}^\infty X_0^{i-1}, \ldots, s_{i-1} = {}^\infty X_{i-1}^{i-1}).$$

The next step is to build yet another model. This third model is made of a single $M/M/1/C_i$ queue with three types of events, arrivals of customers with rate ℓ_i (provided that the number of customers is smaller than C_i), departures with rate μ_i (provided that the number of customers is positive) and null events (with no effect on the queue) with rate $\Lambda - \ell_i - \mu_i$.

For this isolated model, let us introduce the uniformizing probabilities $p = \ell_i/\Lambda$, $q = 1 - p$ and $d = (\Lambda - \ell_i - \mu_i)/\Lambda$. Let F_k be the time to go from state k to state 0 in the isolated system. A one step analysis gives

$$\mathbb{E}[F_k] = 1 + d\mathbb{E}[F_k] + \frac{\ell_i}{\Lambda}\mathbb{E}[F_{(k+1)\wedge C_i}] + \frac{\mu_i}{\Lambda}\mathbb{E}[F_{(k-1)}]$$

$$= \frac{1}{1-d} + p\mathbb{E}[F_{(k+1)\wedge C_i}] + q\mathbb{E}[F_{(k-1)\vee 0}].$$

We get the same equation as (4.14) except for the additional constant which is now $\frac{1}{1-d}$ instead of 1, so that the solution is the same as before up to a multiplicative factor of $\frac{1}{1-d} = \frac{\Lambda}{\ell_i + \mu_i}$. Using Equation (4.16), one gets

(5.2) $$\mathbb{E}[F_{C_i}] = \frac{\Lambda}{\ell_i + \mu_i}\left(\frac{C_i}{q-p} - \frac{p(1 - \left(\frac{p}{q}\right)^{C_i})}{(q-p)^2}\right).$$

LEMMA 5.2. *Under the foregoing notations and assumptions,*

$$^\infty h_{C_i \to 0}(s_0 = {}^\infty X_0^{i-1}, \ldots, s_{i-1} = {}^\infty X_{i-1}^{i-1}) = F_{C_i},$$

in distribution.

Proof. First, using Lemma 5.1 for the new network with infinite queues (except for Q_i), the state $({}^\infty X_0^{i-1}, \ldots, {}^\infty X_{i-1}^{i-1})$ is steady state distributed. Using Burke's Theorem, this implies that the input stream in queue Q_i is Poisson with rate ℓ_i, when one runs a simulation starting in any state in $\mathcal{S} \cap \{s_i = C_i, s_j = {}^\infty X_j^{i-1}, j < i\}$.

Now, during this simulation, one can couple the addition, subtraction et null events for queue Q_i in isolation and for Q_i in the complete network of infinite queues, all of them having the same laws. This implies that the state of queue Q_i in both systems is the same under that coupling. Hence, they reach 0 at the same time: $^\infty h_{C_i \to 0}(s_0 = {}^\infty X_0^{i-1}, \ldots, s_{i-1} = {}^\infty X_{i-1}^{i-1}) = F_{C_i}$ in distribution. □

We are ready to put everything together in expectation.

(5.3) $$\mathbb{E}\tau^b \leqslant_{st} \sum_i \mathbb{E}[h_{C_i \to 0}(s_0 = X_0^{i-1}, \ldots, s_{i-1} = X_{i-1}^{i-1})]$$

(5.4) $$\leqslant \sum_i \mathbb{E}[^\infty h_{C_i \to 0}(s_0 = {}^\infty X_0^{i-1}, \ldots, s_{i-1} = {}^\infty X_{i-1}^{i-1})]$$

(5.5) $$\leqslant \sum_i \mathbb{E}[F_{C_i}].$$

The sequence of inequalities may not hold in distribution because the variables X^i and thus $h_{C_i \to 0}(s_0 = X_0^{i-1}, \ldots, s_{i-1} = X_{i-1}^{i-1})$ are not independent.

Using (5.2),

$$\mathbb{E}\tau^b \leqslant \sum_i \frac{\Lambda}{\ell_i + \mu_i}\left(\frac{C_i}{q-p} - \frac{p(1 - \left(\frac{p}{q}\right)^{C_i})}{(q-p)^2}\right).$$

The result of this part is summarized in the following theorem.

THEOREM 5.3. *In an acyclic stable network of $K+1$ ./M/1/C_i queues with Bernoulli routing and losses in case of overflow, the coupling time from the past* *satisfies in expectaction,*

$$(5.6) \quad \mathbb{E}[\tau^b] \leqslant \sum_{i=0}^{K} \frac{\Lambda}{\ell_i + \mu_i} \left(\frac{C_i}{q-p} - \frac{p(1 - \left(\frac{p}{q}\right)^{C_i})}{(q-p)^2} \right) \leqslant \sum_{i=0}^{K} \frac{\Lambda}{\ell_i + \mu_i}(C_i + C_i^2).$$

Note that this bound on the expectation is ultimately linear in the rate of any event in the system. This behavior is also noticeable for $\mathbb{E}[\tau^b]$ itself.

5.2. Some numerical experiments. In the construction of the bound given in Theorem 5.3, several factors may be responsible for the inaccuracy of the bound.

1. The first factor is the replacement of the max by the sum. We believe that it may be a hard task to get rid of this first approximation because of the intricate dependencies between the queues. Furthermore, experiments reported below show that this may not even be possible in many cases (see Figure 5.2.b).

2. Another factor which may increase the inaccuracy of our bounds is the fact that most events change the states of several queues at the same time, while the bound given here disregards this. In the network studied here, this may add a factor 2 between the true coupling time and the bound given in Theorem 5.3.

3. The most important factor which jeopardizes the quality of the bound is the load issue. If one queues has an heavy load (larger than 1) , the bound provided by Equation (5.6), also called the light traffic bound in Proposition 4.5 is very bad (as seen in Figure 4.3). So far we have not been able to come up with a better bound for queues with large loads. However, when all queues have a small load (smaller than one, and even more so when the load is smaller than 2/3), the bound tends to be more accurate. This is further verified in the experiments reported below.

Computations for the network displayed in Figure 3.1 are reported in Figure 5.2. We have used the following parameters. The input rate is $\lambda_0 = 0.4$. the rates of the other events are $\lambda_1 = 1.4$, $\lambda_2 = 0.6$, $\lambda_3 = 0.8$, $\lambda_4 = 0.5$. The number of simulation runs is 10000. The capacity C is the same in all queues, and we let it vary from 1 to 20. The service rate in the last queue λ_5 takes three values, respectively $0.2, 0.6$ and 0.4.

In the first case (Fig. 5.2.a) , $\lambda_5 = 0.2$ so that queue Q_3 has a load larger than one. Figure 5.2 displays the bound given by formula (5.6) as well as the mean coupling time computed over 10000 simulation runs. As hinted before, the bound is indeed very bad for this system. A ratio larger than 10 w.r.t the true coupling time is reached when $C = 5$. It should also be noticed that our bound is convex in C while the coupling time does not seem to be so.

In the second case (Fig. 5.2.b) , $\lambda_5 = 0.6$, and all queues have a load smaller that 2/3. Figure 5.2.b shows the bound provided by (5.6) and the true coupling time computed by simulation runs. Both curves appear to be almost linear in C (this is true for the bound: when q/p is small, $\mathbb{E}H_{C_i,0}$ is almost linear in C_i) and the ratio is smaller than 1.3. In that case, the curve $\max_{i \in \{0,...K\}} \mathbb{E}H_{C_i,0}$ is also displayed and is below the actual coupling time. This is to be related with the first item in the comments above. The last case (Fig. 5.2.c) is for $\lambda_0 = 0.4$, so that Q_3 has load exactly one. This would correspond to the maximal coupling time for Q_3 if it were alone. Figure 5.2.c displays the backward coupling time and the bound provided by Equation (5.6). For queue Q_3, we use a bound in $C_3 + C_3^2$ which is a bad approximation because of the loss of the factor 2 when compared with the bound for isolated queues. Note that the total gap has a ratio which is almost 2. In that case both the coupling time and the bound exhibit a convex behavior w.r.t. C.

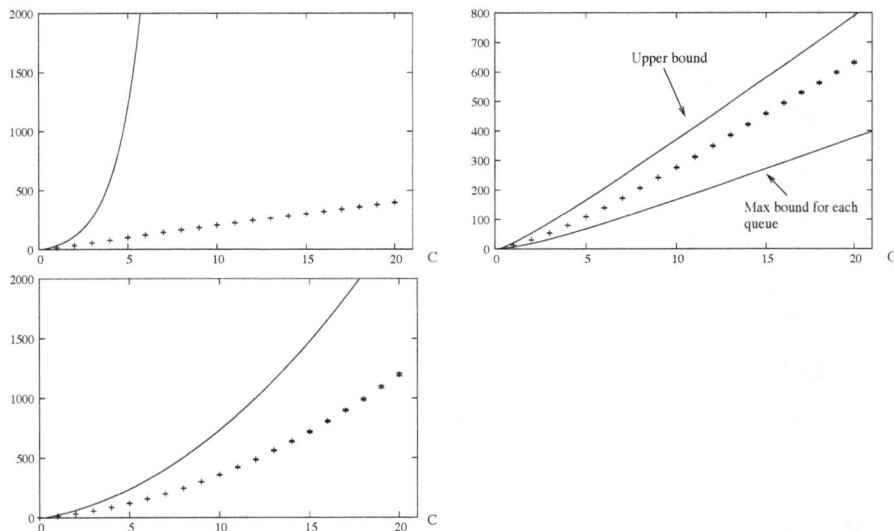

FIG. 5.2. *The capacity C varies from 1 to 20 in all queues. The upper left figure displays the coupling time (dots) with 95% confidence intervals, and the bound given by Equation (5.6) when Queue Q_3 is unstable ($\lambda_5 = 2/10$). In the upper right figure, are given the bound in Equation 5.6, the mean coupling time (dots) with 95% confidence intervals and the maximum over Equations (5.6) for all queues, when Queue Q_3 is stable ($\lambda_5 = 6/10$). The lower figure displays the coupling time (dots) with 95% confidence interval and the bound given by Equation (5.6) when Queue Q_3 is barely unstable ($\lambda_5 = 4/10$)*

A ratio smaller than 2 is indeed interesting because efficient perfect simulation algorithm use a doubling window technique to reduce the complexity and their running time (see Equation (2.4)) so that our bound gives a good estimation of the mean running time of the algorithms.

One should also note that, on a practical point of view, most actual networks which require stationary performance evaluations have small loads.

5.2.1. Extension to more general networks. Actually, extensive simulation runs over many examples show that the bound given in Theorem 5.3 is robust and also holds for more general networks with blocking and with circuits. While we have only been able to show that the light traffic bound holds for each queue, we conjecture that the heavy traffic bound and the critical bound should also hold. This would yield an overall quadratic bound: $\mathbb{E}[\tau^b] \leqslant \sum_{i=0}^{K} \frac{\Lambda}{\ell_i + \mu_i} O(C_i^2)$, for any monotone Markovian network of queues with a finite state space. Furthermore under light or heavy traffic in all queues, the bound should rather be linear: $\mathbb{E}[\tau^b] \leqslant \sum_{i=0}^{K} \frac{\Lambda}{\ell_i + \mu_i} O(C_i)$.

To illustrate this conjecture, we have run simulations for the network displayed in Figure 3.1 with the following parameters. The rates are $\lambda_0 = 0.4, \lambda_1 = 1.4, \lambda_2 = 0.6$, $\lambda_3 = 0.8, \lambda_4 = 0.5$. The capacity is fixed to 10 in all queues and we let λ_5 (the service rate in Q_3) vary from 0 to 4. As long as $\lambda_5 < 0.4$, Q_3 is unstable and our proven bound (B_1) is poor. As soon as λ_5 is large enough our bound becomes acceptable. In Figure 5.3, note that both the bound and the coupling time τ have a linear asymptotic growth in λ_5. The Figure also displays the heavy traffic bound B_2 and the critical bound B_3. Should these two bounds hold, the minimum of B_1, B_2, B_3 (in bold in the figure) would provide a remarkable bound on the coupling time, up to an additional constant. This issue is the subject of our current investigations.

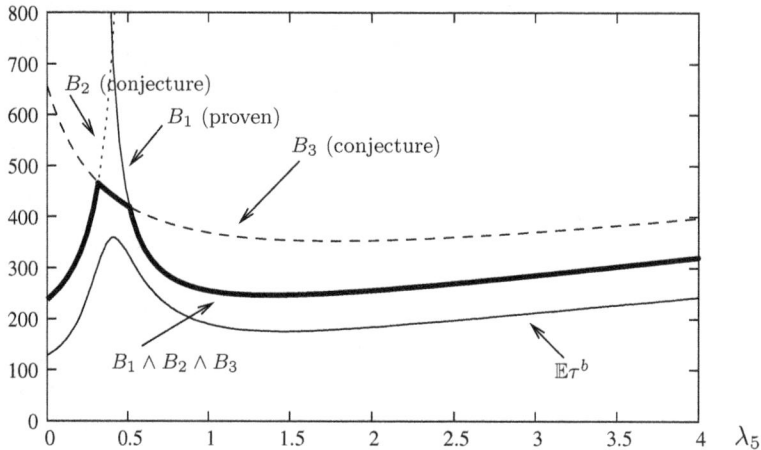

FIG. 5.3. *This figure displays the actual coupling time* $\mathbb{E}\tau^b$ *for the network of Figure 3.1, when the service rate of the last queue ranges from 0 to 5, together with the proven light traffic bound* B_1, *the conjectured heavy traffic bound* B_2, *the conjectured critical bound* B_3 *and the minimum of the three bounds.*

Acknowledgments. The authors would like to thank Jérôme Vienne who partially designed the psi2 perfect simulation software which was used to run all the simulations presented here

REFERENCES

[1] C. ALEXOPOULOS AND D. GOLDSMAN, *To batch or not to batch?*, ACM Trans. Model. Comput. Simul., 14 (2004), pp. 76–114.

[2] J.M. CALVIN, P.W. GLYNN, AND M.K. NAKAYAMA, *The semi-regenerative method of simulation output analysis.* submitted.

[3] ——, *Importance sampling using the semi-regenerative method*, in Proceedings of the Winter Simulation Conference, vol. 1, 2001, pp. 441–450.

[4] M. CRANE AND D.L. IGLEHART, *Simulating stable stochastic systems, iii: Regenerative processes and discrete-event simulation*, Operation Research, 23 (1975), pp. 33–45.

[5] P. GLASSERMAN AND D.D. YAO, *Monotone Structure in Discrete-Event Systems*, Wiley Inter-Science, Series in Probability and Mathematical Statistics, 1994.

[6] O. HÄGGSTRÖM, *Finite Markov Chains and Algorithmic Applications*, Cambridge University Press, 2002.

[7] S.G. HENDERSON AND P.W. GLYNN, *Regenerative steady-state simulation of discrete-event systems*, ACM Trans. Model. Comput. Simul., 11 (2001), pp. 313–345.

[8] J. PROPP AND D. WILSON, *Exact sampling with coupled Markov chains and applications to statistical mechanics*, Random Structures and Algorithms, 9 (1996), pp. 223–252.

[9] S. M. ROSS, *Probability models*, Academic Press, 2003.

[10] W.J. STEWART, *Introduction to the Numerical Solution of Markov Chains*, Princeton, 1994.

[11] J.-M. VINCENT, *Perfect simulation of monotone systems for rare event probability estimation*, in Winter Simulation Conference, Orlando, dec 2005.

[12] ——, *Perfect simulation of queueing networks with blocking and rejection*, in Saint IEEE conference, Trento, 2005, pp. 268–271.

[13] J.-M. VINCENT AND C. MARCHAND, *On the exact simulation of functionals of stationary markov chains*, Linear Algebra and its Applications, 386 (2004), pp. 285–310.

ANALYSIS OF MARKOV REWARD MODELS WITH PARTIAL REWARD LOSS BASED ON A TIME REVERSE APPROACH

GÁBOR HORVÁTH, MIKLÓS TELEK
DEPARTMENT OF TELECOMMUNICATIONS
TECHNICAL UNIVERSITY OF BUDAPEST, 1521 BUDAPEST, HUNGARY
{*HGABOR,TELEK*}@*WEBSPN.HIT.BME.HU*

Abstract. There are effective numerical methods for the analysis of Markov reward models (MRMs) without or with complete reward loss, but the analysis of MRMs with partial reward loss is more complex. This paper presents the analytical description of the distribution and the moments of the accumulated reward of partial increment loss reward models and an effective numerical method to evaluate these measures. The solution method is based on the time reverse process. The time reverse analysis allows to avoid computationally expensive techniques like introducing a supplementary variable or evaluating numerical integrals. Even though, this approach introduces difficulties at the first sight since the reverse process is inhomogeneous, it turns out that with the introduction of a properly chosen performance measure, homogeneous differential equations characterize the desired reward measures. This set of homogeneous differential equations allows one to apply an iterative scheme similar to the randomization (or Jensen's) method with all advantages of that method, i.e., numerical stability, pre-calculated error bound.

Keywords: Time reverse process, Inhomogeneous Markov chain, Reward models, Partial increment loss.

1. Introduction. The extension of discrete state stochastic models with a continuous reward variable results in a very effective modeling tool, the stochastic reward model [9]. Stochastic reward models have been applied for modeling and performance analysis of various engineering systems for a long time. Due to their relative tractability, reward models associated with continuous time Markov chains (CTMC), referred to as Markov reward models (MRM), have gain major attention. The analytical description of Semi-Markov [4] and Markov regenerative reward models [15] are also available, but their numerical analysis is far more complex.

There are two kinds of reward considered in MRMs. During a sojourn of the underlying process in a system state *rate reward* is accumulated continuously. At the state transition of the underlying process *impulse reward* can increase the reward function instantly. The set of reward models with increasing reward function is referred to loss-less or preemptive resume (prs) reward models. Another set of reward models considers a complete or a partial loss of reward at state transitions.

A number of effective numerical methods were developed for the evaluation of MRMs without reward loss [8, 5, 6, 11, 7] or with complete reward loss [3], but the case of partial reward loss was considered only recently. The analytical description of MRM with partial reward loss is provided in [2, 12] and the first numerical procedures for the analysis of larger models in [16]. The most effective numerical method in [16] is restricted to the case when the underlying process is in steady state. We relax this inconvenient restriction here.

This paper focuses on Markov reward models with rate reward accumulation and partial incremental reward loss at state transitions. The rest of the paper is organized as follows. Section 2 defines the set of reward models considered. We summarize the basic properties of the time reverse of the underlying CTMC in section 3 and the analytical description of the accumulated reward based on the time reverse accumulation process in Section 4. Section 5 presents an effective, randomization based numerical method for the analysis of MRM with partial reward loss. Finally, a numerical example demonstrates applicability of the proposed method.

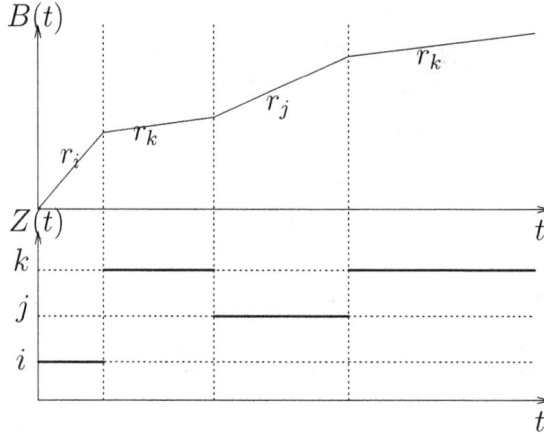

FIGURE 2.1. *Reward accumulation in a preemptive resume model*

2. Model definition.

2.1. Structure state process. Let the (right continuous) structure state process, $\{Z(t), t \geq 0\}$, be an irreducible homogeneous continuous time Markov chain (CTMC) on state space $S = \{1, 2, ..., N\}$ with generator $\mathbf{Q} = \{q_{ij}\}$ and initial probability vector $\underline{\gamma}(0)$. The transient probability vector $\underline{\gamma}(t)$ satisfies

$$(2.1) \qquad\qquad \frac{d}{dt}\underline{\gamma}(t) = \underline{\gamma}(t)\mathbf{Q}$$

with initial condition $\underline{\gamma}(0)$, whose solution is

$$(2.2) \qquad\qquad \underline{\gamma}(t) = \underline{\gamma}(0)e^{\mathbf{Q}t} \ .$$

2.2. Reward accumulation without reward loss. In stochastic reward models with rate reward accumulation each state of the structure state process is assigned with a non-negative constant which describes the rate of the reward accumulation during the sojourn of the structure state process in that state. This constant is commonly referred to as *reward rate*. During the sojourn in state $i \in S$ the amount of accumulated reward $B(t)$ increases according to the reward rate of state i, r_i:

$$\frac{dB(t)}{dt} = r_i$$

$\mathbf{R}$ denotes the diagonal matrix composed of the reward rate of the states in S ($\mathbf{R} = diag < r_i >$). In the case when there is no reward loss at the state transitions of the structure state process the amount of reward accumulated during the interval $(0, T)$ is

$$(2.3) \qquad\qquad B(T) = \int_0^T r_{Z(t)}dt \ .$$

This case is commonly referred to as preemptive resume (prs) case.

Since $Z(t)$ is a CTMC with finite generator ($|q_{ij}| < \infty$, $\forall i, j \in S$) the $Z(t)$ process has a finite number of state transitions in the interval $(0, T)$ with probability one. Let

θ_i ($i \geq 0$, $\theta_0 = 0$) denote the time of the ith state transition and N the number of state transitions in the interval $(0, T)$. The state of the structure state process after the ith state transition is X_i where $X_i = Z(\theta_i)$. The sequence (θ_i, X_i) for $0 \leq i \leq N$ defines a trajectory of the process on the interval $(0, T)$. Having this trajectory the accumulated reward at time T is

$$(2.4) \qquad B(T) = \sum_{i=1}^{N} r_{X_{i-1}}(\theta_i - \theta_{i-1}) + r_{X_N}(T - \theta_N).$$

2.3. Reward accumulation with partial incremental reward loss. During the sojourn of the structure state process in a given state, partial reward loss models accumulate reward in the same way as prs models, except that partial reward loss models might lose a portion of the accumulated reward at the state transitions of the structure state process. When the structure state process undergoes a transition from state i to another state, the fraction $1 - \alpha_i$ of the reward obtained during the last sojourn in state i is lost and only the fraction α_i of the reward (obtained during the last sojourn in i) remains. α_i is a real number, such that $0 \leq \alpha_i \leq 1$. The dynamics of the right continuous reward accumulation process $\{B(t), t \geq 0\}$ ($B(t) = B(t^+)$) is the following (see Figure 2.2):

$$(2.5) \qquad \frac{dB(t)}{dt} = r_{Z(t)} \qquad \text{for} \quad \theta_i < t < \theta_{i+1}$$

$$(2.6) \qquad B(\theta_i) = B(\theta_{i-1}) + \alpha_{Z(\theta_i^-)}[B(\theta_i^-) - B(\theta_{i-1})]$$

Assuming $Z(t)$ follows the (θ_i, X_i) trajectory the reward accumulated in the interval $(0, T)$ is

$$(2.7) \qquad B(T) = \sum_{i=1}^{N} \alpha_{X_{i-1}} r_{X_{i-1}}(\theta_i - \theta_{i-1}) + r_{X_N}(T - \theta_N),$$

since the portion $1 - \alpha_{X_i}$ of the reward accumulated during the first N sojourn is lost, but the reward accumulated in the last state X_n during the (θ_n, T) interval is present at time T.

3. Time reverse of the structure state process. In the subsequent analysis we use the time reverse of the structure state process, $\overleftarrow{Z}(\tau) = Z(T - \tau)$, $0 \leq \tau \leq T$. The time reverse of a CTMC is considered, e.g. in [10, 1]. Here we summarize the main results.

THEOREM 3.1. *If $Z(t)$ is a (homogeneous) CTMC with initial probability $\gamma(0)$ and generator $\mathbf{Q}$, its time reverse process defined by $\overleftarrow{Z}(\tau) = Z(T - \tau)$ ($0 \leq \tau \leq T$) is an inhomogeneous CTMC with initial probability $\overleftarrow{\gamma}(0) = \gamma(T)$ and generator $\overleftarrow{\mathbf{Q}}(\tau) = \{\overleftarrow{q_{ij}}(\tau)\}$ where*

$$\overleftarrow{q_{ij}}(\tau) = \begin{cases} \dfrac{\gamma_j(T - \tau)}{\gamma_i(T - \tau)} q_{ji} & \text{if } i \neq j, \\[3mm] -\displaystyle\sum_{k \in S, k \neq i} \dfrac{\gamma_k(T - \tau)}{\gamma_i(T - \tau)} q_{ki} & \text{if } i = j. \end{cases}$$

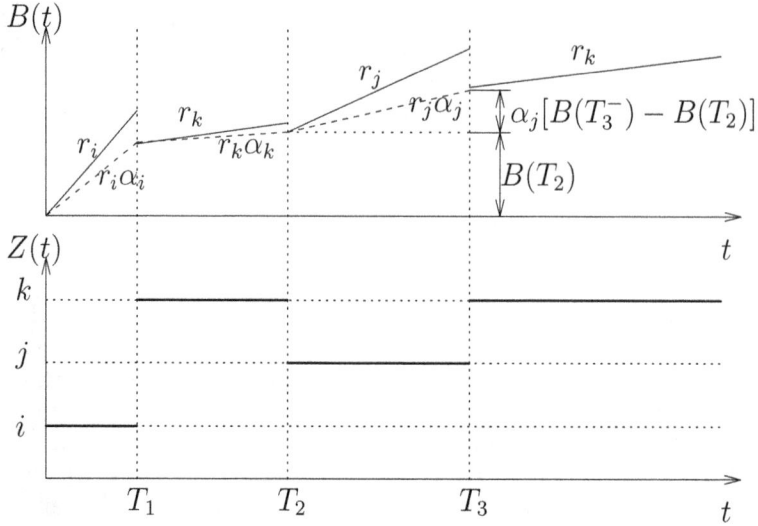

FIGURE 2.2. *Reward accumulation in a partial incremental loss model*

Proof. The generator of a CTMC is defined by the derivatives of the state transition probabilities:

$$\overleftarrow{q_{ij}}(\tau) = \lim_{\Delta \to 0} \frac{1}{\Delta} Pr(\overleftarrow{Z}(\tau + \Delta) = j \mid \overleftarrow{Z}(\tau) = i) \qquad \text{if } i \neq j$$

$$\overleftarrow{q_{ii}}(\tau) = \lim_{\Delta \to 0} \frac{1}{\Delta} \left(Pr(\overleftarrow{Z}(\tau + \Delta) = i \mid \overleftarrow{Z}(\tau) = i) - 1 \right) = \sum_{k \in S, k \neq i} -\overleftarrow{q_{ik}}(\tau)$$

We focus on the first term:

$$Pr(\overleftarrow{Z}(\tau + \Delta) = j \mid \overleftarrow{Z}(\tau) = i) = Pr(Z(T - \tau - \Delta) = j \mid Z(T - \tau) = i) =$$

$$\frac{Pr(Z(T - \tau - \Delta) = j, Z(T - \tau) = i)}{Pr(Z(T - \tau) = i)} =$$

$$\frac{Pr(Z(T - \tau - \Delta) = j)}{Pr(Z(T - \tau) = i)} \, Pr(Z(T - \tau) = i \mid Z(T - \tau - \Delta) = j) \, .$$

Dividing the last expression by Δ and making the $\Delta \to 0$ limit we obtain the statement of the theorem for $i \neq j$. The $i = j$ case is obviously obtained by its definition. $\square$

COROLLARY 3.2. *The transient probability of the reverse process $\overleftarrow{\gamma}(\tau)$ is characterized by the ordinary differential equation*

$$(3.1) \qquad\qquad \frac{d}{d\tau} \overleftarrow{\gamma}(\tau) = \overleftarrow{\gamma}(\tau) \overleftarrow{\mathbf{Q}}(\tau)$$

with initial condition $\overleftarrow{\gamma}(0) = \gamma(T)$ and the solution to this differential equation verifies $\overleftarrow{\gamma}(\tau) = \gamma(T - \tau)$.

Proof. The probability of staying in state j at time τ and $\tau + \Delta$ might change due to a departure from state j or an entrance to state j:

$$Pr(\overleftarrow{Z}(\tau + \Delta) = j) - Pr(\overleftarrow{Z}(\tau) = j) =$$

$$Pr(\overleftarrow{Z}(\tau) \neq j, \overleftarrow{Z}(\tau + \Delta) = j) - Pr(\overleftarrow{Z}(\tau) = j, \overleftarrow{Z}(\tau + \Delta) \neq j) =$$

$$\sum_{k \in S, k \neq j} Pr(\overleftarrow{Z}(\tau) = k, \overleftarrow{Z}(\tau + \Delta) = j) - \sum_{k \in S, k \neq j} Pr(\overleftarrow{Z}(\tau) = j, \overleftarrow{Z}(\tau + \Delta) = k) =$$

$$\sum_{k \in S, k \neq j} Pr(\overleftarrow{Z}(\tau) = k)\, Pr(\overleftarrow{Z}(\tau + \Delta) = j \mid \overleftarrow{Z}(\tau) = k) -$$

$$Pr(\overleftarrow{Z}(\tau) = j) \sum_{k \in S, k \neq j} Pr(\overleftarrow{Z}(\tau + \Delta) = k \mid \overleftarrow{Z}(\tau) = j)$$

Dividing the first and the last expressions by Δ and making the $\Delta \to 0$ limit we obtain eq. (3.1).

The $\overleftarrow{\gamma}(\tau) = \gamma(T - \tau)$ relation holds for $\tau = 0$ by definition. Assuming $\overleftarrow{\gamma}(\tau) = \gamma(T - \tau)$ holds for $0 < \tau < T$ eq. (3.1) describes the reverse of the function defined by eq. (2.1):

$$\begin{aligned}
\text{(3.2)} \qquad \frac{d}{d\tau}\overleftarrow{\gamma}_i(\tau) &= \sum_{k \in S} \overleftarrow{\gamma}_k(\tau)\overleftarrow{q}_{ki}(\tau) = \sum_{k \in S, k \neq i} \overleftarrow{\gamma}_k(\tau)\overleftarrow{q}_{ki}(\tau) + \overleftarrow{\gamma}_i(\tau)\overleftarrow{q}_{ii}(\tau) \\
&= \sum_{k \in S, k \neq i} \overleftarrow{\gamma}_k(\tau)\,\frac{\overleftarrow{\gamma}_i(\tau)}{\overleftarrow{\gamma}_k(\tau)} q_{ik} - \overleftarrow{\gamma}_i(\tau) \sum_{k \in S, k \neq i} \frac{\overleftarrow{\gamma}_k(\tau)}{\overleftarrow{\gamma}_i(\tau)} q_{ki} \\
&= \overleftarrow{\gamma}_i(\tau) \sum_{k \in S, k \neq i} q_{ik} - \sum_{k \in S, k \neq i} \overleftarrow{\gamma}_k(\tau)\, q_{ki} \\
&= -\sum_{k \in S} \overleftarrow{\gamma}_k(\tau)\, q_{ki}
\end{aligned}$$

□

According to Equation (3.2) $\overleftarrow{\underline{\gamma}}(\tau)$ satisfies a homogeneous and an inhomogeneous differential equation at the same time:

$$\frac{d}{d\tau}\overleftarrow{\underline{\gamma}}(\tau) = \overleftarrow{\underline{\gamma}}(\tau)\overleftarrow{\mathbf{Q}}(\tau) = -\overleftarrow{\underline{\gamma}}(\tau)\mathbf{Q},$$

where $\overleftarrow{\mathbf{Q}}(\tau)$ is a proper generator matrix but $-\mathbf{Q}$ is not. The numerical analysis of $\overleftarrow{\underline{\gamma}}(\tau)$ is much easier based on $-\mathbf{Q}$ although it does not have any probabilistic interpretation (with respect to the reverse process). The solution to the homogeneous differential equation is

$$\overleftarrow{\underline{\gamma}}(\tau) = \overleftarrow{\underline{\gamma}}(0)\, e^{-\mathbf{Q}\tau},$$

Note that $\overleftarrow{\mathbf{Q}}(\tau) \neq -\mathbf{Q}$ and $e^{\int_0^\tau \overleftarrow{\mathbf{Q}}(u)du} \neq e^{-\mathbf{Q}\tau}$, only $\overleftarrow{\underline{\gamma}}(\tau)\overleftarrow{\mathbf{Q}}(\tau) = -\overleftarrow{\underline{\gamma}}(\tau)\mathbf{Q}$ and $\overleftarrow{\underline{\gamma}}(0)\, e^{-\mathbf{Q}\tau} = \overleftarrow{\underline{\gamma}}(0)\, e^{\int_0^\tau \overleftarrow{\mathbf{Q}}(u)du}$, where $\overleftarrow{\underline{\gamma}}(\tau) = \underline{\gamma}(T - \tau)$ and $\overleftarrow{\underline{\gamma}}(0) = \underline{\gamma}(T)$.

Based on Corollary 3.2 we often apply the following form of the reverse generator in the sequel:

$$\overleftarrow{q}_{ij}(\tau) = \begin{cases} \dfrac{\overleftarrow{\gamma}_j(\tau)}{\overleftarrow{\gamma}_i(\tau)}\, q_{ji} & \text{if } i \neq j, \\[2ex] -\displaystyle\sum_{k \in S, k \neq i} \dfrac{\overleftarrow{\gamma}_k(\tau)}{\overleftarrow{\gamma}_i(\tau)}\, q_{ki} & \text{if } i = j. \end{cases}$$

THEOREM 3.3. *The elementary probability that the $Z(t)$ process with generator* $\mathbf{Q}$ *and initial distribution $\underline{\gamma}(0)$ follows the trajectory (θ_i, X_i) $(0 \leq i \leq N)$ is identical to the elementary probability that the inhomogeneous $\overleftarrow{Z}(\tau)$ process with generator $\overleftarrow{\mathbf{Q}}(\tau)$ and initial distribution $\overleftarrow{\underline{\gamma}}(0)$ follows the trajectory (ω_i, Y_i) $(0 \leq i \leq N)$ where $\overleftarrow{\underline{\gamma}}(0) = \underline{\gamma}(T)$, $\omega_i = T - \theta_{N+1-i}$, $\omega_0 = 0$ and $Y_i = X_{N-i}$.*

Proof. The proof of the theorem is provided in Appendix A $\square$

4. Analysis of accumulated reward using the time reverse process. Our goal is to apply the time reverse process for the reward analysis of the original structure state process. The preceding detailed analysis of the reverse process and the following theorems prepare the approach.

4.1. Reward accumulation without reward loss. Let $Y_i(t, w)$ denote the joint distribution of the accumulated reward and the system state, i.e. the probability that the accumulated reward is less than w and the CTMC stays in state i at time t,

$$Y_i(t, w) = Pr(B(t) \leq w, Z(t) = i).$$

Let $\underline{Y}(t, w)$ be the row vector composed of $Y_i(t, w)$, i.e., $\underline{Y}(t, w) = \{Y_i(t, w)\}$. $\underline{Y}(t, w)$ is the solution to the partial differential equation

$$\frac{\partial}{\partial t}\underline{Y}(t, w) + \frac{\partial}{\partial w}\underline{Y}(t, w)\mathbf{R} = \underline{Y}(t, w)\mathbf{Q},$$

with initial condition $\underline{Y}(0, w) = \underline{\gamma}(0)$ and $\underline{Y}(t, 0) = \underline{0}$ (with $r_i > 0$) [13, 14]. Further more, let $\overleftarrow{Y}_i(\tau, w)$ denote the distribution of reward accumulated by the reverse process, where

$$\overleftarrow{Y}_i(\tau, w) = Pr(\overleftarrow{B}(\tau) \leq w, \overleftarrow{Z}(\tau) = i)$$

and $\overleftarrow{\underline{Y}}(t, w)$ is the row vector composed of $\overleftarrow{Y}_i(\tau, w)$, i.e., $\overleftarrow{\underline{Y}}(\tau, w) = \{\overleftarrow{Y}_i(\tau, w)\}$. $\overleftarrow{\underline{Y}}(\tau, w)$ is the solution to the partial differential equation

$$(4.1) \qquad \frac{\partial}{\partial \tau}\overleftarrow{\underline{Y}}(\tau, w) + \frac{\partial}{\partial w}\overleftarrow{\underline{Y}}(\tau, w)\mathbf{R} = \overleftarrow{\underline{Y}}(\tau, w)\overleftarrow{\mathbf{Q}}(\tau),$$

with initial condition $\overleftarrow{\underline{Y}}(0, w) = \underline{\gamma}(T)$ and $\overleftarrow{\underline{Y}}(\tau, 0) = \underline{0}$ (with $r_i > 0$) [14].

THEOREM 4.1. *In a prs reward model, the distribution of the reward accumulated by the $Z(t)$ CTMC with generator $\mathbf{Q}$, reward rate matrix $\mathbf{R}$ and initial distribution $\underline{\gamma}(0)$ during the interval $(0, T)$ is identical to the distribution of reward accumulated by the $\overleftarrow{Z}(\tau)$ inhomogeneous CTMC with generator $\overleftarrow{\mathbf{Q}}(\tau)$ (defined in Theorem 3.1), reward rate matrix $\mathbf{R}$ and initial probability $\overleftarrow{\underline{\gamma}}(0) = \underline{\gamma}(T) = \underline{\gamma}(0)e^{\mathbf{Q}T}$ over the interval $(0, T)$. I.e., $\overleftarrow{\underline{Y}}(T, w) \overset{d}{=} \underline{Y}(T, w)$.*

Proof. The accumulated reward of the $Z(t)$ process is characterized by its trajectory via eq. (2.4), which is independent of the order of visited states. The $\overleftarrow{Z}(\tau)$ process follows the inverse trajectory of $Z(t)$ with the same elementary probability (Theorem 3.3), hence the $Z(t)$ and the $\overleftarrow{Z}(\tau)$ processes accumulate the same amount of reward with the same probability. $\square$

To obtain a homogeneous partial differential equation we introduce $\overleftarrow{V}_i(\tau, w)$, the conditional distribution of reward accumulated by the reverse process

$$\overleftarrow{V}_i(\tau, w) = Pr(\overleftarrow{B}(\tau) \leq w \mid \overleftarrow{Z}(\tau) = i)$$

and the row vector $\overleftarrow{V}(\tau, w) = \{\overleftarrow{V}_i(\tau, w)\}$.

THEOREM 4.2. *The distribution of reward accumulated over the interval* $(0, T)$ *is*

$$Pr(B(T) \le w) = \sum_{i \in S} \overleftarrow{V}_i(T, w)\gamma_i(0) \ ,$$

where $\overleftarrow{V}(\tau, w)$ *is the solution to the partial differential equation*

(4.2)
$$\frac{\partial}{\partial \tau}\overleftarrow{V}(\tau, w) + \frac{\partial}{\partial w}\overleftarrow{V}(\tau, w)\mathbf{R} = \overleftarrow{V}(\tau, w)\mathbf{Q}^T \ ,$$

with initial condition $\overleftarrow{V}(0, w) = \underline{e}$ *(where* $\underline{e}$ *is the row vector of ones) and* $\overleftarrow{V}(\tau, 0) = \underline{0}$ *(with* $r_i > 0$*).* $\mathbf{Q}^T$ *is the transpose of* $\mathbf{Q}$*.*

Proof. Since $\overleftarrow{Y}_i(\tau, w) = \overleftarrow{V}_i(\tau, w)\overleftarrow{\gamma}_i(\tau)$, from (4.1) we have

$$\frac{\partial \overleftarrow{V}_i(\tau, w)}{\partial \tau}\overleftarrow{\gamma}_i(\tau) + \frac{\partial \overleftarrow{\gamma}_i(\tau)}{\partial \tau}\overleftarrow{V}_i(\tau, w) + \frac{\partial \overleftarrow{V}_i(\tau, w)}{\partial w}\overleftarrow{\gamma}_i(\tau)r_i =$$

$$\sum_{k \in S} \overleftarrow{V}_k(\tau, w)\overleftarrow{\gamma}_k(\tau)\overleftarrow{q}_{ki}(\tau) =$$

$$\sum_{k \in S, k \ne i} \overleftarrow{V}_k(\tau, w)\overleftarrow{\gamma}_k(\tau)\frac{\overleftarrow{\gamma}_i(\tau)}{\overleftarrow{\gamma}_k(\tau)} \ q_{ik} - \overleftarrow{V}_i(\tau, w)\overleftarrow{\gamma}_i(\tau) \sum_{k \in S, k \ne i} \frac{\overleftarrow{\gamma}_k(\tau)}{\overleftarrow{\gamma}_i(\tau)} \ q_{ki},$$

where $\overleftarrow{q}_{ki}(\tau)$ is replaced in the second step. Dividing both sides by $\overleftarrow{\gamma}_i(\tau)$ and substituting equation (3.2) result eq. (4.2) after some algebra. The distribution of the accumulated reward is obtained as

$$Pr(B(T) \le w) = \sum_{i \in S} Y_i(T, w) = \sum_{i \in S} \overleftarrow{Y}_i(T, w) =$$

$$\sum_{i \in S} \overleftarrow{V}_i(T, w)\overleftarrow{\gamma}_i(T) = \sum_{i \in S} \overleftarrow{V}_i(T, w)\gamma_i(0) \ .$$

□

4.2. Reward accumulation with partial incremental reward loss. Based on eq. (2.7) we can interpret the reward accumulation of a reward model with partial incremental loss as it accumulates reward at rate $\alpha_i r_i$ in the first N visited states and it accumulates reward at rate r_i in the last visited state before time T and there is no reward loss. Using the time reverse of the structure state process we need to evaluate the accumulated reward when the reward rate in the first visited state (of the time reverse process) is r_i and the reward rate in later visited states are $\alpha_i r_i$. We summarize the result of this approach in the following theorem.

THEOREM 4.3. *The distribution of reward accumulated over the interval* $(0, T)$ *by the partial incremental loss Markov reward model with initial distribution* $\underline{\gamma}(0)$, *generator matrix* $\mathbf{Q}$, *reward rate matrix* $\mathbf{R}$ *and reduced reward rate matrix* $\mathbf{R}_\alpha = diag < \alpha_i r_i >$ *is identical to the distribution of reward accumulated over the same interval by the inhomogeneous prs Markov reward model of* $2|S|$ *states with initial distribution* $\pi^*(0)$, *generator* $\overleftarrow{\mathbf{Q}^*}(\tau)$ *and reward rate matrix* $\mathbf{R}^*$, *where*

$$(4.3) \ \pi^*(0) = [\underline{\gamma}(T), \underline{0}], \quad \overleftarrow{\mathbf{Q}^*}(\tau) = \left[\begin{array}{c|c} \overleftarrow{\mathbf{Q}_D}(\tau) & \overleftarrow{\mathbf{Q}}(\tau) - \overleftarrow{\mathbf{Q}_D}(\tau) \\ \hline 0 & \overleftarrow{\mathbf{Q}}(\tau) \end{array}\right], \quad \mathbf{R}^* = \left[\begin{array}{c|c} \mathbf{R} & 0 \\ \hline 0 & \mathbf{R}_\alpha \end{array}\right],$$

$\overleftarrow{\mathbf{Q}}(\tau)$ is defined by Theorem 3.1 and $\overleftarrow{\mathbf{Q}_D}(\tau) = diag\langle\overleftarrow{q_{ii}}(\tau)\rangle$ is the diagonal matrix composed of the diagonal elements of $\overleftarrow{\mathbf{Q}}(\tau)$.

Proof. The structure of the matrices in eq. (4.3) are such that the first $|S|$ states of the inhomogeneous prs Markov reward model represent the reward accumulation in the first visited state (of the time reverse process) and the last $|S|$ states represents the reward accumulation in the later visited states. The theorem is given by the equivalence of the trajectories of the forward and its time reverse process presented in Theorem 3.3 and by the fact that in both cases the accumulated reward is related with the trajectory according to eq. (2.7). □

Theorem 4.3 and eq. (4.1) already provides a computational method. By this method the distribution of the accumulated reward can be calculated using enlarged, possibly inhomogeneous, matrices and vectors with proper initial probability vector.

The numerical solution based on eq. (4.1) could be expensive since $\overleftarrow{\mathbf{Q}}^*(\tau)$ needs to be calculated in each step of the solution method. To overcome this difficulty we look for a partial differential equation description with constant coefficients similar to the one in eq. (4.2).

THEOREM 4.4. *The distribution of reward accumulated over the interval $(0,T)$ is*

$$Pr(B(T) \le w) = \sum_{i \in S}\left(\overleftarrow{X1}_i(T,w) + \overleftarrow{X2}_i(T,w))\right)\gamma_i(0) ,$$

where vectors $\overleftarrow{X1}(\tau,w)$ and $\overleftarrow{X2}(\tau,w)$ are the solution to the partial differential equations

$$(4.4) \qquad \frac{\partial}{\partial\tau}\overleftarrow{X1}(\tau,w) + \frac{\partial}{\partial w}\overleftarrow{X1}(\tau,w)\mathbf{R} = \overleftarrow{X1}(\tau,w)\mathbf{Q}_D ,$$

and

$$(4.5) \quad \frac{\partial}{\partial\tau}\overleftarrow{X2}(\tau,w) + \frac{\partial}{\partial w}\overleftarrow{X2}(\tau,w)\mathbf{R}_\alpha = \overleftarrow{X1}(\tau,w)(\mathbf{Q}-\mathbf{Q}_D)^T + \overleftarrow{X2}(\tau,w)\mathbf{Q}^T ,$$

with initial conditions: $\overleftarrow{X1}(0,w) = \underline{e}$, $\overleftarrow{X2}(0,w) = \underline{0}$, $\overleftarrow{X1}(\tau,0) = \underline{0}$ and $\overleftarrow{X2}(\tau,0) = \underline{0}$ (when $r_i > 0$). In the expressions above $\mathbf{Q}_D$ is the diagonal matrix composed of the diagonal elements of $\mathbf{Q}$ and $\mathbf{Q}_D = diag\langle q_{ii}\rangle$.

Proof. The proof of the theorem is provided in Appendix B. □

Since $\mathbf{Q}_D$ is a diagonal matrix the solution to (4.4) is readable

$$(4.6) \qquad \overleftarrow{X1}_i(\tau,w) = \begin{cases} e^{q_{ii}\tau} & \text{if } w \ge r_i\tau , \\ 0 & \text{if } w < r_i\tau . \end{cases}$$

4.3. Moments of accumulated reward with partial incremental reward loss. The moments of the accumulated reward are given by the following theorem.

THEOREM 4.5. *The nth moment $(n \ge 0)$ of reward accumulated over the interval $(0,T)$ is*

$$E(B(T)^n) = \sum_{i \in S}\left(\overleftarrow{M1}_i^{(n)}(T) + \overleftarrow{M2}_i^{(n)}(T)\right)\gamma_i(0) ,$$

where $\overleftarrow{M1}^{(n)}(\tau)$ and $\overleftarrow{M2}^{(n)}(\tau)$ are the solution to the ordinary differential equations

$$(4.7) \qquad \frac{d}{d\tau}\overleftarrow{M1}^{(n)}(\tau) = n\overleftarrow{M1}^{(n-1)}(\tau)\mathbf{R} + \overleftarrow{M1}^{(n)}(\tau)\mathbf{Q}_D ,$$

and

$$(4.8) \quad \frac{d}{d\tau}\overleftarrow{M2}^{(n)}(\tau) = n\overleftarrow{M2}^{(n-1)}(\tau)\mathbf{R}_\alpha + \overleftarrow{M1}^{(n)}(\tau)(\mathbf{Q} - \mathbf{Q}_D)^T + \overleftarrow{M2}^{(n)}(\tau)\mathbf{Q}^T .$$

with initial conditions $\overleftarrow{M1}^{(0)}(0) = \underline{e}$, $\overleftarrow{M1}^{(i)}(0) = \underline{0}$ *for* $i \geq 1$, *and* $\overleftarrow{M2}^{(i)}(0) = \underline{0}$ *for* $i \geq 0$.

Proof. The proof of the theorem is provided in Appendix C. $\square$

The solution to (4.7) is

$$(4.9) \qquad \overleftarrow{M1}^{(n)}(\tau) = \tau^n \underline{e} \, \mathbf{R}^n \mathbf{E}_D(\tau) ,$$

where $\mathbf{E}_D(\tau)$ is the diagonal matrix $\mathbf{E}_D(\tau) = diag\langle e^{q_{ii}\tau}\rangle$.

5. A randomization based numerical method. The fact that (4.7) and (4.8) are differential equations with constant coefficients suggests that a stable, computationally effective, randomization based numerical algorithm can be constructed to calculate the moments of the accumulated reward. $\overleftarrow{M1}^{(n)}(\tau)$ is explicitly given by (4.9). The solution to $\overleftarrow{M2}^{(n)}(\tau)$ is provided by the following theorem.

THEOREM 5.1. $\overleftarrow{M2}^{(n)}(\tau)$ *can be calculated as*

$$(5.1) \qquad \overleftarrow{M2}^{(n)}(\tau) = n!d^n \sum_{k=0}^{\infty} e^{-\lambda\tau} \frac{(\lambda\tau)^k}{k!} \, \underline{D}^{(n)}(k)$$

where $\lambda = \max_{j\in S} \sum_{i\in S, i\neq j} q_{ij}$ *(maximal aggregated transition rate to a state)*, $d = \max_{i\in S} r_i/\lambda$, $\mathbf{A}_D = \mathbf{Q}_D/\lambda + \mathbf{I}$, $\mathbf{A} = \mathbf{Q}^T/\lambda + \mathbf{I}$, $\mathbf{S} = \mathbf{R}/(\lambda d)$, $\mathbf{S}_\alpha = \mathbf{R}_\alpha/(\lambda d)$ *and*

$$(5.2) \quad \underline{D}^{(n)}(k) = \begin{cases} \underline{e}\,(\mathbf{I} - \mathbf{A}_D^k) & n = 0 \\[2mm] 0 & k \leq n, n \geq 1 \\[2mm] \underline{D}^{(n-1)}(k-1)\mathbf{S}_\alpha + \underline{D}^{(n)}(k-1)\mathbf{A}+ & \\[1mm] \quad \binom{k-1}{n} \underline{e}\,\mathbf{S}^n\mathbf{A}_D^{k-1-n}(\mathbf{A} - \mathbf{A}_D) & k > n, n \geq 1 \end{cases}$$

Proof. Substituting (4.9) into (4.8) results

$$(5.3)\frac{d}{d\tau}\overleftarrow{M2}^{(n)}(\tau) = n\overleftarrow{M2}^{(n-1)}(\tau)\mathbf{R}_\alpha + \tau^n\underline{e}\mathbf{R}^n\mathbf{E}_D(\tau)(\mathbf{Q} - \mathbf{Q}_D)^T + \overleftarrow{M2}^{(n)}(\tau)\mathbf{Q}^T$$

with initial conditions $\overleftarrow{M2}^{(n)}(0) = \underline{0}$ and (from the definition of $\overleftarrow{M2}^{(n)}(\tau)$) $\overleftarrow{M2}^{(0)}(\tau) = \underline{e}(\mathbf{I} - \mathbf{E}_D(\tau))$. Substituting $\mathbf{A}, \mathbf{A}_D, \mathbf{S}, \mathbf{S}_\alpha, \mathbf{E}_D(\tau) = \sum_{k=0}^{\infty} e^{-\lambda\tau}\frac{(\lambda\tau)^k}{k!}\mathbf{A}_D^k, \mathbf{I} - \mathbf{E}_D(\tau) =$

$$\sum_{k=0}^{\infty} e^{-\lambda\tau} \frac{(\lambda\tau)^k}{k!} \left(\mathbf{I} - \mathbf{A}_D^k\right) \text{ and (5.1) into (5.3) results}$$

$$n!d^n\lambda \sum_{i=0}^{\infty} e^{-\lambda\tau} \frac{(\lambda\tau)^i}{i!} \left(\underline{D}^{(n)}(i+1) - \underline{D}^{(n)}(i)\right) =$$

$$n(n-1)!d^{n-1} \sum_{i=0}^{\infty} e^{-\lambda\tau} \frac{(\lambda\tau)^i}{i!} \underline{D}^{(n-1)}(i)\lambda d\ \mathbf{S}_\alpha +$$

$$\frac{1}{\lambda^n} \underline{e}\lambda^n d^n \mathbf{S}^n \sum_{i=n}^{\infty} e^{-\lambda\tau} \frac{(\lambda\tau)^i}{i!} \mathbf{A}_D^i\lambda(\mathbf{A} - \mathbf{A}_D) +$$

$$n!d^n \sum_{i=0}^{\infty} e^{-\lambda\tau} \frac{(\lambda\tau)^i}{i!} \underline{D}^{(n)}(i)\lambda(\mathbf{A}_D - \mathbf{I})$$

The recursive formulae of $\underline{D}^{(n)}(i)$ ensures the equality for each $e^{-\lambda\tau}(\lambda\tau)^i$ terms.

□

The error caused by truncating the infinite sum in (5.1) can be bounded as follows.

THEOREM 5.2. $\overleftarrow{\underline{M2}}^{(n)}(\tau)$ can be calculated as a finite sum and an error vector, $\underline{\xi}(G)$, such that all elements of the error vector is less than ε:

$$(5.4) \qquad \overleftarrow{\underline{M2}}^{(n)}(\tau) = n!d^n \sum_{k=0}^{G-1} e^{-\lambda\tau} \frac{(\lambda\tau)^k}{k!} \underline{D}^{(n)}(k) + \underline{\xi}(G)$$

where $\underline{D}^{(n)}(k)$ is given by (5.2) and

$$G = \min_{g>n} \left((\lambda\tau)^{n+1}d^n \sum_{k=g-n-1}^{\infty} e^{-\lambda\tau} \frac{(\lambda\tau)^k}{k!} < \varepsilon\right)$$

Proof. The error term, $\underline{\xi}(G)$ satisfies

$$\begin{aligned}
\underline{\xi}(G) &= n!d^n \sum_{k=G}^{\infty} e^{-\lambda\tau} \frac{(\lambda\tau)^k}{k!} \underline{D}^{(n)}(k) \\
&\leq n!d^n \sum_{k=G}^{\infty} e^{-\lambda\tau} \frac{(\lambda\tau)^k}{k!} \frac{k!}{n!\ (k-n-1)!} \underline{e} \qquad \text{due to Corollary D.2} \\
&= (\lambda\tau)^{n+1}d^n \sum_{k=G-n-1}^{\infty} e^{-\lambda\tau} \frac{(\lambda\tau)^k}{k!} \underline{e} \\
&< \varepsilon\ \underline{e} \qquad\qquad\qquad\qquad\qquad\qquad\quad \text{due to the definition of } G
\end{aligned}$$

□

6. Numerical behaviour: complexity and stability. The two main sources of numerical errors in computations with floating point numbers are the summation of numbers with different orders of magnitude and the presence of substractions. The methods based on multiplications and summations of positive numbers between 0 and 1 usually have nice numerical properties.

In our case, the introduction of matrices $\mathbf{A}, \mathbf{A}_D, \mathbf{S}$ and $\mathbf{S}_\alpha$ ensures that the general iteration step ((5.2), case $k > n$) involves only summations of vectors and vector-matrix multiplications where the elements of the vectors and the matrices are between 0 and 1 (e.g., matrix $\mathbf{A} - \mathbf{A}_D$ has this property as well).

The only step where a real substraction takes place is the initialization step at $n = 0$ (5.2). In this step we calculate the diagonal matrix $\mathbf{I} - \mathbf{A}_D^k$, whose elements are non-negative between 0 and 1. Hence this substraction does not affect the numerical stability of the proposed algorithm.

The core iteration step, the computation of $\underline{D}^{(n)}(k)$ based on (5.2), is implemented using vector-matrix multiplications. The complexity of multiplications of vectors with full matrices is $o\left(N^2\right)$ and vectors with diagonal matrices is $o\left(N\right)$. There are only 2 real vector-matrix multiplications in an iteration step (computing $\underline{e}\mathbf{S}^n\mathbf{A}_D^{k-1-n}(\mathbf{A} - \mathbf{A}_D)$ and $\underline{D}^{(n)}(k-1)\mathbf{A}$). The computational complexity of the algorithm is dominated by these vector-matrix multiplications. The core iteration step requires $o\left(N^2\right) \sim 2N^2$ floating point multiplications and summations. To compute the nth moment, we have to compute moments $0 \ldots n - 1$ as well. The overall computational complexity of one iteration step is $o\left(n \cdot N^2\right) \sim 2(n + 1)N^2$ and G iterations are required to reach the prescribed precision.

7. Numerical example. We implemented (in C) a numerical method that provides the moments of the accumulated reward in reward models with partial incremental loss. Matrices are stored in a sparse manner. With this implementation it is possible to handle very large models, with 10^6 states.

As an example, let consider the following system. The Markov chain describing the model is shown in Figure 7.1. The state numbering reflects the number of working servers. State M is the maintenance state. The reward rates r_i and α_i corresponding to the states are indicated on the dashed line. If there are k working servers, the reward rate is $r_k = k \cdot r$.

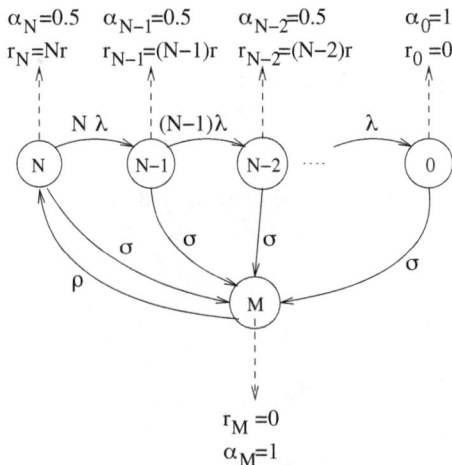

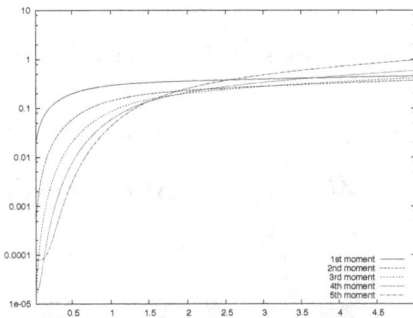

FIGURE 7.1. *Structure of the Markov chain*

FIGURE 7.2. *Moments of the accumulated reward*

Figure 7.2 shows the first 5 moments of the accumulated reward as the function of time. Figure 7.3 and 7.4 compares the moments of the accumulated reward while changing parameter α to 0.25, 0.5, 0.75, accordingly.

Appendix A. Proof of Theorem 3.3.

To prove the theorem we need the following corollary.

Number of servers:	$N = 500000$
Server break down rate:	$\lambda = 0.000004$
System inter-maintenance rate:	$\sigma = 1.5$
Inverse of system maintenance time:	$\rho = 0.1$
Reward accumulation of one server:	$r = 0.000002$
Incremental reward loss at failures:	$\alpha = 0.5$

TABLE 7.1

Parameters used in the example

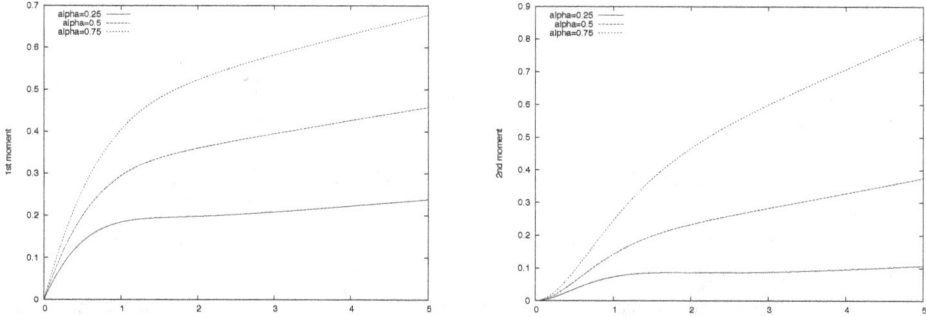

FIGURE 7.3. *The effect of α on the 1st and 2nd moments*

COROLLARY A.1. *The short term behaviour of $Pr(\overleftarrow{Z}(\tau) = j)$ satisfies*

$$Pr(\overleftarrow{Z}(\tau) = j)\left(1 + \overleftarrow{q}_{jj}(\tau)\Delta + \sigma(\Delta)\right) = Pr(\overleftarrow{Z}(\tau + \Delta) = j)\left(1 + q_{jj}\Delta + \sigma(\Delta)\right).$$

Proof. We start from the same relation as in the proof of Corollary 3.2:

$$Pr(\overleftarrow{Z}(\tau + \Delta) = j) - Pr(\overleftarrow{Z}(\tau) = j) =$$

$$Pr(\overleftarrow{Z}(\tau) \neq j, \overleftarrow{Z}(\tau + \Delta) = j) - Pr(\overleftarrow{Z}(\tau) = j, \overleftarrow{Z}(\tau + \Delta) \neq j) =$$

$$\sum_{k \in S, k \neq j} Pr(\overleftarrow{Z}(\tau) = k, \overleftarrow{Z}(\tau + \Delta) = j) - \sum_{k \in S, k \neq j} Pr(\overleftarrow{Z}(\tau) = j, \overleftarrow{Z}(\tau + \Delta) = k) =$$

$$Pr(\overleftarrow{Z}(\tau + \Delta) = j) \sum_{k \in S, k \neq j} Pr(\overleftarrow{Z}(\tau) = k \mid \overleftarrow{Z}(\tau + \Delta) = j) -$$

$$Pr(\overleftarrow{Z}(\tau) = j) \sum_{k \in S, k \neq j} Pr(\overleftarrow{Z}(\tau + \Delta) = k \mid \overleftarrow{Z}(\tau) = j) =$$

$$Pr(\overleftarrow{Z}(\tau + \Delta) = j)\left(- q_{jj}\Delta + \sigma(\Delta)\right) - Pr(\overleftarrow{Z}(\tau) = j)\left(- \overleftarrow{q}_{jj}(\tau)\Delta + \sigma(\Delta)\right).$$

Rearranging the first and the last expressions provides the corollary. □

Based on Corollary A.1 the proof of Theorem 3.3 proceeds as follow.

The elementary probability that the $Z(t)$ process follows the trajectory (θ_i, X_i)

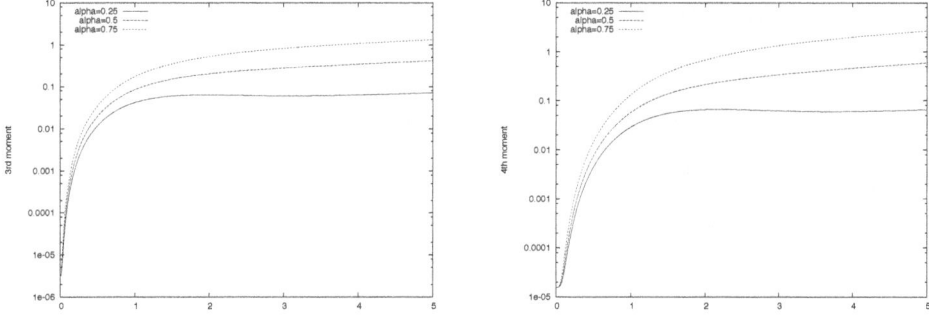

FIGURE 7.4. *The effect of α on the 3rd and 4th moments*

is

$$\underbrace{\gamma_{X_0}(0)}_{\text{init. prob.}} \prod_{i=0}^{N-1} \left(\underbrace{-q_{X_i X_i} \, e^{q_{X_i X_i}(\theta_{i+1}-\theta_i)}}_{\text{sojourn time}} \quad \underbrace{\frac{q_{X_i X_{i+1}}}{-q_{X_i X_i}}}_{\text{state tr. prob.}} \right) \underbrace{e^{q_{X_N X_N}(T-\theta_N)}}_{\text{last sojourn}} =$$

$$\gamma_{X_0}(0) \prod_{i=0}^{N-1} \left(e^{q_{X_i X_i}(\theta_{i+1}-\theta_i)} \, q_{X_i X_{i+1}} \right) e^{q_{X_N X_N}(T-\theta_N)} \, ,$$

and the elementary probability that the $\overleftarrow{Z}(\tau)$ process follows the trajectory (ω_i, Y_i) is

$$\underbrace{\overleftarrow{\gamma}_{Y_0}(0)}_{\text{init. prob.}} \prod_{i=0}^{N-1} \left(\underbrace{-\overleftarrow{q}_{Y_i Y_i}(\omega_{i+1}) \, e^{\int_{\omega_i}^{\omega_{i+1}} \overleftarrow{q}_{Y_i Y_i}(\tau)d\tau}}_{\text{sojourn time}} \quad \underbrace{\frac{\overleftarrow{q}_{Y_i Y_{i+1}}(\omega_{i+1})}{-\overleftarrow{q}_{Y_i Y_i}(\omega_{i+1})}}_{\text{state tr. prob.}} \right) \underbrace{e^{\int_{\omega_N}^{T} \overleftarrow{q}_{Y_N Y_N}(\tau)d\tau}}_{\text{last sojourn}} =$$

$$\overleftarrow{\gamma}_{Y_0}(0) \prod_{i=0}^{N-1} \left(e^{\int_{\omega_i}^{\omega_{i+1}} \overleftarrow{q}_{Y_i Y_i}(\tau)d\tau} \, \overleftarrow{q}_{Y_i Y_{i+1}}(\omega_{i+1}) \right) e^{\int_{\omega_N}^{T} \overleftarrow{q}_{Y_N Y_N}(\tau)d\tau} \, .$$

Let us consider the $\overleftarrow{\gamma}_{Y_i}(\omega_i) \, e^{\int_{\omega_i}^{\omega_{i+1}} \overleftarrow{q}_{Y_i Y_i}(\tau)d\tau}$ term first:

$$\overleftarrow{\gamma}_{Y_i}(\omega_i) \, e^{\int_{\omega_i}^{\omega_{i+1}} \overleftarrow{q}_{Y_i Y_i}(\tau)d\tau} = \overleftarrow{\gamma}_{Y_i}(\omega_i) \lim_{M \to \infty} e^{\sum_{k=0}^{M-1} h \overleftarrow{q}_{Y_i Y_i}(\omega_i + kh)} =$$

$$\lim_{M \to \infty} \overleftarrow{\gamma}_{Y_i}(\omega_i) \prod_{k=0}^{M-1} e^{h \overleftarrow{q}_{Y_i Y_i}(\omega_i + kh)} = \lim_{M \to \infty} \overleftarrow{\gamma}_{Y_i}(\omega_i) \prod_{k=0}^{M-1} \left(1 + h \overleftarrow{q}_{Y_i Y_i}(\omega_i + kh) + \sigma(h) \right) =$$

$$\lim_{M \to \infty} \overleftarrow{\gamma}_{Y_i}(\omega_i) \left(1 + h \overleftarrow{q}_{Y_i Y_i}(\omega_i) + \sigma(h) \right) \prod_{k=1}^{M-1} \left(1 + h \overleftarrow{q}_{Y_i Y_i}(\omega_i + kh) + \sigma(h) \right) =$$

$$\lim_{M \to \infty} \left(1 + h q_{Y_i Y_i} + \sigma(h) \right) \overleftarrow{\gamma}_{Y_i}(\omega_i + h) \prod_{k=1}^{M-1} \left(1 + h \overleftarrow{q}_{Y_i Y_i}(\omega_i + kh) + \sigma(h) \right) = \ldots$$

$$\lim_{M \to \infty} \prod_{k=0}^{M-1} \left(1 + h q_{Y_i Y_i} + \sigma(h) \right) \overleftarrow{\gamma}_{Y_i}(\omega_{i+1}) = \overleftarrow{\gamma}_{Y_i}(\omega_{i+1}) \, e^{q_{Y_i Y_i}(\omega_{i+1}-\omega_i)} \, ,$$

where $h = \frac{\omega_{i+1}-\omega_i}{M}$ and Corollary A.1 is applied in the fifth step. Using this relation the elementary probability of the reverse trajectory is

$$\overleftarrow{\gamma}_{Y_0}(\omega_0) \prod_{i=0}^{N-1} \left(e^{\int_{\omega_i}^{\omega_{i+1}} \overleftarrow{q}_{Y_i Y_i}(\tau)d\tau} \overleftarrow{q}_{Y_i Y_{i+1}}(\omega_{i+1}) \right) e^{\int_{\omega_N}^{T} \overleftarrow{q}_{Y_N Y_N}(\tau)d\tau} =$$

$$\overleftarrow{\gamma}_{Y_0}(\omega_0) e^{\int_{\omega_0}^{\omega_1} \overleftarrow{q}_{Y_0 Y_0}(\tau)d\tau} \overleftarrow{q}_{Y_0 Y_1}(\omega_1) \;\; \ldots \;\; e^{\int_{\omega_N}^{T} \overleftarrow{q}_{Y_N Y_N}(\tau)d\tau} =$$

$$e^{q_{Y_0 Y_0}(\omega_1 - \omega_0)} \overleftarrow{\gamma}_{Y_0}(\omega_1) \frac{\overleftarrow{\gamma}_{Y_1}(\omega_1)}{\overleftarrow{\gamma}_{Y_0}(\omega_1)} q_{Y_1 Y_0} \;\; \ldots \;\; e^{\int_{\omega_N}^{T} \overleftarrow{q}_{Y_N Y_N}(\tau)d\tau} =$$

$$e^{q_{Y_0 Y_0}(\omega_1 - \omega_0)} q_{Y_1 Y_0} \overleftarrow{\gamma}_{Y_1}(\omega_1) \;\; \ldots \;\; e^{\int_{\omega_N}^{T} \overleftarrow{q}_{Y_N Y_N}(\tau)d\tau} = \ldots =$$

$$\prod_{i=0}^{N-1} \left(e^{q_{Y_i Y_i}(\omega_{i+1} - \omega_i)} q_{Y_{i+1} Y_i} \right) e^{q_{Y_N Y_N}(T - \omega_N)} \overleftarrow{\gamma}_{Y_N}(T)$$

Substituting $Y_i = X_{N-i}$, $\overleftarrow{\gamma}_{Y_N}(T) = \gamma_{X_0}(0)$, $\omega_{i+1} - \omega_i = \theta_{N+1-i} - \theta_{N-i}$ and $T - \omega_N = \theta_1$ results the elementary probability of the forward trajectory, which concludes the proof of the theorem.

Appendix B. Proof of Theorem 4.4.

The proof follows the same pattern as the proof of Theorem 4.2, but here we exploit the block structure of $\pi^*(0)$, $\overleftarrow{\mathbf{Q}^*}(\tau)$ and $\mathbf{R}^*$. $\pi^*(\tau)$ is decomposed as $\pi^*(\tau) = [\overleftarrow{\gamma 1}(\tau), \overleftarrow{\gamma 2}(\tau)]$ where

$$(B.1) \qquad \overleftarrow{\gamma 1}_i(\tau) = Pr(\overleftarrow{Z}(T) = i, \text{no state transition in } (0,T)) = \overleftarrow{\gamma}_i(\tau) e^{q_{ii}\tau},$$

$$(B.2) \quad \overleftarrow{\gamma 2}_i(\tau) = Pr(\overleftarrow{Z}(T) = i, \text{state transition in } (0,T)) = \overleftarrow{\gamma}_i(\tau)(1 - e^{q_{ii}\tau}) \,.$$

Substituting $\pi^*(0)$, $\overleftarrow{\mathbf{Q}^*}(\tau)$ and $\mathbf{R}^*$ into eq. (4.1), utilizing the specific block structure provided in (4.3) and completing similar steps as in Theorem 4.2 results:

$$(B.3) \qquad \frac{\partial}{\partial \tau}\overleftarrow{V1}_i(\tau, w) + \frac{\partial}{\partial w}\overleftarrow{V1}_i(\tau, w)r_i = 0 \,,$$

and

$$\frac{\partial}{\partial \tau}\overleftarrow{V2}_i(\tau, w) + \frac{\partial}{\partial w}\overleftarrow{V2}_i(\tau, w)r_i \alpha_i =$$

$$\overleftarrow{V2}_i(\tau, w)\frac{\overleftarrow{\gamma}_i(\tau)}{\overleftarrow{\gamma 2}_i(\tau)} q_{ii} + \sum_{k \in S, k \neq i} \left(\overleftarrow{V1}_k(\tau, w)\frac{\overleftarrow{\gamma 1}_k(\tau)}{\overleftarrow{\gamma 2}_i(\tau)} + \overleftarrow{V2}_k(\tau, w)\frac{\overleftarrow{\gamma 2}_k(\tau)}{\overleftarrow{\gamma 2}_i(\tau)} \right) \frac{\overleftarrow{\gamma}_i(\tau)}{\overleftarrow{\gamma}_k(\tau)} q_{ik} =$$

$$\overleftarrow{V2}_i(\tau, w)\frac{\overleftarrow{\gamma}_i(\tau)}{\overleftarrow{\gamma 2}_i(\tau)} q_{ii} + \sum_{k \in S, k \neq i} \left(\overleftarrow{V1}_k(\tau, w)\frac{\overleftarrow{\gamma 1}_k(\tau)}{\overleftarrow{\gamma}_k(\tau)} + \overleftarrow{V2}_k(\tau, w)\frac{\overleftarrow{\gamma 2}_k(\tau)}{\overleftarrow{\gamma}_k(\tau)} \right) \frac{\overleftarrow{\gamma}_i(\tau)}{\overleftarrow{\gamma 2}_i(\tau)} q_{ik}$$

$$(B.4)$$

Using (B.1) and (B.2) we have

$$\frac{\partial}{\partial \tau}\overleftarrow{V2}_i(\tau,w) + \frac{\partial}{\partial w}\overleftarrow{V2}_i(\tau,w)r_i\alpha_i =$$

$$\overleftarrow{V2}_i(\tau,w)\,\frac{q_{ii}}{1-e^{q_{ii}\tau}} + \sum_{k\in S, k\neq i}\left(\overleftarrow{V1}_k(\tau,w)e^{q_{kk}\tau} + \overleftarrow{V2}_k(\tau,w)\,(1-e^{q_{kk}\tau})\right)\frac{q_{ik}}{1-e^{q_{ii}\tau}} =$$

$$\sum_{k\in S}\overleftarrow{V2}_k(\tau,w)\frac{q_{ik}}{1-e^{q_{ii}\tau}} + \sum_{k\in S, k\neq i}\left(\overleftarrow{V1}_k(\tau,w) - \overleftarrow{V2}_k(\tau,w)\right)\frac{e^{q_{kk}\tau}\,q_{ik}}{1-e^{q_{ii}\tau}}$$

(B.5)

The vector-matrix versions of eq. (B.3) and (B.5) are

(B.6)
$$\frac{\partial}{\partial \tau}\underline{\overleftarrow{V1}}(\tau,w) + \frac{\partial}{\partial w}\underline{\overleftarrow{V1}}(\tau,w)\mathbf{R} = \underline{0}\,,$$

and

$$\frac{\partial}{\partial \tau}\underline{\overleftarrow{V2}}(\tau,w) + \frac{\partial}{\partial w}\underline{\overleftarrow{V2}}(\tau,w)\mathbf{R}_\alpha =$$

(B.7)
$$\left(\underline{\overleftarrow{V2}}(\tau,w)\mathbf{Q}^T + \left(\underline{\overleftarrow{V1}}(\tau,w) - \underline{\overleftarrow{V2}}(\tau,w)\right)\mathbf{E}_D(\tau)(\mathbf{Q}-\mathbf{Q}_D)^T\right)\left(\mathbf{I}-\mathbf{E}_D(\tau)\right)^{-1},$$

with initial conditions: $\underline{\overleftarrow{V1}}(0,w) = \underline{e}$, $\underline{\overleftarrow{V2}}(0,w) = \underline{e}$, $\underline{\overleftarrow{V1}}(\tau,0) = \underline{0}$ and $\underline{\overleftarrow{V2}}(\tau,0) = \underline{0}$ (when $r_i > 0$).

The stochastic meaning of $\overleftarrow{V1}_i(T,w)$ and $\overleftarrow{V2}_i(T,w)$ are

$$\overleftarrow{V1}_i(T,w) = Pr(\overleftarrow{B}(T) < w \mid \overleftarrow{Z}(T) = i, \text{no state transition in } (0,T))\,,$$

$$\overleftarrow{V2}_i(T,w) = Pr(\overleftarrow{B}(T) < w \mid \overleftarrow{Z}(T) = i, \text{state transition in } (0,T))\,,$$

hence

$$Pr(B(T) \leq w) = \sum_{i\in S}\overleftarrow{V1}_i(T,w)\,Pr(\overleftarrow{Z}(T) = i, \text{no state transition in } (0,T))+$$
$$\sum_{i\in S}\overleftarrow{V2}_i(T,w)\,Pr(\overleftarrow{Z}(T) = i, \text{state transition in } (0,T))\,,$$

To simplify (B.6) and (B.7) we introduce

$$\overleftarrow{X1}_i(\tau,w) = \overleftarrow{V1}_i(\tau,w)e^{q_{ii}\tau} \text{ and } \overleftarrow{X2}_i(\tau,w) = \overleftarrow{V2}_i(\tau,w)(1-e^{q_{ii}\tau})\,.$$

The stochastic interpretation of $\overleftarrow{X1}_i(\tau,w)$ and $\overleftarrow{X2}_i(\tau,w)$ are

$$\overleftarrow{X1}_i(\tau,w) = Pr(\overleftarrow{B}(\tau) < w, \text{ no state transition in } (0,\tau) \mid \overleftarrow{Z}(\tau) = i)\,,$$

and

$$\overleftarrow{X2}_i(\tau,w) = Pr(\overleftarrow{B}(\tau) < w, \text{ state transition in } (0,\tau) \mid \overleftarrow{Z}(\tau) = i)\,,$$

from which

$$Pr(B(T) \leq w) = \sum_{i\in S}\left(\overleftarrow{X1}_i(T,w) + \overleftarrow{X2}_i(T,w)\right)\gamma_i(0)\,.$$

From (B.3) and (B.5) we have

(B.8) $$\frac{\partial}{\partial \tau}\overleftarrow{X1}_i(\tau,w) + \frac{\partial}{\partial w}\overleftarrow{X1}_i(\tau,w)r_i = \overleftarrow{X1}_i(\tau,w)q_{ii} ,$$

and

(B.9) $$\frac{\partial}{\partial \tau}\overleftarrow{X2}_i(\tau,w) + \frac{\partial}{\partial w}\overleftarrow{X2}_i(\tau,w)r_i\alpha_i = \sum_{k\in S, k\neq i}\overleftarrow{X1}_k(\tau,w)q_{ik} + \sum_{k\in S}\overleftarrow{X2}_k(\tau,w)q_{ik} .$$

Eq. (4.4) and (4.5) are the vector-matrix versions of eq. (B.8) and (B.9), respectively.

Appendix C. Proof of Theorem 4.5.

The Laplace-Stieltjes transform of $\overleftarrow{X1}_i(T,w)$ and $\overleftarrow{X2}_i(X,w)$ with respect to w is denoted by $\overleftarrow{X1^*}_i(T,v)$ and $\overleftarrow{X2^*}_i(T,v)$, where v is the transform variable.

The nth derivative of $\overleftarrow{X1^*}_i(T,v)$ and $\overleftarrow{X2^*}_i(T,v)$, at $v=0$ are

$$\overleftarrow{M1}_i^{(n)}(T) = (-1)^n \frac{d^n}{dv^n}\overleftarrow{X1^*}_i(T,v)|_{v=0}$$

and

$$\overleftarrow{M2}_i^{(n)}(T) = (-1)^n \frac{d^n}{dv^n}\overleftarrow{X2^*}_i(T,v)|_{v=0} .$$

The stochastic meaning of $\overleftarrow{M1}_i^{(n)}(T)$ and $\overleftarrow{M2}_i^{(n)}(T)$ are

$$\overleftarrow{M1}_i^{(n)}(T) = E(\overleftarrow{B}(T)^n, \text{no state transition in } (0,T) \mid \overleftarrow{Z}(T) = i) ,$$

$$\overleftarrow{M2}_i^{(n)}(T) = E(\overleftarrow{B}(T)^n, \text{state transition in } (0,T) \mid \overleftarrow{Z}(T) = i) ,$$

and based on these quantities the nth moment ($n \geq 0$) of the reward accumulated over the interval $(0,T)$ is

$$E(B(T)^n) = \sum_{i\in S}\left(\overleftarrow{M1}_i^{(n)}(T) + \overleftarrow{M2}_i^{(n)}(T)\right)\gamma_i(0) ,$$

The LST of Eq. (B.8) and (B.9) with respect to w (denoting the transform variable by v) are:

(C.1) $$\frac{\partial}{\partial \tau}\underleftarrow{X1^*}(\tau,v) + v\underleftarrow{X1^*}(\tau,v)\mathbf{R} - \underbrace{\underleftarrow{X1}(\tau,0)\mathbf{R}}_{0} = \underleftarrow{X1^*}(\tau,v)\mathbf{Q}_D ,$$

and

$$\frac{\partial}{\partial \tau}\underleftarrow{X2^*}(\tau,v) + v\underleftarrow{X2^*}(\tau,v)\mathbf{R}_\alpha - \underbrace{\underleftarrow{X2}(\tau,0)\mathbf{R}_\alpha}_{0} = \underleftarrow{X1^*}(\tau,v)(\mathbf{Q} - \mathbf{Q}_D)^T + \underleftarrow{X2^*}(\tau,v)\mathbf{Q}^T$$
(C.2)

The nth derivative of Eq. (C.1) and (C.2) with respect to v are:

(C.3) $$\frac{\partial}{\partial \tau}\underleftarrow{X1^*}^{(n)}(\tau,v) + n\underleftarrow{X1^*}^{(n-1)}(\tau,v)\mathbf{R} + v\underleftarrow{X1^*}^{(n)}(\tau,v)\mathbf{R} = \underleftarrow{X1^*}^{(n)}(\tau,v)\mathbf{Q}_D ,$$

and

$$
\text{(C.4)} \quad \frac{\partial}{\partial \tau} \overleftarrow{X2^*}^{(n)}(\tau, v) + n \overleftarrow{X2^*}^{(n-1)}(\tau, v) \mathbf{R}_\alpha + v \overleftarrow{X2^*}^{(n)}(\tau, v) \mathbf{R}_\alpha =
$$
$$
\overleftarrow{X1^*}^{(n)}(\tau, v)(\mathbf{Q} - \mathbf{Q}_D)^T + \overleftarrow{X2^*}^{(n)}(\tau, v)\mathbf{Q}^T .
$$

Substituting $v = 0$ provides the theorem.

Appendix D. Corollaries utilized in the proof of Theorem 5.2.

COROLLARY D.1. *The $\underline{D}^{(n)}(k)$ terms are upper bounded by*

$$
\text{(D.1)} \quad
\begin{aligned}
\underline{D}^{(0)}(k) &\leq \underline{e} & &\text{if } n = 0, k > 0 \\
\underline{D}^{(n)}(k) &\leq \frac{(k-1)! \, (n+1+nk)}{(n+1)! \, (k-n-1)!} \, \underline{e} & &\text{if } n > 0, k > n
\end{aligned}
$$

Proof. Based on their definition $\mathbf{A}_D$, $\mathbf{S}$, $\mathbf{S}_\alpha$ and $\mathbf{A}$ are matrices of non-negative elements whose row-sum are between 0 and 1, hence the following inequalities hold element-wise:

$$
\underline{0} \leq \underline{e}\,(\mathbf{I} - \mathbf{A}_D^k) \leq \underline{e}
$$

$$
\underline{0} \leq \underline{D}^{(n-1)}(k-1)\mathbf{S}_\alpha \leq \underline{D}^{(n-1)}(k-1)
$$

$$
\underline{0} \leq \underline{D}^{(n)}(k-1)\mathbf{A} \leq \underline{D}^{(n)}(k-1)
$$

$$
\underline{0} \leq \underline{e}\,\mathbf{S}^n \mathbf{A}_D^{k-1-n}(\mathbf{A} - \mathbf{A}_D) \leq \underline{e}
$$

The first inequality shows that the corollary holds for $n = 0$. Assuming that the corollary holds for $\underline{D}^{(n)}(k-1)$ and $\underline{D}^{(n-1)}(k-1)$ we have

$$
\underline{D}^{(n)}(k) = \underline{D}^{(n-1)}(k-1)\mathbf{S}_\alpha + \underline{D}^{(n)}(k-1)\mathbf{A} + \binom{k-1}{n} \underline{e}\,\mathbf{S}^n \mathbf{A}_D^{k-1-n}(\mathbf{A} - \mathbf{A}_D) \leq
$$
$$
\underline{D}^{(n-1)}(k-1) + \underline{D}^{(n)}(k-1) + \binom{k-1}{n} \underline{e} \leq
$$
$$
\frac{(k-2)! \, (n + (n-1)(k-1))}{n! \, (k-n-1)!} \, \underline{e} + \frac{(k-2)! \, (n+1+n(k-1))}{(n+1)! \, (k-n-2)!} \, \underline{e} + \binom{k-1}{n} \underline{e} =
$$
$$
\frac{(k-1)! \, (n+1+nk)}{(n+1)! \, (k-n-1)!} \, \underline{e}.
$$

COROLLARY D.2. *The $\underline{D}^{(n)}(k)$ terms for $k \geq n+1$ are further upper bounded by*

$$
\text{(D.2)} \quad \underline{D}^{(n)}(k) \leq \frac{(k-1)! \, (n+1+nk)}{(n+1)! \, (k-n-1)!} \, \underline{e} \leq \frac{k!}{n! \, (k-n-1)!} \, \underline{e}
$$

Proof. Multiplying (D.2) by $\dfrac{n! \, (k-n-1)!}{(k-1)!}$ we have $1 + \dfrac{nk}{n+1} \leq k$, which holds for $k \geq n+1$.

REFERENCES

[1] W. J. Anderson. *Continuous-Time Markov Chains*. Springer-Verlag, New York, 1991.
[2] A. Bobbio, V. G. Kulkarni, and M. Telek. Partial loss in reward models. In *2nd Int. Conf. on Mathematical Methods in Reliability*, pages 207–210, Bordeaux, France, July 2000.
[3] A. Bobbio and M. Telek. The task completion time in degradable sytems. In B. R. Haverkort, R. Marie, G. Rubino, and K. S. Trivedi, editors, *Performability Modelling: Techniques and Tools*, chapter 7, pages 139–161. Wiley, 2001.
[4] G. Ciardo, R.A. Marie, B. Sericola, and K.S. Trivedi. Performability analysis using semi-Markov reward processes. *IEEE Transactions on Computers*, 39:1252–1264, 1990.
[5] E. de Souza e Silva and H.R. Gail. Calculating cumulative operational time distributions of repairable computer systems. *IEEE Transactions on Computers*, C-35:322–332, 1986.
[6] E. de Souza e Silva and R. Gail. An algorithm to calculate transient distributions of cumulative rate and impulse based reward. *Commun. in Statist. – Stochastic Models*, 14(3):509–536, 1998.
[7] L. Donatiello and V. Grassi. On evaluating the cumulative performance distribution of fault-tolerant computer systems. *IEEE Transactions on Computers*, 1991.
[8] W.K. Grassmann. Means and variances of time averages in markovian environment. *European Journal of Operational Research*, 31:132–139, 1987.
[9] R.A. Howard. *Dynamic Probabilistic Systems, Volume II: Semi-Markov and Decision Processes*. John Wiley and Sons, New York, 1971.
[10] J. G. Kemeny, J. L. Snell, and A. W. Knapp. *Denumerable Markov Chains*. Springer-Verlag, New York, 2nd edition, 1976.
[11] H. Nabli and B. Sericola. Performability analysis: a new algorithm. *IEEE Transactions on Computers*, 45:491–494, 1996.
[12] V.F. Nicola, R. Martini, and P.F. Chimento. The completion time of a job in a failure environment and partial loss of work. In *2nd Int. Conf. on Mathematical Methods in Reliability (MMR'2000)*, pages 813–816, Bordeaux, France, July 2000.
[13] A. Reibman, R. Smith, and K.S. Trivedi. Markov and Markov reward model transient analysis: an overview of numerical approaches. *European Journal of Operational Research*, 40:257–267, 1989.
[14] M. Telek, A. Horváth, and G. Horváth. Analysis of inhomogeneous markov reward models. In *International Conference on Numerical Solution of Markov Chains – NSMC 2003*, pages 305–322, Urbana, IL, USA, Sept 2003.
[15] M. Telek and A. Pfening. Performance analysis of Markov Regenerative Reward Models. *Performance Evaluation*, 27&28:1–18, 1996.
[16] M. Telek and S. Rácz. Analysis of partial loss reward models and its application. In *Measuring, Modelling and Evaluation of Computer and Communication Systems (MMB)*, pages 25–40, Aachen, Germany, Sept 2001. RWTH Aachen.

THE FIVE GREATEST APPLICATIONS
OF
MARKOV CHAINS

PHILIPP VON HILGERS[*] AND AMY N. LANGVILLE[†]

Abstract. One hundred years removed from A. A. Markov's development of his chains, we take stock of the field he generated and the mathematical impression he left. As a tribute to Markov, we present what we consider to be the five greatest applications of Markov chains.

Key words. Markov chains, Markov applications, stationary vector, PageRank, HIdden Markov models, performance evaluation, Eugene Onegin, information theory

AMS subject classifications. 60J010, 60J20, 60J22, 60J27, 65C40

1. Introduction. The five applications that we have selected are presented in the boring but traditional chronological order. Of course, we agree that this ordering scheme is a bit unfair and problematic. Because it appears first chronologically, is A. A. Markov's application of his chains to the poem Eugeny Onegin the most important or the least important of our five applications? Is A. L. Scherr's application of Markov chains (1965) to the performance of computer systems inferior to Brin and Page's application to web search (1998)? For the moment, we postpone such difficult questions. Our five applications, presented in Sections 2-6, appear in the admittedly unjust chronological order. In Section 7, we right the wrongs of this imposed ordering and unveil our proper ordering, ranking the applications from least important to most important. We conclude with an explanation of this ordering. We hope you enjoy this work, and further, hope this work is discussed, debated, and contested.

2. A. A. Markov's Application to Eugeny Onegin. Any list claiming to contain the five greatest applications of Markov chains must begin with Andrei A. Markov's own application of his chains to Alexander S. Pushkin's poem "Eugeny Onegin." In 1913, for the 200th anniversary of Jakob Bernoulli's publication [4], Markov had the third edition of his textbook [19] published. This edition included his 1907 paper, [20], supplemented by the materials from his 1913 paper [21]. In that edition he writes, "Let us finish the article and the whole book with a good example of dependent trials, which approximately can be regarded as a simple chain." In what has now become the famous first application of Markov chains, A. A. Markov, studied the sequence of 20,000 letters in A. S. Pushkin's poem "Eugeny Onegin," discovering that the stationary vowel probability is $p = 0.432$, that the probability of a vowel following a vowel is $p_1 = 0.128$, and that the probability of a vowel following a consonant is $p_2 = 0.663$. In the same article, Markov also gave the results of his other tests; he studied the sequence of 100,000 letters in S. T. Aksakov's novel "The Childhood of Bagrov, the Grandson." For that novel, the probabilities were $p = 0.449$, $p_1 = 0.552$, and $p_2 = 0.365$.

At first glance, Markov's results seem to be very specific, but at the same time his application was a novelty of great ingenuity in very general sense. Until that time, the theory of probability ignored temporal aspects related to random events. Mathematically speaking, no difference was made between the following two events: a die

---

[*]Max Planck Institute for History of Science, Berlin, Germany (philgers@mpiwg-berlin.mpg.de)

[†]Department of Mathematics, College of Charleston, Charleston, SC 29424, (langvillea@cofc.edu)

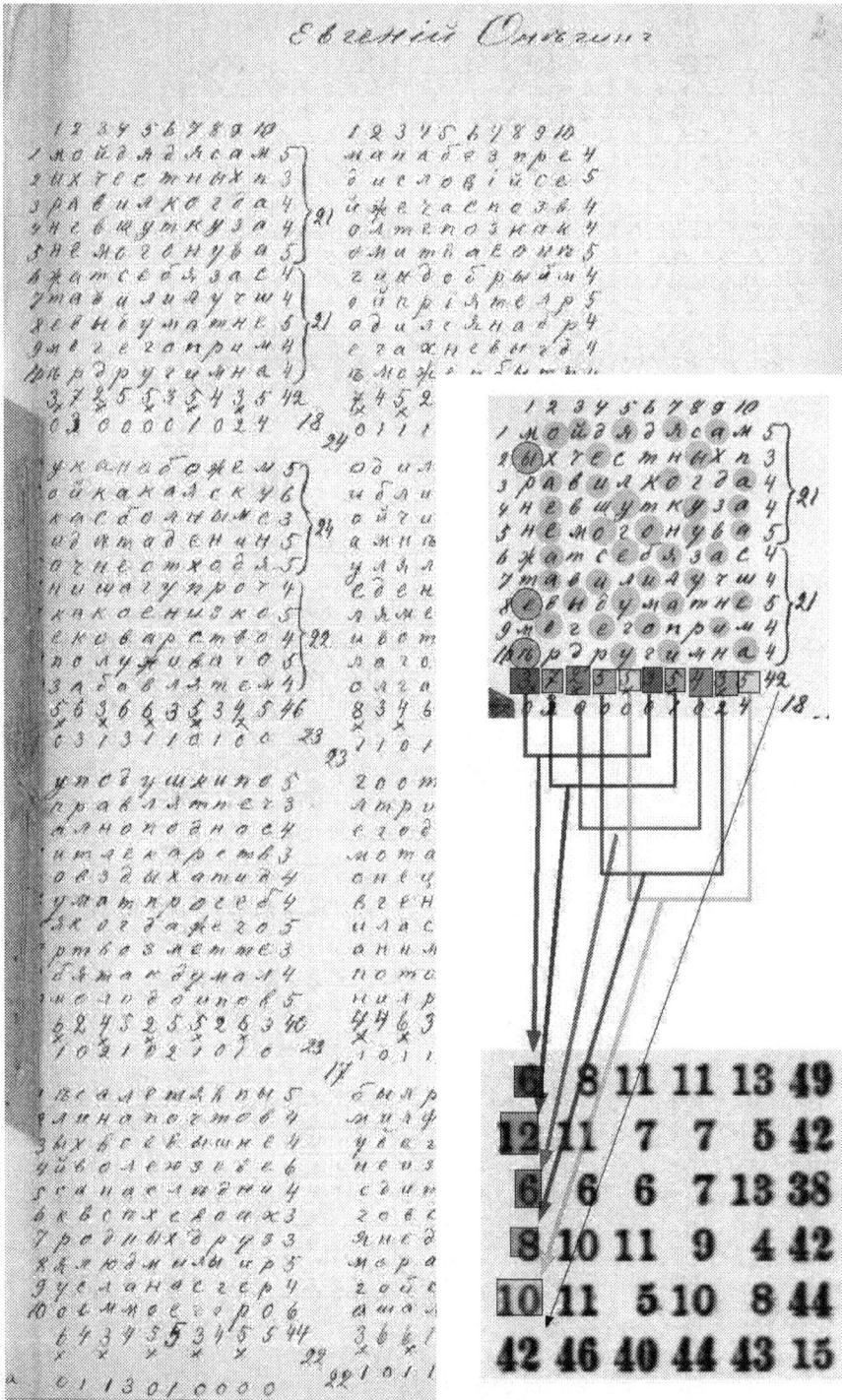

FIG. 2.1. *Left background: The first 800 letters of 20,000 total letters compiled by Markov and taken from the first one and a half chapters of Pushkin's poem "Eugeny Onegin." Markov omitted spaces and punctuation characters as he compiled the cyrillic letters from the poem. Right foreground: Markov's count of vowels in the first matrix of 40 total matrices of 10 × 10 letters. The last row of the 6 × 6 matrix of numbers can be used to show the fraction of vowels appearing in a sequence of 500 letters. Each column of the matrix gives more information. Specifically, it shows how the sums of counted vowels are composed by smaller units of counted vowels. Markov argued that if the vowels are counted in this way, then their number proved to be stochastically independent.*

thrown a thousand times versus a thousand dice thrown once each. Even dependent random events do not necessarily imply a temporal aspect. In contrast, a temporal aspect is fundamental in Markov's chains. Markov's novelty was the notion that a random event can depend only on the most recent past. When Markov applied his model to Pushkin's poem, he compared the probability of different distributions of letters taken from the book with probabilities of sequences of vowels and consonants in term of his chains. The latter models a stochastic process of reading or writing while the former is simply a calculation of the statistical properties of a distribution of letters. Figure 2.1 shows Markov's original notes in computing the probabilities needed for his Pushkin chain. In doing so, Markov demonstrated to other scholars a method of accounting for time dependencies. This method was later applied to the diffusion of gas molecules, Mendel's genetic experiments, and the random walk behavior of certain particles.

The first response to Markov's early application was issued by a colleague at the Academy of Sciences in St. Petersburg, the philologist and historian Nikolai A. Morozov. Morozov enthusiastically credited Markov's method as a "new weapon for the analysis of ancient scripts" [24]. To demonstrate his claim Morozov himself provided some statistics that could help identify the style of some authors. In his typical demanding, exacting, and critical style [3], Markov found few of Morozov's experiments to be convincing. Markov, however, did mention that a more advanced model and an extended set of data might be more successful at identifying an author solely by mathematical analysis of his writings [22].

3. C. E. Shannon's Application to Information Theory. When Claude E. Shannon introduced "A Mathematical Theory of Communication" [30] in 1948, his intention was to present a general framework for communication based on the principles of the new digital media. Shannon's information theory gives mathematically formulated answers to questions such as how analog signals could be transformed into digital ones, how digital signals then could be coded in such way that noise and interference would not do harm to the original message represented by such signals, and how an optimal utilization of a given bandwidth of a communication channel could be ensured. A famous *entropy* formula associated with Shannon's information theory is $H = -(p_1 \log_2 p_1 + p_2 \log_2 p_2 + \ldots + p_n \log_2 p_n)$, where H is the amount of information and p_i is probability of occurrence of the states in question. This formula is the entropy of a source of discrete events. In Shannon's words this formula "gives values ranging from zero—when one of the two events is certain to occur (i.e., probability 1) and all others are certain not to occur (i.e., probability 0)—to a maximum value of $\log_2 N$ when all events are equally probable (i.e., probability $\frac{1}{N}$). These situations correspond intuitively to the minimum information produced by a particular event (when it is already certain what will occur) and the greatest information or the greatest prior uncertainty of the event" [31].

It is evident that if something is known about a message beforehand, then the receiver in a communication system should somehow be able to take advantage of this fact. Shannon suggested that any source transmitting data is a Markov process. This assumption leads to the idea of determining a priori the transition probabilities of communication symbols, i.e., the probability of a symbol following another symbol or group of symbols. If, for example, the information source consists of words of the English language and excludes acronyms, then the transition probability of the letter "u" following the letter "q" is 1.

Shannon applied Markov's mathematical model in a manner similar to Markov's own first application of "chains" to the vowels and consonants in Alexander Pushkin's poem. It is interesting to pause here to follow the mathematical trail from Markov to Shannon. For further details about the theory of Markov chains, Shannon referred to a 1938 book by Maurice Fréchet [7]. While Fréchet only mentions Markov's own application very briefly, he details an application of Markov chains to genetics. Beyond Fréchet's work, within the mathematical community Markov chains had become a prominent subject in their own right since the early 1920s, especially in Russia. Most likely Fréchet was introduced to Markov chains by the great Russian mathematician Andrei Kolmogorov. In 1931 Kolmogorov met Fréchet in France. At this time Kolmogorov came to France after he had visited Göttingen, where he had achieved fame particularly due to his axiomatization of probability theory [14].

Kolmogorov speculated that if physics in Russia had reached the same high level as it had in some Western European countries, where advanced theories of probability tackled distributions of gas and fluids, Markov would not have picked Pushkin's book from the shelf for his own experiments [33]. Markov's famous linguistic experiment might have been a physics experiment instead. It is significant that Kolmogorov himself contributed an extension to the theory of Markov chains by showing that "it is a matter of indifference which of the two following assumptions is made: either the time variable t runs through all real values, or only through the integers" [15].[1] In other words, Kolmogorov made the theory suitable not only for discrete cases, but also for all kinds of physical applications that includes continuous cases. In fact, he explicitly mentioned Erwin Schrödinger's wave mechanics as an application for Markov chains.

With this pattern of prior applications of Markov theory to physical problems, it is somewhat ironic that Shannon turned away from physics and made great use of Markov chains in the new domain of communication systems, processing "symbol by symbol" [30] as Markov was the first to do. However, Shannon went beyond Markov's work with his information theory application. Shannon used Markov chains not solely as a method for analyzing stochastic events but also to generate such events.

Shannon demonstrated that a sequence of letters, which are generated by an increasing order of overlapping groups of letters, is able to reflect properties of a natural language. For example, as a first-order approximation to an English text, an arbitrary letter is added to its predecessor, with the condition that this specific letter be generated according to its relative frequency in the written English language in general. The probability distribution of each specific letter can, of course, be acquired only by empirical means. A second-order approximation considers not only the distribution of a single letter but a bigram of two letters. Each subsequent letter becomes the second part of the next bigram. Again, the transition probabilities of these overlapping bigrams are chosen to match a probability distribution that has been observed empirically. By applying such a procedure Shannon generated a sequence of order 3 by using trigrams. Shannon's first attempt at using Mar! kov chains to produce English sentences resulted in "IN NO IST LAT WHEY CRATICT FOURE BIRS GROCID" [30]. Although this line makes no sense, nevertheless, some similarities with written English are evident. With this start, Shannon felt that communication systems could be viewed as Markov processes, and their messages analyzed by means of Markov

---

[1]Kolmogorov saw the potential of Markov theory in the physical domain. And later, he was among the first to promote information theory and further develop it into an algorithmic information theory. In addition, due to translation issues, Kolmogorov's contribution to the mathematization of poetry is scarcely known outside of Russia.

theory. In studying artificial languages Shannon distinguished between ergodic and non-ergodic processes. Shannon used the phrase "ergodic process" to refer to the property that all processes originating from the same source have the same statistical properties.

While Shannon introduced his generative model for technical reasons, other scholars became convinced that Markov models could even play a far more general role in the sciences and arts. In fact, Shannon, himself helped popularize information theory in other diverse fields. For instance, Shannon and his colleague David W. Hagelbarger created a device that could play the game of "Even and Odd." It is reported that Shannon's so-called "Mind-Reading Machine" won most of the games against visitors at the Bell Labs who dared to challenge the machine to a match. Despite the whimsical nature of Shannon's machine, it deserves credit as being the first computer to have implemented a Markov model [32]. At this same time, the ever well-informed French psychoanalyst Jacques Lacan introduced Markov chains as the underlying mechanism for explaining the process by which unconscious choices are made. Lacan hints that Shannon's machine was the model for his theory [16, 17].

Later researchers continued Shannon's trend of using computers and Markov chains to generate objects from text to music to pictures. Artists such as musician Iannis Xenakis developed "Free Stochastic Music" based on Markov chains, and early media artists and computer experts such as Frieder Nake plotted pictures generated by Markov models. In fact, a group of influential linguists claimed that the modus operandi of language is a Markov process [12]. Such a general assumption provoked a controversial debate between Roman Jakobson and Noam Chomsky. Chomsky argued that language models based on Markov chains do not capture some nested structures of sentences, which are quite common in many languages such as English [10]. We now recognize that Chomsky's account of the limitations of Markovian language models was too harsh. The early 1980s saw a resurgence of the success of Markovian language models in speech recognition. Section 5 treats this era briefly.

4. A. L. Scherr's Application to Computer Performance Evaluation. In 1965 Allan L. Scherr completed his thesis, "An Analysis of Time-Shared Computer Systems," and received his Ph.D. in electrical engineering from M.I.T. At the time, the Compatible Time-Sharing System was new to the M.I.T. campus and allowed 300 users to interactively access the computer and its software. The goal of Scherr's thesis was to characterize the system's usage. He conducted simulation studies to predict the system's usage and wanted to compare this with real user data from similar systems. He found that no such data existed, so he conducted his own comprehensive measurements of system performance. Scherr declared his analysis of time-shared systems complete after he obtained his own real data and compared this with his simulation results.

At this point in the story, many of us can thank one of Scherr's thesis advisors for our current careers and research field. This particular advisor complained that Scherr's thesis was not yet complete as it wasn't quite academic enough. Scherr recalls that "there weren't enough mathematical formulas in it" [8]. So in response to this complaint, Scherr hobbled together a "very quick and dirty" mathematical analysis by applying a method from a recent operations research course he had taken. He used a continuous-time Markov chain to model M.I.T's Compatible Time-Sharing System. The chain not only added enough mathematics, it also led Scherr to a surprising result. Scherr's quick and dirty measure gave a very good approximation to system performance. According to Scherr, this was surprising because "this very simple, analytic model that ignored 99 percent of the details of the system was just as accurate in predicting performance and capacity as the very e! laborate, very accurate simulation model" [8].

Scherr's single-server, single-queue Markov chain captured the dominating aspects of the system. The tridiagonal transition rate matrix $\mathbf{A}$ appearing in Scherr's thesis is below. The states of the chain are the number of users interacting with the system, and thus, are labeled $\{0, 1, \ldots, n\}$.

$$
\mathbf{A} =
\begin{pmatrix}
-\frac{n}{T} & \frac{n}{T} & & & & & \\
\frac{1}{P} & -(\frac{1}{P} + \frac{n-1}{T}) & \frac{n-1}{T} & & & & \\
 & \frac{1}{P} & -(\frac{1}{P} + \frac{n-2}{T}) & \frac{n-2}{T} & & & \\
 & & \ddots & \ddots & \ddots & & \\
 & & & \frac{1}{P} & -(\frac{1}{P} + \frac{2}{T}) & \frac{2}{T} & \\
 & & & & \frac{1}{P} & -(\frac{1}{P} + \frac{1}{T}) & \frac{1}{T} \\
 & & & & & \frac{1}{P} & -\frac{1}{P}
\end{pmatrix}.
$$

Scherr defined n as the number of users in the system. When a user interacted with the time-shared computer system, Scherr defined T as the mean time for the console portion of the interaction and P as the mean processor time per interaction. Scherr then solved the system $\boldsymbol{\pi}^T \mathbf{A} = \mathbf{0}$ for the stationary probability vector $\boldsymbol{\pi}$. Using standard queuing theory techniques, Scherr reported π_0, the long-run probability that the M.I.T. shared processor was idle, $\bar{Q} = \sum_{i=1}^{n} i\pi_i$, the average number of users waiting for service, and W, the mean response time of the system. Scherr concluded that "the think times of the users and the service times of the equipment were really all you needed to know to do a pretty accurate characterization of the response time and capacity of the system" [8].

In the years after his dissertation, researchers argued over the validity of Scherr's simple Markov analysis. Were Scherr's results—that a simple Markov model came to within 4-5 percent accuracy of the very detailed simulation—merely one atypical study or was simple analysis preferable to detailed simulation? A few decades after his dissertation, Scherr, then at IBM, had an answer to that question, an answer arrived at through practice. Scherr discovered that system updates and upgrades are the primary reason for the superiority of the simplified analytical model over detailed simulation. The simple analytical models give a better return on investment. By the time a detailed simulation is finally complete, the system is usually updated, and a new simulation required. Also, Scherr found that the simple Markov models force designers to get to the heart of the system, because only the essential information can be modeled; all extraneous information must be identified and omitted.

By the way, ten years later the performance evaluation community, which evolved as a result of Scherr's 1965 thesis [29], noted Scherr's ground-breaking contribution. In 1975, his thesis was awarded the Grace Murray Hopper Award by the Association for Computing Machinery. Each year this award is presented to a young computer professional, aged 35 or less, who has made a "single significant technical or service contribution" to the computer field. At present, there are dozens of journals and conferences devoted to computer performance evaluation. Papers from these journals and proceedings demonstrate the advancement achieved since Scherr's development of this new field. These papers describe Markov models for computer performance evaluation that have expanded far beyond Scherr's famous single-server, single queue model.

5. L. E. Baum's Application to Hidden Markov Models. When Lawrence R. Rabiner introduced Hidden Markov Models (HMMs) in the widely read proceedings of the IEEE in 1989, he stated that even though most of their principles had been known for twenty years, engineers had failed to read mathematical journals to grasp their potential and mathematicians had failed to "provide sufficient tutorial material for most readers to understand the theory and to be able to apply it to their own research" [28]. The purpose of Rabiner's paper was to rectify this under-appreciation of an important tool.

Rabiner might have overstated the case a bit as HMMs were receiving some appreciation prior to 1989. In 1980 a symposium "On the application of Hidden Markov Models to Text and Speech" was held in Princeton for the reason that it "was thought fitting to gather researchers from across the country" for one day [9]. It was at this gathering, that Lee P. Neuwirth coined the phrase "Hidden Markov Model," instead of calling it the bit unwieldy alternative of "probabilistic functions of Markov chains" [25, 26]. It is worth mentioning that Lee P. Neuwirth was the longtime director of the Communications Research Division within the Institute for Defense Analysis (IDA-CRD) in Princeton, New Jersey. Some called this institute "the most secret of the major think tanks" of the day [1].

Today, now that speech recognition software is available off the shelf, it is no longer a secret that, thanks to HMMs, "word spotting" in stream of spoken language is done by a computer—even in the presence of noise. In addition, speech cannot only be recognized algorithmically with great accuracy, but the individual voice of the speaker can be identified as well. It stands to reason that such features are of great interest, especially for intelligence services, in order to scan communication channels for key words. Moreover, HMMs enable the extraction of significant information from the acoustical pattern of a spoken language without requiring any semantic knowledge of the language in question.

John D. Ferguson, also a mathematician at the IDA, specified the theoretical impact of HMMs by solving three fundamental problems in the proceedings of that 1980 IDA symposium.

1. Compute the probability of an observed sequence based on a given model.
2. Maximize the probability of an observed sequence by adjusting the parameters using, for instance, the "Baum-Welsh" algorithm.
3. Select the most likely hidden states, given the observation, and the model.

By listing these problems Ferguson provided a building block for later tutorials on HMMs [9, 26, 28]. The meaning of these rather abstract instructions might become clearer with an example taken from Alan B. Poritz's instructive paper "Hidden Markov Models: A Guided Tour" [26]. This example is illustrated in Figure 5.1. Imagine three mugs, each containing its own mixture of stones. The stones are marked either "state 1" or "state 2." Somebody randomly draws a stone from mug "0." If the stone is marked "state 1," as shown in the illustrated example, a ball is drawn from urn "1." Otherwise, a ball is drawn from the other urn. The two urns are filled with a different mixture of black and white balls. After every draw, the ball will be replaced in the urn from which it was selected. An observer only gets to know the outcome of the drawing in terms of a sequence of symbols: "BWBWWB", where "B" stands for a black ball and "W" for a white ball.[2] Notice that the particular urn from which a ball

---

[2]It is interesting that Ferguson and Neuwirth were not the first scientists to consider urn problems in the Markov context. Markov and his colleague A. A. Chuprov had a heated debate over the notion of "double probability," the phrase Markov gave to a random selection between two or more urns for

is drawn is not available to the observer, only the color is. Hence the label "hidden" in the HMM acronym. An HMM is valuable because it helps uncover this hidden information, or at least gives a reasonable approximation to it.

Let's start with most difficult problem, recovering a mixture model from the observed tokens, thereby solving the second problem. In the given example, the mixture model corresponds to guessing each urn's mixture. In the late 1960s and particularly the early 1970s, Leonard E. Baum, supported by the work of other scholars, demonstrated that the underlying model can be recovered from a sufficiently long observation sequence by an iterative procedure, which maximizes the probability of the observation sequence. This procedure is called Baum-Welsh algorithm. It is based on the auxiliary Q-function, which is denoted $Q(\lambda, \bar{\lambda}) = \sum_{s \in S} P_\lambda(O, s) \log P_{\bar{\lambda}}(O, s)$, where s is a element out of set of states S and $\bar{\lambda}$ is an exstimated model of λ. In an iterative procdure $\bar{\lambda}$ becomes λ until there is no further improvement in terms of a measurement of increased probabiltiy. Baum and his collegues proved that maximization of $Q(\lambda, \bar{\lambda})$ increased likelihood of the estimated $\bar{\lambda}$ and a hidden true model, i.e. $max_{\bar{\lambda}}[Q(\lambda, \bar{\lambda})] \Rightarrow P(Q|\bar{\lambda} \geq P(Q|\lambda)$ [2, 26, 28]. Second, from this model the hidden state sequence can then be estimated, thereby solving the third problem. A technique for doing this is called the Viterbi Algorithm. This algorithm estimates the best state sequence, $Q = (q_1 q_2 \cdots q_T)$, for the given observation sequence $O = (O_1 O_2 \cdots O_T)$ by defining the quantity $\delta(i) = max_{q_1, q_2, \cdots, q_{t-1}} P[q_1 q_2 \cdots q_t = i, O_1 O_2 \cdots O_t | \lambda]$. In our example, the outcome of the unobserved draws of stones from the mug generated the hidden state sequence. Eventually, once the parameters of the hidden model denoted by the vector $\lambda = (A, B, \pi)$ are known, the probability of the observed sequence can be computed, thereby solving the first problem. Instead of calculating the probability of the observed sequence by finding each possible sequence of the hidden states and summing these probabilities, a short cut exists. It is called the forward algorithm and minimizes computational efforts from exponential growth to linear growth by calculating partial probabilities at every time step in a recursive manner [26].

To summarize, hidden Markov models require the solution of some fundamental, interrelated problems of modeling and analyzing stochastic events. Several mathematicians (only a few are mentioned in this brief survey) defined the main problems and developed different techniques to solve them. Once tutorial material spread among scholars from other fields, HMMs quickly demonstrated their potential through real-life applications. For instance, in 1989 Gary A. Churchill used HMMs to separate genomic sequences into segments [11]. Since then, HMMs have become an off-the-shelf technology in biological sequence analysis.

We close this section by returning once again to Andrei A. Markov. Even though Markov died in St. Petersburg in 1922, his work on his famous chains played a prominent role in 1980 in Princeton's secret think tank. This prominence is apparent in the bibliographies of the proceedings of the IDA symposium that year. In fact, the proceedings open with an article by Lee P. Neuwirth and Robert L. Cave that states explicitly their intention to examine the relation of HMMs to the English language "in the spirit of Markov's application"[25], referring to Eugeny Onegin. The first result of their experiment showed that the most simple model of English text provides only two states: vowel and consonant. The HMM's ability to distinguish vowels from consonants was their main result. In this experiment, Neuwirth and Cave analyzed the letters of an English text without letting the HMM know which were vowels and

---

the drawing of lots. Markov was more skeptical than Chuprov that such a scheme could lead to new results [27].

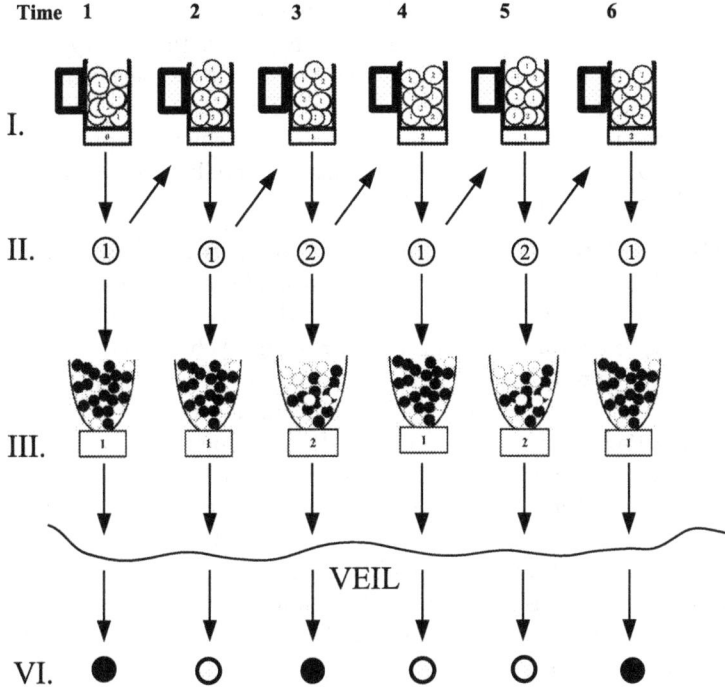

FIG. 5.1. *Sampling of the Urns veils the sampling of a sequence of Mugs.*

I. Let a_{01} be the fraction of stones with the state 1 marking in the first mug, and a_{02} the fraction of stones with the state 2 marking so that $a_{01} + a_{02} = 1$. Similarly, the fraction of stones in the other two mugs are denoted a_{21}, a_{22} in one case and a_{31}, a_{32} in the other.

II. At each time $t = 1, 2, \ldots, T$, a stone is selected. The number of selections in the example above is $T = 6$, while the outcome of these unobserved random selections determines the sequence of states s=(1,1,2,1,2,1).

III. Say $b_1(B)$ is the fration of white balls, $b_1(W)$ the fraction of black balls in one of the two urns and analog $b_2(B)$, $b_2(W)$ in case of the other urn so that for each state s, the vector $b = (b_s(1), b_s(2), \cdots b_s(K))$ is the output probability vector of a HMM with finite output alphabet with K elements. In this example the number of elements in the alphabet is $K = 2$.

VI. The T-long observation sequence is O=(B,W,B,W,W), while $|B| = 3$ is the number of drawn balls that are black and $|W| = 3$ is the number of drawn balls that are white.

The parameter vector which describes the HMM is
$\lambda = (a_{01}, a_{02}, a_{21}, a_{22}, a_{31}, a_{32}, b_1(B), b_1(W), b_2(B), b_2(W))$. P_λ stands for probability density associated with the model. It also has become common to denote the parameter model in general by the vector $\lambda = (A, B, \pi)$, where π is the vector of the initial state probabilities, and thus, is equivalent to a_{01}, a_{02}.

which were consonants. In hindsight, Markov has to be admired for his intuition to focus on the vowel-consonant distinction, which proved to be very significant four score later. Markov was forced to stop his letter-counting experiments, when he had nearly completely lost his sight due to glaucoma. Even if Markov had had more time and better eyesight to carry his experiments further, such extensions would have been very difficult to complete, given the precomputer era he lived in, when computational efforts had to be paid in man-years.

6. S. Brin and L. Page's Application to Web Search. In the late 1990s, Sergey Brin and Larry Page, then graduate students at Stanford University, were working on their PageRank project to organize the World Wide Web's information. Their 1998 paper, "PageRank: Bringing Order to the Web" [6], contributed to the order Google brought to web search. Brin and Page saw the potential of their PageRank idea, took a leave of absence from Stanford, and formed Google in 1998.

By Brin and Page's own admission, PageRank, which is the stationary vector of an enormous Markov chain, is the driving force behind Google's success in ranking webpages. In the PageRank context, the web and its hyperlinks create an enormous directed graph. Brin and Page's vision of a web surfer taking a random walk on this graph led to the formulation of the world's largest Markov chain [23]. The Google Markov chain $\mathbf{G}$ is defined as a convex combination of two other chains: $\mathbf{H}$, a reducible Markov matrix defined by the web's hyperlinks, and $\mathbf{E}$, a completely dense rank-one Markov matrix. Specifically,

$$\mathbf{G} = \alpha\mathbf{H} + (1 - \alpha)\mathbf{E},$$

where $0 < \alpha < 1$ is a parameter that Brin and Page originally set to .85, $H_{ij} = 1/O_i$ if page i links to page j, and 0, otherwise, and O_i is the number of outlinks from page i. $\mathbf{E} = \mathbf{e}\mathbf{v}^T$, where $\mathbf{v}^T > \mathbf{0}$ is the so-called personalization vector. The idea is that for $\alpha = .85$, 85% of the time surfers follow the hyperlink structure of the web and 15% of the time they jump to a new page according to the distribution in the personalization vector $\mathbf{v}^T$. For the 6-node web example in Figure 6.1, using $\alpha = .85$ and $\mathbf{v}^T = 1/6\,\mathbf{e}^T$, the $\mathbf{H}$ and $\mathbf{G}$ matrices are below.

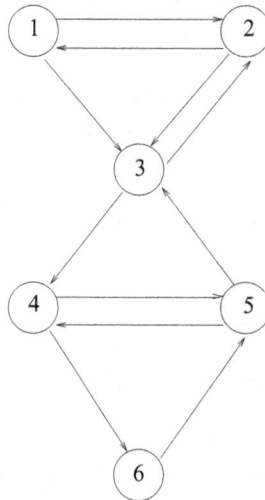

FIG. 6.1. *6-node web graph*

$$\mathbf{H} = \begin{pmatrix} 0 & 1/2 & 1/2 & 0 & 0 & 0 \\ 1/2 & 0 & 1/2 & 0 & 0 & 0 \\ 0 & 1/2 & 0 & 1/2 & 0 & 0 \\ 0 & 0 & 0 & 0 & 1/2 & 1/2 \\ 0 & 0 & 1/2 & 1/2 & 0 & 0 \\ 0 & 0 & 0 & 0 & 1 & 0 \end{pmatrix} \quad \text{and}$$

$$\mathbf{G} = \begin{pmatrix} 0.025 & 0.450 & 0.450 & 0.025 & 0.025 & 0.025 \\ 0.450 & 0.025 & 0.450 & 0.025 & 0.025 & 0.025 \\ 0.025 & 0.450 & 0.025 & 0.450 & 0.025 & 0.025 \\ 0.025 & 0.025 & 0.025 & 0.025 & 0.450 & 0.450 \\ 0.025 & 0.025 & 0.450 & 0.450 & 0.025 & 0.025 \\ 0.025 & 0.025 & 0.025 & 0.025 & 0.875 & 0.025 \end{pmatrix}.$$

$\mathbf{G}$ has been carefully formed so that it is an aperiodic, irreducible Markov chain, which means that its stationary vector π^T exists and is unique. Element π_i represents the long-run proportion of the time that the random surfer spends in page i. A page with a large stationary probability must be important as it means other important pages point to it, causing the random surer to return there often. The webpages returned in response to a user query are ranked in large part by their π_i values [5]. In the above example, the pages would be ordered from most important to least important as $\{3, 5, 4, 2, 6, 1\}$ since $\pi^T = (\begin{array}{cccccc} 0.092 & 0.158 & 0.220 & 0.208 & 0.209 & 0.113 \end{array})$. This ranking idea is so successful that nearly all search engines use some hyperlink analysis to augment their retrieval systems.

π^T was originally computed using the simple power method [6], but since then a variety of methods for accelerating this computation have been proposed [18]. Judging by the hundreds of papers produced in the last few years on PageRank computation, without doubt PageRank presents a fun computational challenge for our community.

7. Ranked List. It's now time to present these five great Markov applications in their proper order. Employing a common trick among late night television hosts, we present the list in reverse order of importance, as this serves to build the anticipation and excitement inherent in all top k lists.

---

  5. Scherr's application to Computer Performance Evaluation
  4. Brin and Page's application to PageRank and Web Search
  3. Baum's application to Hidden Markov Models[3]
  2. Shannon's application to Information Theory
  1. Markov's application to Eugeny Onegin

How did we arrive at this ordering? Good question. We considered several popular schemes for ordering elements in a set. We've already discussed the drawbacks of the chronological system. For similar reasons, the alphabetical system seemed equally unfair and unjustified. For instance, should we alphabetize by author of the application or name of the application? And what if an application was produced by two authors, as in the PageRank case? Do we order by Brin or by Page?

Having ruled out such elementary systems, we moved on to more advanced systems, such as the "Markov number" system and the "impact factor" system. The Markov number system is an analogy to the Erdos number system. In this system,

---

[3]Actually, there was a tie between Shannon and Baum for second place. See the next page for details.

A. A. Markov has a Markov number of 0. Markov applications produced by Markov's coauthors would receive a Markov number of 1, coauthors of coauthors, a 2, and so on. This means that Markov's own Eugeny Onegin application would be the most important of our five chosen applications, because Markov numbers are like golf scores, the lower, the better. This ranking system seemed reasonable, but it wasn't without its faults. For instance, could we find a coauthor of Scherr's who coauthored a paper with Markov? Maybe, but we admit that we didn't try very hard. This left us with a Markov number for only Markov and no Markov numbers for Shannon, Scherr, Baum, Brin, or Page.

So we proceeded to the "impact factor" system of ranking our five applications. The idea here is to rank each application according to its perceived impact on society. A laudable goal, but practically difficult. Should applications that created entire new fields, such as Shannon's information theory and Scherr's computer performance evaluation receive the highest impact factor? If so, where does that leave PageRank? While the PageRank application hasn't created a new field (yet), it does affect billions of people each day as they access Google to surf the web.

Then inspiration came to us. The proper ranking system should have been self-evident. PageRank is the most famous system for ranking items, so why not use PageRank to rank our five applications? However, this idea gave cause for caution. Could PageRank be used to rank items, one of which is PageRank itself? The idea sounded dangerous, and worth avoiding. So we sidestepped the problem by employing a PageRank-like system, called HITS [13], to rank our five applications. In order to apply the HITS system for ranking items we needed to create a graph that represented the relationships between our five Markov applications. We produced the graph in Figure 7.1. We created a link between applications if we perceived some connection between the two topics. For instance, there is a link between Information Theory and Eugeny Onegin because Information Theory uses Markov chains to capture the stochastic relationship between letters and words and Eugeny Onegin uses chains to capture the stochastic relationship between vowels and consonants. Similarly, the link from PageRank to Performance Evaluation represents the fact that Scherr's models can be used to evaluate the massive parallel machines used to compute PageRank.

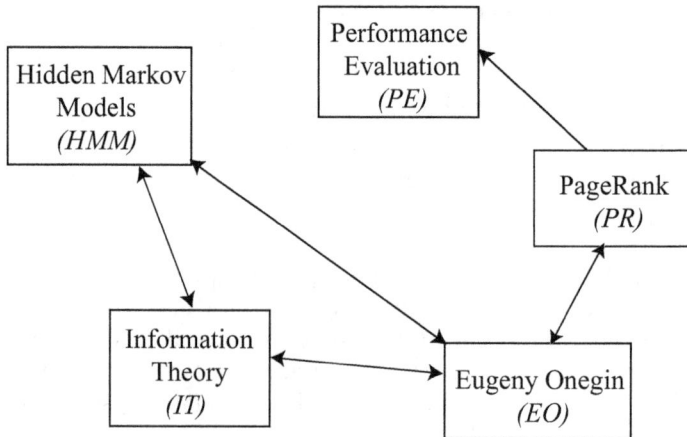

FIG. 7.1. *Graph of relationships between five Markov applications*

The adjacency matrix $\mathbf{L}$ associated with Figure 7.1 is

$$
\mathbf{L} = \begin{array}{c} \\ EO \\ IT \\ PE \\ HMM \\ PR \end{array}
\begin{array}{ccccc}
EO & IT & PE & HMM & PR \\
\left(\begin{array}{ccccc}
0 & 1 & 0 & 1 & 1 \\
1 & 0 & 0 & 1 & 0 \\
0 & 0 & 0 & 0 & 0 \\
1 & 1 & 0 & 0 & 0 \\
1 & 0 & 1 & 0 & 0
\end{array} \right)
\end{array},
$$

and the HITS authority vector $\mathbf{a}$ (the dominant eigenvector of $\mathbf{L}^T\mathbf{L}$) is

$$
\mathbf{a} = \begin{array}{ccccc}
EO & IT & PE & HMM & PR \\
(\;.64 & .50 & .17 & .50 & .26\;),
\end{array}
$$

meaning that the most authoritative of our five applications is Markov's application to Eugeny Onegin and the least authoritative is Scherr's application to computer performance evaluation.

Now that the mystery behind our ranking of the five greatest applications of Markov chains has been revealed, do you agree with us? Since you are all fellow Markov enthusiasts, we expect little argument about A. A. Markov's claim to the first position. It is, after all, his 150th birthday that we are celebrating. But what about the other four positions. We await your suggestions and complaints by email. Regardless, Happy 150th Markov!

REFERENCES

[1] James Bamford. The Puzzle Palace. A Report on America's Most Secret Agency. Houghton Mifflin Company, Boston, 1982.

[2] Leonard E. Baum, Ted Petrie, George Soules, and Norman Weiss. A maximization technique occuring in the statistical analysis of probabilistic functions of Markov chains. *Ann. Math. Statist.*, 41:164-171, 1970.

[3] Gely P. Basharin, Amy N. Langville, and Valeriy A. Naumov. The Life and Work of A. A. Markov. *Linear Algebra and its Applications*, 386: 3-26, 2004.

[4] Jakob Bernoulli. *Ars Conjectandi*, Opus Posthumum, Accedit Tractatus de Seriebus infinitis, et Epistola Gallice scripta de ludo Pilae recticularis, Basileae, 1713 (Ch. 1-4 translated into English by B. Sung, *Ars Conjectandi*, Technical Report No. 2, Dept. of Statistics, Harvard University, 1966).

[5] Nancy Blachman, Eric Fredricksen, Fritz Schneider. *How to Do Everything with Google*. McGraw-Hill, 2003.

[6] Sergey Brin, Lawrence Page, R. Motwami, Terry Winograd. The PageRank citation ranking: Bringing order to the Web. Technical Report 1999-0120, Computer Science Department, Stanford University, 1999.

[7] Maurice Fréchet. Méthode des fonctions arbitraires. Théorie des énénements en chaine dans les cas d'un nombre fini d'états possibles. Gauthier-Villars, Paris, 1938.

[8] Karen A. Frenkel. Allan L. Scherr: Big Blue's Time-Sharing Pioneer. *Communications of the ACM*, 30(10): 824-828, 1987.

[9] John D. Ferguson. Hidden Markov Models for Language. *Symposium on the Application of Hiddden Markov Models to Text and Speech*, IDA-CRD, Princeton, 1980, 1-7.

[10] Noam Chomsky. Three models for the description of language. *IRE Transaction of Information Theory*, 2(3):113-124, 1956.

[11] Gary A. Churchill. Stochastic models for heterogeneous DNA sequences. *Bull Math Biol.* 51, 1989, 79-94.

[12] Colin Cherry, Morris Halle, and Roman Jakobson. Toward the logicaldescription of languages in their phonemic aspect. *Language*, 29 (1): 34-46, 1953.

[13] Jon Kleinberg. Authoritative sources in a hyperlinked environment. *Journal of the ACM.* 46, 1999.

[14] Andrey N. Kolmogorov. Grundbegriffe der Wahrscheinlichkeitsrechnung. Springer, Berlin, 1933.

[15] Andrey N. Kolmogorov. Zur Theorie der Markoffschen Ketten. *Math. Ann.* 112, 1936, 155-160. (Translated into English by G. Lindquist, *Selected Works of A.N. Kolmogorov*, Vol. 2, Probability Theory and Mathematical Statistics. Kluwer, Dordrecht, Boston, London,1986, 182-192.

[16] Jacques Lacan. Écrits. Le champ freudien. (Ch. 1 *Le séminare sur "la Lettre volée"*) Éditions du Seuil, Paris, 1966.

[17] Jacques Lacan. Le Séminare. Livre II. Le moi dans la théorie de Freud et dans la technique de la psychanalyse. (Ch. 15 *Pair ou impair? Au-delà de l'intersubjectivitivé.*) Éditions du Seuil, Paris, 1978.

[18] Amy N. Langville and Carl D. Meyer. *Google's PageRank and Beyond: The Science of Search Engine Rankings.* Princeton University Press, Princeton, 2006.

[19] Andrei A. Markov. *Ischislenie veroyatnostej*, SPb, 1900; 2-e izd., SPb, 1908, Translated into German, *Wahrscheinlichkeitsrechnung*, Teubner, Leipzig-Berlin, 1912; 3-e izd., SPb, 1913; 4-e izd., Moskva, 1924.

[20] Andrei A. Markov. Issledovanie zamechatel'nogo sluchaya zavisimyh ispytanij, *Izvestiya Akademii Nauk*, SPb, VI seriya, tom 1, 9 3, 1907, 61-80 (Translated into French, Recherches sur un cas remarquable d'epreuves dependantes, *Acta mathematica*, Stockholm, 33, 1910, 87-104).

[21] Andrei A. Markov. Primer statisticheskogo issledovaniya nad tekstom "Evgeniya Onegina", illyustriruyuschij svyaz' ispytanij v cep', *Izvestiya Akademii Nauk*, SPb, VI seriya, tom 7, 9 3, 1913, 153-162.

[22] Andrei A. Markov. Ob odnom primenenii statisticheskogo metoda (On some application of statistical method).*Izvestija Imp. Akademii nauk*, serija VI,4:239-242, 1916.

[23] Cleve Moler. The world's largest matrix computation. *Matlab News and Notes.* Oct. (10): 12-13, 2002.

[24] N. A. Morozov, N. A. Lingvisticheskie spektry (Linguistic spectra).*Izvestija Akademii Nauk*, Section of Russian Language,20:1-4, 1915.

[25] Robert L. Cave, Lee P. Neuwirth. Hidden Markov Models for English. *Symposium on the Application of Hiddden Markov Models to Text and Speech*, IDA-CRD, Princeton, 1980, 1-7.

[26] Alan B. Poritz. Hidden Markov Models: A Guided Tour. *Proc. from ICASSP – International Conference on Acoustics, Speech, and Signal Processing*, 1988, 7-13.

[27] Kh. O. Ondar (Ed.). The Correspondence Between A.A. Markov and A.A. Chuprov on the Theory of Probability and Mathematical Statistics. Springer, New York, Heidelberg, Berlin, 1982.

[28] Lawrence R. Rabiner. A Tutorial on Hidden Markov Models and Selected Applications in Speech Recognition. *Proc. of the IEEE*, Vol. 77, 2, 1989, 257-286.

[29] Allan Lee Scherr. An Analysis of Time-Shared Computer Systems. Ph.D. thesis, Massachusetts Institute of Technology, 1962.

[30] Claude E. Shannon. A Mathematical Theory of Communication. *The Bell System Technical Jorurnal*, Vol. 27, 1948, July, 379-423, October, 623-656. Also included in *Collected Papers*, ed. by N.J. Sloane and Aaron D. Wyner. IEEE Press, Piscataway, 1993, 5-83.

[31] Claude E. Shannon. Information Theory. *Encyclopædia Britiannica*, Vol 12, 1953, 350-353.

[32] Claude E. Shannon. A Mind-Readinng(?)Machine. Bell Laboratories Memorandum, March 18, 1953. 4 pp. Also included in *Collected Papers*, ed. by N.J. Sloane and Aaron D. Wyner. IEEE Press, Piscataway, 1993, 688-690

[33] Oscar B. Sheynin. A.A. Markov's Work on Probability. *Archive for History of Exact Sciences* 39:337-375, 1988/89.

BOUNDING THE MEAN CUMULATED REWARD
UP TO ABSORPTION

ANA PAULA COUTO DA SILVA* AND GERARDO RUBINO[†]

Abstract.
Markov models are the most used tools when analyzing complex systems from the performance, dependability or performability points of view. They can capture most of the important aspects of the systems, at the price however of a possible combinatorial explosion in the number of states of the models. In such a case, the many available numerical methods are often of no use, and one of the possible answers to the problem is to design methods that are able to provide bounds of the needed metrics, at moderate costs. This paper proposes such a method when the model is an absorbing Markov process. The method is designed to be accurate in the case of models of highly reliable systems, and can handle the case of models with rewards. The paper illustrates its properties with different numerical examples.

Key words. Large Reward Markov Models, Stochastic Complementation, Path-based Approach

AMS subject classifications. 60J22,60J50

1. Introduction. The scientific literature is rich in methods developed for analyzing, in many different ways, performance, dependability or performability measures using models. Looking for analytical methods for evaluating these measures has been an active area of research for a long time ([14], [13], [9], [17], [1]) and a number of tools have been built for the specification of system models and to make the solving methods more accessible ([10], [8], [23], [14]). Increased complexity of the target systems and the sophistication of the performability, reliability and availability measures that are of interest continue to challenge our modeling skills.

Among performability measures, transient ones relate to the behavior of the system over a finite period of time, e.g., the duration of a mission. An example is the mean number of packets in a router during a specific amount of time, e.g., the RTT (round trip time). This type of measure is more appropriate for the analysis of mission-oriented systems, e.g., a space mission or an aircraft flight. Steady-state measures are associated with limiting behavior and therefore are appropriate for systems with a lifetime that for modeling purposes is viewed as infinite in length.

Markov models, under different forms, are the most widely used mathematical tools in the area. The usefulness of Markov chains is due to the power of the theory and the set of efficient algorithms associated with. However, such a power has a price. There are two major drawbacks when using Markov models. The first one is the fact that, to be able to represent the more and more complex systems built nowadays, the state spaces are larger and larger, making in many cases that the numerical procedures become useless. The second one is the so-called "rare event" problem, meaning that in many cases, the interesting events have very low probabilities, making problematic the use of Monte Carlo techniques. This is usually related to high numerical values of the ratios between different transition rates of the chain, which leads also to numerical problems in the exact analysis of the models (*stiffness*). Often, the two problems appear simultaneously. As far as Monte Carlo techniques are considered, specific

---
*Federal University of Rio de Janeiro, COPPE/PESC/CS Department, Rio de Janeiro, Brazil (anapaula@land.ufrj.br). A.P.C. da Silva was supported by a fellowship from CAPES during her visit at IRISA.

[†]INRIA/IRISA,Campus de Beaulieu, Rennes, France (rubino@irisa.fr)

algorithms are available to deal with rare events (the size of the model is not a problem here). In this paper, we are concerned with numerical evaluation. In this case, a natural way to deal with a large state space is via state reduction techniques, the most common of which involve truncation of the state space, aggregation of states (see [2], [24], [25] and references therein), stochastic complementation theory ([16] and references therein) and path-based approaches ([3], [20], [19]). A major difficulty with all these methods is to provide a bound on the error that is introduced. On these lines, several works in the literature are concerned with providing bounds for a set of measures ([18], [15] and references therein, [7], [25]).

This paper deals with the evaluation of the expected cumulated reward for systems with a finite life-time. We describe a bounding technique that not only provides a state space reduction for numerical analysis but also error bounds. Bounding techniques have been mainly developed for cases in which, at the same time, the model is large and *stiff*. This second aspect will actually help us to deal with the first one. The intuition behind the approach is quite simple. When the model is *stiff*, the stochastic process spends most of its time in a (small) part of the state space. It is then a natural attempt to try to approximate the interesting measures by working basically with that subset of states. The approach presented here merges the idea of state space reduction using stochastic complementation with a path-based technique. We will use the so-called *slow-fast approximation method*, partially following ideas presented in [19], which in turn is based on a state space classification proposed in [2].

The paper is organized as follows: Section 2 presents some related works in the literature that study efficient methods for calculating some measures of interest. These methods use state reduction techniques and bounding procedures in order to calculate dependability or performability metrics. Sections 3 and 4 explain the method proposed, showing how the bounds are built. In Section 5, we give examples illustrating the procedure. Section 6 concludes the paper.

2. Related Works. The major problem when applying numerical solution methods to Markov chains is the state space explosion. For several realistic models having very large space state sets, it quickly becomes prohibitively expensive to compute a numerically exact solution, unless there is some special structure that can be exploited. We know that as computing power increases the size of the models that can be solved grows. However, state space sizes increase exponentially with the natural parameters characterizing the models, and thus there will always be problems of interest that are out of reach. We can often relax the requirement for an exact solution and be satisfied with bounds or approximations, or use Monte Carlo techniques, thus being able to quantitatively handle larger models than those that can be solved exactly.

Concerning approximations and bounds, there are several approaches that have appeared in the literature. An established and widely used technique for bounding the stationary distribution of Markov process has been published by Courtois and Semal ([5], [6]), 20 years ago. The method is defined for discrete time Markov processes, for which lower or upper bounds for the transition probabilities are known. The stationary distribution is shown to belong to a polyhedron of vectors defined by the normalized rows of the inverse of a matrix computed from the lower or upper bounds of the transition probabilities. An important problem with this approach is that the bounds often become looser when the dimension of the matrix increases. The method presented in [4] is an extension of [5] and avoids the dependence between the quality of the bounds and the matrix probability dimension, with some additional computational cost.

In [18] the authors showed how to compute bounds for the asymptotic expected reward, avoiding the evaluation of several sub-models in order to establish the conditional state probabilities as in the approach presented in [5]. The main idea is to partition the state space into classes $C_0, C_1, \ldots$ such that the higher the index, the less probable the set. Then, starting with some index k, each of the classes $C_i, i \geq k$ is lumped into a single state. The new state space is much smaller than the original one, and the effort in the paper is to define properly new processes on the reduced space which provide the bounds of the availability metric which was defined in the original model. One of the conditions necessary for this approach to work is that for any state in any class C_i, a repair must always be possible. This can be very restrictive in some cases. In [15] the authors propose an improvement of the technique of [18], allowing to relax the previous assumption. The method is also more accurate in the cases where both procedures can be used. This has however a price: in some cases, different and possibly large linear systems must still be solved. It must be added that the technique in [15] can, in some cases, deal with infinite state spaces as well.

Most of previous work deals with irreducible processes. In this paper, we are interested in absorbing ones, and specifically in the calculation of transient measures where we want to study the behavior until the system reaches a given subset of states. Among these measures, a basic one is the mean time the modeled system is *up*, that is, works as specified in its design. More generally, if a reward model is used to calculate the expected gain when using the system until the end of its life-time (or the expected cost, depending on the semantics of the rewards used), the corresponding metric is the mean expected reward up to absorption. The main objectives of this paper are (i) to be able to work on large state spaces, and (ii) to compute mathematical bounds of this expectation. The method uses a state space reduction, based on the so-called stochastic complementation theory (see [16] and references therein), and on the concept of *slow* and *fast* transitions and states, going back to [2]. In that paper, the authors present an aggregation technique for transient analysis of *stiff* Markov Chains. Concerning this classification, our work follows some ideas and results presented in [19], where a bounding scheme is developed for transient Markov chain analysis. The intuition behind this is that the system spends most of its time on the set of slow states, and that the cardinality of this set is usually much smaller than the cardinality of the whole state space.

It must be pointed out that for some models, especially for large ones, calculating the stochastic complement of a chain is not possible. In this paper we are specifically interested in this (general) case, and to deal with this problem, we follow a path-based approach that allows computing an approximated stochastic complement without generating all the state space. This type of approach was used in [3], [20], [19], for different purposes.

3. Preliminaries.

3.1. The Model. The system is represented by an homogeneous continuous time Markov chain $X = \{X(t), t \geq 0\}$ with finite state space $\{1, \cdots, M, M+1\}$ and infinitesimal generator (Q_{ij}). State $M+1$ is absorbing, and states $1, \cdots, M$ are transient: from any $i \leq M$ there is a path to state $M+1$, that is, a sequence of states $(i, i_1, i_2, \cdots, i_k, M+1)$ (reduced to $(i, M+1)$ if $k = 0$) with $Q_{ii_1} Q_{i_1 i_2} \cdots Q_{i_k M+1} > 0$. With these assumptions, the chain will be absorbed after a random finite time T (called the absorption time): $T = \inf\{t \geq 0 : X(t) = M+1\}$ and $\Pr(T < \infty) = 1$. We denote $q_i = \sum_{j \neq i} Q_{ij}$ the rate out of state i.

It will be useful to give a name to the set of transient states. We set $\Omega = \{1, 2, \cdots, M\}$.

Associated with each state $i \leq M$ there is a reward r_i, which is a real number interpreted as the rate at which the system generates a benefit while in state i (or at which the system cumulates a cost). Thus, the Cumulated Reward up to Absorption is the random variable

$$R_\infty = \int_0^\infty r_{X_t} \, dt.$$

When for all $i \leq M$ we set $r_i = 1$, we have $R_\infty = T$. When X represents the state of a system subject to failures and repairs, if states $1, \cdots, M$ represent the system working as specified and $M + 1$ represents a failed system, $\mathrm{E}(T)$ is also called the Mean Time To Failure of the system (MTTF).

The goal of the analysis developed here is the evaluation of the expectation of the random variable R_∞, called the Expected Cumulated Reward up to Absorption (ECRA):

$$\mathrm{ECRA} = \mathrm{E}(R_\infty).$$

3.2. Exact Analysis of the ECRA Metric. Let us denote by P the restriction to Ω of the transition probability matrix (t.p.m.) of the discrete Markov chain canonically embedded at the transition epochs of X (that is, if Q is the restriction of (Q_{ij}) to Ω and if Λ is the diagonal matrix $\Lambda = \mathrm{diag}(q_i, \ i \leq M)$, then $P = \Lambda^{-1}Q$). Matrix P is strictly sub-stochastic (that is, the sum of the elements of at least one of its rows is strictly less than 1), otherwise state $M + 1$ would not be reachable from Ω.

Let r be the column vector indexed on Ω whose ith entry is r_i/q_i. Let $\varrho_i = \mathrm{E}_i(R_\infty)$, where $\mathrm{E}_i()$ denotes the expectation operator conditioned to the fact that the initial state is state $i \in \Omega$, and denote by ϱ the vector indexed on Ω whose ith entry is ϱ_i. Since P is strictly sub-stochastic, matrix $I - P$ is not singular and $\varrho = (I - P)^{-1}r$, that is, ϱ is the solution to the linear system $(I - P)\varrho = r$, with size $|\Omega|$. If $\alpha_i = \Pr(X(0) = i)$ and if α is the vector indexed on Ω whose ith entry is α_i, then $\mathrm{ECRA} = \sum_{i=1}^M \alpha_i \varrho_i$.

So, computing the ECRA metric is a linear problem. This paper proposes a scheme allowing to compute lower and upper bounds of this expectation by working with linear systems much smaller than $|\Omega|$ in size. The method is designed for the case of $|\Omega| \gg 1$ and when the chain is *stiff* (this typically happens when X models a highly dependable system).

3.3. Fast and Slow Transitions and States. We say that the transition (i, j) is a *fast* one if $Q_{ij} \geq \theta > 0$ where θ is a given positive threshold. It is a *slow* one if $0 < Q_{ij} < \theta$. A state $i \neq M + 1$ is a *fast state* if at least one transition out of it is fast; otherwise, i is a *slow state* [2]. This defines a partition of the state space of X into three subsets: F, the set of fast states, S, the set of slow ones, and $\{M + 1\}$. We assume that $F \neq \emptyset$ and $S \neq \emptyset$. We also assume that the distribution of $X(0)$ has its support in S.

Let the transition rate matrix (Q_{ij}) be reordered so that the states belonging to S are numbered from 1 to $|S|$ and the states belonging to F from $|S|+1$ to $|S|+|F| = M$. Its sub-matrix Q can be then partitioned according to (S, F), as well as any other matrix indexed on Ω, and the same for vectors. For instance, we write

$$Q = \begin{pmatrix} Q_S & Q_{SF} \\ Q_{FS} & Q_F \end{pmatrix}, \qquad d = \begin{pmatrix} d_S \\ d_F \end{pmatrix}, \qquad \text{etc.}$$

Assumption: We assume that in the graph of chain X there is no circuit composed only of fast transitions. This is the usual situation in dependability models, where fast transitions are associated with repairs. In some models, however, the choice of the θ threshold can make this assumption invalid. The reader can see [2] for an approximation used to handle this situation.

3.4. Uniformization. Let us denote by Y the discrete time Markov chain associated with X, obtained by uniformization of the latter with respect to the uniformization rate η, $\eta \geq \max_{1 \leq i \leq M}\{q_i\}$. We denote by $U = (U_{ij})$ the restriction of the t.p.m. of Y to Ω (that is, $U = I + Q/\eta$). As for matrix Q, we similarly decompose U into the corresponding blocks using the analogous notation U_S, U_{SF}, U_{FS} and U_F.

The number ECRA defined on X can be computed by dividing the corresponding metric on Y by η. Let us write an exponent Y to explicitly indicate that we work on Y. Then,

$$R_\infty^Y = \sum_{k=0}^\infty r_{Y(k)},$$

$$\mathrm{E}(R_\infty^Y) = \alpha(I - U)^{-1}r$$

and

$$\mathrm{ECRA} = \frac{1}{\eta}\mathrm{E}(R_\infty^Y) = \frac{1}{\eta}\alpha(I - U)^{-1}r.$$

In the previously introduced vector notation, $\varrho = \varrho^Y/\eta$. All this means that we can work with chain Y to derive the ECRA metric defined on X.

3.5. Decomposing the Cumulated Reward. Let us write

$$R_\infty^{Y,S} = \sum_{k=0}^\infty r_{Y(k)}1_{Y(k)\in S}, \qquad R_\infty^{Y,F} = \sum_{k=0}^\infty r_{Y(k)}1_{Y(k)\in F}.$$

That is, $R_\infty^{Y,S}$ (resp. $R_\infty^{Y,F}$) is the reward cumulated on the slow states (resp. on the fast ones) by Y until absorption. Then obviously $R_\infty^Y = R_\infty^{Y,S} + R_\infty^{Y,F}$ and the same relation holds for the respective expectations.

If $\varrho_i^Y = \mathrm{E}(R_\infty^Y \mid Y(0) = i)$, $\varrho_i^{Y,S} = \mathrm{E}(R_\infty^{Y,S} \mid Y(0) = i)$ and $\varrho_i^{Y,F} = \mathrm{E}(R_\infty^{Y,F} \mid Y(0) = i)$, we also have $\varrho_i^Y = \varrho_i^{Y,S} + \varrho_i^{Y,F}$. As stated before, we assume that the initial state is a slow one (the natural situation). Correspondingly, we define the column vectors ϱ^Y, $\varrho^{Y,S}$ and $\varrho^{Y,F}$ indexed on S, with entry i respectively equal to ϱ_i^Y, $\varrho_i^{Y,S}$ and $\varrho_i^{Y,F}$. Again, $\varrho^Y = \varrho^{Y,S} + \varrho^{Y,F}$.

3.6. Stochastic Complement. Let us denote by $\widetilde{Y}$ the chain obtained by stochastically complementing [16] the subset of states $S \cup \{M+1\}$ in Y. This means considering the chain obtained by copying Y while it lives inside $S \cup \{M+1\}$ and by freezing it in state $i \in S$ if after visiting i the chain goes to some state in F; in this case, if Y enters again $S \cup \{M+1\}$ by state k, $\widetilde{Y}$ jumps from i to k. Process $\widetilde{Y}$ is also called the *reduction* of Y with respect to $S \cup \{M+1\}$. Its t.p.m., indexed on $S \cup \{M+1\}$, is

$$\begin{pmatrix} \widetilde{U} & \widetilde{u} \\ 0 & 1 \end{pmatrix}$$

where

$$(3.1) \qquad \widetilde{U} = U_S + U_{SF}(I - U_F)^{-1}U_{FS}$$

and $\widetilde{u} = 1 - \widetilde{U}1$, 1 denoting the column vector, indexed on S, with all its entries equal to 1.

As for Y, we define the cumulated reward until absorption $R_\infty^{\widetilde{Y}}$ and the vector $\widetilde{\varrho}$ indexed on S whose ith entry is the conditional expectation $\widetilde{\varrho}_i = \mathrm{E}(R_\infty^{\widetilde{Y}} \mid \widetilde{Y}(0) = i)$. A key observation now is that, by definition of stochastic complement,

$$(3.2) \qquad \widetilde{\varrho} = \varrho^{Y,S}.$$

4. Bounding Procedure. In this Section we describe our bounding procedure. We first derive a lower bound of $\varrho^Y = \varrho^{Y,S} + \varrho^{Y,F}$, by computing a lower bound of $\widetilde{\varrho} = \varrho^{Y,S}$. Then, an upper bound of ϱ^Y is computed in two parts: first, an upper bound of $\widetilde{\varrho} = \varrho^{Y,S}$ is obtained; then, an upper bound of $\varrho^{Y,F}$ ends the bounding scheme.

In the sequel, all the inequalities between vectors or between matrices mean that the inequality holds element per element: for instance, if $a = (a_i)$ and $b = (b_i)$ are vectors with the same dimension, then $a \leq b \iff a_i \leq b_i$ for all index value i.

4.1. Lower Bound of ECRA. We provide here a lower bound of vector $\widetilde{\varrho}$. We had that

$$(4.1) \qquad \widetilde{\varrho} = (I - \widetilde{U})^{-1}r.$$

In general, matrix $\widetilde{U}$ is unknown. Note however that in some specific models, it can be exactly computed with a moderate effort, in which case we can evaluate $\widetilde{\varrho}$ exactly and then use it as our lower bound of ϱ^Y (see the numerical examples). Assume here that we have no access to $\widetilde{U}$. In this case, we can use the following approach. Fix some integer $N \geq 2$. For all states $i, j \in S$, we build the set $\mathcal{P}_N$ of all the paths of Y having the form $\pi = (i, f_1, f_2, \cdots, f_{N-1}, j)$, where $f_1, f_2, \cdots, f_{N-1} \in F$. For such a path its probability is

$$p(\pi) = U_{i f_1} U_{f_1 f_2} \ldots U_{f_{N-1} j}.$$

Then,

$$\sum_{\pi \in \mathcal{P}_N} p(\pi) = (U_{SF} U_F^{N-2} U_{FS})_{ij}.$$

Define matrix U' by

$$(4.2) \qquad U' = U_S + U_{SF}\left(I + U_F + U_F^2 + \cdots + U_F^{N-2}\right)U_{FS}.$$

Clearly, $U' \leq \widetilde{U}$, and if ϱ' is the solution to the system

$$(4.3) \qquad (I - U')\varrho' = r,$$

then

$$\varrho' \leq \widetilde{\varrho} = \varrho^{Y,S} \leq \varrho^Y.$$

For the first inequality, just observe that $\widetilde{\varrho} - \varrho' = U_{SF}\left(\sum_{k \geq N-1} U^k\right)U_{FS}\,r \geq 0$.
Vector ϱ', solution to linear system (4.3) with size $|\widetilde{\Omega}|$, is our lower bound of ϱ^Y.
To build the matrix of the system, we need to explore the set of fast states up to a distance $N-2$ from S. As it can be seen in the examples, we get accurate bounds in all cases with small values of N.

4.2. Upper Bound. For upper bounding ϱ^Y, we first find an upper bound of $\varrho^{Y,S}$, then an upper bound of $\varrho^{Y,F}$.

4.2.1. Upper Bound of $\varrho^{\mathbf{Y},\mathbf{S}}$. Define $U'' = \widetilde{U} - U' \geq 0$. We have

$$U'' = U_{SF}\left(U_F^{N-1} + U_F^N + \cdots\right)U_{FS}$$
$$= U_{SF}U_F^{N-1}(I - U_F)^{-1}U_{FS}.$$

Define

$$V = (I - U_F)^{-1}.$$

The number V_{ij} is the mean number of visits of Y to $j \in F$ starting from $i \in F$, before leaving F. The main idea now is to build an upper bound of matrix U''.

For this purpose, we use the following result, adapted from [19]:

LEMMA 4.1. *Since there is no circuit in the graph of X composed of fast transitions, we can partition the states of F into D classes $C_1, \cdots, C_D$, where C_1 is the set of states directly connected to S by a fast transition, C_2 is the set of states directly connected to C_1 by a fast transition, etc. Integer D is the maximal number of fast transitions from any state of F to set S. Then, if we define*

$$\lambda = \max_{f \in F} \sum_{g \in F,\ (f,g)\ \text{slow}} Q_{fg}$$

and

$$\mu = \min_{f \in F} \sum_{g \in F,\ (f,g)\ \text{fast}} Q_{fg},$$

and if we assume that $\lambda < \mu$ (this is the natural situation), any sojourn of Y on F has a duration less, in average, than σ, where

(4.4)
$$\sigma = \frac{\eta}{\mu}\frac{1 - (1 - \psi)^D}{\psi(1 - \psi)^D},$$

and $\psi = \lambda/\mu < 1$.

Proof:
    We assume here that $Y(0) \in F$. Let us transform $S \cup \{M + 1\}$ into a single absorbing state 0. Let us denote by $I(k)$ the index of the class where $Y(k)$ is, $I(k) = d \iff Y(k) \in C_d$, with the convention $I(k) = 0$ if $Y(k) = 0$.
    We will use an auxiliary discrete time Markov chain Z defined on the set of indexes $\{0, 1, 2, \cdots, D\}$. State 0 is absorbing and the remaining states are transient. From any $d > 0$, transitions are possible to $d - 1$ with probability $1 - \psi$ and to D with probability ψ.

The proof consists of first showing that Y visits less classes than Z visits states. This comes from the fact that for any $k \geq 0$, the random variable $I(k)$ is stochastically smaller than $Z(k)$, which is obtained by recurrence on k. The claim is obviously true for $k = 0$ since, by construction, $I(0) = Z(0)$. The recurrence assumption is that $I(k) \leq_{\mathrm{st}} Z(k)$. We must prove that $\Pr(I(k + 1) \geq l) \leq \Pr(Z(k + 1) \geq l)$ for all possible l. Write

$$\Pr(I(k + 1) \geq l) = \sum_{d=0}^{D} p(l, d) \Pr(I(k) = d),$$

where $p(l, d) = \Pr(I(k + 1) \geq l \,|\, I(k) = d)$, and similarly,

$$\Pr(Z(k + 1) \geq l) = \sum_{d=0}^{D} q(l, d) \Pr(Z(k) = d),$$

where $q(l, d) = \Pr(Z(k + 1) \geq l \,|\, Z(k) = d)$. We have $q(0, d) = q(1, d) = \cdots = q(d - 1, d) = 1$, $q(d, d) = q(d + 1, d) = \cdots = q(D, d) = \psi$.

We see immediately that $p(l, d) \leq q(l, d) = 1$ for $l - 0, 1, \cdots, d - 1$. For $l \geq d$,

$$p(l, d) = \sum_{f \in C_d} \Pr(I(k) \geq l \,|\, Y(k) = f) \Pr(Y(k) = f \,|\, Y(k) \in C_d).$$

The conditional probability $\Pr(I(k) \geq l \,|\, Y(k) = f)$ can be written

$$\Pr(I(k) \geq l \,|\, Y(k) = f) = \frac{u_f}{u_f + v_f} = \frac{1}{1 + \dfrac{v_f}{u_f}},$$

where

$$u_f = \frac{1}{\eta} \sum_{g \in C_l \cup C_{l+1} \cup \cdots} Q_{fg}, \qquad v_f = \frac{1}{\eta} \sum_{g \in C_{d-1} \cup C_d \cup \cdots \cup C_{l-1}} Q_{fg}.$$

Observe that u_f is a "small" probability and v_f is a "large" one, since the former comes from transition rates of slow transitions and the latter includes rates of fast ones. Then, by definition of λ and μ,

$$\frac{1}{1 + \dfrac{v_f}{u_f}} \leq \frac{1}{1 + \dfrac{\mu/\eta}{\lambda/\eta}} = \frac{\psi}{1 + \psi}.$$

So, for $l \geq d$,

$$p(l, d) \leq \frac{\psi}{1 + \psi} \leq \psi = q(l, d).$$

Resuming, we have in all cases $p(l, d) \leq q(l, d)$.

Now, look at $q(l, d)$ as a function of d: $p(l, d) = g(d)$. Observe that $g()$ is increasing. Then,

$$\Pr(I(k+1) \geq l) \leq \sum_{d_0}^{D} g(d) \Pr(I(k) = d) = \mathrm{E}(g(I(k))) \leq \mathrm{E}(g(Z(k))) = \Pr(Z(k+1) \geq l).$$

The last inequality in the derivation comes from the recurrence assumption and the fact that $g()$ is an increasing function. We have thus proved that $I(k) \leq_{\mathrm{st}} Z(k)$ for all $k \geq 0$. This means that the number of states visited by Z before absorption is stochastically larger than the number of classes visited by Y.

It remains to look at the time Y spends in each of the classes. We prove here that, on the average, this time is upperbounded by the ratio μ/η. Denote by π_f the probability of starting a sojourn in C_d by state $f \in C_d$, and by π the vector whose fth entry is π_f, indexed on C_d. Let U_d be the restriction to C_d of matrix U, also indexed on that class of states. If S is the length of the sojourn time of Y in C_d when the first visited state is chosen according to distribution π, then we have $\Pr(S > k) = \pi^{\mathrm{T}} U_d^k 1$, where π^{T} denotes the transposed of the column vector π [21]. The expected length of S is

$$\mathrm{E}(S) = \sum_{k \geq 0} \pi^{\mathrm{T}} U_d^k 1 \quad \left(= \pi^{\mathrm{T}} (I - U_d)^{-1} 1\right).$$

Write $U_d 1 = u$. Entry f of u is $u_f = 1 - \sum_{g \notin C_d} (U_d)_{fg}$. For any f in any of the D classes,

$$u_f \leq 1 - q, \qquad q = \frac{\mu}{\eta} \quad (\leq 1)$$

which follows by definition of μ (and η). So, we have $u \leq (1 - q)1$. In the same way, $U_d^2 1 = U_d u \leq (1 - q) U_d 1 \leq (1 - q)^2 1$. By recurrence it is then immediate to check that $U_d^k 1 \leq (1 - q)^k 1$, $k \geq 0$. This leads to

$$\mathrm{E}(S) = \sum_{k \geq 0} \pi^{\mathrm{T}} U_d^k 1 \leq \sum_{k \geq 0} \pi^{\mathrm{T}} (1 - q)^k 1 = \sum_{k \geq 0} (1 - q)^k = \frac{1}{q} = \frac{\eta}{\mu}.$$

The final step is to prove that the expected absorption time of Z, when the initial state is state D, is given by

$$\frac{1 - (1 - \psi)^D}{\psi(1 - \psi)^D},$$

which can easily done by recurrence on D. This concludes the proof. _____ •

The preceding Lemma means that for any $i, j \in F$, $V_{ij} \leq \sigma$, and thus,

(4.5) $$U'' \leq \sigma U_{SF} U_F^{N-1} U_{FS} = B.$$

Then, if $\sum_{n=0}^{\infty} (U' + B)^n < \infty$,

$$\varrho^{Y,S} = \widetilde{\varrho}$$
$$= \sum_{n=0}^{\infty} \widetilde{U}^n r$$
$$= \sum_{n=0}^{\infty} (U' + U'')^n r$$
$$\leq \sum_{n=0}^{\infty} (U' + B)^n r$$
$$= \varphi.$$

Now, since matrix U_F is strictly sub-stochastic, we have $\lim_{N \to \infty} B = 0$. This means that, for N large enough, matrix $U' + B$ is strictly sub-stochastic, and thus, that vector φ is well defined. Observe that φ is obtained as the solution to the linear system

$$(4.6) \qquad\qquad (I - U' - B)\varphi = r.$$

4.2.2. Upper Bound of $\varrho^{Y,F}$. For $i, j \in S$, denote by W_{ij} the mean number of visits that Y makes to state j, starting at i, and by ν_i the mean number of visits that Y makes to the whole subset F, again conditioned to $Y(0) = i$. Observe that

$$\nu_i = \sum_{j \in S} W_{ij} U_{jF}$$

where $U_{jF} = \sum_{f \in F} U_{jf}$. Also observe that this number ν_i is computable on the reduced process: if $W = (W_{ij})$, we have

$$W = (I - \widetilde{U})^{-1}.$$

The ith entry of vector $\varrho^{Y,F}$ can be upper-bounded by $\sigma \nu_i r^*$ where

$$r^* = \max\{r_f, \ f \in F\}.$$

If we define vectors u_F and ν, indexed on S, by $(u_F)_i = U_{iF}$ and $\nu = (\nu_i)$, then

$$(4.7) \qquad\qquad \nu = W u_F.$$

Now,

$$\nu = W u_F = \sum_{n=0}^{\infty} \widetilde{U}^n u_F \leq \sum_{n=0}^{\infty} (U' + B)^n u_F = \tau.$$

Observe that τ is the solution to the linear system

$$(4.8) \qquad\qquad (I - U' - B)\tau = u_F.$$

So, the upper bound of $\varrho^{Y,F}$ is

$$\varrho^{Y,F} \leq \sigma r^* \tau.$$

Finally, the upper bound of ϱ is:

$$\varrho \leq \sigma r^* \tau + \varphi$$

4.3. The Case of Stochastic Complement Available. In some cases, it is possible to compute the exact stochastic complement (without any matrix inversion). This happens when the system has some special structure, as a single return state in S, a single exit state from F, or in the cases where all non-zeros rows in U_{FS} are proportional to each other. For instance, consider the special case where there is a single return state in S, that is, a single state $s \in S$ such that for some $f \in F$, $U_{fs} > 0$, and that for all $g \in F$ and $z \in S, z \neq s$, $U_{gz} = 0$. Note that this simply means that in matrix U_{FS} all columns are composed of 0 except one. In these cases, matrix $(I - U_F)^{-1} U_{FS}$ can be evaluated easily, with small computational complexity. For more details, see [16] and also [22].

Assume then that $\widetilde{U}$ is known. In this case, the lower bound of ϱ is simply given by $\widetilde{\varrho}$, after solving the linear system

$$(4.9) \qquad\qquad (I - \widetilde{U})\widetilde{\varrho} = r.$$

The obtained bound is clearly better than the lower bound given by ϱ'. The upper bound is obtained by summing $\sigma r^* \nu$ and $\widetilde{\varrho}$. The first term is computed after solving the linear system

$$(4.10) \qquad\qquad (I - \widetilde{U})\nu = u_F.$$

Again, this bound is better than in the general case.

In the example described in 5.2 this situation is illustrated.

4.4. Summary. Let us summarize the proposed technique in algorithmic form: one for the general case, where $\widetilde{U}$ is unknown and the second one for the case where the exact computation of the stochastic complement is possible.

ECRA Bounding Technique – General Case ()
1   Choose θ
2   Choose η
3   Choose N
4   Build U' (equation (4.2))
5   Compute ϱ' (linear system (4.3))
6   Compute σ (equation (4.4))
7   Compute B (equation (4.5))
8   Compute φ (linear system (4.6))
9   Compute τ (linear system (4.8))
10 Bounds: $\varrho' \le \varrho \le \sigma r^* \tau + \varphi$

Remark: after step 7 we can check if matrix $U' + B$ is strictly sub-stochastic. If it is not, a larger value of N must be tried, repeating step 7. In our examples we always used small values of N and we never found this situation.

ECRA Bounding Technique – Exact Stochastic Complement Case ()
1   Choose θ
2   Choose η
3   Compute $\widetilde{U}$ (equation (3.1))
4   Compute $\widetilde{y}$ (linear system (4.1))
5   Compute σ (equation (4.4))
6   Compute ν (linear system (4.10))
7   Bounds: $\widetilde{\varrho} \le \varrho \le \sigma r^* \nu$

4.5. An immediate extension. In many dependability models of multicomponent systems, there is only one slow state, the initial one, where all the components are up. More generally, suppose we have an example where $|S|$ is small. In these cases, we can develop the same bounding numerical scheme, by replacing S by some $S' = S \cup F'$ where F' is a subset of F, and correspondingly F by $F - F'$. This puts more information into the models that will be effectively solved, and leads to a better precision. The example of Subsection 5.3 is used to illustrate this case.

4.6. Computational Complexity. Concerning the cost of the procedure, observe than in the general case, the cost is dominated by the resolution of 3 linear systems having the size of S, usually much less than the size of the whole state space. Moreover, some supplementary gain can be obtained using the fact that two of the linear systems to solve correspond to the same matrix. The rest of the cost is concentrated on the computation of matrices U' and B (the latter is "almost" a sub-product of the former). In most of the examples we have considered here, $N \ll |S|$, which explains that the linear systems' resolution dominates the global cost. Since N will be always small or moderate in applications, the case of $|S| \le N$ is not a very important one, since in that case, S will be small and the global procedure will run very quickly (or we will use a superset S' of S, and $|S'| \gg N$).

When the stochastic complement is available, the bounds are better, as explained before, and also less expensive to find since we have now only two linear systems to solve, both with the same matrix, and no path-based part.

5. Examples. This section illustrates the behavior of our method through a few examples. In the first one we present the evaluation of a multicomponent system subject to failures and repairs. The second example studies the behavior of a system with a set of identical components. The third one presents a grid system based on [12], used for studying the case where the original slow set has cardinality equal to 1 and a superset of S is used in order to calculate a more efficient approximated solution, but keeping a reduced total computational cost. The last example presents a database model, where we are interested in studying the economical loss due to failures. An important feature of the illustrations is that they are built in such a way that the exact solution will be always available (even when the state spaces are huge).

In the examples, the results are presented in two types of tables: the first one gives the set of parameters characterizing the example and the numerical results: the exact solution (ECRA), the lower and upper bounds (L.B. and U.B. respectively) and the relative error (R.E.) defined by $|(\text{L.B.} + \text{U.B.})/2 - \text{ECRA}|/\text{ECRA}$. The second table type gives information related to the main computational gains, which includes the reduction of the state space size and the parameters related to the path-based approach. For the three first examples, the ECRA measure is a basic reliability measure: the Mean Time to Failure (MTTF), that is, the reward of a state is 1 if the system is operational when in that state, 0 otherwise. The last example presents a computation with more general rewards.

In the tables, we use the following type of notation: $2.0e6$ means 2.0×10^6, and $3.0e{-}4$ means 3.0×10^{-4}.

5.1. A multicomponent system with deferred repairs. Consider a system composed of C identical components subject to failures and repairs. The repair facility starts working only when there are H failed components and it has only one server. Once started, the repair facility remains active until there is no failed unit in the system (hysteresis phenomenon). The system starts with all its C components up. The time to failure of any component is exponentially distributed with parameter ϕ. The system is considered operational if there are at least K components up. We are interested in the MTTF measure, that is, the expected operational time until the system has only $K - 1$ components up. Observe that for this description to make sense, we need $C \ge H + K$. Finally, components' life-times and repair times are independent of each other.

All states that have a repair exit rate are fast states here. This means that the total number of slow states is equal to H.

TABLE 5.1

Numerical results for the system with $C = 1000$ components. We use $\gamma = 1$ for all cases. L.B.: Lower Bound, U.B.: Upper Bound, R.E.: Relative Error.

H	K	ϕ	L.B.	ECRA	U.B.	R.E.
3	1	$1e{-}4$	$2.5378e16$	$2.5466e16$	$2.6268e16$	$1.4e{-}2$
100	10	$1e{-}5$	$2.7186e16$	$2.7187e16$	$2.7189e16$	$1.8e{-}5$
100	10	$1e{-}4$	$2.5381e16$	$2.5471e16$	$2.5484e16$	$1.5e{-}3$

TABLE 5.2

State space reduction and N parameter for the system with $C = 1000$ components.

orig. state size	new state size	reduction rate (ratio)	N
1003	3	334	15
1100	100	11	106
1100	100	11	106

In the first set of results the repair time is exponentially distributed with parameter γ. For a total number of components $C = 1000$ (think of a network) we performed some experiments varying the values of the failure rate as well as the parameters H and K.

The space state can be written initially composed of states $i = C, C - 1, \ldots, C - H, \ldots, K, \ldots, 1, 0$ plus the "doubles" $(C - 1)', (C - 2)', \ldots, (C - H + 1)'$ necessary because of the hysteresis phenomenon. We can then collapse states $K - 1, \ldots, 1, 0$ into a single absorbing state, say $0'$. This last state space has $C + H - K + 1$ states, one absorbing, H slow and $C - K$ fast. In this particular model, there is only one exit state from S and there is only one entering state into F. Obviously, this simple topology allows for a numerical analysis of the example, for virtually any values of its parameters.

Table 5.1 shows the main parameters and numerical results. Table 5.2 shows the computational gains and the path-based approach parameters (following the order of the results presented in Table 5.1). For the cases analyzed, the bounding method is efficient even when a small number of states is effectively generated.

Now, let us consider a system with the same characteristics and a total number of components $C = 5000$. The parameter varied was H. Table 5.3 shows the numerical results and Table 5.4 presents the additional information about the sizes of the corresponding state spaces.

The parameter N used in all cases, in this example, is at least equal to H. The choice of this number is based on the particular model structure and the idea of generating at least one path into the set of fast states. This special topology makes that the fact of generating at least one path into F does not significantly increase the computational cost and better bounds can be obtained. It is worth nothing that the idea of generating at least one path into F is not a necessary assumption for the method to work, given that all bounds are computed based on having a lower bound for the exact stochastic complement. This is a choice that can be made during the dependability analysis and is based on the tradeoff between better bounds and computational cost.

Let us introduce a special way of increasing the original state space used for the

TABLE 5.3

Numerical results for the system with $C = 5000$ components. In these cases, we use $\gamma = 1$, $K = 1$ and $\phi = 1e{-}4$. We use $\gamma = 1$ for all cases. L.B.: Lower Bound, U.B.: Upper Bound, R.E.: Relative Error.

H	L.B.	ECRA	U.B.	R.E.
20	$2.5032e16$	$2.5471e16$	$2.7480e16$	$3.0e{-}2$
50	$2.5447e16$	$2.5472e16$	$2.5777e16$	$5.5e{-}3$
100	$2.5458e16$	$2.5472e16$	$2.5629e16$	$2.8e{-}3$
500	$2.5441e16$	$2.5472e16$	$2.5478e16$	$4.9e{-}4$

TABLE 5.4

State space reduction and N parameter for the system with $C = 5000$ components.

orig. state size	new state size	reduction rate (ratio)	N
5020	20	251	20
5050	50	101	60
5100	100	51	108
5500	500	11	516

bounding method and at the same time keeping a very small one in order to obtain the exact MTTF solution. For the exact solution, the repair rate remains exponentially distributed with parameter γ. However, for applying the bounding method let us consider that after a failure, the components enter in a repair facility with one server and that the repair time is distributed according to a Coxian distribution with mean γ. The parameters of the Coxian distribution are then chosen in such a way that the distribution is equivalent to an exponential one. If v_s is the parameter of the sth stage and if l_s is the probability that phase s is the last one, then, for all phase s, we have $v_s l_s = 1/\gamma$. Technically, we put ourselves in a *strong lumpability* situation, which allows us to obtain the exact solution from the previous case, and to test our bounding technique in the case of a large state space.

Consider the system with $C = 10000$ components, for which we know the exact MTTF value. Using the Coxian distribution version, the bounding method is performed for the cases where the number of phases is $s = 10$ and $s = 100$. The state space cardinality increases 10 and 100 times, respectively, with respect to the exponential repairs case. Table 5.5 shows the numerical results, as well as the model parameters. Table 5.6 shows the second set of the numbers related to the bounding method.

The numbers on Table 5.5 deserve a special comment related to the upper bound results. For the case where $s = 10$, the bounds are tighter than for the case where $s = 100$. This can be explained by the way the time spent on the fast set is bounded. One of the parameters is D, the largest distance between any state in the fast set to the slow set. In this very special case, D is the cardinality of the fast set. The other point is that ϕ is not much smaller than γ. However, even for this extreme case, the approximation is good enough with a high reduction rate and with a small number of states generated.

5.2. Several identical components system. This example is based on [19], where a system is composed of C identical and independent components. The evolu-

TABLE 5.5

Numerical results for the system with $C = 10000$ components, using a Coxian service distribution. The values for the parameters are: $\gamma = 1$, $K = 1$ and $\phi = 1e-5$. L.B.: Lower Bound, U.B.: Upper Bound, R.E.: Relative Error.

H	s	L.B.	ECRA	U.B.	R.E.
10	10	$2.7170e16$	$2.7187e16$	$2.7227e16$	$4.2e{-}4$
10	100	$2.7186e16$	$2.7187e16$	$3.3415e16$	$1.5e{-}1$

TABLE 5.6

State space reduction and N parameter for the system with $C = 10000$ components. In the first row, the Coxian distribution has $s = 10$ phases; in the second, $s = 100$.

orig. state size	new state size	reduction rate (ratio)	N
100100	10	10010	20
1001000	10	100100	30

tion of any of them is modeled by the Markov chain shown in Fig. 5.1. The failure rate is ϕ, the *coverage factor* is c and the repair rate is γ. We will calculate the MTTF measure.

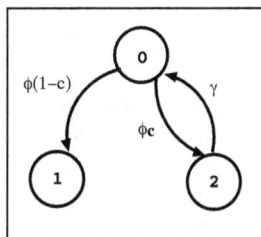

FIG. 5.1. *Evolution of any of the components of the system. There are C components. Failure rate: ϕ; coverage factor: c. Repair rate: γ. State 0 means that the component is up; in state 1, it is down and not-repairable; in 2 it is down and repairable.*

For a first group of systems, with a total number of components C into the set $\{10, 30, 40, 50\}$, we will study how the method performs when the exact stochastic complement is used (this is possible in this case). All the systems were modeled exploring the symmetries in the Markov chain which results in a state space reduction (due to the *strong lumpability* property). A state is classified as fast if it has at least one exit transition rate greater or equal to γ ($\theta = \gamma$). Table 5.7 shows the numerical results. Table 5.8 shows the reduction rate reached by applying the method.

Let us now model all the systems *without* exploiting the strong lumpability property. The idea is to generate a set of models with much larger state space and to apply the bounding method on it. This allows us to work on large models and still have access to the exact value of the MTTF. Table 5.9 shows the numerical results and Table 5.10 shows the reduction rate and the used value of N. We can point out that tight bounds can be provided even when large reduction rates occur.

As a last study case, consider that the system has $C = 11$ components. Here, we apply the bounding technique for two different values of N: $N = 3$ and $N = 6$. Table 5.11 shows the numerical results. As expected, increasing the total number of

TABLE 5.7

Numerical results for the system with several identical components, using the symmetries in the Markov Chain (strong lumpability). The values for the parameters are $\phi c 10^{-6}$, $\phi(1-c) = 10^{-7}$ and $\gamma = 1$ in all cases. L.B.: Lower Bound, U.B.: Upper Bound, R.E.: Relative Error.

C	L.B.	ECRA	U.B.	R.E.
10	$2.9285e7$	$2.9285e7$	$2.9285e7$	0
30	$3.9932e7$	$3.9933e7$	$3.9933e7$	$1.2e{-}5$
40	$4.2765e7$	$4.2765e7$	$4.2767e7$	$2.3e{-}5$
50	$4.4959e7$	$4.4959e7$	$4.4961e7$	$2.2e{-}5$

TABLE 5.8

State space reduction for the system with several identical components, using the symmetries in the Markov Chain (strong lumpability).

orig. state size	new state size	reduction rate (ratio)
66	10	6
496	30	16
861	40	21
1326	50	26

possible explored paths yields a smaller relative error measure. However, even for a case where the maximum size path is $N = 3$ a good approximation is reached.

5.3. Grid. Consider the system presented in [12] used for studying the sensitivity of system reliability/availability with respect to the coverage parameter c and look at it as a grid composed by C processors. We will assume that only one node up is enough for consider the system as operational. Each processor fails at rate $\phi = 10^{-10}$ and is repaired at rate $\gamma = 1$, and we consider a large number of connected elements in the grid: $C = 5000$. Any failure is repairable with probability c. That is, for any state $i = 1, \cdots, C$ there is a transition to the absorbing state, with rate $i\phi(1-c)$. The state transition diagram is show in Fig. 5.2.

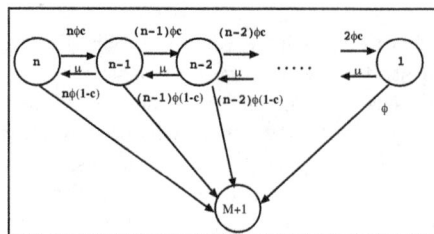

FIG. 5.2. *Grid State Diagram.*

Applying the slow state set definition introduced in section 3.3, this system has only one slow state (the initial one, where all processors are working). The numerical results are show in Table 5.12. Even for the case where we have just one slow state, the relative error is small and the computational savings are very significant.

Let us now illustrate the idea of redefining the slow states set, considering some states that belong to the fast states as slow ones. Or, saying this in a different way,

TABLE 5.9

Numerical results for the system with several identical components, without using the symmetries in the Markov Chain. The exact values of the MTTF are obtained exploiting the strong lumpability property. The values for the parameters are $\phi c = 10^{-6}$, $\phi(1 - c) = 10^{-7}$ and $\gamma = 1$, for all cases. L.B.: Lower Bound, U.B.: Upper Bound, R.E.: Relative Error.

C	L.B.	ECRA	U.B.	R.E.
11	$3.0116e7$	$3.0177e7$	$3.0206e7$	$5.3e$–4
12	$3.0925e7$	$3.1026e7$	$3.1031e7$	$1.5e$–3
18	$3.4943e7$	$3.4951e7$	$3.4997e7$	$5.4e$–4
20	$3.5977e7$	$3.5977e7$	$3.5990e7$	$1.8e$–4

TABLE 5.10

State space reduction and N parameter for the system with several components without using the strong lumpability property.

C	orig. state size	new state size	reduction rate (ratio)	N
11	177147	2048	86	6
12	531441	4096	130	6
18	382742098	262144	1460	5
20	3486784401	1048576	3325	5

let us perform the stochastic complement (and all the analysis) using S' instead of S, where $S \subset S'$. We do this by "adding 1000 states to S", that is, by using a set S' where $|S'| = 1001$, in order to reach a better approximation. Table 5.13 shows the new numerical results. In this case, increasing the total number of states in the slow set results not only in a better lower bound, as expected, but also in a better upper bound. This happens due to the way $\rho^{Y,F}$ is upper bounded, given that we consider the worst case for each visit in F.

5.4. Database System. In this last example, we will study a database system composed by CPUs, hard disks and memory units. Each component in the system has a similar behavior as the component presented in Section 5.2: failures can be repaired with rate ϕc and are not repairable with rate $\phi(1 - c)$. The repair rate is γ.

First, suppose that we are interested in the CPU and hard disk units performance. Let us study a system with 3 CPUs and 5 hard disks, and let us calculate the MTTF measure. In order to increase the original state space and at the same time to be able to compute the exact solution, we will apply the same approach used in Section 5.1. Each component has a repair time distributed according to a *s-stage* Coxian distribution. Table 5.14 shows the numerical results for this first database system. Table 5.15 shows the reduction rate reached by applying the method. For the exact solution, we model the same system using the *strong lumpability* property.

Now, let us also consider the memory performance. In this last study case, our system has 4 CPU, 5 disk and 3 memory units. Suppose that for each component failure without repair, we have an economical loss $L = \$500$ for the CPU component, $L = \$5000$ for the disk component and $L = \$1000$ for the memory component and for each failure with repair, we have an economical loss $L = \$100$ for the three types of components. We will calculate the total amount of money lost until the system breakdown (state where all components are failed). This scenario can represent a

TABLE 5.11
Numerical results for the system with several identical components, without using the symmetries in the Markov Chain. There are $C = 11$ components here. The exact values of the MTTF are obtained exploiting the strong lumpability property. The values for the parameters are $\phi c = 1e^{-4}$, $\phi(1-c) = 1e^{-5}$ and $\gamma = 1$, for all cases. We vary N in this illustration. L.B.: Lower Bound, U.B.: Upper Bound, R.E.: Relative Error.

N	L.B.	ECRA	U.B.	R.E.
3	3.0199e5	3.0202e5	3.0566e5	6.0e–3
6	3.0191e5	3.0202e5	3.0558e5	5.7e–3

TABLE 5.12
Numerical results for the multiprocessor system with $C = 5000$ processors. L.B.: Lower Bound, U.B.: Upper Bound, R.E.: Relative Error.

L.B.	ECRA	U.B.	R.E.
1.9984e11	2.0000e11	2.0034e11	4.5e–4

commercial Web site, where the lost of performance affects the total amount of capital cumulated. Table 5.16 shows the numerical result for this database system. The values are negative due to the fact that the system is loosing capital.

6. Conclusions. We have developed a method for determining bounds for the ECRA measure. The results are based on a classification of the states and transitions as slow and fast and on the reduction of the state space using the stochastic complementation, performed by means of a path-based approach. We showed that by reducing the original stochastic process that represents the system under study with respect to the state classification presented and by using the path-based approach, tight bounds can be obtained with a much lower computational cost than the one necessary to solve the original problem.

One important relevant property for better bounds and computational savings is that the system spends most of the time in a few (slow) states. It is reasonable to expect that this same property will hold in other cases as those considered here (for instance, when analyzing the mean number of users before reaching a total capacity of N clients or processes in some Web service systems).

The ongoing research work for improving our bounding method follows the following lines:

1. A better upper bound for the total time spent on the fast states set, by using different λ and μ rates for each specific class of states inside the set F. This improvement looks interesting, for instance in cases where the distance D is large (as in the case studied in Section 5.1).

2. In this paper, we used a Breath First Search for going from S into F up to depth N. However, for the cases where we have a large value D this algorithm is not very efficient (see Example 5.1, where we had to use a parameter $N = H$ in order to have at least one path explored inside F). Our technique also works if a different strategy for exploring F is used. We are looking at this possibility now. Ideas about different criteria used for exploring the fast states can be found in papers like [11] and references therein.

TABLE 5.13

Numerical results for the grid with $C = 5000$ processors, using a superset of S in the bounds computation.

L.B.	ECRA	U.B.	R.E.
$1.9999e11$	$2.0000e11$	$2.0000e11$	$2.5e{-}5$

TABLE 5.14

Numerical results for the Database system. We are interested in the CPU and hard disk units' behavior. The exact values of the MTTF are obtained exploiting the strong lumpability property. The values for the parameters are: $CPU = \phi c = 1e^{-5}$, $\phi(1 - c) = 1e^{-6}$ and $\gamma = 1$; $Disk = \phi c = 1e^{-3}$, $\phi(1 - c) = 1e^{-4}$ and $\gamma = 5$. L.B.: Lower Bound, U.B.: Upper Bound, R.E.: Relative Error.

s	L.B.	ECRA	U.B.	R.E.
3	$1.8139e6$	$1.8333e6$	$1.8354e6$	$4.7e{-}3$
10	$1.8150e6$	$1.8333e6$	$1.8870e6$	$9.6e{-}3$

7. Acknowledgments. The authors would like to thank Flavio Pimentel Duarte and Fernando Silveira Filho for helping with the programs used for generating the subsets of the state space using the path-based approach.

REFERENCES

[1] VIKRAM ADVE, ADNAN AGBARIA, MATTI HILTUNEN, RAVI IYER, KAUSTUBH JOSHI, ZBIGNIEW KALBAREZYK, RYAN LEFEVER, RAYMOND PLANTE, WILLIAM H. SANDERS, AND RICHARD D. SCHLICHTING, *A Compiler-Enabled Model- and Measurement-Driven Adaptation Environment for Dependability and Performance*, in Next Generation Software (NGS) Workshop at the International Parallel & Distributed Processing Symposium (IPDPS), 2005.

[2] ANDREA BOBBIO AND KISHOR S. TRIVEDI, *An Aggregation Technique for the Transient Analysis of Stiff Markov Chains*, IEEE Trans. Computers, C-35 (1986), pp. pp. 803–814.

[3] MARC BOUISSOU AND YANNICK LEFEBVRE, *A Path-Based Algorithm to Evaluate Asymptotic Unavailability for Large Markov Models*, in Annual Reliability and Maintainability Symposium, 2002, pp. 32–39.

[4] P. BUCHHOLZ, *An improved method for bounding stationary measures*, Performance Evaluation, 62 (2005), pp. 349–365.

[5] P.J. COURTOIS AND P. SEMAL, *Bounds for the Positive Eigenvectors of Nonnegative Matrices and for their Approximation by Decomposition*, J. ACM, 31 (1984), pp. 804–825.

[6] ———, *On Polyhedra of Perron-Frobenius Eigenvectors*, Linear Algebra Applications, 65 (1985), pp. 157–170.

[7] DAVID DALY, PETER BUCHHOLZ, AND WILLIAM H. SANDERS, *An Approach for Bounding Reward Measures Markov Models using Aggregation*, tech. report, University of Illinois at Urbana-Chmapaign, 2004.

[8] DAVID DALY, DANIEL D. DEAVOURS, JAY M. DOYLE, PATRICK G. WEBSTER, AND WILLIAM H. SANDERS, *Mobius: An Extensible Tool For Performance and Dependability Modeling*, in Computer Peformance Evaluation - Modelling Techniques and Tools - 11$^{\underline{th}}$ International Conference (TOOLS2000), vol. 1786, Springer, Maro 2000, pp. 332–336.

[9] EDMUNDO DE SOUZA E SILVA AND H. RICHARD GAIL, *Calculating Availability and Performability Measures of Repairable Computer Systems Using Randomization*, Journal of the Association for Computing Machinery, 36 (1989), pp. 171–193.

[10] EDMUNDO DE SOUZA E SILVA AND ROSA M. M. LEO, *The TANGRAM-II Environment*, in Computer Peformance Evaluation - Modelling Techniques and Tools - 11$^{\underline{th}}$ International Conference (TOOLS2000), vol. 1786, Springer, March 2000, pp. 366–369.

[11] EDMUNDO DE SOUZA E SILVA AND PEDRO MEJA OCHOA, *State Space Exploration in Markov Models*, Performance Evaluation Review, 20 (1992), pp. pp. 152–166.

[12] JOANNE BECHTA DUGAN AND KISHOR S. TRIVEDI, *Covarage Modeling for Dependability Analysis of Fault-Tolerant Systems*, IEEE Trans. Computers, 38 (1989), pp. pp. 775–787.

TABLE 5.15

State space reduction and N parameter for the database system composed by CPU and hard disk units.

C	orig. state size	new state size	reduction rate (ratio)	N
3	390625	256	1525	7
10	429998169	256	1679680	7

TABLE 5.16

Numerical results for the Database system, considering the memory units' behavior. The exact value of the expected cumulated economical loss is obtained exploiting the strong lumpability property. The values for the parameters are: $CPU = \phi c = 2e^{-6}$, $\phi(1 - c) = 1e^{-5}$; $Disk = \phi c = 1e^{-5}$, $\phi(1 - c) = 1e^{-5}$; $Memory = \phi c = 1e^{-5}$, $\phi(1 - c) = 1e^{-4}$ and $\gamma = 1$ for all components.

L.B.	ECRA	U.B.	R.E.
$-4.4740e10$	$-4.4741e10$	$-4.4753e10$	$1.2e$–4

[13] E.DE SOUZA E SILVA AND H. R. GAIL, *The Uniformization Method in Performability Analysis*, in Performability Modelling: Techniques and Tools, G. Rubino B. R. Haverkort, R. Marie and K. S. Trivedi, eds., Wiley, 2001, ch. 3, pp. 31–58.

[14] B. R. HAVERKORT, R. MARIE, G. RUBINO, AND K. S. TRIVEDI, *Performability Modelling: Techniques and Tools*, Wiley, 2001.

[15] STEPHANIE MAHEVAS AND GERARDO RUBINO, *Bound Computation of Dependability and Performance Measures*, IEEE Trans. Computers, 50 (2001), pp. 399–413.

[16] C. D. MEYER, *Stochastic Complementation, Uncoupling Markov Chains and the Theory of Nearly Reducible Systems*, Siam Review, 31 (1989), pp. 240–272.

[17] JOHN F. MEYER AND WILLIAM H. SANDERS, *Specification and Construction of Performability Models*, in Proceedings of the Second International Workshop on Performability Modeling of Computer and Communication Systems, 1993.

[18] RICHARD R. MUNTZ, EDMUNDO DE SOUZA E SILVA, AND AMBUJ GOYAL, *Bounding Availability of Repairable Computer Systems*, IEEE Trans. Computers, 38 (1989), pp. 1714–1723.

[19] OLIVIER POURRET, *The Slow-Fast Approximation for Markov Processes in Reliability (text in French)*, PhD thesis, University of Paris-Sud Orsay, 1998.

[20] M. AKBER QURESHI AND WILLIAM H. SANDERS, *A New Methodology for Calculating Distributions of Reward Accumulated During a Finite Interval*, in Proceedings of the 26th Annual International Symposion on Fault-Tolerant Computing, 1996, pp. 116–125.

[21] G. RUBINO AND B. SERICOLA, *Sojourn times in Markov processes*, Journal of Applied Probability, (1989), pp. 744–756.

[22] ——, *Interval availability analysis using operational periods*, Performance Evaluation, 14 (1992), pp. 257–272.

[23] R. SAHNER AND KISHOR S. TRIVEDI, *Reliability Modeling using SHARPE*, IEEE Trans. Reliability, 36 (1987), pp. pp. 186–193.

[24] H. SIMON AND A. ANDO, *Aggregation of Variables in Dynamic Systems*, Econometrica, 29 (1961), pp. 111–138.

[25] W.J. STEWART, ed., *Numerical Solution of Markov Chains*, Marcel Dekker, 1991.

INCREASING CONVEX MONOTONE MARKOV CHAINS: THEORY, ALGORITHM AND APPLICATIONS[*]

MOUAD BEN MAMOUN[†‡], ANA BUŠIĆ[†], JEAN-MICHEL FOURNEAU[†], AND NIHAL PEKERGIN[†§]

Abstract. We develop theoretical and algorithmic aspects of discrete-time Markov chain comparison with the increasing convex order. This order is based on the variability of the process and it is expected that one can get more accurate bounds with such an order although the monotonicity property is more complex. We give a characterization for finite state space to obtain an algebraic description which is suitable for an algorithmic framework. We develop an algorithm and we introduce some applications related to the worst case stochastic analysis when some high level information is known, but not the complete structure of the chain.

Key words. Markov chains, stochastic bounds

AMS subject classifications. 60E15, 60J22, 15A51

1. Introduction. Comparison techniques have gained an increasing popularity in the study of stochastic processes [23]. These techniques may be related to various mathematical theory (stochastic ordering, polyhedral theory, Markov chain decision process, stochastic recurrence equation).

In the context of numerical analysis of Markov chains, the first idea was to analyze systems which are too difficult for numerical analysis. One can compare a chain of the model with another one which is simpler to solve. A recent survey [16] presents several solutions that we can group into two key ideas: reduction of the state space or using an ad hoc structure, the numerical analysis of which is simpler. The first approach was shown to be very efficient [25]. The stochastic approach was developed using projection or functions of Markov chains [13] and an algorithmic derivation of smaller chains based on strong or weak lumpability [28, 21, 15] was proposed. Other algorithms to obtain upper Hessenberg or single input macro state chains were also proposed in [16, 8]. A tool providing all these algorithms was also demonstrated [14].

However all these approaches are based on the strong stochastic ordering ($\preceq_{st}$-ordering) among random variables. This order is quite natural because it is associated with sample-paths and coupling. Nevertheless, many other stochastic orders have been studied. For instance, the variability orders ($\preceq_{icx}$ and $\preceq_{icv}$) have been used to compare different queueing models when we change the variability of arrival or service distributions. To the best of our knowledge, very little was done to construct bounding Markov chains with these orders. Vincent's pioneering work [1] was the only reference we could find. Note that the lack of algorithms for the increasing convex order ($\preceq_{icx}$) has precluded to compare the accuracy of these orderings when we bound Markov chains. However, $\preceq_{icx}$-ordering provides more accurate bounds when we compare random variables, as $\preceq_{st}$-comparison implies $\preceq_{icx}$-comparison.

Another completely different application of bounds was recently proposed by P. Buchholz [7]. The main assumption is that the modelers do not know the real transi-

---

[*]This work was supported by project Sure-Paths from ACI Sécurité and EURO-NGI network of excellence.

[†]PRiSM, Université de Versailles Saint-Quentin-en-Yvelines, 45, Av. des Etats-Unis, 78035 Versailles, France. Email: {mobe, abusic, jmf, nih}@prism.uvsq.fr

[‡]Université Mohammed V, B.P 1014, Rabat, Maroc

[§]Centre Marin Mersenne, Université Paris 1, 75013 Paris, France

tion probabilities. Thus, one wants to model a system by a family of Markov chains where the transition probabilities belong to an interval of probabilities. One has to derive the worst case (or the best case) for all the matrices in the set. The theoretical arguments rely on Courtois's polyhedral approach. The algorithms are very accurate as the bounds can be reached by a matrix in the set. Unfortunately the complexity is quite high. Very recently a similar problem was solved independently by Haddad and Moreaux [18]. Again, one has to find a bound for a set of matrices. However, Haddad and Moreaux's approach is quite different and relies on stochastic comparison with $\preceq_{st}$ order. The set is given through componentwise extremal matrices. Note that these matrices are useless for a direct computation as they are not stochastic. The authors derive an algorithm to find an upper bounding monotone matrix for all elements in the set according to the $\preceq_{st}$ order. This algorithm is very simple and its complexity is relatively small (less than quadratic). However the method seems to be less accurate than Buchholz's method. Note that to the best of our knowledge there is no comparison between these two new methods.

The techniques we use in this paper are quite different. Here we improve the theory of $\preceq_{icx}$-ordering for finite discrete time Markov chains. This order is known for a long time but very little was known in the context of Markov chains. Unlike the $\preceq_{st}$ order, this stochastic order imposes difficult constraints for the monotonicity property and, until recently, it was an open problem to build $\preceq_{icx}$-monotone bounding matrices. In [4] an algorithm has been designed to construct $\preceq_{icx}$-monotone bounding matrices that belong to a class of matrices denoted as class $\mathcal{C}$. However the algorithm relies on the specific monotonicity characterization of class $\mathcal{C}$ Markov chains. Here we develop a general algorithm to obtain an $\preceq_{icx}$-monotone upper bound for a given stochastic matrix. We also show that the $\preceq_{icx}$ order is more accurate than the $\preceq_{st}$ order when we derive some worst case stochastic process for which only the expected value and a pattern of nonzero transitions are known. This problem is somehow related to the problem considered by Buchholz, Haddad and Moreaux. We do not assume that the set of Markov chains we must bound is defined by intervals on the elements, but we know the pattern of nonzero transitions and the average, and we build the processes which provides extremal distributions.

The remaining of the paper is organized as follows. In Section 2 we present a brief overview of the comparison of random variables and Markov chains under some stochastic order when the state space is endowed with a total order. We also stress that the stochastic comparison approach is much more versatile than the polyhedral technique developed by Courtois [10, 11] even if it is often less accurate. Then we develop in Section 3 the theory to compare finite discrete time Markov chains (DTMCs) with the $\preceq_{icx}$ order. Section 4 is devoted to an algorithm to build an $\preceq_{icx}$-monotone upper bound of a Markov chain. Finally in Section 5 we present two applications of our approach. The first application consists in deriving the worst case of a family of Markov chains where the transitions are defined by their expectation and a pattern for nonzero transitions. The second one is related to the absorption time of a DTMC with one absorbing state and can be used to bound a phase type (PH) distribution modeling a general service time. By doing so, the complexity in two level modeling formalisms can be significantly reduced.

2. A brief presentation. Here we state some basic definitions and results on the stochastic comparison approach. We refer to [24, 26] for further details. First we give the definition of stochastic comparison of two random variables taking values on a totally ordered space $\mathcal{E}$. Let $\mathcal{F}_{st}$ denote the class of all increasing real functions on

$\mathcal{E}$ and $\mathcal{F}_{icx}$ the class of all increasing and convex real functions on $\mathcal{E}$. We denote by $\preceq_{\mathcal{F}}$ the stochastic order relation, where $\mathcal{F}$ can be replaced by st, icx to be associated respectively with the class of functions $\mathcal{F}_{st}$, $\mathcal{F}_{icx}$. Throughout the paper, $\leq$ denotes the componentwise comparison when comparing two vectors or matrices.

DEFINITION 2.1. *Let X and Y be two random variables taking values on a totally ordered space $\mathcal{E}$,*

$$X \preceq_{\mathcal{F}} Y \iff Ef(X) \leq Ef(Y), \quad \forall f \in \mathcal{F}$$

whenever the expectations exist.

Remark that, since $\mathcal{F}_{icx} \subset \mathcal{F}_{st}$, the $\preceq_{st}$-comparison is stronger than the $\preceq_{icx}$-comparison, i.e.

$$X \preceq_{st} Y \implies X \preceq_{icx} Y.$$

Notice that for discrete random variables X and Y with probability vectors p and q, the notations $p \preceq_{\mathcal{F}} q$ and $X \preceq_{\mathcal{F}} Y$ are used interchangeably. Stochastic comparison of discrete random variables according to $\preceq_{st}$ and $\preceq_{icx}$ orders can also be defined through matrices (see [20, 22]). We assume here that $\mathcal{E} = \{1, \ldots, n\}$, but the following statements may be extended to the infinite case. We denote by $\boldsymbol{K}_{\mathcal{F}}$ the matrix related to the $\preceq_{\mathcal{F}}$ order, $\mathcal{F} \in \{st, icx\}$:

$$(2.1) \quad \boldsymbol{K}_{st} = \begin{pmatrix} 1 & 0 & 0 & \ldots & 0 \\ 1 & 1 & 0 & \ldots & 0 \\ 1 & 1 & 1 & \ldots & 0 \\ \vdots & \vdots & \vdots & \ddots & \vdots \\ 1 & 1 & 1 & \ldots & 1 \end{pmatrix}, \quad \boldsymbol{K}_{icx} = \begin{pmatrix} 1 & 0 & 0 & \ldots & 0 \\ 2 & 1 & 0 & \ldots & 0 \\ 3 & 2 & 1 & \ldots & 0 \\ \vdots & \vdots & \vdots & \ddots & \vdots \\ n & n-1 & n-2 & \ldots & 1 \end{pmatrix}.$$

PROPOSITION 2.2. *Let X and Y be two random variables with probability vectors $p = (p_i)_{i=1}^n$ and $q = (q_i)_{i=1}^n$ ($p_i = P(X = i)$ and $q_i = P(Y = i)$, $1 \leq i \leq n$). Then*

$$X \preceq_{\mathcal{F}} Y \iff p\boldsymbol{K}_{\mathcal{F}} \leq q\boldsymbol{K}_{\mathcal{F}}.$$

For the $\preceq_{st}$ and $\preceq_{icx}$ orders, this can be given as follows:

$$(2.2) \quad X \preceq_{st} Y \iff \sum_{k=i}^n p_k \leq \sum_{k=i}^n q_k, \quad \forall i \in \{1, \ldots, n\},$$

$$(2.3) \quad X \preceq_{icx} Y \iff \sum_{k=i}^n (k-i+1) p_k \leq \sum_{k=i}^n (k-i+1) q_k, \quad \forall i \in \{1, \ldots, n\}.$$

In the following we only compare discrete time Markov chains (DTMCs). Continuous time models can be considered after uniformization.

It is shown in Theorem 5.2.11 of [24, p. 186] that monotonicity and comparability of the probability transition matrices of time-homogeneous Markov chains yield sufficient conditions to compare stochastically the underlying chains. We first define the monotonicity and comparability of stochastic matrices and then state this fundamental theorem.

DEFINITION 2.3. *Let $\boldsymbol{P}$ be a stochastic matrix. $\boldsymbol{P}$ is said to be stochastically $\preceq_{\mathcal{F}}$-monotone if for any probability vectors p and q,*

$$p \leq_{\mathcal{F}} q \implies p\boldsymbol{P} \leq_{\mathcal{F}} q\boldsymbol{P}.$$

DEFINITION 2.4. *Let P and Q be two stochastic matrices. Q is said to be an upper bounding matrix of P in the sense of the $\preceq_{\mathcal{F}}$ order ($P \preceq_{\mathcal{F}} Q$) if*

$$P \, K_{\mathcal{F}} \leq Q \, K_{\mathcal{F}}.$$

Let us remark that this is equivalent to saying that $P \preceq_{\mathcal{F}} Q$, if

$$P_{i,*} \preceq_{\mathcal{F}} Q_{i,*}, \quad \forall i \in \{1, \ldots, n\}$$

where $P_{i,*}$ denotes the ith row of matrix P.

DEFINITION 2.5. *Let $\{X_k\}_{k \geq 0}$ and $\{Y_k\}_{k \geq 0}$ be two homogeneous DTMCs. We say that the chain $\{X_k\}$ is smaller than the chain $\{Y_k\}$ in the sense of $\preceq_{\mathcal{F}}$ order,*

$$\{X_k\} \preceq_{\mathcal{F}} \{Y_k\}$$

if

$$X_k \preceq_{\mathcal{F}} Y_k \text{ for all } k \geq 0.$$

THEOREM 2.6. *Two homogeneous Markov chains $\{X_k\}_{k \geq 0}$ and $\{Y_k\}_{k \geq 0}$ with the transition matrices P and Q satisfy $\{X_k\} \preceq_{\mathcal{F}} \{Y_k\}$ if*
   (i) *$X_0 \preceq_{\mathcal{F}} Y_0$,*
   (ii) *there exists an $\preceq_{\mathcal{F}}$-monotone transition matrix R such that*

$$P \preceq_{\mathcal{F}} R \preceq_{\mathcal{F}} Q.$$

A special case of Theorem 2.6 is the comparison of two chains when at least one of two transition matrices P or Q is $\preceq_{\mathcal{F}}$-monotone.

COROLLARY 2.7. *Let $\{X_k\}_{k \geq 0}$ and $\{Y_k\}_{k \geq 0}$ be two homogeneous DTMCs with the transition matrices P and Q. If*
   (i) *$X_0 \preceq_{\mathcal{F}} Y_0$,*
   (ii) *$P \preceq_{\mathcal{F}} Q$,*
   (iii) *at least one of the transition matrices P or Q is $\preceq_{\mathcal{F}}$-monotone,*
then

$$\{X_k\} \preceq_{\mathcal{F}} \{Y_k\}.$$

COROLLARY 2.8. *Let P and Q be two transition matrices such that there exists an $\preceq_{\mathcal{F}}$-monotone matrix R satisfying $P \preceq_{\mathcal{F}} R \preceq_{\mathcal{F}} Q$. If the steady-state distributions $(\pi_P, \ \pi_Q)$ exist, then*

$$\pi_P \preceq_{\mathcal{F}} \pi_Q.$$

Let us now consider the comparison of two absorbing chains. The comparison of two chains in the sense of $\preceq_{\mathcal{F}}$ order (we remind that $\preceq_{\mathcal{F}}$ stands for $\preceq_{st}$ or $\preceq_{icx}$ order) provides $\preceq_{st}$-comparison of their absorption times.

PROPOSITION 2.9. *Let $\{X_k\}_{k \geq 0}$ and $\{Y_k\}_{k \geq 0}$ be two homogeneous Markov chains with an absorbing state n (the last one), and let $T_a(X)$ and $T_a(Y)$ denote respectively the absorption times into n for the two chains. If*

$$\{X_k\} \preceq_{\mathcal{F}} \{Y_k\},$$

then

$$T_a(Y) \preceq_{st} T_a(X).$$

Notice that the $\preceq_{st}$-comparison of absorption times is now on random variables T_a defined on $\mathbb{N}_0$ (dates), and not on states.

Proof. We have $X_k \preceq_{\mathcal{F}} Y_k$, for all $k \geq 0$. Particularly, for both $\preceq_{st}$ and $\preceq_{icx}$ order, this implies that

(2.4) $$P(X_k = n) \leq P(Y_k = n), \quad \text{for all } k \geq 0.$$

Thus, $P(T_a(X) \leq k) = P(X_k = n) \leq P(Y_k = n) = P(T_a(Y) \leq k)$, which gives $T_a(Y) \preceq_{st} T_a(X)$.   □

Remark that we obtain the $\preceq_{st}$-comparison of absorption times even if we compare the DTMCs in the $\preceq_{icx}$-ordering sense. In fact, the ordering relation needs only to satisfy (2.4).

Like the polyhedral approach, the stochastic comparison approach enables the comparison of steady-state distributions. The Courtois's approach is often more accurate because the polyhedral constraints are usually weaker than the monotonicity constraints. But the stochastic comparison approach is much more versatile. It can provide bounds on the distribution at any time time t, it can give a stochastic bound for the absorption time and for several measures of interest in performance evaluation, reliability modeling, stochastic model checking. For instance, it has been shown that path properties which are studied by stochastic model checking can be simplified using the stochastic comparison approach [5].

3. Increasing convex ordering of finite DTMCs.

The monotonicity is an important property for comparison of Markov chains (Theorem 2.6). However, it is obvious that checking the monotonicity of a transition matrix using Definition 2.3 is not tractable since we must check the implication for all comparable probability vectors. Thus, an algorithmic characterization of monotonicity is mandatory.

For the usual stochastic order $\preceq_{st}$, a matrix characterization of monotonicity has been established [20, 24]: $\boldsymbol{P}$ is $\preceq_{st}$-monotone if and only if all the entries of the matrix $\boldsymbol{K}_{st}^{-1}\boldsymbol{P}\boldsymbol{K}_{st}$ are non negative. This is valid for both finite and infinite state space cases. In [22] an equivalent characterization for the $\preceq_{icx}$-monotonicity has been provided when the state space is infinite (for the chains taking values in $\mathbb{Z}$): $\boldsymbol{P}$ is $\preceq_{icx}$-monotone if and only if all the entries of the matrix $\boldsymbol{K}_{icx}^{-1}\boldsymbol{P}\boldsymbol{K}_{icx}$ are non negative. The finite case has been first studied by Vincent in his pioneering work and the above condition was assumed to be sufficient [1]. Note that if the condition is necessary and sufficient for infinite state space, it is only sufficient for finite state space.

In this section we complete Vincent's work and we obtain a complete matrix characterization of the $\preceq_{icx}$-monotonicity. First, we state in the following proposition a necessary and sufficient condition for the $\preceq_{icx}$-monotonicity in terms of transition probabilities $P_{i,j}$. For a transition matrix $\boldsymbol{P}$ on the state space $\mathcal{E} = \{1, \ldots, n\}$, for all i and j, we will denote by $f_{i,j}(\boldsymbol{P})$,

(3.1) $$f_{i,j}(\boldsymbol{P}) = \sum_{k=j}^{n} (k - j + 1) P_{i,k}.$$

Notice that

(3.2) $$f_{i,j}(\boldsymbol{P}) = (\boldsymbol{P}\boldsymbol{K}_{icx})_{i,j}, \quad \forall i, j \in \{1, \ldots, n\}.$$

REMARK 3.1. *For a vector $x \in \mathbb{R}^n$ we will write $x \in \mathcal{F}$ if and only if vector x, seen as a real function on $\mathcal{E} = \{1, \ldots, n\}$, is in $\mathcal{F}$:*

(i) *$x \in \mathcal{F}_{st}$ if and only if vector x is increasing, i.e.*

$$x \in \mathcal{F}_{st} \iff x_i \le x_{i+1}, \forall i \in \{1, \ldots, n-1\}.$$

(ii) *$x \in \mathcal{F}_{icx}$ if and only if vector x is increasing and convex, i.e.*

$$x \in \mathcal{F}_{icx} \iff x_1 \le x_2 \text{ and } 2x_i \le x_{i-1} + x_{i+1}, \forall i \in \{2, \ldots, n-1\}.$$

Notice that $x_1 \le x_2$ and $2x_i \le x_{i-1} + x_{i+1}, 1 < i < n-1$ imply $x_i \le x_{i+1}, \forall i < n$.

PROPOSITION 3.2. *A stochastic matrix $\boldsymbol{P}$ is $\preceq_{icx}$-monotone if and only if, for all $j \in \{2, \ldots, n\}$, the vectors*

$$f_{*,j}(\boldsymbol{P}) = (f_{i,j}(\boldsymbol{P}))_{i=1}^n,$$

defined by (3.1) are increasing and convex, i.e.

$$f_{1,j}(\boldsymbol{P}) \le f_{2,j}(\boldsymbol{P}) \text{ and } 2f_{i,j}(\boldsymbol{P}) \le f_{i-1,j}(\boldsymbol{P}) + f_{i+1,j}(\boldsymbol{P}), \forall i \in \{2, \ldots, n-1\}.$$

Proof. We first show the following relation for two given probability vectors p and q

$$(3.3) \qquad p\boldsymbol{P} \preceq_{icx} q\boldsymbol{P} \iff \sum_{k=1}^n p_k f_{k,j}(\boldsymbol{P}) \le \sum_{k=1}^n q_k f_{k,j}(\boldsymbol{P}), \quad \forall j.$$

Using Proposition 2.2, characterizing the $\preceq_{icx}$-comparison of two probability vectors, and (3.2) we have:

$$p\boldsymbol{P} \preceq_{icx} q\boldsymbol{P} \iff (p\boldsymbol{P})\boldsymbol{K}_{icx} \le (q\boldsymbol{P})\boldsymbol{K}_{icx} \iff p(\boldsymbol{P}\boldsymbol{K}_{icx}) \le q(\boldsymbol{P}\boldsymbol{K}_{icx})$$

$$\iff \sum_{k=1}^n p_k f_{k,j}(\boldsymbol{P}) \le \sum_{k=1}^n q_k f_{k,j}(\boldsymbol{P}), \quad \forall j.$$

$\Leftarrow$ *Necessary condition:* suppose that vectors $f_{*,j}(\boldsymbol{P})$ are increasing and convex and show that $\boldsymbol{P}$ is $\preceq_{icx}$-monotone. For that let us consider two probability vectors p and q such that $p \preceq_{icx} q$ and prove that $p\boldsymbol{P} \preceq_{icx} q\boldsymbol{P}$. According to Definition 2.1, $p \preceq_{icx} q$ implies that $\sum_{k=1}^n p_k h(k) \le \sum_{k=1}^n q_k h(k), \forall h \in \mathcal{F}_{icx}$. For each $j \in \mathcal{E}$, let us denote by g_j a function on $\mathcal{E}$ defined by

$$g_j(i) = f_{i,j}(\boldsymbol{P}), \forall i.$$

Since $f_{*,j}(\boldsymbol{P})$ are increasing and convex for all $j \ge 2$, functions g_j belong to $\mathcal{F}_{icx}, \forall j \ge 2$ by hypothesis of the proposition. As $f_{i,1}(\boldsymbol{P}) = 1 + f_{i,2}(\boldsymbol{P}), \forall i$, g_1 belongs also to $\mathcal{F}_{icx}$. Thus, $\sum_{k=1}^n p_k f_{k,j}(\boldsymbol{P}) \le \sum_{k=1}^n q_k f_{k,j}(\boldsymbol{P}), \forall j \in \mathcal{E}$, so $p\boldsymbol{P} \preceq_{icx} q\boldsymbol{P}$ follows from (3.3).

$\Rightarrow$ *Sufficient condition:* Suppose that $\boldsymbol{P}$ is $\preceq_{icx}$-monotone, and show that vectors $f_{*,j}(\boldsymbol{P}) \in \mathcal{F}_{icx}$, for all $j \ge 2$.
Let us define the probability vectors $p^{(i)}, 1 \le i \le n$:

$$p^{(i)} = (p_1^{(i)} = 0, \ldots, p_i^{(i)} = 1, \ldots, p_n^{(i)} = 0).$$

Using (2.3), it is obvious that $p^{(i)} \preceq_{icx} p^{(i+1)}$, $1 \leq i \leq n - 1$. Matrix P is $\preceq_{icx}$-monotone, thus $p^{(i)}P \preceq_{icx} p^{(i+1)}P$. It follows from (3.3) that

$$f_{i,j}(P) = \sum_{k=1}^{n} p_k^{(i)} f_{k,j}(P) \leq \sum_{k=1}^{n} p_k^{(i+1)} f_{k,j}(P) = f_{i+1,j}(P), \forall j \in \mathcal{E}.$$

Hence, vectors $f_{*,j}(P)$ are increasing.

Let us now define the probability vectors $q^{(i)}$, $2 \leq i \leq n - 1$ as follows:

$$q^{(i)} = (q_1^{(i)} = 0, \ldots, q_{i-1}^{(i)} = \frac{1}{2}, q_i^{(i)} = 0, q_{i+1}^{(i)} = \frac{1}{2}, \ldots, q_n^{(i)} = 0).$$

It can be easily shown that $p^{(i)} \preceq_{icx} q^{(i)}$, $2 \leq i \leq n - 1$. Thus,

$$f_{i,j}(P) = \sum_{k=1}^{n} p_k^{(i)} f_{k,j}(P) \leq \sum_{k=1}^{n} q_k^{(i)} f_{k,j}(P) = \frac{1}{2} f_{i-1,j}(P) + \frac{1}{2} f_{i+1,j}(P), \forall j.$$

Therefore, $2f_{i,j}(P) \leq f_{i-1,j}(P) + f_{i+1,j}(P)$ and vectors $f_{*,j}(P)$ are convex for all $j \in \mathcal{E}$. □

The above proposition can also be proved from Stoyan's theorem 5.2.3 [24] which provides conditions ensuring monotonicity of transition matrices in the general case of an integral stochastic order $\preceq$, and an arbitrary state space. However, this requires to introduce some notions like maximal generators used in this theorem. For the sake of simplicity and to make the paper self contained, we preferred to give a direct proof based only on the definition of monotonicity.

In the following proposition we give the matrix characterization for the $\preceq_{icx}$-monotonicity.

PROPOSITION 3.3. *A stochastic matrix P is $\preceq_{icx}$-monotone if and only if*

(3.4) $$Z_{icx} P K_{icx} \geq 0,$$

where K_{icx} is the matrix given by (2.1) and

(3.5) $$Z_{icx} = \begin{pmatrix} 1 & 0 & 0 & \cdots & 0 \\ -1 & 1 & 0 & \cdots & 0 \\ 1 & -2 & 1 & \cdots & 0 \\ \vdots & \ddots & \ddots & \ddots & \vdots \\ 0 & \cdots & 1 & -2 & 1 \end{pmatrix}$$

Proof. Let us denote by $A = (A_{i,j})_{i,j=1}^n$ the matrix PK_{icx}, then

$$Z_{icx} A \geq 0 \iff \forall j \in \{1, \ldots, n\}, \quad \begin{cases} A_{1,j} \geq 0 \\ -A_{1,j} + A_{2,j} \geq 0 \\ A_{i-2,j} - 2A_{i-1,j} + A_{i,j} \geq 0, \quad \forall i \geq 3 \end{cases}$$

Notice that the inequality $A_{1,j} \geq 0$ is always satisfied and that the conditions for $j = 1$ follow from the conditions for $j = 2$ ($A_{i,1} = A_{i,2} + 1$). Since $A_{i,j} = \sum_{k=j}^{n} (k - j + 1) P_{i,k} = f_{i,j}(P)$, by Proposition 3.2 it follows that P is $\preceq_{icx}$-monotone if and only if (3.4) holds. □

Let us give some remarks concerning the matrix characterization of Proposition 3.3. First this characterization is not unique. Indeed, the values of the first row of Z_{icx} can be replaced by any non negative values and several set of values are possible for the first column of K_{icx}. However, the values of the other rows of Z_{icx} and columns of K_{icx} are necessary. For instance, in the case where all values of the first column of K_{icx} are replaced by 1, we obtain the matrix $K_{icx}(1)$:

$$K_{icx}(1) = \begin{pmatrix} 1 & 0 & 0 & \cdots & 0 \\ 1 & 1 & 0 & \cdots & 0 \\ 1 & 2 & 1 & \cdots & 0 \\ \vdots & \vdots & \vdots & \vdots & \vdots \\ 1 & n-1 & n-2 & \cdots & 1 \end{pmatrix}$$

and we have the following characterization:

$$P \text{ is } \preceq_{icx} \text{-monotone} \iff K_{icx}(1)^{-1} P K_{icx}(1) \geq 0.$$

In fact it can be seen that $K_{icx}(1)^{-1} = Z_{icx}$.

Let us emphasize that the matrix $K_{icx}(1)$ allows also to define the $\preceq_{icx}$ order. For two probability vectors p and q defined on $\mathcal{E} = \{1, \ldots, n\}$, it can easily be seen that

$$p\, K_{icx} \leq q\, K_{icx} \iff p\, K_{icx}(1) \leq q\, K_{icx}(1).$$

Indeed the sum of the first two columns of $K_{icx}(1)$ gives the first column of K_{icx}.

Contrary to the $\preceq_{st}$ order, in the case of the finite state space $\mathcal{E} = \{1, \ldots, n\}$ the condition $K_{icx}^{-1} P K_{icx} \geq 0$ is not equivalent to the $\preceq_{icx}$-monotonicity of the matrix P. We show this by a counter example but we give before K_{icx}^{-1} which is the inverse of the matrix K_{icx} given by (2.1).

$$K_{icx}^{-1} = \begin{pmatrix} 1 & 0 & 0 & \cdots & 0 \\ -2 & 1 & 0 & \cdots & 0 \\ 1 & -2 & 1 & \cdots & 0 \\ \vdots & \ddots & \ddots & \ddots & \vdots \\ 0 & \cdots & 1 & -2 & 1 \end{pmatrix}$$

Note that the only difference between the matrices Z_{icx} and K_{icx}^{-1} is the value of the first element of the second row. Let us consider the matrix P:

$$P = \begin{pmatrix} 0.5 & 0.1 & 0.4 \\ 0.4 & 0.15 & 0.45 \\ 0.3 & 0.2 & 0.5 \end{pmatrix}$$

we have

$$K_{icx}^{-1} P K_{icx} = \begin{pmatrix} 1.9 & 0.9 & 0.4 \\ -1.75 & -0.75 & -0.35 \\ 0 & 0 & 0 \end{pmatrix} \quad Z_{icx} P K_{icx} = \begin{pmatrix} 1.9 & 0.9 & 0.4 \\ 0.15 & 0.15 & 0.15 \\ 0 & 0 & 0 \end{pmatrix}$$

Hence, the condition $K_{icx}^{-1} P K_{icx} \geq 0$ is not satisfied while P is $\preceq_{icx}$-monotone as $Z_{icx} P K_{icx} \geq 0$. In fact, the condition $K_{icx}^{-1} P K_{icx} \geq 0$ is sufficient but not necessary for the $\preceq_{icx}$-monotonicity.

We want to emphasize here the importance of this counter-example and the matrix characterization of Proposition 3.3. Indeed, the condition $K_{icx}^{-1} P K_{icx} \geq 0$ has very important consequences. It implies that the first and last states are absorbing ($P_{1,1} = P_{n,n} = 1$) as it has been shown in [1]. Fortunately, the condition $Z_{icx} P K_{icx} \geq 0$

given in Proposition 3.3 is weaker than the condition $K_{icx}^{-1} P K_{icx} \geq 0$ and it can be easily proven that $K_{icx}^{-1} P K_{icx} \geq 0 \implies Z_{icx} P K_{icx} \geq 0$. Moreover, as we will see in the following section, it is possible to construct $\preceq_{icx}$ bounding monotone matrices without absorbing states.

In [1], the authors studied the irreducibility of monotone bounding matrices. They state that if all sub-diagonal entries of the matrix K characterizing the considered order are positive such as for K_{st} and K_{icx}, then the bounding matrix has only one recurrent class. Let us emphasize that this result is independent on the algorithm of construction of the bounding matrix. We recall this result in the case of the $\preceq_{icx}$ order which is of particular interest for us in this work.

PROPOSITION 3.4. *Let* $P = (P_{i,j})_{i,j=1}^n$ *be an irreducible stochastic matrix. A stochastic matrix* $Q = (Q_{i,j})_{i,j=1}^n$ *such that* $P \preceq_{icx} Q$ *and* Q *is* $\preceq_{icx}$*-monotone, has only one recurrent class. This class contains the state* n.

Proof. We show that $\forall i < n$, $\exists j > i$ such that $Q_{i,j} > 0$. Indeed, this implies that for each state $i < n$, there exist a path between state i and state n and consequently there is only one recurrent class which contains necessarily state n.

By contradiction, if we suppose that $\exists i < n$, such that $\forall j > i$, $Q_{i,j} = 0$, then $\sum_{j=i+1}^n (j-i) Q_{i,j} = 0$. Using the $\preceq_{icx}$-monotonicity of Q (Proposition 3.2), we have: $\forall k \leq i, \sum_{j=i+1}^n (j-i) Q_{k,j} = 0$. On the other hand, since $P \preceq_{icx} Q$, then $\forall k \leq i, \sum_{j=i+1}^n (j-i) P_{k,j} \leq \sum_{j=i+1}^n (j-i) Q_{k,j} = 0$. This implies that $\forall k \leq i, \forall j > i, P_{k,j} = 0$, which means that the set $\{1, \ldots, i\}$ is absorbing. This is impossible because P is irreducible. $\qquad\square$

Recall that $\preceq_{st}$-comparison implies $\preceq_{icx}$-comparison. However, we cannot compare $\preceq_{st}$ and $\preceq_{icx}$-monotonicity property.

EXAMPLE 3.5. *Let us consider the following two matrices*

$$P = \begin{pmatrix} 0.5 & 0.4 & 0.1 \\ 0.3 & 0.3 & 0.4 \\ 0.1 & 0.4 & 0.5 \end{pmatrix} \text{ and } Q = \begin{pmatrix} 0.2 & 0.5 & 0.3 \\ 0.3 & 0.3 & 0.4 \\ 0.2 & 0.3 & 0.5 \end{pmatrix}.$$

Matrix P *is an* $\preceq_{st}$*-monotone matrix that is not* $\preceq_{icx}$*-monotone. On the other hand,* Q *is an example of an* $\preceq_{icx}$*-monotone matrix that is not* $\preceq_{st}$*-monotone.*

4. Algorithm for an $\preceq_{icx}$ bound. Let us suppose that we have a stochastic matrix P. We would like to compute an $\preceq_{icx}$-monotone matrix Q such that $P \preceq_{icx} Q$. Then by Corollary 2.7 we have $\preceq_{icx}$-comparison of underlying chains.

Contrary to the case of $\preceq_{st}$ order (see [1]), in the case of $\preceq_{icx}$ order it is not generally possible to find an optimal $\preceq_{icx}$-monotone upper bound Q, i.e. a matrix Q such that

(i) $P \preceq_{icx} Q$,
(ii) Q is $\preceq_{icx}$-monotone, and
(iii) for each $\preceq_{icx}$-monotone transition matrix U, $P \preceq_{icx} U \Rightarrow Q \preceq_{icx} U$.

EXAMPLE 4.1. *We will consider the transition matrix*

$$P = \begin{pmatrix} 0.5 & 0.4 & 0.1 \\ 0.3 & 0.3 & 0.4 \\ 0.1 & 0.4 & 0.5 \end{pmatrix}.$$

Let us suppose now that there is an optimal $\preceq_{icx}$*-monotone upper bound* Q *for* P. *Both*

$$\bar{U} = \begin{pmatrix} 0.5 & 0.4 & 0.1 \\ 0.3 & 0.3 & 0.4 \\ 0.1 & 0.2 & 0.7 \end{pmatrix} \text{ and } \hat{U} = \begin{pmatrix} 0.5 & 0.2 & 0.3 \\ 0.3 & 0.3 & 0.4 \\ 0.1 & 0.4 & 0.5 \end{pmatrix}$$

are $\preceq_{icx}$-monotone upper bounds for matrix $\boldsymbol{P}$. Thus, $\boldsymbol{Q}$ should satisfy $\boldsymbol{Q} \preceq_{icx} \bar{\boldsymbol{U}}$ and $\boldsymbol{Q} \preceq_{icx} \hat{\boldsymbol{U}}$. If we consider only the last column, this implies $Q_{*,3} = (0.1, 0.4, 0.5)$, which is not convex. This is in contradiction with the fact that $\boldsymbol{Q}$ is $\preceq_{icx}$-monotone (see Proposition 3.2).

However, for each matrix $\boldsymbol{P}$ we can find an $\preceq_{icx}$-monotone upper bound. A trivial one is given by

$$Q_{i,j} = \begin{cases} 1, j = n, \\ 0, j < n. \end{cases}$$

In this section we discuss the compatibility of $\preceq_{icx}$-monotonicity and comparison constraints, and we propose an algorithm to derive a non-trivial $\preceq_{icx}$-monotone upper bound for an arbitrary finite transition matrix $\boldsymbol{P}$.

Let us remind that, for a transition matrix $\boldsymbol{P}$ of size n, $f_{i,j}(\boldsymbol{P})$ denotes $f_{i,j}(\boldsymbol{P}) = \sum_{k=j}^{n}(k - j + 1)P_{i,j}, \forall i, j$ (see (3.1)). Similarly we will define $s_{i,j}(\boldsymbol{P})$ as

$$s_{i,j}(\boldsymbol{P}) = \sum_{k=j}^{n} P_{i,j}, \forall i, j.$$

It can be easily shown that

$$(4.1) \qquad f_{i,j}(\boldsymbol{P}) = f_{i,j+1}(\boldsymbol{P}) + s_{i,j}(\boldsymbol{P}), \ \forall i, \ \forall j < n.$$

Notice that $Q_{i,j} \geq 0, \ \forall i, j$ if and only if

$$s_{i,n}(\boldsymbol{Q}) \geq 0, \ \forall i \text{ and } s_{i,j}(\boldsymbol{Q}) \geq s_{i,j+1}(\boldsymbol{Q}), \ \forall i, \forall j < n.$$

The constraints on $\boldsymbol{Q}$ can be then given in terms of $s_{i,j}(\boldsymbol{Q})$ and $f_{i,j}(\boldsymbol{Q})$ as follows:
  1. Comparison ($\boldsymbol{P} \preceq_{icx} \boldsymbol{Q}$)

$$(4.2) \qquad f_{i,j}(\boldsymbol{Q}) \geq f_{i,j}(\boldsymbol{P}), \ \forall i, \forall j \geq 2.$$

  2. Monotonicity (see Proposition 3.2)

$$(4.3) \qquad \text{vectors } f_{*,j}(\boldsymbol{Q}) \text{ are increasing and convex for all } j \geq 2.$$

  3. $\boldsymbol{Q}$ is a stochastic matrix

$$(4.4) \qquad \begin{aligned} &0 \leq s_{i,n}(\boldsymbol{Q}) \leq 1, \ \forall i, \\ &s_{i,j+1}(\boldsymbol{Q}) \leq s_{i,j}(\boldsymbol{Q}) \leq 1, \ \forall i, \forall j < n, \\ &s_{i,1}(\boldsymbol{Q}) = 1, \forall i. \end{aligned}$$

We will compute the entries of the bounding matrix decreasingly by columns. Remark that we need to compute $n - 1$ columns as the first one is completely determined by the fact that the matrix $\boldsymbol{Q}$ is stochastic.

The constraints for the last column can be written as follows:

$$\begin{aligned} &Q_{i,n} \geq P_{i,n}, \ \forall i \\ &Q_{*,n} \text{ is increasing and convex} \\ &Q_{n,n} \leq 1. \end{aligned}$$

Remark that $Q_{*,n} \in \mathcal{F}_{icx}$ and $Q_{n,n} \leq 1$ imply $Q_{i,n} \leq 1, \forall i$. Let $\mathbf{0}, \mathbf{1} \in \mathbb{R}^n$ denote the vectors $\mathbf{0} = (0, \ldots, 0)$ and $\mathbf{1} = (1, \ldots, 1)$. The last column of a bounding matrix $\mathbf{Q}$ is then a solution of Problem 4.2 for $a = P_{*,n}$ and $b = \mathbf{1}$. Moreover, we will show in §4.1 (Proposition 4.4) that the computation of each column of an $\preceq_{icx}$-monotone bounding matrix can be seen as Problem 4.2 with different vectors a and b that can be easily computed.

PROBLEM 4.2. *Let a and b be two vectors such that $\mathbf{0} \leq a \leq b$. Find an increasing and convex vector $x \in \mathbb{R}^n$ such that $a \leq x \leq b$.*

Let us remark that Problem 4.2 does not always have a solution. Take for instance $a = b$ and $a_1 > a_2$. In our particular problem of last column computation, $b = \mathbf{1}$ is trivially increasing and convex, so there is at least one solution of Problem 4.2, $x = b$. This trivial solution is not satisfying in our case, as this would result by a bounding matrix $\mathbf{Q}$ with all the elements of the last column equal to 1. Intuitively, we are interested in finding a solution x of Problem 4.2 that is as closest as possible to the vector a. Notice that if vector a is increasing and convex, we can simply take $x = a$. If a is not increasing and convex it is generally not possible to find an optimal solution x of Problem 4.2 in the following sense:

$$\text{for each } y, \text{ a solution of Problem 4.2, } x \leq y.$$

EXAMPLE 4.3. *Consider for example vectors*

$$a = (0.1, 0.4, 0.5) \text{ and } b = (1, 1, 1).$$

The vectors $\bar{y} = (0.1, 0.4, 0.7)$ and $\hat{y} = (0.3, 0.4, 0.5)$ are both solutions of Problem 4.2. Thus the optimal solution x of Problem 4.2 should satisfy $x \leq \bar{y}$ and $x \leq \hat{y}$ which implies $x = a$. This is not possible as vector a is not increasing and convex ($a_3 - a_2 = 0.1 < 0.3 = a_2 - a_1$).

Some heuristics for Problem 4.2 will be given in §4.2.

4.1. Basic algorithm. We will present here an algorithm to construct an $\preceq_{icx}$-monotone bounding matrix for a given transition matrix $\mathbf{P}$. If we suppose that we know the last $n - j$ columns of the bounding matrix $\mathbf{Q}$, then the sufficient and necessary conditions on column j are given by the following proposition.

PROPOSITION 4.4. *Let us suppose that we have already computed the columns n to $j + 1, 1 < j < n$ of matrix $\mathbf{Q}$, satisfying conditions (4.2), (4.3) and (4.4). If we fix the last $n - j$ already computed columns of $\mathbf{Q}$, then $f_{*,j}(\mathbf{Q})$ must be a solution of Problem 4.2 with vectors a and b such that*

$$(4.5) \qquad a_i = \max(f_{i,j}(\mathbf{P}), f_{i,j+1}(\mathbf{Q}) + s_{i,j+1}(\mathbf{Q})), \quad b_i = f_{i,j+1}(\mathbf{Q}) + 1, \forall i.$$

Problem 4.2 with the above vectors a and b always has a solution. Furthermore, each solution of Problem 4.2 with the above vectors a and b can be taken as $f_{*,j}(\mathbf{Q})$.

Proof. We will first show that Problem 4.2 with vectors a and b given by (4.5) always has a solution. Notice that vector b given by (4.5) is increasing and convex by the hypothesis of the proposition (condition (4.3)). Therefore, if $a \leq b$, Problem 4.2 always has at least one trivial solution b. By the hypothesis of the proposition we have $f_{i,j+1}(\mathbf{P}) \leq f_{i,j+1}(\mathbf{Q})$ (condition (4.2)) and $s_{i,j+1}(\mathbf{Q}) \leq 1$ (condition (4.4)), so $f_{i,j}(\mathbf{P}) = f_{i,j+1}(\mathbf{P}) + s_{i,j}(\mathbf{P}) \leq f_{i,j+1}(\mathbf{Q}) + 1 = b_i, \forall i$ and $f_{i,j+1}(\mathbf{Q}) + s_{i,j+1}(\mathbf{Q}) \leq b_i, \forall i$. Thus $a \leq b$.

Let x be now any solution of Problem 4.2 with vectors a and b given by (4.5), and let us take this solution as vector $f_{*,j}(\mathbf{Q})$. We will show that $f_{*,j}(\mathbf{Q})$ satisfies the conditions (4.2), (4.3) and (4.4). In other words, that any solution of Problem 4.2 with vectors a and b given by (4.5) can be taken as $f_{*,j}(Q)$. Condition (4.3) follows directly from the fact that any solution of Problem 4.2 is increasing and convex. Condition (4.2) follows from $f_{*,j}(Q) \geq a$. It remains us to show that $s_{i,j+1}(Q) \leq s_{i,j}(Q) \leq 1, \forall i$. We know that $f_{*,j}(Q) \geq a$, thus

$$s_{i,j}(Q) = f_{i,j}(Q) - f_{i,j+1}(Q) \geq s_{i,j+1}(Q), \forall i.$$

On the other hand, $f_{*,j}(Q) \leq b$ implies $s_{i,j}(Q) \leq 1, \forall i$. Therefore, any solution of Problem 4.2 with vectors a and b given by (4.5) can be taken as $f_{*,j}(Q)$.

Finally, we will show that this is also a necessary condition for $f_{*,j}(Q)$. Let us suppose that $f_{*,j}(Q)$ satisfies conditions (4.2), (4.3) and (4.4). Condition (4.2) implies $f_{i,j}(Q) \geq f_{i,j}(P), \forall i$, and condition (4.4) implies $s_{i,j}(Q) \leq s_{i,j+1}(Q), \forall i$. We have

$$f_{i,j}(Q) = f_{i,j+1}(Q) + s_{i,j}(Q) \leq f_{i,j+1}(Q) + s_{i,j+1}(Q), \ \forall i.$$

Thus, $f_{*,j}(Q) \geq a$. On the other hand, (4.4) implies $s_{i,j}(Q) \leq 1, \forall i$ which gives $f_{i,j}(Q) = f_{i,j+1}(Q) + s_{i,j}(Q) \leq b_i, \forall i$. Finally, (4.3) implies that $f_{*,j}(Q)$ is increasing and convex. Therefore, $f_{*,j}(Q)$ is a solution of Problem 4.2 with vectors a and b given by (4.5). □

Proposition 4.4 allows us to build the bounding matrix Q decreasingly by columns. As there is generally no optimal $\preceq_{icx}$-monotone upper bounding matrix (see Example 4.3), we prefer to give first the general algorithm. Algorithm 4.5 computes an $\preceq_{icx}$-monotone upper bounding matrix Q for an arbitrary finite transition matrix P. We reduce the problem of computing an $\preceq_{icx}$-monotone upper bound to Problem 4.2. Then, in §4.2 we present some heuristics to solve Problem 4.2.

ALGORITHM 4.5. *Let P be an arbitrary transition matrix of size n. An $\preceq_{icx}$-monotone upper bound Q for matrix P can be obtained as follows:*

1. *Solve Problem 4.2 with $a = P_{*,n}$ and $b = \mathbf{1}$. Let x_n denote the obtained solution. Set $q_{*,n} = s_{*,n}(Q) = f_{*,n}(Q) = x_n$.*

2. *For each $j = n - 1$ to 2:*
Solve Problem 4.2 with vectors a and b as in Proposition 4.4, i.e.

$$a_i = \max(f_{i,j}(P), f_{i,j+1}(Q) + s_{i,j+1}(Q)), \ b_i = f_{i,j+1}(Q) + 1, \forall i.$$

Denote the solution by x_j.
Set $f_{i,j}(Q) = x_j$, $s_{i,j}(Q) = f_{i,j}(Q) - f_{i,j+1}(Q)$, $Q_{i,j} = s_{i,j}(Q) - s_{i,j+1}(Q)$.

3. *$f_{i,1}(Q) = f_{i,2}(Q) + 1$, $s_{i,1}(Q) = 1$, $q_{i,1} = 1 - s_{i,2}(Q)$.*

THEOREM 4.6. *Matrix Q obtained by Algorithm 4.5 is an $\preceq_{icx}$-monotone matrix such that $P \preceq_{icx} Q$.*

Proof. Follows directly from definition of Problem 4.2 and from Proposition 4.4 by induction on j. □

Notice that Theorem 4.6 does not depend on how Problem 4.2 is solved in Algorithm 4.5.

4.2. Solving Problem 4.2. Let us remark first that in all the $n - 1$ instances of Problem 4.2 in Algorithm 4.5, the vector b is increasing and convex. In this case, Problem 4.2 always has a trivial solution $x = b$. We remind that we are interested in finding an increasing and convex vector x that is as close as possible to the vector a. In terms of Algorithm 4.5, this corresponds to the local optimization for a current column. Recall that, generally, a global optimal $\preceq_{icx}$-monotone upper bounding matrix does not exist (Example 4.3). Additionally, we need solutions that can be easily computed. More precisely, we will consider here only the algorithmic constructions with complexity of $O(n)$ for Problem 4.2.

Vector a does not need to be increasing or convex. However, as we have $a \leq b$, and b is an increasing vector, we know that $r(a) \leq b$, where $r(a)$ denotes the vector of local maxima of vector a,

$$r(a)_i = \max_{k \leq i} a_k.$$

In the following, we will suppose that the vector a is increasing. If this is not the case, we can simply take $r(a)$ instead of a. Notice that $r(a)$ can be easily computed as

$$r(a)_1 = a_1, \ r(a)_i = \max(r(a)_{i-1}, a_i), \ i > 1.$$

Therefore, we will only consider Problem 4.2 with an increasing vector a, and an increasing and convex vector b.

Furthermore, we will distinguish a special case where the vector b is a constant vector. We will show that the general case, where vector b is an arbitrary increasing and convex vector, can be reduced to this special case.

PROPOSITION 4.7. *Let a and b be two increasing vectors such that $a \leq b$. Let the vector b be additionally convex. Denote by δ the maximal distance between a and b, and let d be the vector with all the entries equal to δ,*

$$\delta = \max_{1 \leq i \leq n} \{b_i - a_i\}, \quad d = (\delta, \ldots, \delta).$$

If y is a solution of Problem 4.2 with vectors $a + d - b$ and d, then $x = y + b - d$ is a solution of Problem 4.2 with vectors a and b.

Proof. Both b and y are increasing and convex vectors, thus x is increasing and convex. As $a + d - b \leq y \leq d$, we have $a \leq x \leq b$. $\qquad \square$

In the following we present some heuristics for Problem 4.2, far from being exhaustive. We remind that we focus only on linear time complexity algorithms.

We have the following constraints:
1. $a_i \leq x_i \leq b_i$,
2. $x_2 \leq x_1, \ x_i \geq 2x_{i-1} - x_{i-2}, \forall i \geq 3$,

where a and b are increasing vectors and b is additionally convex.

Forward computation. The very simple idea is to order the above inequalities increasingly in row index and to take equalities instead of inequalities. We obtain

$$\begin{cases} x_1 = a_1, \\ x_2 = a_2, \\ x_i = \max\{2x_{i-1} - x_{i-2}, \ a_i\}, \ \forall i \geq 3. \end{cases}$$

However, this can yield $x \not\leq b$. For example, for $a = (0.1, 0.5, 0.5, 0.7)$ and $b = (1, 1, 1, 1)$, we obtain $x = (0.1, 0.5, 0.9, 1.3)$. We can notice that we cannot guarantee that vector x will be a solution of Problem 4.2 even in the special case when b is a constant vector.

Backward computation. If we use the same basic idea as above, but we compute the entries in decreasing order, in the case of a constant vector b, x is a solution of Problem 4.2.

(4.6) $$\begin{cases} x_n = a_n, \\ x_{n-1} = a_{n-1}, \\ x_i = \max\{2x_{i+1} - x_{i+2}, \ a_i\}, \ \forall i \leq n - 2. \end{cases}$$

PROPOSITION 4.8. *Let a be an increasing vector and let $b = (\beta, \ldots, \beta)$ be a constant vector such that $a \leq b$. Then vector x computed by backward computation (4.6) is a solution of Problem 4.2, i.e. x is an increasing and convex vector such that $a \leq x \leq b$.*

Proof. $x \geq a$ and x is convex are trivial. It remains us to show that $x_i \leq x_{i+1}, \forall i < n$. Then we have $x_i \leq x_n = a_n \leq \beta$, so x is an increasing convex vector such that $a \leq x \leq b$. We will show that $x_i \leq x_{i+1}, \forall i < n$ by induction on i. For $i = n - 1$, we have $x_n = a_n \geq a_{n-1} = x_{n-1}$. Let us suppose now that $x_k \leq x_{k+1}, \forall k, i < k < n$. Then, $x_{i+1} - x_i = x_{i+1} - \max\{2x_{i+1} - x_{i+2}, a_i\} = \min\{x_{i+2} - x_{i+1}, x_{i+1} - a_i\} \geq 0$, since $x_{i+1} \geq a_{i+1} \geq a_i$. Thus, $x_i \leq x_{i+1}, \forall i < n$. ☐

Notice that, if there are i and j such that $i < j$ and $a_i = a_j$, then backward computation yields $a_k = a_j$ for all $k \leq j$.

EXAMPLE 4.9. *For vectors $a = (0, 0, 0.2, 0.2, 0.45, 0.6, 0.6, 0.9)$ and $b = \mathbf{1}$, the solution of Problem 4.2 obtained by backward computation is $x = (0.6, 0.6, 0.6, 0.6, 0.6, 0.6, 0.6, 0.9)$.*

To avoid this problem, we propose the following heuristic for Problem 4.2.

Modified backward computation. We suppose that a is not convex (otherwise, take simply $x = a$). Particularly, a is not a constant vector. Let us denote by $\{i_1 > 1, \ldots, i_s\}$ indices for which vector a strictly increases, i.e.

$$\text{for all } j < s, a_{i_j} < a_{i_{j+1}} \text{ and } a_k = a_{i_j}, \ i_j \leq k < i_{j+1},$$
$$a_k = a_{i_s}, \ k \geq i_s.$$

ALGORITHM 4.10 (Modified backward computation).
  1. *If $i_s = n$, then $x_n = a_n$, $d_s = 1$,*
else,

$$\begin{cases} x_n = b_n - \delta, \ where \ 0 \leq \delta \leq b_n - a_{i_s}. \\ d_s = \frac{x_n - a_{i_s}}{n - i_s}, \\ x_k = x_n - (n - k)d_s, \ i_s \leq k < n. \end{cases}$$

  2. *For $t = s - 1$ to 1,*

$$\begin{cases} d_t = \min\left\{d_{t+1}, \frac{x_{i_{t+1}} - a_{i_t}}{i_{t+1} - i_t}\right\}, \\ x_k = x_{i_{t+1}} - (i_{t+1} - k)d_t, \ i_t \leq k < i_{t+1}. \end{cases}$$

  3. *Computation of entries x_k, $1 \leq k < i_1$:*

$$d \leftarrow d_1,$$
$$\text{For } k = i_1 - 1, \ldots, 1, \begin{cases} x_k = \max\{x_{k+1} - d, a_1\}, \\ d \leftarrow x_{k+1} - x_k. \end{cases}$$

PROPOSITION 4.11. *Let a be an increasing vector and let $b = (\beta, \ldots, \beta)$ be a constant vector such that $a \leq b$. Then vector x computed by Algorithm 4.10 is a solution of Problem 4.2, i.e. x is an increasing and convex vector such that $a \leq x \leq b$.*

Proof. We will first show that $a \leq x \leq b$. For x_n, either $i_s = n$ and $x_n = a_{i_s} = a_n$ or $x_n = \beta - \delta$, where $0 \leq \delta \leq \beta - a_{i_s} = \beta - a_n$. Therefore, $a_n \leq x_n \leq \beta$. If $i_s < n$, then for x_k, $i_s \leq k < n$, we have $x_k = x_n - (n - k)d_s \leq x_n$, as $d_s \geq 0$. On the other hand,

$$x_k = x_n - \frac{n - k}{n - i_s}(x_n - a_{i_s}) \geq a_{i_s} = a_k.$$

Thus, $a_k \leq x_k \leq x_n \leq \beta$, $i_s \leq k < n$. Similarly, by induction on t, we can show that $d_t \geq 0$, for each t such that $1 \leq t \leq s - 1$. Thus, $x_k \leq x_{i_{t+1}}$, $i_t \leq k < i_{t+1}$. By induction on t, $x_k \leq \beta$, $k \geq i_1$. On the other hand,

$$x_k = x_{i_{t+1}} - (i_{t+1} - k)d_t \geq x_{i_{t+1}} - \frac{i_{t+1} - k}{i_{t+1} - i_t}(x_{i_{t+1}} - a_{i_t}) \geq a_{i_t} \geq a_k, \ i_t \leq k < i_{t+1}.$$

Thus, $a_k \leq x_k \leq \beta$, $k \geq i_1$. For $k < i_1$, using the same arguments as in the proof of Proposition 4.8, it can be easily shown that $a_1 = a_k \leq x_k \leq x_{i_1} \leq \beta$.

It remains us to show that vector x is increasing and convex. It can be easily seen that $0 \leq d_1 \leq \ldots \leq d_s$. We have $x_{k+1} - x_k = d_t$, $i_t \leq k < i_{t+1}$. Thus, $x_{k+1} - x_k \leq x_{k+2} - x_{k+1}$, $i_1 \leq k \leq n$. Using the same arguments as in the proof of Proposition 4.8, it can be shown that $x_{k+1} - x_k \leq x_{k+2} - x_{k+1}, k < i_1$, and that $x_2 \geq x_1$. Therefore, vector x is increasing and convex. $\square$

EXAMPLE 4.12. *For vectors a and b from Example 4.9, the solution of Problem 4.2, obtained by modified backward computation is vector x given below:*

$$x = (0, 0.075, 0.2, 0.325, 0.45, 0.6, 0.75, 0.9).$$

Remark that heuristics we presented here do not use the fact that b is a convex vector. Therefore, it is not surprising that we can actually guarantee that they always yield a solution for Problem 4.2 only in the case of a constant vector b. Heuristics that guarantee the solution in the case when b is an arbitrary increasing and convex vector should exploit the fact that we can start with an initial solution. This can be either b, or a solution obtained by means of simple heuristics described above and Proposition 4.7. Describing those heuristics is not in the scope of this paper. However, notice that, if we know a solution x of Problem 4.2, then it can be locally improved in the following way. Suppose $\hat{x}_i = x_i - \epsilon$, $\hat{x}_j = x_j$, $j \neq i$. Then,

$$\epsilon \leq \min\{x_i - a_i, x_i - x_{i-1}, x_i - 2x_{i-1} + x_{i-2}, x_i - 2x_{i+1} + x_{i+2}\}.$$

For $i = n - 1, n$, we have $\epsilon \leq \min\{x_i - a_i, x_i - x_{i-1}, x_i - 2x_{i-1} + x_{i-2}, \}$, for $i = 2$, $\epsilon \leq \min\{x_2 - a_2, x_2 - x_1, x_2 - 2x_3 + x_4\}$, and for $i = 1$, $\epsilon \leq \min\{x_1 - a_1, x_1 - 2x_2 + x_3\}$. This local improvement is interesting only for the entries of a solution x where the slope changes, i.e. where

$$x_{i-1} - x_{i-2} < x_i - x_{i-1} \text{ and } x_{i+1} - x_i < x_{i+2} - x_{i+1}.$$

However, this idea can be generalized to take into account the intervals of constant slope.

4.3. Example. We will illustrate Algorithm 4.5 on a small matrix P.

$$P = \begin{pmatrix} 0.2 & 0 & 0.4 & 0.4 & 0 \\ 0.1 & 0.5 & 0.2 & 0.1 & 0.1 \\ 0.25 & 0.25 & 0 & 0.3 & 0.2 \\ 0.2 & 0.1 & 0 & 0.3 & 0.4 \\ 0.1 & 0 & 0.35 & 0 & 0.55 \end{pmatrix}$$

We will denote by x^t the transposed vector of x. If we use the backward computation algorithm for the last column, we obtain $(Q_{*,5})^t = (0, 0.1, 0.25, 0.4, 0.55)$. Let us consider now column 4. We have

$$\begin{aligned} a &= (\max(f_{*,4}(P), f_{*,5}(Q) + s_{*,5}(Q)))^t &= (0.4, 0.3, 0.7, 1.1, 1.1), \\ b &= (f_{*,5}(Q) + 1)^t &= (1.0, 1.1, 1.25, 1.4, 1.55). \end{aligned}$$

Thus, in order to compute column 4, we have to solve Problem 4.2 with vectors $r(a) = (0.4, 0.4, 0.7, 1.1, 1.1)$ and b. Note that our vector b is not a constant vector, so Proposition 4.8 does not apply. Indeed, the backward computation yields $x = (1.1, 1.1, 1.1, 1.1, 1.1) \not\leq b$ so x is not a solution. By means of Proposition 4.7, we have $\delta = \max_{1 \leq i \leq n} \{b_i - a_i\} = 0.7$, $d = (0.7, \ldots, 0.7)$, and $a + d - b = (0.1, 0, 0.15, 0.4, 0.25)$. Thus, using backward computation for Problem 4.2 with vectors $a + d - b$ and d, we have $y = (0.4, 0.4, 0.4, 0.4, 0.4)$, and

$$(f_{*,4}(\boldsymbol{Q}))^t = y + b - d = (0.7, 0.8, 0.95, 1.1, 1.25).$$

For column 3 we have

$$
\begin{aligned}
a &= (\max(f_{*,3}(\boldsymbol{P}), f_{*,4}(\boldsymbol{Q}) + s_{*,4}(\boldsymbol{Q})))^t &&= (1.4, 1.5, 1.65, 1.8, 2.0), \\
b &= (f_{*,5}(\boldsymbol{Q}) + 1)^t &&= (1.7, 1.8, 1.95, 2.1, 2.25).
\end{aligned}
$$

We can notice that a is increasing and convex. Thus, we can take

$$(f_{*,3}(\boldsymbol{Q}))^t = a = (1.4, 1.5, 1.65, 1.8, 2.0).$$

Finally, for column 2 we have

$$a = (\max(f_{*,2}(\boldsymbol{P}), f_{*,3}(\boldsymbol{Q}) + s_{*,3}(\boldsymbol{Q})))^t = (2.1, 2.2, 2.35, 2.6, 2.9),$$

which is increasing and convex. Thus, $f_{*,2}(\boldsymbol{Q})) = a^t$. The bounding matrix $\boldsymbol{Q}$ is given below. The matrix $\boldsymbol{Q}'$ is obtained by Algorithm 4.5 and modified backward computation.

$$
\boldsymbol{Q} = \begin{pmatrix}
0.3 & 0 & 0 & 0.7 & 0 \\
0.3 & 0 & 0 & 0.6 & 0.1 \\
0.3 & 0 & 0 & 0.45 & 0.25 \\
0.2 & 0.1 & 0 & 0.3 & 0.4 \\
0.1 & 0.15 & 0.05 & 0.15 & 0.55
\end{pmatrix}
\qquad
\boldsymbol{Q}' = \begin{pmatrix}
0.2 & 0.1 & 0.2 & 0.5 & 0 \\
0.3 & 0.1 & 0 & 0.5 & 0.1 \\
0.25 & 0.1 & 0 & 0.4 & 0.25 \\
0.2 & 0.1 & 0 & 0.3 & 0.4 \\
0.15 & 0.1 & 0 & 0.2 & 0.55
\end{pmatrix}
$$

The steady-state distributions are respectively

$$
\begin{aligned}
\pi_{\boldsymbol{P}} &= (0.1663, 0.1390, 0.1982, 0.1998, 0.2966), \\
\pi_{\boldsymbol{Q}} &= (0.1955, 0.0872, 0.0172, 0.3553, 0.3447), \\
\pi_{\boldsymbol{Q}'} &= (0.1951, 0.1000, 0.0390, 0.3293, 0.3366).
\end{aligned}
$$

with expectations $E(\pi_{\boldsymbol{P}}) = 3.3213$, $E(\pi_{\boldsymbol{Q}}) = 3.5665$, and $E(\pi_{\boldsymbol{Q}'}) = 3.5122$. Notice that $\preceq_{st}$-monotone upper bound obtained by Vincent's algorithm [1] is given by

$$
\boldsymbol{R} = \begin{pmatrix}
0.2 & 0 & 0.4 & 0.4 & 0 \\
0.1 & 0.1 & 0.4 & 0.3 & 0.1 \\
0.1 & 0.1 & 0.3 & 0.3 & 0.2 \\
0.1 & 0.1 & 0.1 & 0.3 & 0.4 \\
0.1 & 0 & 0.2 & 0.15 & 0.55
\end{pmatrix}
$$

with $\pi_{\boldsymbol{R}} = (0.1111, 0.0544, 0.2302, 0.2594, 0.3449)$ and $E(\pi_{\boldsymbol{R}}) = 3.6726$. We can notice that, for this example, $E(\pi_{\boldsymbol{Q}'}) < E(\pi_{\boldsymbol{Q}}) < E(\pi_{\boldsymbol{R}})$. Numerical experimentations have shown that this is not always the case.

5. Applications. We will not develop here a complete algorithm to reduce the state space or the complexity of numerical resolution such as the algorithms presented in [15] for the $\preceq_{st}$ order. Indeed, the $\preceq_{st}$-ordering constraints are consistent with ordinary lumpability [28] and with some matrix structures which allow a simpler resolution technique [16]. Clearly, it is more complex to build a monotone upper bound matrix for the $\preceq_{icx}$ order. Thus, it is difficult to generalize the various algorithms presented in [16] which design in only one step a monotone bound simpler to solve.

Instead, we advocate a two step approach to design $\preceq_{icx}$ lumpable bounds. In the first step we obtain using one of the algorithms described in Section 4 an $\preceq_{icx}$-monotone upper bound B of matrix A. Then we use a simpler algorithm to design a lumpable $\preceq_{icx}$ bound (say C) of B. C is not monotone (it may be but it is not enforced by the method). This is a direct consequence of Theorem 2.6. We do not detail here how we can build a lumpable upper bound. Instead, we present two applications. The first one consists in the worst case analysis of models which are not completely specified, while the second one is related to a more traditional use of bounds to reduce the complexity of numerical computation. Formally, in both cases we are interested in finding $\preceq_{st}$ and $\preceq_{icx}$ bounds in a family of distributions.

5.1. Worst case arrivals in a Batch/D/1/N queue. We consider a queue with a single server, finite capacity N, batch arrivals and deterministic service. Let $A = (a_0, \ldots, a_K)$ denote the distribution of the batch arrivals. We assume that we only know the average batch size $\alpha = E(A)$. Note that the average batch size is closely related to the load which is quite simple to measure. Assume that $N >> K$ and $\alpha < 1$. The exact values of a_i ($0 \leq i \leq K$) are unknown. A natural question when we analyze the average queue size of such a system is to find the worst batch distribution. We also describe the model by the set of possible nonzero transitions. Obviously, we know the abstract transition matrix of the Markov chain:

$$P = \begin{pmatrix} a_0 & a_1 & \cdots & a_K & 0 & \cdots & 0 \\ a_0 & a_1 & \cdots & a_K & 0 & \cdots & 0 \\ 0 & a_0 & a_1 & \cdots & a_K & \cdots & 0 \\ \vdots & \ddots & \ddots & \ddots & \ddots & \ddots & \vdots \\ 0 & \ddots & 0 & a_0 & a_1 & \cdots & a_K \\ \vdots & & & \ddots & \ddots & \ddots & \vdots \\ 0 & \cdots & \cdots & \cdots & 0 & a_0 & \sum_{i=1}^{K} a_i \end{pmatrix}$$

Let $\mathcal{F}_\alpha$ be the family of all distributions on the space $\mathcal{E} = \{0, \cdots, n\}$ having the same mean α. In the two following properties, we study the existence of a maximal element in the set $\mathcal{F}_\alpha$.

PROPERTY 5.1. *(See [26, Theorem 2.A.9]) The worst case distribution in the sense of the $\preceq_{icx}$ order is given by $q = (\frac{n-\alpha}{n}, 0, \ldots, 0, \frac{\alpha}{n})$, i.e.*

$$q \in \mathcal{F}_\alpha \quad and \quad p \preceq_{icx} q, \; \forall p \in \mathcal{F}_\alpha.$$

The non existence of the $\preceq_{st}$ bound in the same set (next property) is simply related to the fact that if $Y \preceq_{st} X$ and $E(X) = E(Y)$ then X and Y have the same distribution.

PROPERTY 5.2. *For the $\preceq_{st}$ order, there is no distribution r satisfying:*

$$r \in \mathcal{F}_\alpha \quad and \quad p \preceq_{st} r, \; \forall p \in \mathcal{F}_\alpha$$

We proceed in the following way to obtain matrix B which is an $\preceq_{icx}$-monotone upper bounding matrix for P:

(i) First, we obtain an upper bound Q which is not $\preceq_{icx}$-monotone. We apply Proposition 5.1 at each row to construct matrix Q. For the sake of brevity, we cannot detail the whole process here (see [9] for a complete derivation). Note that we must adapt Property 5.1 because the pattern of nonzero transitions changes at every row. For rows $i \leq N - K + 1$, the probability mass is concentrated in transitions to states $i - 1$ and $i - 1 + K$. Then the bounding distribution is $q = (\bar{b}, 0, \ldots, 0, b)$ where $b = \frac{m}{K}$ and $\bar{b} = 1 - b$. For the other rows the probability mass is concentrated in transitions to states $i - 1$ and N. When $i \geq N - K + 2$, $Q_{i,i-1} = 1 - b_i$ and $Q_{i,N} = b_i$ with $b_i = \frac{m}{N-i+1}$.

Matrix Q is not $\preceq_{icx}$-monotone. Indeed the last row is not convex. We can see that $Q_{N-K,N} = 0$, $Q_{N-K+1,N} = \frac{m}{K}$, and $Q_{N-K+2,N} = \frac{m}{K-1}$. Moreover, it is not sufficient to make the last row convex.

(ii) We now apply the polynomial transform $t_\delta(Q) = \delta Q + (1 - \delta)Id$, where Id is the identity matrix. This transform does not change the steady-state distribution and it is known to increase the accuracy of the $\preceq_{st}$ bounds when it is used as a preprocessing [12]. Here it allows to move some probability mass to the diagonal elements.

(iii) Then we apply the forward algorithm to the last row of the transformed matrix, $t_\delta(Q)$.

(iv) Finally we change some diagonal and sub-diagonal elements to make the matrix $\preceq_{icx}$-monotone and we obtain matrix B:

$$B = \begin{cases} B_{0,0} = 1 - \delta b & B_{0,K} = \delta b \\ i = 1, \cdots, N - K + 1 : \\ B_{i,i-1} = \delta(1 - b) & B_{i,i} = 1 - \delta & B_{i,i+K-1} = \delta b \\ i = N - K + 2, \cdots, N - 1 : \\ B_{i,i-1} = f & B_{i,i} = e & B_{i,N} = \delta b(i - N + K) \\ B_{N,N-1} = \delta(1 - m) & B_{N,N} = 1 - \delta + \delta m \end{cases}$$

where

$$e = 1 - \delta + \delta m - \delta m \frac{(i - N + K)(N - i + 1)}{K} \text{ and } f = 1 - e - \delta b(i - N + K).$$

Let U be the maximum value of $\frac{(i-N+K)(N-i+1)}{K}$. We have proved in [9] the following property:

PROPERTY 5.3. *If $\delta \leq \frac{1}{1+mU}$ then matrix B is irreducible and stochastic. Furthermore, B is $\preceq_{icx}$-monotone and provides an upper bound of the steady state distribution of P.*

Matrix B has the same structure as $t_\delta(Q)$ and its stationary distribution can be easily computed using an elimination algorithm. This matrix provides an upper bound for the distribution of the population in the queue.

5.2. Absorption time. Several high level modeling approaches combine a hierarchy of submodels. For instance, PEPA nets [17] are based on Petri nets and the descriptions of places and transitions use PEPA, a Stochastic Process Algebra. This is an explicit two level model. For Stochastic Automata Network the hierarchy is implicit [27]: the automata describe local transitions and the interaction between them is modeled by synchronized transitions and functions which are carried by the labeled transitions. This hierarchy is appealing to model complex structures but it is not always useful to solve the model because the classical technique considers the global state space. Note that even if we represent the transition matrix by a tensor representation [27], we consider the global state space during the resolution process. So we want to develop new methods which can use the submodels during the resolution and combine them in an efficient way. Here we present an application of $\preceq_{icx}$ bounds when the model of the low level is an absorbing Markov chain while the first level of the hierarchy exhibits some structural properties.

We just introduce the approach which will be developed in a sequel paper. We assume here that the high level formalism is a precedence graph but the approach can be easily generalized to other formalisms. There are n nodes representing tasks to complete according to the synchronization constraints defined by the arcs of the precedence graph. Each task is modeled by a DTMC with one absorbing state and the service time (holding time) of task i, d_i is the the absorption time of this Markov chain. We assume that the precedence graph has an unique input node (1) and an unique end node (n). The overall completion time is defined as the duration between the beginning of node 1 and the termination of node n. We are interested in computing the distribution of the completion time for the underlying graph. Note that if a return arc from the end node to the beginning node is added, similar techniques can be used to compute the distribution of the cycle time or the throughput of the system.

The service time of each node follows indeed a discrete PHase type (PH) distribution. A discrete PH distribution is defined by the initial distribution (say τ) and the transition probability matrix Q. Let X be the absorption time of the corresponding chain when the initial distribution is τ. Without loss of generality we assume that there exists only one absorbing state which is the last one. We also assume that the initial distribution τ is $(1, 0, \ldots, 0)$. Indeed, a general distribution can be considered by adding an extra state at the beginning. The global description of the model is Markovian but the state space is huge. Typically we must consider a subset of the product of all the PH distributions for task service times. The main idea here consists in the reduction of the complexity of a PH distribution taking into account the properties of both levels in the model.

It is known that precedence graphs exhibit (max,+) linear equations for their sequence of daters (i.e. the triggering instant of a transition) [2]. Let t_i be the completion time of node i, and $P(i)$ be the set of predecessors of i in the precedence graph. Since node i is trigged (executed) as soon as all its predecessors have been completed, its completion time is defined as follows:

$$t_i = d_i + \max_{j \in P(i)} t_j$$

Thus we obtain linear equations with two operators: the addition and the maximum. Remark that in the case $t_1 = d_1$, the overall completion time is t_n. Such linear (max,+) equations have been extensively studied as they allow new types of analytical or numerical methods which are not based on exponential delays or embedded Markov chains. They also allow an important reduction of complexity for Markovian models.

A fundamental property of increasing convex ordering is the compatibility with these operators. Thus if we can build upper bounds on service times of nodes, we obtain upper bounds on the node completion times, so an upper bound on the completion time of the global system. Note that service times of nodes are supposed to be independent.

PROPERTY 5.4. *If for some* i, $d_i \preceq_{st} m$ *(res.* $d_i \preceq_{icx} m$*), then* $t_n \preceq_{st} \tilde{t}_n$ *(res.* $t_n \preceq_{icx} \tilde{t}_n$*), where* $\tilde{t}_n$ *denotes the completion time of node* n *in the system where* d_i *has been replaced by* m.

In our application example, we aim to bound PH type service times of nodes. Let us first define a family of random variables related to a well known set in reliability modeling [3].

DEFINITION 5.5 (Discrete New Better than Used (in Expectation) (DNBU(E))). *Let* X_t *be an integer valued random variable modeling the residual time of* X, *given that* $X > t$, *i.e.* $X_t = [X - t | X > t]$. X *is said to be [19]:*
   (i) *DNBUE if* $E(X_t) \le E(X)$ *for all* t *integer,*
   (ii) *DNBU if* $X_t \preceq_{st} X$ *for all* t *integer.*
Notice that DNBU $\Rightarrow$ *DNBUE.*

The main result we use is the $\preceq_{icx}$-comparison for any DNBUE random variable with a geometric one.

PROPERTY 5.6. *[19] If* X *is DNBUE of mean* m, *then* X *is smaller in the* $\preceq_{icx}$ *sense than a geometric distributed random variable of mean* m.

Since we only need a one state model to generate a geometric distribution, if we can bound the low level model (PH distributions for service times of nodes) by a DNBUE distribution, the global state space would be largely reduced.

Of course a PH distribution is not in general DNBUE. However it is simple to bound some set of PH distributions by DNBUE ones. We show in the sequel how DNBUE bounds can be constructed for acyclic PH distributions. These distributions have received considerable attention as they are sufficient to approximate general distributions (see [6] for the theory and a fitting algorithm). The following property is a direct consequence of the memoryless property of the geometric distribution.

PROPERTY 5.7. *A sum of independent but not necessarily identically distributed geometric distributions is DNBUE. Remark that this distribution is the uniformization of a hypoexponential distribution.*

PROPERTY 5.8. *An arbitrary acyclic discrete PH distribution is upper bounded in the* $\preceq_{st}$ *sense by a PH distribution defined as a sum of independent but not necessarily identically distributed geometric distributions.*

Recall that $\preceq_{st}$-comparison implies $\preceq_{icx}$-comparison. Thus, this property can be combined with Properties 5.6 and 5.7 to derive an upper bound for an acyclic PH distribution. Let us now illustrate how we can algorithmically construct this upper bound. Suppose that the acyclic PH distribution X is given by its upper triangular transition matrix denoted by $\boldsymbol{P}$. We only need to find a lower bounding (in the $\preceq_{st}$ or $\preceq_{icx}$ sense) matrix $\boldsymbol{Q}$ which has nonzero entries only on the diagonal and first upper diagonal. Note that a matrix of this form is always $\preceq_{st}$ and $\preceq_{icx}$-monotone. Let us denote the corresponding PH distribution by Y. Then $X \preceq_{st} Y$ follows directly from Proposition 2.9.

It can be easily shown that the greatest lower bound of the above form for matrix $\boldsymbol{P}$ in the $\preceq_{st}$ (res. $\preceq_{icx}$) sense is the matrix $\boldsymbol{A}$ (res. $\boldsymbol{B}$), where

$$(5.1) \qquad A_{i,i+1} = s_{i,i+1}(\boldsymbol{P}) = \sum_{j=i+1}^{n} P_{i,j}, \quad B_{i,i+1} = \min(f_{i,i+1}(\boldsymbol{P}), 1), \; \forall i.$$

Note that for each matrix $\boldsymbol{P}$ we obtain $\boldsymbol{A} \preceq_{st} \boldsymbol{B}$, since $s_{i,i+1}(\boldsymbol{P}) \le \min(f_{i,i+1}(\boldsymbol{P}), 1)$ (see (4.1)). Thus, in this simple case the $\preceq_{icx}$ order provides better bounds.

Let us now illustrate this on a small numerical example. Let P be the matrix of an acyclic PH distribution X,

$$P = \begin{pmatrix} 0.3 & 0.5 & 0.2 \\ 0 & 0.3 & 0.7 \\ 0 & 0 & 1 \end{pmatrix}.$$

Then the matrices A and B given by (5.1) are as follows:

$$A = \begin{pmatrix} 0.3 & 0.7 & 0 \\ 0 & 0.3 & 0.7 \\ 0 & 0 & 1 \end{pmatrix}, B = \begin{pmatrix} 0.1 & 0.9 & 0 \\ 0 & 0.3 & 0.7 \\ 0 & 0 & 1 \end{pmatrix}.$$

Let Y and Z denote PH distributions (starting at state 1) corresponding to A and B. We have $E(X) = 2.449$, $E(Y) = 2.857$, and $E(Z) = 2.540$. Clearly, the $\preceq_{st}$ bound (A) is less accurate than the $\preceq_{icx}$ bound (B). Furthermore, from Property 5.6 it follows that $X \preceq_{icx} Geom(p)$, where $1/p = E(Z) = 2.540$.

We can also easily check if an arbitrary PH distribution (starting at state 1) is DNBUE. The following property provides only a sufficient condition.

PROPERTY 5.9. *Let $a_i, i = 1, \ldots, n-1$ denote the mean residual time X_t of PH distribution X (with transition matrix P), given that $X_t = i$. Suppose that $P_{i,i} < 1, \forall i$. Then a_i can be found by solving the linear system:*

$$a_i = \frac{1}{1 - P_{i,i}}(1 + \sum_{j \neq i} P_{i,j} a_j).$$

If $a_1 \geq \max_{i>1} a_i$, then X (starting at 1) is DNBUE.

Finally, we can use the following property to build an $\preceq_{st}$-lower bound that is DNBUE for an arbitrary PH distribution and combine this result with Property 5.6 in order to reduce the state space of our model.

PROPERTY 5.10. *An arbitrary PH distribution starting at state 1 with the transition matrix that is $\preceq_{st}$ or $\preceq_{icx}$-monotone is DNBU (and, consequently, DNBUE).* The proof of this property uses similar arguments as the proof of Proposition 2.9.

6. Conclusion. The stochastic comparison has been largely applied in different areas of applied probability. Recently, algorithmic stochastic comparison in the sense of the $\preceq_{st}$ order has been developed for Markovian analysis. In this paper, we have considered theoretical and algorithmic issues of the $\preceq_{icx}$ order. We hope that these will open new horizons for Markov chain analysis approaches. Our aim was not to compare $\preceq_{icx}$ bounds with $\preceq_{st}$ bounds. We found that it is generally not possible to stochastically compare the distributions obtained by both methods. We will see in the future which approach, if any, is the more accurate and we will develop the complexity issues about these new algorithms, especially for sparse matrices. We presented some applications for the worst case analysis which is an important issue in performance evaluation when the complete specification of the underlying model is unknown.

REFERENCES

[1] O. ABU-AMSHA AND J.-M. VINCENT, *An algorithm to bound functionals of Markov chains with large state space*, in 4th INFORMS Conference on Telecommunications, Boca Raton, FL, 1998.

[2] F. BACCELLI, G. COHEN, G. J. OLSDER, AND J.-P. QUADRAT, *Synchronization and Linearity: An Algebra for Discrete Event Systems*, Willey, New York, 1992.

[3] R. E. BARLOW AND F. PROSCHAN, *Statistical Theory of Reliability and Life Testing*, Holt, Rinehart and Winston, New York, 1975.

[4] M. BEN MAMOUN AND N. PEKERGIN, *Closed-form stochastic bounds on the stationary distribution of Markov chains*, Probability in the Engineering and Informational Sciences, 16 (2002), pp. 403–426.

[5] M. BEN MAMOUN, N. PEKERGIN, AND SANA YOUNES, *Model checking of continuous-time Markov chains by closed-form bounding distributions*. Submitted.

[6] A. BOBBIO, A. HORVATH, M. SCARPA, AND M. TELEK, *Acyclic discrete phase type distributions: properties and a parameter estimation algorithm*, Performance Evaluation, 54 (2003), pp. 1–32.

[7] P. BUCHHOLZ, *An improved method for bounding stationary measures of finite Markov processes*, Performance Evaluation, 62 (2005), pp. 349–365.

[8] A. BUSIC AND J.-M. FOURNEAU, *A matrix pattern compliant strong stochastic bound*, in 2005 IEEE/IPSJ International Symposium on Applications and the Internet Workshops (SAINT 2005 Workshops), Trento, Italy, IEEE Computer Society, 2005, pp. 256–259.

[9] A. BUSIC, J.-M. FOURNEAU, AND N. PEKERGIN, *Worst case analysis of batch arrivals with the icx ordering*. Submitted.

[10] P.-J. COURTOIS AND P. SEMAL, *Bounds for the positive eigenvectors of nonnegative matrices and for their approximations by decomposition*, Journal of the ACM, 31 (1984), pp. 804–825.

[11] ———, *Computable bounds for conditional steady-state probabilities in large Markov chains and queueing models*, IEEE Journal on Selected Areas in Communications, 4 (1986), pp. 926–937.

[12] T. DAYAR, J.-M. FOURNEAU, AND N. PEKERGIN, *Transforming stochastic matrices for stochastic comparison with the st-order*, RAIRO Operations Research, 37 (2003), pp. 85–97.

[13] M. DOISY, *Comparaison de Processus Markoviens*, PhD thesis, Université de Pau et des Pays de l'Adour, 1992.

[14] J.-M. FOURNEAU, M. LE COZ, N. PEKERGIN, AND F. QUESSETTE, *An open tool to compute stochastic bounds on steady-state distributions and rewards*, in 11th International Workshop on Modeling, Analysis, and Simulation of Computer and Telecommunication Systems (MASCOTS 2003), Orlando, FL, IEEE Computer Society, 2003, p. 219.

[15] J.-M. FOURNEAU, M. LE COZ, AND F. QUESSETTE, *Algorithms for an irreducible and lumpable strong stochastic bound*, Special Issue on the Conference on the Numerical Solution of Markov Chains 2003, Linear Algebra and its Applications, 386 (2004), pp. 167–185.

[16] J.-M. FOURNEAU AND N. PEKERGIN, *An algorithmic approach to stochastic bounds*, in Performance Evaluation of Complex Systems: Techniques and Tools, Performance 2002, Tutorial Lectures, M. Calzarossa and S. Tucci, eds., vol. 2459 of Lecture Notes in Computer Science, Springer, 2002, pp. 64–88.

[17] S. GILMORE, J. HILLSTON, L. KLOUL, AND M. RIBAUDO, *PEPA nets: a structured performance modelling formalism*, Performance Evaluation, 54 (2003), pp. 79–104.

[18] S. HADDAD AND P. MOREAUX, *Sub-stochastic matrix analysis for bounds computation: Theoretical results*. To appear in European Journal of Operational Research.

[19] T. HU, M. MA, AND A. K. NANDA, *Moment inequalities for discrete ageing families*, Communications in Statistics, 32 (2003), pp. 61–90.

[20] J. KEILSON AND A. KESTER, *Monotone matrices and monotone Markov processes*, Stochastic Processes and their Applications, 5 (1977), pp. 231–241.

[21] J. LEDOUX AND L. TRUFFET, *Markovian bounds on functions of finite Markov chains*, Advances in Applied Probability, 33 (2001), pp. 505–519.

[22] H. LI AND M. SHAKED, *Stochastic convexity and concavity of Markov processes*, Mathematics of Operations Research, 19 (1994), pp. 477–493.

[23] K. MOSLER AND M. SCARSINI, *Stochastic Orders and Applications: A Classified Bibliography*, vol. 401 of Lecture Notes in Economics and Mathematical Systems, Springer-Verlag, 1993.

[24] A. MULLER AND D. STOYAN, *Comparison Methods for Stochastic Models and Risks*, Wiley, New York, NY, 2002.

[25] N. PEKERGIN, *Stochastic performance bounds by state space reduction*, Performance Evaluation, 36-37 (1999), pp. 1–17.

[26] M. SHAKED AND J. G. SHANTIKUMAR, *Stochastic Orders and their Applications*, Academic Press, San Diego, CA, 1994.

[27] W. J. STEWART, *Introduction to the Numerical Solution of Markov Chains*, Princeton University Press, New Jersey, NJ, 1994.

[28] L. TRUFFET, *Reduction technique for discrete time Markov chains on totally ordered state space using stochastic comparisons*, Journal of Applied Probability, 37 (2000), pp. 795–806.

POLYNOMIALS OF A STOCHASTIC MATRIX AND STRONG STOCHASTIC BOUNDS*

TUĞRUL DAYAR[†], JEAN-MICHEL FOURNEAU[‡], NIHAL PEKERGIN[§], AND JEAN-MARC VINCENT[¶]

Abstract. Bounding by stochastic comparison is a promising technique for performance evaluation since it enables the verification of performance measures efficiently. To improve the accuracy of this technique, preprocessing of a stochastic matrix before computing a strong stochastic bound is considered. Using results from linear algebra, it is shown that some polynomials of the stochastic matrix give more accurate bounds. Numerical results are presented to illustrate the ideas, and a stochastic interpretation is provided.

Key words. Polynomials of a stochastic matrix, stochastic comparison, strong stochastic order, bounding

AMS subject classifications. 60J10, 15A51, 15A45, 60J22, 65F30

1. Introduction. Stochastic comparison is often presented as a powerful technique in various areas of applied probability (see, for instance, the books by Stoyan [20], Shaked and Shantikumar [18], and Kijima [10]). When models are too complex to be analyzed efficiently, the technique enables the study of models that guarantee upper or lower bounds on performance measures of the original model, which remain unknown. Recently, several applications of this technique in practical problems of telecommunications engineering [13, 14, 16] and stochastic model checking [11, 17] are reported.

There are different stochastic ordering relations and the most widely used is the strong stochastic (i.e., st) ordering, which defines a partial order on probability distributions. Sufficient conditions for the stochastic comparability (under a stochastic ordering relation) of two time-homogeneous Markov chains (MCs) are given by the stochastic monotonicity and bounding properties of their one step stochastic transition probability matrices [9, 12]. In [21], this idea is used to devise an algorithm which constructs an optimal st-monotone (upper) bounding MC for a given MC. Later in [1], this algorithm is generalized so as to compute stochastic bounds on performance measures of a reduced model obtained from the original one by aggregation. The ideas in [1] are extended to nearly completely decomposable MCs in [22] and to ordinary and exact lumpability in [23]. Algorithmic derivations of stochastic bounds obtained by stochastic comparison of MCs can be found in [6, 7, 16], and a tool is presented in [8]. In passing, we remark that stochastic comparison may also be applied to continuous-time MCs through uniformization.

The key problem in the stochastic comparison approach is to obtain accurate

---

*This work was supported by a TÜBİTAK-CNRS joint grant and project Sure-Paths from ACI Sécurité.

†Department of Computer Engineering, Bilkent University, TR-06800 Bilkent, Ankara, Turkey; Tel: (+90) (312) 290-1981; Fax: (+90) (312) 266-4047 (tugrul@cs.bilkent.edu.tr).

‡PR*i*SM, Université de Versailles-St.Quentin, 45 av. des Etats Unis, 78035 Versailles, France; Tel: (+33) 1 39254000; Fax: (+33) 1 39254057 (Jean-Michel.Fourneau@prism.uvsq.fr).

§PR*i*SM, Université de Versailles-St.Quentin, 45 av. des Etats Unis, 78035 Versailles, France and Centre Marin Mersenne, Université Paris 1, 90 rue Tolbiac, 75013 Paris, France; Tel: (+33) 1 39254220; Fax: (+33) 1 39254057 (Nihal.Pekergin@prism.uvsq.fr).

¶Laboratoire Informatique et Distribution, 51, avenue Jean Kuntzmann, 38330 Montbonnot, France; Tel: (+33) 4 76612055; Fax: (+33) 4 76612099 (Jean-Marc.Vincent@imag.fr).

bounds. That is, a bounding probability distribution close to the exact probability distribution in the stochastic ordering sense is sought. It is observed that some pre-processing of the stochastic matrix may lead to more accurate bounds, and to that effect, two types of preprocessing are considered. First, states of the MC may be re-ordered using some heuristics [4, 16]. The results in [16] are important in showing that there are problems for which such a bounding technique coupled with aggregation can give accurate bounds in a smaller amount of time than it would take to compute the exact distribution up to machine precision using state-of-the-art numerical methods alone. Furthermore, as indicated in [3], reorderings are especially beneficial if the MC has some structural properties that can be exploited. The advantage of reordering is that it is still possible to obtain bounds on transient measures. Second, the MC may be transformed to another MC so that the steady-state probability distribution remains invariant. In particular, it is shown in [5] that the accuracy of the bounds improve in the strong stochastic sense if a very simple linear transformation is em-ployed. The transformation is so simple that the computation required is proportional to the number of nonzeros in the stochastic matrix, which is roughly the computation required by a vector-matrix multiplication. Furthermore, the transformation can be computed on the fly while the matrix is generated from a higher level specification and does not require any specific structure. Clearly, this approach provides bounds only on steady-state measures since it is not evident how the transient distributions are related. The objective of this paper is not to discuss the benefits of aggregation or to undertake a numerical study exploiting different reorderings and structural properties with stochastic bounding. Here, we consider the second type of preprocessing using higher degree polynomials, which will be more expensive to compute, but which will improve the accuracy of the steady-state bounds further. However, we also extend the result to improving the bound for the absorption time in a transient MC with one absorbing state.

The paper is organized as follows. In section 2, we briefly provide some back-ground information and introduce the algorithm in [1]. Section 3 is devoted to the analysis of the proposed transformations including the improvement of the bound for the absorption time. In section 4, we give some numerical examples and introduce two classes of MCs for which results are readily available. In section 5, we show the convexity of a desirable set of polynomials, while in section 6, we present a formulation based on stochastic processes for some of our results. In section 7, we conclude.

2. Background.

First, we recall the definition of st-ordering used in this paper. For further information on the stochastic comparison method, we refer the reader to [20]. The first definition is quite general; the second one applies only to discrete random variables and it leads to an algebraic statement of st-comparison. In the following, when used between two vectors or two matrices, $\leq$ has the usual meaning of elementwise less than or equal to. However, $p[j]$ refers to element j of the vector p and $P[i, j]$ refers to element (i, j) of the matrix P. Although unconventional, this notation is adopted since parentheses and subscripts are reserved for compositions and dependencies.

DEFINITION 2.1. *Let X and Y be random variables taking values on a totally ordered space. Then X is said to be less than Y in the strong stochastic sense, that is, $X \leq_{st} Y$ if and only if $E[f(X)] \leq E[f(Y)]$ for all nondecreasing functions f whenever the expectations exist.*

DEFINITION 2.2. *Let X and Y be random variables taking values on the state space $\mathcal{S} = \{1, 2, \ldots, n\}$. Let p and q be probability distribution vectors such that*

$$p[j] = Pr(X = j) \quad and \quad q[j] = Pr(Y = j) \quad \forall j \in \mathcal{S}.$$

Then X is said to be less than Y in the strong stochastic sense, that is, $X \leq_{st} Y$, if and only if

$$\sum_{j=k}^{n} p[j] \leq \sum_{j=k}^{n} q[j] \quad \forall k \in \mathcal{S}.$$

Hence, the st-ordering defines a partial order on probability distributions, meaning the relation is reflexive, antisymmetric, and transitive. The reflexivity implies that $X \leq_{st} X$ for all random variables X; however, the antisymmetricity does not require all pairs of random variables X and Y to be related as $X \leq_{st} Y$ or $Y \leq_{st} X$, thus the partialness of the order.

For the sake of completeness, we recall the well known fundamental theorem (see for instance Theorem 3.4 of [12, p. 355]) which states for two time-homogeneous MCs that the st-comparability of their initial probability distributions, the st-monotonicity of one of them, and their st-comparability yield sufficient conditions for their strong stochastic comparison:

THEOREM 2.3. *Let P and Q be stochastic matrices of order n respectively characterizing the time-homogeneous MCs $X(t)$ and $Y(t)$ for $t \in \mathbb{N}$ on the state space $\mathcal{S} = \{1, 2, \ldots, n\}$. Then $\{X(t)\}_{t \in \mathbb{N}} \leq_{st} \{Y(t)\}_{t \in \mathbb{N}}$ (meaning, $X(t) \leq_{st} Y(t)$ for $\forall t \in \mathbb{N}$) if*

(i) $X(0) \leq_{st} Y(0)$,

(ii) st-monotonicity of at least one of the matrices holds; that is,

$$either \quad P[i, *] \leq_{st} P[j, *] \quad or \quad Q[i, *] \leq_{st} Q[j, *] \quad \forall i, j \in \mathcal{S} \quad such \ that \quad i \leq j,$$

(iii) st-comparability of the matrices holds; that is, $P[i, *] \leq_{st} Q[i, *] \ \forall i \in \mathcal{S}$, where $P[i, *]$ refers to row i of P.

COROLLARY 2.4. *Let P and Q be stochastic matrices of order n respectively characterizing the time-homogeneous MCs $X(t)$ and $Y(t)$ on the state space $\mathcal{S} = \{1, 2, \ldots, n\}$. If $\{X(t)\}_{t \in \mathbb{N}} \leq_{st} \{Y(t)\}_{t \in \mathbb{N}}$, $\lim_{t \to +\infty} X(t)$ and $\lim_{t \to +\infty} Y(t)$ exist, and π_P and π_Q are respectively the steady-state distributions of P and Q, then $\pi_P \leq_{st} \pi_Q$.*

We remark that π_P and π_X are used interchangeably throughout the text.

Unrolling the constraints in parts (ii) and (iii) of Theorem 2.3, one can devise the following algorithm due to Vincent [1]:

VINCENT'S ALGORITHM. Constructs an optimal st-monotone upper bounding stochastic matrix Q corresponding to the stochastic matrix P of order n.

$Q[1, n] = P[1, n]$;
for $i = 2, 3, \ldots, n$,
    $Q[i, n] = \max\{Q[i - 1, n], P[i, n]\}$;
for $l = n - 1, n - 2, \ldots, 1$,
    $Q[1, l] = P[1, l]$;
    for $i = 2, 3, \ldots, n$,
        $Q[i, l] = \max\{\sum_{j=l}^{n} Q[i - 1, j], \sum_{j=l}^{n} P[i, j]\} - \sum_{j=l+1}^{n} Q[i, j]$;

Let us now introduce two operators on nonnegative matrices of order n which provide a simple formulation for Vincent's algorithm as in [5, pp. 92–93].

DEFINITION 2.5. *Let $\mathcal{B}$ be the set of stochastic matrices of order n on the state space $\mathcal{S} = \{1, 2, \ldots, n\}$, and let $P \in \mathcal{B}$. Then:*

(i) r is the summation operator used in st-comparison and defined on P as

$$r(P)[i,j] = \sum_{k=j}^{n} P[i,k] \qquad \forall i, j \in \mathcal{S}.$$

In general, $r(P)$ is not a stochastic matrix. Now, let $\mathcal{A} = \{r(P) \mid P \in \mathcal{B}\}$ and $Z \in \mathcal{A}$. Then $r^{-1}(Z)$ is well defined and given by

$$r^{-1}(Z)[i,j] = \begin{cases} Z[i,n], & \forall i \in \mathcal{S} \quad \text{and} \quad j = n \\ Z[i,j] - Z[i,j+1], & \forall i \in \mathcal{S} \quad \text{and} \quad 1 \le j \le n-1 \end{cases}.$$

(ii) v is an operator which transforms $P \in \mathcal{B}$ into a matrix in $\mathcal{A}$ and is defined by

$$v(P)[i,j] = \max_{m \le i}\{r(P)[m,j]\} \qquad \forall i, j \in \mathcal{S}.$$

The next result follows from Definition 2.5 and appears in [9].

COROLLARY 2.6. *The summation operation r on the stochastic matrix $P \in \mathcal{B}$ is equivalent to postmultiplying P by the lower-triangular matrix*

$$L = \begin{pmatrix} 1 & & & \\ 1 & 1 & & \\ \vdots & \vdots & \ddots & \\ 1 & 1 & \cdots & 1 \end{pmatrix}$$

of order n. That is, $r(P) = PL$. The inverse of the summation operation on the matrix $Z \in \mathcal{A}$ is equivalent to postmultiplying Z by the lower-bidiagonal matrix

$$L^{-1} = \begin{pmatrix} 1 & & & \\ -1 & 1 & & \\ & \ddots & \ddots & \\ & & -1 & 1 \end{pmatrix}.$$

That is, $r^{-1}(Z) = ZL^{-1}$.

PROPERTY 2.7. *The st-comparability of the probability distribution vectors p, q of length n, the st-comparability of $P, Q \in \mathcal{B}$, and the st-monotonicity of Q can be defined through the operator r and the lower-triangular matrix L as follows:*

(i) st-comparability $p \le_{st} q \iff pL \le qL$;

(ii) st-comparability $P \le_{st} Q \iff r(P) \le r(Q) \iff PL \le QL$;

(iii) st-monotonicity of $Q \iff r(Q)[i,j] \le r(Q)[i+1,j]$
$$1 \le i \le n-1 \quad \text{and} \quad \forall j \in \mathcal{S}$$
$$\iff L^{-1}QL \ge 0.$$

A careful look at Vincent's algorithm shows that it is related to both operators introduced in Definition 2.5.

PROPOSITION 2.8. *The operator corresponding to Vincent's algorithm is $r^{-1}v$.*

Some important properties have been proved about Vincent's algorithm one of which is that the st-monotone upper bounding stochastic matrix Q may be reducible, even if the stochastic matrix P is irreducible. Yet, when Q is irreducible, it has a single subset of irreducible states and therefore a steady-state distribution (see section 4 in [15]). A new algorithm, named IMSUB, is introduced in [6] to fix this problem. Nevertheless, Vincent's algorithm possesses an important optimality property [21]:

PROPERTY 2.9. *Let $U \in \mathcal{B}$ be another st-monotone upper bounding stochastic matrix for $P \in \mathcal{B}$. Then $Q \in \mathcal{B}$ obtained by Vincent's algorithm is optimal in the sense that $Q \leq_{st} U$.*

However, the optimality property does not imply that the bound on the state probability distribution of P cannot be improved. In [5], the idea of first preprocessing P with a transformation, which does not affect its steady-state distribution, and then applying Vincent's algorithm is studied. Specifically, a linear transformation of the form

$$(2.1) \qquad \Gamma_\delta(P) = \delta I + (1 - \delta)P,$$

where I is the identity matrix and $\delta \in [0, 1)$, is considered. Therein, it is proved for all δ that

$$\pi_P \leq_{st} \pi_{r^{-1}v(\Gamma_\delta(P))} \leq_{st} \pi_{r^{-1}v(P)},$$

where π_P, $\pi_{r^{-1}v(\Gamma_\delta(P))}$, and $\pi_{r^{-1}v(P)}$ are respectively the steady-state distributions of P, $r^{-1}v(\Gamma_\delta(P))$, and $r^{-1}v(P) = Q$. Observe that the equalities are attained if P is already st-monotone; otherwise, it is a better idea to transform P before applying Vincent's algorithm. In [5], it is also shown that $\delta = 1/2$ provides the most accurate bounds among linear transformations. However, this should not be interpreted to mean that the bounding probability distribution always improves with this transformation. The transformation may simply generate the same bounding distribution with that of the untransformed stochastic matrix, yet equality also satisfies st-ordering.

In the next section, we study higher degree polynomials which leave the steady-state distribution of P invariant and give more accurate bounds. This provides a trade-off between computational cost and accuracy.

3. Stochastic comparison: an algebraic approach. We first provide a new proof of Property 2.9 using the operators in Definition 2.5 so as to aid the ideas developed in this section. Then, through algebraic techniques, we introduce and prove results for the st-comparison of a special set of polynomials of a stochastic matrix.

3.1. A new proof of optimality. Let $\mathcal{B}$ be the set of stochastic matrices of order n as in Definition 2.5.

LEMMA 3.1. *If $P \in \mathcal{B}$ and $r^{-1}v(P) = Q$, then $Q \in \mathcal{B}$ is the smallest st-monotone stochastic matrix such that $P \leq_{st} Q$. That is, if $U \in \mathcal{B}$ is another stochastic matrix such that*

(i) U is st-monotone

(ii) $P \leq_{st} U$,

then $Q \leq_{st} U$.

Proof. Observe that $r^{-1}v(P) = Q$ may be written as $v(P) = r(Q)$. Then it follows from the st-comparability of Q and U in part (ii) of Property 2.7 and the operator for Vincent's algorithm in Proposition 2.8 that we must prove for each row

$i \in \mathcal{S}$ the inequality

$$v(P)[i,j] \leq r(U)[i,j] \qquad \forall j \in \mathcal{S}.$$

The proof is based on an induction argument over the row indices. For row 1, the operators r and v are equivalent (see Definition 2.5), and therefore, from part (ii) of Property 2.7 we have

$$v(P)[1,j] = r(P)[1,j] \leq r(U)[1,j] \qquad \forall j \in \mathcal{S}.$$

This is the basis of induction. Now, suppose the statement to be proved is true for row k; this is the inductive hypothesis. Let us prove it for row $(k+1)$. From part (ii) of Definition 2.5, the inductive hypothesis, and the st-monotonicity of U in part (i) of the lemma for row k, we have

$$v(P)[k,j] = \max_{m \leq k}\{r(P)[m,j]\} \leq r(U)[k,j] \leq r(U)[k+1,j] \qquad \forall j \in \mathcal{S}.$$

On the other hand, since $P \leq_{st} U$ from part (ii) of the lemma, for row $(k+1)$ we have

$$r(P)[k+1,j] \leq r(U)[k+1,j] \qquad \forall j \in \mathcal{S}.$$

Using part (ii) of Definition 2.5 for row $(k+1)$ and combining the previous two statements, we obtain

$$v(P)[k+1,j] = \max_{m \leq k+1}\{r(P)[m,j]\} \leq r(U)[k+1,j] \qquad \forall j \in \mathcal{S},$$

which completes the proof. $\square$

3.2. Polynomials of a stochastic matrix. Now, let us introduce a special set of polynomials of stochastic matrices and two subsets derived from it.

DEFINITION 3.2. *Let $\mathcal{D}(P)$ be the set of polynomials $\Phi(P)$ of stochastic matrices $P \in \mathcal{B}$ such that $\Phi(I) = I$, $\Phi(P) \neq I$ as long as $P \neq I$, and the coefficients of $\Phi(P)$ are nonnegative.*

Hence, $\mathcal{D}(P)$ is the set of polynomials of $P \in \mathcal{B}$ with nonnegative coefficients summing up to 1 and different than I. Next we define the nested subsets $\mathcal{Z}_k(P)$ of degree 1 through k of $\mathcal{D}(P)$.

DEFINITION 3.3. *Let $\mathcal{D}_k(P) \subset \mathcal{D}(P)$ consist of those polynomials in $\mathcal{D}(P)$ of degree k and let $\mathcal{Z}_k(P) = \cup_{l=1}^{k} \mathcal{D}_l(P)$ for $k \geq 1$.*

The subset of $\mathcal{Z}_k(P)$ defined next will be of interest in bounding steady-state distributions.

DEFINITION 3.4. *Let $\mathcal{G}_k(P) \subseteq \mathcal{Z}_k(P)$ be the set of good polynomials of degree at most k for $P \in \mathcal{B}$ and $k \geq 1$ such that $\Psi(P) \in \mathcal{G}_k(P)$ if and only if $\pi_{r^{-1}v(\Psi(P))} \leq_{st} \pi_{r^{-1}v(\Phi(P))}$ for all $\Phi(P) \in \mathcal{Z}_k(P)$.*

In other words, the set good polynomials $\mathcal{G}_k(P)$ are those polynomials of degree at most k in $\mathcal{Z}_k(P)$ which yield the most accurate strong stochastic bound computed by Vincent's algorithm on the steady-state distribution π_P. For instance, $\Gamma_{1/2}(P) \in \mathcal{G}_1(P)$ (see (2.1)).

LEMMA 3.5. *If $P \in \mathcal{B}$ is irreducible and aperiodic, then $\Phi(P) \in \mathcal{D}(P)$ is also an irreducible and aperiodic stochastic matrix, and has the same steady-state distribution as P.*

Proof. Without loss of generality, let $\Phi(P) = \sum_{k=0}^{K} a_k P^k$ be a Kth degree polynomial of P with coefficients $a_k \geq 0$ for $k \in \{0, 1, \ldots, K\}$. Since P is assumed to be irreducible and aperiodic, P^k must be an irreducible and aperiodic stochastic matrix for all $k \in \{1, 2, \ldots, K\}$. Furthermore, $\Phi(P) \neq I$ in Definition 3.2 implies that there exists $k \in \{1, 2, \ldots, K\}$ for which $a_k > 0$. On the other hand, $\Phi(I) = I$ in Definition 3.2 implies $\sum_{k=0}^{K} a_k = 1$. Combining these two results, we have $\Phi(P) = \sum_{k=0}^{K} a_k P^k$ as a convex sum of irreducible and aperiodic stochastic matrices (and possibly I). Hence, $\Phi(P)$ must be an irreducible and aperiodic stochastic matrix. As for the second part, since P is assumed to be an irreducible and aperiodic stochastic matrix, it must have a positive steady-state distribution. Let this distribution be π_P. Since π_P is the steady-state distribution of P, it must be the unique fixed-point of P satisfying $\pi_P = \pi_P P$. Hence, from

$$\pi_P \Phi(P) = \pi_P \sum_{k=0}^{K} a_k P^k = \sum_{k=0}^{K} a_k \pi_P P^k = \left(\sum_{k=0}^{K} a_k\right)\pi_P = \pi_P,$$

we obtain π_P as the fixed-point of $\Phi(P)$. Since $\Phi(P)$ is an irreducible and aperiodic stochastic matrix, it must have a positive steady-state distribution. This cannot be any other distribution than its positive fixed-point π_P. □

3.2.1. The polynomial P^2 in $\mathcal{D}$. Before continuing with polynomials of higher degree, let us consider the second degree polynomial $\Phi(P) = P^2$, where $P \in \mathcal{B}$.

LEMMA 3.6. *If $P, Q \in \mathcal{B}$ satisfy $P \leq_{st} Q$ and Q is st-monotone, then*

(i) Q^2 *is st-monotone,*

(ii) $P^2 \leq_{st} Q^2$.

Proof. Part (i) follows from part (iii) of Property 2.7 after expressing the st-monotonicity of $Q^2 \in \mathcal{B}$ as $L^{-1}Q^2 L \geq 0$, where L is the lower-triangular matrix of ones in Corollary 2.6. This condition can be rewritten as $(L^{-1}QL)(L^{-1}QL) \geq 0$. Now, since Q is st-monotone by assumption, we have $L^{-1}QL \geq 0$ from part (iii) of Property 2.7, which proves the result. Regarding the proof of part (ii), note that from the st-comparability of the stochastic matrices $P^2, Q^2 \in \mathcal{B}$ in part (ii) of Property 2.7, we must show that $r(P^2) \leq r(Q^2)$, or equivalently $P^2 L \leq Q^2 L$. By assumption, we have $r(P) \leq r(Q)$ from the st-comparability of $P, Q \in \mathcal{B}$ from part (ii) of Property 2.7, which may be written as $PL \leq QL$. Since both sides of this inequality are nonnegative and P is also nonnegative, we can premultiply both sides with P to obtain $P^2 L \leq PQL$. Hence, it remains to show that $P(QL) \leq Q(QL)$. But, this is equivalent to showing that $PL(L^{-1}QL) \leq QL(L^{-1}QL)$. Since Q is st-monotone, we have $L^{-1}QL \geq 0$. Postmultiplying both sides of $PL \leq QL$ with $L^{-1}QL$, which is nonnegative, proves the result. □

It is now possible to conclude for this simple case:

THEOREM 3.7. *If $P \in \mathcal{B}$ and $\Phi(P) = P^2$, then*

$$r^{-1}v(\Phi(P)) \leq_{st} \Phi(r^{-1}v(P)).$$

Proof. Applying part (ii) of Lemma 3.6 with $Q = r^{-1}v(P)$, we obtain $\Phi(P) \leq_{st} \Phi(r^{-1}v(P))$, where the stochastic matrix $\Phi(r^{-1}v(P))$ is st-monotone from part (i) of Lemma 3.6. Now, consider the application of operator $r^{-1}v$ to the stochastic matrix $\Phi(P) \in \mathcal{B}$. According to Lemma 3.1, $r^{-1}v(\Phi(P))$ is the smallest st-monotone stochastic matrix such that $\Phi(P) \leq_{st} \Phi(r^{-1}v(P))$. Since $\Phi(r^{-1}v(P))$ is another st-monotone upper bounding stochastic matrix for $\Phi(P)$, it must be that

$$\Phi(P) \leq_{st} r^{-1}v(\Phi(P)) \leq_{st} \Phi(r^{-1}v(P)).$$

□

Theorem 3.7 essentially says that it is better to apply the polynomial transformation $\Phi(P) = P^2$ before Vincent's algorithm.

3.2.2. Arbitrary polynomials in $\mathcal{D}$. Now we consider an arbitrary polynomial in $\mathcal{D}(P)$ and generalize Lemma 3.6.

LEMMA 3.8. *If $P, Q \in \mathcal{B}$ satisfy $P \leq_{st} Q$, Q is st-monotone, and $\Phi(P) \in \mathcal{D}(P)$, then*

(i) *$\Phi(Q)$ is st-monotone,*
(ii) *$\Phi(P) \leq_{st} \Phi(Q)$.*

Proof. Without loss of generality, let $\Phi(Q) = \sum_{k=0}^{K} a_k Q^k$ be a Kth degree polynomial of Q with coefficients $a_k \geq 0$ for $k \in \{0, 1, \ldots, K\}$. Since Q is st-monotone, we have $L^{-1}QL \geq 0$ from part (iii) of Property 2.7. But this implies $(L^{-1}QL)^k = Q^k \geq 0$, that is, Q^k is st-monotone for $k \in \{0, 1, \ldots, K\}$. Then part (i) follows directly from the last statement in Theorem 1.2 in [9, p. 234] by virtue of the fact that $a_k \geq 0$ and $\sum_{k=0}^{K} a_k = 1$, and $\Phi(Q)$ is the convex sum of st-monotone stochastic matrices. The proof of part (ii) is purely algebraic and follows in the same way as in the proof of Lemma 3.6. □

Since the lemma is established, the general theorem may be given as follows:

THEOREM 3.9. *If $P \in \mathcal{B}$ and $\Phi(P) \in \mathcal{D}(P)$, then*

$$r^{-1}v(\Phi(P)) \leq_{st} \Phi(r^{-1}v(P)).$$

Proof. The proof follows in the same way as that of Theorem 3.7 with the difference that Lemma 3.8 is used instead of Lemma 3.6. □

Hence, for an arbitrary polynomial in $\mathcal{D}(P)$, it is better to apply the polynomial transformation before Vincent's algorithm. Clearly, st-comparability between the stochastic matrices $r^{-1}v(\Phi(P))$ and $\Phi(r^{-1}v(P))$ implies st-comparability between their steady-state distributions from Corollary 2.4.

COROLLARY 3.10. *If $P \in \mathcal{B}$ is irreducible, aperiodic, and $\Phi(P) \in \mathcal{D}(P)$, then*

$$\pi_P \leq_{st} \pi_{r^{-1}v(\Phi(P))} \leq_{st} \pi_{r^{-1}v(P)}.$$

Proof. Recall from Lemma 3.5 that $\Phi(P) \in \mathcal{B}$ is an irreducible and aperiodic stochastic matrix. Hence, it is possible to use Theorem 3.9 to compare the matrices $r^{-1}v(\Phi(P))$ and $\Phi(r^{-1}v(P))$. Then, since $r^{-1}v(\Phi(P))$ is st-monotone by construction, we can use Corollary 2.4 to obtain

$$\pi_{r^{-1}v(\Phi(P))} \leq_{st} \pi_{\Phi(r^{-1}v(P))}.$$

Furthermore, from Lemma 3.5 the matrices $\Phi(r^{-1}v(P))$ and $r^{-1}v(P)$ have the same steady-state distribution. Hence,

$$\pi_{r^{-1}v(\Phi(P))} \leq_{st} \pi_{r^{-1}v(P)},$$

which proves half of the result. On the other hand, $r^{-1}v(\Phi(P))$ is an st-monotone upper bounding stochastic matrix for $\Phi(P)$, implying

$$\pi_{\Phi(P)} \leq_{st} \pi_{r^{-1}v(\Phi(P))}$$

from Corollary 2.4. But this is equivalent to

$$\pi_P \leq_{st} \pi_{r^{-1}v(\Phi(P))}$$

since $\pi_{\Phi(P)} = \pi_P$ from Lemma 3.5, and the remaining half of the proof is completed.
□

In general, the accuracy of bounds improves with higher degree polynomial transformations in $\mathcal{D}(P)$ for $P \in \mathcal{B}$ and is better than those obtained by the linear transformation introduced in [5]. In section 4, we provide evidence to that effect. However, higher degree polynomial transformations mean more computation. For instance, if the degree of the polynomial is two, one must consider a sparse matrix-matrix multiplication for the generation of P^2 instead of P from a higher level specification. This is not so difficult because the matrices considered are usually extremely sparse. If the order of the stochastic matrix is n and the average number of transitions per state is d, the sparse matrix-matrix multiplication has a time complexity of $O(nd^2)$. This results from the fact that the sparse matrix has about nd nonzeros and each nonzero gets multiplied by d nonzeros when computing the product. On the other hand, the generation of P from a higher level specification based on states and events has time a complexity of $O(ne)$, whereas the generation of P^2 requires only $O(ne^2)$ steps, where e is the number of events.

3.2.3. Composition of polynomials in $\mathcal{D}(P)$. In [5], the linear transformation $\Gamma_\delta(P) = \delta I + (1 - \delta)P$, where $P \in \mathcal{B}$ and $\delta \in [0, 1)$, is considered, implying $\Gamma_\delta(P) \in \mathcal{Z}_1(P)$, and it is proved that the accuracy of bounds increases with δ until the transformed matrix becomes row diagonally dominant (RDD). Since the transformed matrix is clearly RDD for all P when $\delta = 1/2$, the transformation $\Gamma_{1/2}(P)$ provides the most accurate bounds among linear transformations (i.e., $\Gamma_{1/2}(P) \in \mathcal{G}_1(P)$).

Here, we state two results obtained through the composition of polynomials that follow from Corollary 3.10 by the fact that each polynomial yields a stochastic matrix on which another polynomial transformation can be applied before Vincent's algorithm.

COROLLARY 3.11 (Composition with $\Gamma_{1/2}(P)$). *If $P \in \mathcal{B}$ is irreducible, aperiodic, and $\Phi(P) \in \mathcal{D}(P)$, then the transformation $\Gamma_{1/2}(\Phi(P)) = (\Phi(P) + I)/2$ yields*

$$\pi_P \leq_{st} \pi_{r^{-1}v(\Gamma_{1/2}(\Phi(P)))} \leq_{st} \pi_{r^{-1}v(\Phi(P))}.$$

COROLLARY 3.12 (General Composition). *If $P \in \mathcal{B}$ is irreducible, aperiodic, and $\Phi(P), \Psi(P) \in \mathcal{D}(P)$, then*

$$\pi_P \leq_{st} \pi_{r^{-1}v(\Psi(\Phi(P)))} \leq_{st} \pi_{r^{-1}v(\Phi(P))}.$$

Hence, the composition of polynomials improves the st-bounds.

3.3. Transformations of a stochastic matrix. Finally, we can generalize the stated results to a theorem for an arbitrary transformation assuming that it satisfies the two basic properties considered in Lemmas 3.5 and 3.8.

THEOREM 3.13. *Let $P \in \mathcal{B}$ be irreducible and aperiodic, and let $\Xi(\cdot)$ be a (not necessarily polynomial) transformation such that*

(i) *$\Xi(P)$ is an irreducible and aperiodic stochastic matrix and has the same steady-state distribution as P;*

(ii) *if P is st-monotone, so is $\Xi(P)$;*

   if $P \leq_{st} Q$, then $\Xi(P) \leq_{st} \Xi(Q)$.

Then $r^{-1}v(\Xi(P)) \leq_{st} \Xi(r^{-1}v(P))$ and $\pi_P \leq_{st} \pi_{r^{-1}v(\Xi(P))} \leq_{st} \pi_{\Xi(r^{-1}v(P))}$.

Now, we turn to transient MCs with one absorbing state and show how we can use the simple quadratic polynomial to bound the absorption time.

3.4. Improving the bound for absorption time. In [2], two corollaries of Theorem 2.3 are proven. Here, we restate them so that they can be used to improve the bound for time to absorption in transient MCs with one transient state.

COROLLARY 3.14. *Let $P, Q \in \mathcal{B}$ be respectively characterizing the time-homogeneous MCs $X(t), Y(t)$ for $t \in \mathbb{N}$. If P or Q is st-monotone and if $P \leq_{st} Q$, then*

$$Pr(X(t) = n|X(0) = i) \leq Pr(Y(t) = n|Y(0) = i) \qquad \forall i \in \mathcal{S} \ \ and \ \ \forall t \in \mathbb{N}.$$

Now, let $P, Q \in \mathcal{B}$ be transient with only one absorbing state, n, and let $T_i(P)$, $T_i(Q)$ be random variables defined respectively as the time to be absorbed in state n of P, Q given that the initial state is $i \in \mathcal{S}$. Clearly,

$$Pr(T_i(P) \leq t) = Pr(X(t) = n|X(0) = i) \qquad \forall i \in \mathcal{S},$$
$$Pr(T_i(Q) \leq t) = Pr(Y(t) = n|Y(0) = i) \qquad \forall i \in \mathcal{S},$$

and we can state the second corollary.

COROLLARY 3.15. *Let $P, Q \in \mathcal{B}$ be respectively characterizing the time-homogeneous transient MCs $X(t), Y(t)$ for $t \in \mathbb{N}$ with absorbing state n. If P or Q is st-monotone and if $P \leq_{st} Q$, then*

$$T_i(Q) \leq_{st} T_i(P) \qquad \forall i \in \mathcal{S}.$$

Note that st-comparison is now between the random variables $T_i(P)$ and $T_i(Q)$, which are defined on time rather than states, and it is based on the st-ordering of the respective cumulative distribution functions (cdfs). This is a useful tool and implies that we can use the algorithms in [21] to bound absorption times in transient MCs with only one absorbing state. Now, we relate this to the polynomial approach. However, instead of a general theory, we present results for the simple polynomial $\Phi(P) = P^2$.

Assuming that $P \in \mathcal{B}$ is transient with one absorbing state, n, let $X(t)$, $Y(t)$, $Z(t)$, $W(t)$, and $U(t)$ for $t \in \mathbb{N}$ denote the time-homogeneous MCs defined by P, $\Phi(P)$, $r^{-1}v(P)$, $\Phi(r^{-1}v(P))$, and $r^{-1}v(\Phi(P))$, respectively. First, we show a relation between $T_i(P)$ and $T_i(\Phi(P))$. Clearly,

$$Pr(X(2t) = n|X(0) = i) = Pr(Y(t) = n|Y(0) = i) \qquad \forall i \in \mathcal{S} \ \ and \ \ \forall t \in \mathbb{N}.$$

Therefore, $Pr(T_i(P) \leq 2t) = Pr(T_i(\Phi(P)) \leq t)$ for $t \in \mathbb{N}$. So, it is possible to obtain the cdf of $T_i(P)$ using the cdf of $T_i(\Phi(P))$, but only for even time steps. However, we can build, $F(t)$, an upper bound on the cdf $Pr(T_i(P) \leq t)$ for $t \in \mathbb{N}$ using the cdf of $T_i(\Phi(P))$ as

$$F(t) = \begin{cases} Pr(T_i(\Phi(P)) \leq k), & t = 2k, \ \forall k \in \mathbb{N} \\ Pr(T_i(\Phi(P)) \leq k+1), & t = 2k+1, \ \forall k \in \mathbb{N} \end{cases}.$$

This is possible, since if t is even (i.e., $t = 2k$ for $k \in \mathbb{N}$), then $F(t)$ and the cdf of $T_i(P)$ are equal. So, F is an upper bound. Otherwise, if t is odd (i.e., $t = 2k+1$ for $k \in \mathbb{N}$), then

$$F(t) = F(t+1) = Pr(T_i(\Phi(P)) \leq k+1) = Pr(T_i(P) \leq 2k+2) \geq Pr(T_i(P) \leq 2k+1).$$

So, again we obtain an upper bound because the cdf is a nondecreasing function. Similarly, we can build, $G(t)$, a lower bound on the cdf $Pr(T_i(P) \leq t)$ for $t \in \mathbb{N}$ using the cdf of $T_i(\Phi(P))$ as

$$G(t) = \begin{cases} Pr(T_i(\Phi(P)) \leq k), & t = 2k, \ \forall k \in \mathbb{N} \\ Pr(T_i(\Phi(P)) \leq k), & t = 2k+1, \ \forall k \in \mathbb{N} \end{cases}.$$

Let us now return to the st-comparison idea. We know from Theorem 3.7 that $r^{-1}v(\Phi(P)) \leq_{st} \Phi(r^{-1}v(P))$. This implies $T_i(\Phi(r^{-1}v(P))) \leq_{st} T_i(r^{-1}v(\Phi(P)))$ from Corollary 3.15. From the same corollary, we also have $T_i(r^{-1}v(\Phi(P))) \leq_{st} T_i(\Phi(P))$. Now, using the shorter notation

$$Pr(X_i = t) = Pr(X(t) = n|X(0) = i)$$

for $X(t)$ and similarly for $Y(t)$, $Z(t)$, $W(t)$, $U(t)$, we can compare the cdfs of the random variables X_i, Y_i, Z_i, W_i, U_i at time $t = 2k$ for $k \in \mathbb{N}$ and obtain

$$Pr(W_i \leq k) = Pr(Z_i \leq 2k), \quad Pr(Y_i \leq k) = Pr(X_i \leq 2k),$$

$$\text{and} \quad Pr(W_i \leq k) \geq Pr(U_i \leq k) \geq Pr(Y_i \leq k).$$

Thus at time $t = 2k$ for $k \in \mathbb{N}$, the cdf of U_i provides a more accurate bound than that of Z_i and W_i for $i \in S$. Hence, to compute a lower bound for the absorption time, it is more accurate to form $\Phi(P)$ and then apply Vincent's algorithm to obtain the bounding matrix. Of course, we do not obtain a bound for the complete cdf, but only for even time steps.

In the next section, we discuss various examples to illustrate how some of the ideas developed in this section can be put to work.

4. Illustrative examples and special classes of stochastic matrices. We first give an example of a second degree polynomial to show how the st-bounds on the steady-state distribution of a MC can be improved. Then we introduce two special classes of matrices for which results are readily available from an st-bounding point of view.

4.1. Improving the quality of st-bounds using a second degree polynomial. We consider the (4×4) stochastic matrix

$$P = \begin{pmatrix} 0.1 & 0.2 & 0.4 & 0.3 \\ 0.2 & 0.3 & 0.2 & 0.3 \\ 0.1 & 0.5 & 0.4 & 0 \\ 0.2 & 0.1 & 0.3 & 0.4 \end{pmatrix}$$

and study the effects of using the two polynomial transformations $\Gamma_{1/2}(P) = (I+P)/2$ and $\Phi(P) = (I + P^2)/2$. Recall that $\Gamma_{1/2}(P)$ provides the best transformation among polynomials of degree one.

First we compute $\Gamma_{1/2}(P)$ and $\Phi(P)$:

$$
\Gamma_{1/2}(P) = \begin{pmatrix} 0.55 & 0.1 & 0.2 & 0.15 \\ 0.1 & 0.65 & 0.1 & 0.15 \\ 0.05 & 0.25 & 0.7 & 0 \\ 0.1 & 0.05 & 0.15 & 0.7 \end{pmatrix}, \Phi(P) = \begin{pmatrix} 0.575 & 0.155 & 0.165 & 0.105 \\ 0.08 & 0.63 & 0.155 & 0.135 \\ 0.075 & 0.185 & 0.65 & 0.09 \\ 0.075 & 0.13 & 0.17 & 0.625 \end{pmatrix}.
$$

Then we apply the operator $r^{-1}v$ corresponding to Vincent's algorithm to obtain the st-monotone upper bounding stochastic matrices

$$
r^{-1}v(P) = \begin{pmatrix} 0.1 & 0.2 & 0.4 & 0.3 \\ 0.1 & 0.2 & 0.4 & 0.3 \\ 0.1 & 0.2 & 0.4 & 0.3 \\ 0.1 & 0.2 & 0.3 & 0.4 \end{pmatrix}, r^{-1}v(\Gamma_{1/2}(P)) = \begin{pmatrix} 0.55 & 0.1 & 0.2 & 0.15 \\ 0.1 & 0.55 & 0.2 & 0.15 \\ 0.05 & 0.25 & 0.55 & 0.15 \\ 0.05 & 0.1 & 0.15 & 0.7 \end{pmatrix}
$$

and

$$
r^{-1}v(\Phi(P)) = \begin{pmatrix} 0.575 & 0.155 & 0.165 & 0.105 \\ 0.08 & 0.63 & 0.155 & 0.135 \\ 0.075 & 0.185 & 0.605 & 0.135 \\ 0.075 & 0.13 & 0.17 & 0.625 \end{pmatrix}.
$$

Finally, we compute the steady-state distributions of the three st-monotone upper bounding stochastic matrices and P in four decimal digits of precision to be

$$
\pi_{r^{-1}v(P)} = (0.1000, 0.2000, 0.3667, 0.3333),
$$
$$
\pi_{r^{-1}v(\Gamma_{1/2}(P))} = (0.1259, 0.2587, 0.2821, 0.3333),
$$
$$
\pi_{r^{-1}v(\Phi(P))} = (0.1530, 0.2997, 0.2916, 0.2557),
$$
$$
\pi_P = (0.1530, 0.3025, 0.3167, 0.2278).
$$

Clearly,

$$
\pi_P \leq_{st} \pi_{r^{-1}v(\Phi(P))} \leq_{st} \pi_{r^{-1}v(\Gamma_{1/2}(P))} \leq_{st} \pi_{r^{-1}v(P)},
$$

and hence, bounds obtained using the polynomial transformation $\Phi(P)$ are more accurate since $\Phi(P)$ is the composition of $\Gamma_{1/2}(P)$ and P^2 (i.e., $\Phi(P) = \Gamma_{1/2}(P^2)$).

4.2. The classes of st-monotone and st-antimonotone stochastic matrices. We first remark that Vincent's algorithm does not modify an st-monotone stochastic matrix.

PROPERTY 4.1. *If $P \in \mathcal{B}$ is st-monotone, then $r^{-1}v(P) = P$.*

Now, recall the set of polynomials $\mathcal{D}(P)$ for $P \in \mathcal{B}$ in Definition 3.2, and observe that the next property is a simple generalization of the result in Theorem 1.2 of [9, p. 234].

PROPERTY 4.2. *If $P \in \mathcal{B}$ is st-monotone, then $\Phi(P)$ is st-monotone for all $\Phi(P) \in \mathcal{D}(P)$.*

Using these two properties, we can state the next lemma.

LEMMA 4.3. *Let $P \in \mathcal{B}$ be st-monotone, then $\pi_{r^{-1}v(P)} = \pi_{r^{-1}v(\Phi(P))}$ for all $\Phi(P) \in \mathcal{D}(P)$.*

Proof. Due to Property 4.1, we have $\pi_{r^{-1}v(P)} = \pi_P$, and because of property 4.2, we obtain $\pi_{\Phi(P)} = \pi_{r^{-1}v(\phi(P))}$. Since $\pi_{\Phi(P)} = \pi_P$ from Lemma 3.5, the lemma is proved. $\square$

For the set of good polynomials $\mathcal{G}_k(P)$ for $P \in \mathcal{B}$ in Definition 3.4, we have:

REMARK 4.4. *If $P \in \mathcal{B}$ is st-monotone, then $\mathcal{G}_k(P) = \mathcal{Z}_k(P)$ for $k \geq 1$.*

Let us now consider the case of an st-antimonotone[1] stochastic matrix.

DEFINITION 4.5. *A stochastic matrix $P \in \mathcal{B}$ is st-antimonotone if and only if $vP \leq_{st} uP$ for all probability vectors u and v such that $u \leq_{st} v$.*

The class of st-antimonotone stochastic matrices can be characterized further as in the next two lemmas.

LEMMA 4.6. *If $P \in \mathcal{B}$ is st-antimonotone, then P^2 is st-monotone.*

Proof. If P is st-antimonotone, then for $u \leq_{st} v$, where $u \geq 0$, $ue = 1$, $v \geq 0$, $ve = 1$, and e is a column vector of ones, we have $vP \leq_{st} uP$. Now, let $vP = y$ and $uP = z$. Since $P \geq 0$ and $Pe = e$, we obtain $y \geq 0$, $ye = 1$, $z \geq 0$, and $ze = 1$. Then, from $y \leq_{st} z$ we must have $zP \leq_{st} yP$, since P is st-antimonotone. But this is equivalent to $uP^2 \leq_{st} vP^2$. Hence, we have $u \leq_{st} v$ implying $uP^2 \leq_{st} vP^2$, which means P^2 is st-monotone. $\square$

LEMMA 4.7. *The stochastic matrix $P \in \mathcal{B}$ is st-antimonotone if and only if the rows of P are nonincreasing according to the st-ordering.*

Proof. First note that the rows of P are nonincreasing if and only if $L^{-1}PL \leq 0$ for L in Corollary 2.6. For the first part of the proof, assume that the rows are nonincreasing according to st-ordering. Then it must be that $L^{-1}PL \leq 0$ (see Property (2.7)). Hence, if $u \leq_{st} v$, we must prove that $vP \leq_{st} uP$. Let us rewrite $u \leq_{st} v$ as $(v - u)L \geq 0$ and $vP \leq_{st} uP$ as $(v - u)PL \leq 0$ from part (i) of Property 2.7. Inserting $I = LL^{-1}$ between the vector $(v - u)$ and the matrix PL in the latter, we obtain $(v - u)LL^{-1}PL \leq 0$. But, $(v - u)L \geq 0$; therefore, we must have $L^{-1}PL \leq 0$. As for the second part, assume that P is st-antimonotone. Now, let e_j represent the jth column of I, and consider the vectors e_1^T and e_2^T. Clearly, we have $e_1^T \leq_{st} e_2^T$. Thus according to the definition of st-antimonotonicity, $e_2^T P \leq_{st} e_1^T P$. As $e_1^T P$ and $e_2^T P$ are respectively equal to the first and the second rows of P, we have proved that $P[2, *] \leq_{st} P[1, *]$. The rest of the proof for the other rows of P follows similarly. $\square$

The following property is a simple consequence of this characterization of st-antimonotone stochastic matrices.

PROPERTY 4.8. *If $P \in \mathcal{B}$ is st-antimonotone, then all the rows of $r^{-1}v(P)$ are equal.*

We now show that we can find very easily good polynomials for st-antimonotone matrices as suggested in the next property, which follows from Lemma 4.6 and Property 4.1.

PROPERTY 4.9. *If $P \in \mathcal{B}$ is st-antimonotone, then $\Phi(P) = P^2$ is st-monotone and $r^{-1}v(\Phi(P)) = \Phi(P)$. Thus, $\Phi(P)$ provides the most accurate st-monotone upper bounding stochastic matrix as it yields the exact steady-state distribution:*

$$\pi_{r^{-1}v(\Phi(P))} = \pi_{\Phi(P)} = \pi_P$$

Furthermore, $\Phi(P) \in \mathcal{G}_k(P)$ for $k \geq 2$ since it is not possible to improve the bound. Clearly, all polynomials with coefficients of zero for the odd degree terms are in $\mathcal{G}_k(P)$ for $k \geq 2$.

---

[1]This notion is not claimed to be equivalent to the antimonotone property studied for Monte Carlo Markov Chains algorithms.

Let us now consider a higher degree polynomial of odd degree. For instance, $\Psi(P) = P^3$.

PROPERTY 4.10.   *If $P \in \mathcal{B}$ is st-antimonotone, then $\Psi(P) = P^3$ is st-antimonotone. Clearly, the property holds for P^{2k+1}, $k \in \mathbb{N}$.*

We know that $\Phi(P) = P^2 \in \mathcal{G}_3(P)$ from Property 4.9. Thus, the more accurate bound is the exact result. If a polynomial provides a bound which is not equal to the exact result, it cannot be in $\mathcal{G}_3(P)$.

Consider, for instance, the (4×4) stochastic matrix

$$P = \begin{pmatrix} 0.5 & 0.1 & 0.2 & 0.2 \\ 0.6 & 0 & 0.3 & 0.1 \\ 0.6 & 0.2 & 0.2 & 0 \\ 0.6 & 0.3 & 0.1 & 0 \end{pmatrix} \quad \text{with} \quad \pi_P = (0.5455, 0.1314, 0.2009, 0.1222).$$

The rows are decreasing according to the st-ordering; thus, P is st-antimonotone. The st-monotone upper bounding matrix corresponding to P computed with Vincent's algorithm is

$$r^{-1}v(P) = \begin{pmatrix} 0.5 & 0.1 & 0.2 & 0.2 \\ 0.5 & 0.1 & 0.2 & 0.2 \\ 0.5 & 0.1 & 0.2 & 0.2 \\ 0.5 & 0.1 & 0.2 & 0.2 \end{pmatrix} = e\pi_{r^{-1}v(P)} \quad \text{with} \quad \pi_{r^{-1}v(P)} = (0.5, 0.1, 0.2, 0.2).$$

Now let us compute P^2, P^3, and the st-monotone upper bounding stochastic matrices provided by Vincent's algorithm for these polynomials:

$$P^2 = \begin{pmatrix} 0.55 & 0.15 & 0.19 & 0.11 \\ 0.54 & 0.15 & 0.19 & 0.12 \\ 0.54 & 0.1 & 0.22 & 0.14 \\ 0.54 & 0.08 & 0.23 & 0.15 \end{pmatrix} \quad \text{and} \quad P^3 = \begin{pmatrix} 0.545 & 0.126 & 0.204 & 0.125 \\ 0.546 & 0.128 & 0.203 & 0.123 \\ 0.546 & 0.14 & 0.196 & 0.118 \\ 0.546 & 0.145 & 0.193 & 0.116 \end{pmatrix}.$$

We remark that P^2 is st-monotone, P^3 is st-antimonotone, and Vincent's algorithm yields

$$r^{-1}v(P^2) = \begin{pmatrix} 0.55 & 0.15 & 0.19 & 0.11 \\ 0.54 & 0.15 & 0.19 & 0.12 \\ 0.54 & 0.1 & 0.22 & 0.14 \\ 0.54 & 0.08 & 0.23 & 0.15 \end{pmatrix},$$

$$r^{-1}v(P^3) = \begin{pmatrix} 0.545 & 0.126 & 0.204 & 0.125 \\ 0.545 & 0.126 & 0.204 & 0.125 \\ 0.545 & 0.126 & 0.204 & 0.125 \\ 0.545 & 0.126 & 0.204 & 0.125 \end{pmatrix}.$$

Finally, we compute the st upper bounds on the steady-state distribution as

$$\pi_{r^{-1}v(P^2)} = (0.5455, 0.1314, 0.2009, 0.1222),$$
$$\pi_{r^{-1}v(P^3)} = (0.545, 0.126, 0.204, 0.125).$$

Thus, P^3 is not a good polynomial for P (i.e., $P^3 \notin \mathcal{G}_3(P)$). In fact, all polynomials are not good polynomials for an st-antimonotone matrix. This example shows that it is not always a good idea to increase the degree of the polynomial since the bounds provided by P^3 are less accurate than the bounds provided by P^2.

Furthermore, we also have in $\mathcal{G}_2(P)$ some polynomials with positive coefficients. For instance, consider the polynomial $0.49P^2 + 0.01P + 0.5$, which yields

$$\begin{pmatrix} 0.7745 & 0.0745 & 0.0951 & 0.0559 \\ 0.2706 & 0.5735 & 0.0961 & 0.0598 \\ 0.2706 & 0.051 & 0.6098 & 0.0686 \\ 0.2706 & 0.0422 & 0.1137 & 0.5735 \end{pmatrix}.$$

Furthermore, some polynomials in $\mathcal{G}_2(P)$ even result in a non-RDD stochastic matrix. Check, for instance, the polynomial $0.75P^2 + 0.05P + 0.2$.

5. On the set of good polynomials. In this section, we prove for $P \in \mathcal{B}$ that if $\mathcal{G}_k(P) \neq \emptyset$ for some $k \geq 1$, then $\mathcal{G}_k(P)$ is convex.

First, we state and prove two lemmas regarding Vincent's algorithm which will aid the proof of the main result.

LEMMA 5.1. *If $P, Q \in \mathcal{B}$, then for all $a, b > 0$ such that $a + b = 1$, we have*

$$r^{-1}v(aP + bQ) \leq_{st} ar^{-1}v(P) + br^{-1}v(Q).$$

Proof. Since st-comparison is established by means of the sum operator r as in part (ii) of Property 2.7, it suffices to show that

$$v(aP + bQ) \leq av(P) + bv(Q).$$

By writing the effect of the v operator elementwise using part (ii) of Definition 2.5, we obtain

$$v(aP + bQ)[i, j] = \max_{m \leq i} \sum_{k=j}^{n} (aP + bQ)[m, k] \quad \forall i, j$$

$$\leq \max_{m \leq i} a \sum_{k=j}^{n} P[m, k] + \max_{m \leq i} b \sum_{k=j}^{n} Q[m, k] \quad \forall i, j$$

$$\leq av(P)[i, j] + bv(Q)[i, j] \quad \forall i, j.$$

□

LEMMA 5.2. *If $P, Q \in \mathcal{B}$ have the same steady-state distribution, then any convex sum of P and Q also has the same steady-state distribution.*

Proof. Let $C = aP + (1-a)Q$, where $0 < a < 1$, $\pi P = \pi$, $\pi Q = \pi$, and $\pi e = 1$. Then $\pi C = a\pi P + (1-a)\pi Q = a\pi + (1-a)\pi = \pi$. □

Now, recall that $\Gamma_{1/2}(P) = (I + P)/2$ is

LEMMA 5.3. *If $\Phi(P) \in \mathcal{G}_k(P)$ for some $k \geq 1$, then $\Gamma_{1/2}(\Phi(P)) \in \mathcal{G}_k(P)$.*

Proof. Assume that $\mathcal{G}_k(P) \neq \emptyset$ and let $\Phi(P) \in \mathcal{G}_k(P)$. Clearly, $\Phi(P)$ is a stochastic matrix from Lemma 3.5. We apply $\Gamma_{1/2}$ to the matrix $\Phi(P)$. Because $\Gamma_{1/2}(P) \in \mathcal{G}_1(P)$, we have

$$\pi_{r^{-1}}v(\Gamma_{1/2}(\Phi(P))) \leq_{st} \pi_{r^{-1}}v(\Phi(P)).$$

But, if $\Phi(P) \in \mathcal{G}_k(P)$, then also $\Phi(P) \in Z_k(P)$, and clearly, $\Gamma_{1/2}(\Phi(P)) \in \mathcal{Z}_k(P)$. Thus, we can apply the definition of the set $\mathcal{G}_k(P)$, and remark that, if a polynomial provides a more accurate bound, it must give the same bound:

$$\pi_{r^{-1}}v(\Gamma_{1/2}(\Phi(P))) = \pi_{r^{-1}}v(\Phi(P))$$

Hence, $\Gamma_{1/2}(\Phi(P)) \in \mathcal{G}_k(P)$. □

Now let us turn to the result we want to prove about the set of good polynomials.

THEOREM 5.4. *If* $\mathcal{G}_k(P) \neq \emptyset$ *for some* $k \geq 1$, *then* $\mathcal{G}_k(P)$ *is convex.*

Proof. Let $\mathcal{G}_k(P) \neq \emptyset$ for some $k \geq 1$ and consider two polynomials $\Phi(P), \Psi(P) \in \mathcal{G}_k(P)$ and two real numbers $a, b > 0$ such that $a + b = 1$. According to Lemma 5.1, we have

$$r^{-1}v(a\Phi(P) + b\Psi(P)) \leq_{st} ar^{-1}v(\Phi(P)) + br^{-1}v(\Psi(P)),$$

since $\Phi(P)$ and $\Psi(P)$ are stochastic matrices for $P \in \mathcal{B}$ from Lemma 3.5. Thus,

$$\pi_{r^{-1}v(a\Phi(P)+b\Psi(P))} \leq_{st} \pi_{ar^{-1}v(\Phi(P))+br^{-1}v(\Psi(P))}.$$

Recall that $\Phi(P), \Psi(P) \in \mathcal{G}_k(P)$. Thus, $r^{-1}v(\Phi(P))$ and $r^{-1}v(\Phi(P))$ have the same steady-state distribution (i.e., they yield the most accurate st-bounds):

$$\pi_{r^{-1}v(\Phi(P))} = \pi_{r^{-1}v(\Psi(P))}$$

Applying Lemma 5.2, we obtain

$$\pi_{ar^{-1}v(\Phi(P))+br^{-1}v(\Psi(P))} = \pi_{r^{-1}v(\Phi(P))} = \pi_{r^{-1}v(\Psi(P))}.$$

Combining all the relations, we get

$$\pi_{r^{-1}v(a\Phi(P)+b\Psi(P))} \leq_{st} \pi_{r^{-1}v(\Phi(P))}.$$

But, since $\pi_{r^{-1}v(\Phi(P))}$ yields the most accurate bounds on the steady-state distribution, we must have $\pi_{r^{-1}v(a\Phi(P)+b\Psi(P))} = \pi_{r^{-1}v(\Phi(P))}$. Thus, $a\Phi(P) + b\Psi(P) \in \mathcal{G}_k(P)$, and therefore $\mathcal{G}_k(P)$ is convex. □

The next section includes a stochastic interpretation of some of the results developed in this paper.

6. Stochastic comparison: a stochastic approach. Consider the time-homogeneous MC $\{X(t)\}_{t \in \mathbb{N}}$ which takes values in $\mathcal{S}$. Let this MC be characterized by the irreducible and aperiodic stochastic matrix P with steady-state distribution π_X.

Consider now an independent and identically distributed random sequence of integers $\{\tau(t)\}_{t \in \mathbb{N}}$ stochastically independent of $\{X(t)\}_{t \in \mathbb{N}}$. Define the subchain $\{X^\tau(t)\}_{t \in \mathbb{N}}$ of $\{X(t)\}_{t \in \mathbb{N}}$ by

$$X^\tau(0) = X(0) \quad \text{and} \quad X^\tau(t) = X(\tau(1) + \tau(2) + \cdots + \tau(t)).$$

Provided that $Pr(\tau(i) > 0) > 0$, the stochastic process $\{X^\tau(t)\}_{t \in \mathbb{N}}$ is a MC. Its probability transition matrix is simply defined by P^τ. It is clear that this chain is irreducible and aperiodic because

$$\lim_{t \to +\infty} Pr(X^\tau(t) = j | X^\tau(0) = i) = \lim_{t \to +\infty} Pr(X(t) = j | X(0) = i) = \pi_X[j].$$

The left limit is the limit of a subsequence of the converging sequence $Pr(X(t) = j | X(0) = i)$. This ensures the divergence of the potential matrix $\sum_{t \geq 0}(P^\tau)^t$ of $\{X^\tau(t)\}_{t \in \mathbb{N}}$. Hence, P^τ has the same steady-state distribution as P (i.e., $\pi_{X^\tau} = \pi_X$).

Recall that a MC can also be constructed from a uniform random sequence $\{U(t)\}_{t\in\mathbb{N}}$ on $[0,1)$. It is sufficient to define for each state $i \in \mathcal{S}$, the cdf $F(i,j) = Pr(X(1) \leq j|X(0) = i)$. Then the generalized inverse of F at (i,u) for $u \in [0,1)$ is given by

$$F^{-1}(i,u) = \max\{j : F(i,j) < u\}.$$

Hence, the stochastic process $\{\overline{X(t)}\}_{t\in\mathbb{N}}$ defined by the recurrence relation

$$\overline{X(t)} = F^{-1}(\overline{X(t-1)}, U(t+1))$$

is a MC. Furthermore, $\{X(t)\}_{t\in\mathbb{N}}$ and $\{\overline{X(t)}\}_{t\in\mathbb{N}}$ have the same distribution.

Strassen's theorem [19] bases the stochastic ordering $\leq_{st}$ on a coupling argument. Following this idea, we observe that a MC is st-monotone if and only if $F^{-1}(i,u)$ is an increasing function of i for all $u \in [0,1)$. Furthermore, a MC $\{Y(t)\}_{t\in\mathbb{N}}$ with cdf G is an st-monotone upper bound if and only if

$$F^{-1}(i,u) \leq G^{-1}(i,u) \leq G^{-1}(i+1,u) \leq G^{-1}(i+2,u) \leq \cdots \quad \forall i \in \mathcal{S} \quad \text{and} \quad u \in [0,1).$$

Now associate with the MC $\{X(t)\}_{t\in\mathbb{N}}$ the set $\mathcal{M}_X$ of its st-monotone upper bounds. It is clear that $(\mathcal{M}_X, \wedge)$ is lattice structured, where $\wedge$ is the minimum operator on functions. More precisely, if $\{Y_1(t)\}_{t\in\mathbb{N}}, \{Y_2(t)\}_{t\in\mathbb{N}}\} \in \mathcal{M}_X$ respectively have the cdfs G_1, G_2, define G by $G^{-1}(i,u) = \min(G_1^{-1}(i,u), G_2^{-1}(i,u))$ and $\{Y(t)\}_{t\in\mathbb{N}}$ as the corresponding coupled MC. Then $\{Y(t)\}_{t\in\mathbb{N}}$ is st-monotone, $\{Y(t)\}_{t\in\mathbb{N}} \leq_{st} \{Y_1(t)\}_{t\in\mathbb{N}}$, $\{Y(t)\}_{t\in\mathbb{N}} \leq_{st} \{Y_2(t)\}_{t\in\mathbb{N}}$, and $\{X(t)\}_{t\in\mathbb{N}} \leq_{st} \{Y(t)\}_{t\in\mathbb{N}}$, so $\{Y(t)\}_{t\in\mathbb{N}} \in \mathcal{M}_X$. Because $\mathcal{M}_X$ is bounded from below, there exists a unique minimum element in $\mathcal{M}_X$ denoted by $\{X_*(t)\}_{t\in\mathbb{N}}$, which is the corresponding coupled MC. Observing that $\{(X_*)^\tau(t)\}_{t\in\mathbb{N}}$ is also st-monotone because the composition of increasing functions is an increasing function, we deduce that $\{(X_*)^\tau(t)\}_{t\in\mathbb{N}}$ belongs to $\mathcal{M}_{X^\tau}$ and is greater than the minimal st-monotone upper bound $\{(X^\tau)_*(t)\}_{t\in\mathbb{N}}$ of $\{X^\tau(t)\}_{t\in\mathbb{N}}$. At steady-state, this yields

$$\pi_X \leq_{st} \pi_{(X^\tau)_*} \leq_{st} \pi_{(X_*)^\tau} = \pi_{X_*}.$$

7. Conclusion. The theory for Markov chain comparison is well stated since many years. However, the algorithmic aspects of this theory are still under development; yet, it is possible to automatically derive bounds in a software tool [8]. Improving the accuracy of bounds is one of the major problems since the quality of bounds may vary largely due to several choices made during the modeling process. We must now find some rule-of-thumbs to help the users during model design and analysis. The results presented in this paper have a lot of algorithmic appeal and will most likely be stepping stones for larger improvements in the future.

REFERENCES

[1] O. ABU-AMSHA AND J.-M. VINCENT, *An algorithm to bound functionals of Markov chains with large state space*, in 4th INFORMS Conference on Telecommunications, Boca Raton, Florida, 8-11 March 1998. Available as Rapport de recherche MAI No. 25, IMAG, Grenoble, France, 1996.
[2] M. BEN MAMMOUN, A. BUŠIĆ, J. M. FOURNEAU, AND N. PEKERGIN, *Increasing convex monotone Markov chains: theory, algorithm and applications*, in Proceedings of the A. A. Markov Anniversary Meeting, A. N. Langville and W. J. Stewart, eds., Boson Books, 2006, pp. 189–210.

[3] A. BUŠIĆ AND J. M. FOURNEAU, *Bounds for point and steady-state availability: an algorithmic approach based on lumpability and stochastic ordering*, in Formal Techniques for Computer Systems and Business Processes, Lecture Notes in Computer Science 3670, M. Bravetti, L. Kloul, and G. Zavattaro, eds., Springer Verlag, 2005, pp. 94–108.

[4] T. DAYAR AND N. PEKERGIN, *Stochastic comparison, reorderings, and nearly completely decomposable Markov chains*, in Numerical Solution of Markov Chains, B. Plateau, W. J. Stewart, and M. Silva, eds., Prensas Universitarias de Zaragoza, Zaragoza, Spain, 1999, pp. 228–246.

[5] T. DAYAR, J.-M. FOURNEAU, AND N. PEKERGIN, *Transforming stochastic matrices for stochastic comparison with the st-order*, RAIRO Operations Research, 37 (2003), pp. 85–97.

[6] J. M. FOURNEAU, M. LECOZ, AND F. QUESSETTE, *Algorithms for an irreducible and lumpable strong stochastic bound*, Linear Algebra and Its Applications, 386 (2004), pp. 167–185.

[7] J.-M. FOURNEAU AND N. PEKERGIN, *An algorithmic approach to stochastic bounds*, in Performance Evaluation of Complex Systems: Techniques and Tools, Lecture Notes in Computer Science 2459, M. Calzarossa and S. Tucci, eds., Springer Verlag, 2002, pp. 64–88.

[8] J.-M. FOURNEAU, M. LE COZ, N. PEKERGIN, AND F. QUESSETTE, *An open tool to compute stochastic bounds on steady-state distributions and rewards*, in Proceedings of the 11th IEEE/ACM International Symposium on Modeling, Analysis, and Simulation of Computer Telecommunication Systems, G. Kotsis, ed., IEEE CS-Press, 2003, pp. 219–224.

[9] J. KEILSON AND A. KESTER, *Monotone matrices and monotone Markov processes*, Stochastic Processes and Their Applications, 5 (1977), pp. 231–241.

[10] M. KIJIMA, *Markov Processes for Stochastic Modeling*, Chapman & Hall, London, England, 1997.

[11] M. KWIATKOWSKA, G. NORMAN, AND A. PACHECO, *Model checking CSL until formulae with random time bounds*, in Process Algebra and Probabilistic Methods, Lecture Notes in Computer Science 2399, H. Hermanns and R. Segala, eds., Springer Verlag, 2002, pp. 152–168.

[12] W. A. MASSEY, *Stochastic orderings for Markov processes on partially ordered spaces*, Mathematics of Operations Research, 12 (1987), pp. 350–367.

[13] N. PEKERGIN, *Stochastic delay bounds on fair queueing algorithms*, in Proceedings of the 18th Annual Joint Conference of the IEEE Computer and Communication Societies, IEEE Press, 1999, pp. 1212–1220.

[14] N. PEKERGIN, *Stochastic performance bounds by state reduction*, Performance Evaluation, 36–37 (1999), pp. 1–17.

[15] N. PEKERGIN, T. DAYAR, AND D. N. ALPARSLAN, *Componentwise bounds for nearly completely decomposable Markov chains using stochastic comparison and reordering*, Technical Report BU-CE-0202, Department of Computer Engineering, Bilkent University, Ankara, Turkey, 2002.
    Available at http://www.cs.bilkent.edu.tr/tech-reports/2002/BU-CE-0202.ps.gz
    Last accessed on January 9, 2006.

[16] N. PEKERGIN, T. DAYAR, AND D. N. ALPARSLAN, *Componentwise bounds for nearly completely decomposable Markov chains using stochastic comparison and reordering*, European Journal of Operational Research, 165 (2005), pp. 810–825.

[17] N. PEKERGIN AND S. YOUNÈS, *Stochastic model checking with stochastic comparison*, in Formal Techniques for Computer Systems and Business Processes, Lecture Notes in Computer Science 3670, M. Bravetti, L. Kloul, and G. Zavattaro, eds., Springer Verlag, 2005, pp. 109–123.

[18] M. SHAKED AND J. G. SHANTIKUMAR, *Stochastic Orders and Their Applications*, Academic Press, San Diego, California, USA, 1994.

[19] V. STRASSEN, *The existence of probability measures with given marginals*, The Annals of Mathematical Statistics, 36 (1965), pp. 423–439.

[20] D. STOYAN, *Comparison Methods for Queues and Other Stochastic Models*, John Wiley & Sons, Berlin, Germany, 1983.

[21] M. TREMOLIERES, J.-M. VINCENT, AND B. PLATEAU, *Determination of the optimal upper bound of a Markovian generator*, Technical Report 106, LGI-IMAG, Grenoble, France, 1992.

[22] L. TRUFFET, *Near complete decomposability: bounding the error by stochastic comparison method*, Advances in Applied Probability, 29 (1997), pp. 830–855.

[23] L. TRUFFET, *Reduction techniques for discrete-time Markov chains on totally ordered state space using stochastic comparisons*, Journal of Applied Probability, 37 (2000), pp. 795–806.

UPDATING MARKOV CHAINS

AMY N. LANGVILLE[*] AND CARL D. MEYER[†]

1. Introduction. Suppose that the stationary distribution vector

$$\phi^T = (\phi_1, \phi_2, \ldots, \phi_m)$$

for an m-state homogeneous irreducible Markov chain with transition probability matrix $\mathbf{Q}_{m \times m}$ is known, but the chain requires updating by altering some of its transition probabilities or by adding or deleting some states. Suppose that the updated transition probability matrix $\mathbf{P}_{n \times n}$ is also irreducible. The updating problem is to compute the updated stationary distribution $\pi^T = (\pi_1, \pi_2, \ldots, \pi_n)$ for $\mathbf{P}$ by somehow using the components in ϕ^T to produce π^T with less effort than that required by starting from scratch.

2. The Power Method. For the simple case in which the updating process calls for perturbing transition probabilities in $\mathbf{Q}$ to produce the updated matrix $\mathbf{P}$ without creating or destroying states, restarting the power method is a possible updating technique. In other words, simply iterate with the new transition matrix but use the old stationary distribution as the initial vector

$$(1) \qquad \mathbf{x}_{j+1}^T = \mathbf{x}_j^T \mathbf{P} \quad \text{with} \quad \mathbf{x}_0^T = \phi^T.$$

Will this produce an acceptably accurate approximation to π^T in fewer iterations than are required when an arbitrary initial vector is used? To some degree this is true, but intuition generally overestimates the extent, even when updating produces a $\mathbf{P}$ that is close to $\mathbf{Q}$. For example, if the entries of $\mathbf{P} - \mathbf{Q}$ are small enough to ensure that each component π_i agrees with ϕ_i in the first significant digit, and if the goal is to compute the update π^T to twelve significant places by using (1), then about $11/R$ iterations are required, whereas starting from scratch with an initial vector containing no significant digits of accuracy requires about $12/R$ iterations, where $R = -\log_{10} |\lambda_2|$ is the asymptotic rate of convergence with λ_2 being the subdominant eigenvalue of $\mathbf{P}$. In other words, the effort is reduced by about 8% for each correct significant digit that is built into $\mathbf{x}_0^T$ [22]. In general, the restarted power method is not particularly effective as an updating technique, even when the updates represent small perturbations.

3. Rank-One Updating. When no states are added or deleted, the updating problem can be formulated in terms of updating $\mathbf{Q}$ one row at a time by adapting the Sherman–Morrison rank-one updating formula [29] to the singular matrix $\mathbf{A} = \mathbf{I} - \mathbf{Q}$. The mechanism for doing this is by means of the group inverse $\mathbf{A}^{\#}$ for $\mathbf{A}$, which is often involved in questions concerning Markov chains [6, 9, 11, 14, 23, 25, 27, 28, 31, 32, 38]. $\mathbf{A}^{\#}$ is the unique matrix satisfying $\mathbf{A}\mathbf{A}^{\#}\mathbf{A} = \mathbf{A}$, $\mathbf{A}^{\#}\mathbf{A}\mathbf{A}^{\#} = \mathbf{A}^{\#}$, and $\mathbf{A}\mathbf{A}^{\#} = \mathbf{A}^{\#}\mathbf{A}$.

THEOREM 3.1. [31]. *If the ith row $\mathbf{q}^T$ of $\mathbf{Q}$ is updated to produce $\mathbf{p}^T = \mathbf{q}^T - \delta^T$, the ith row of $\mathbf{P}$, and if ϕ^T and π^T are the respective stationary probability distribu-*

---
[*]Department of Mathematics, College of Charleston, Charleston, SC 29424, (langvillea@cofc.edu).
[†]Department of Mathematics, North Carolina State University, Raleigh, NC 27695-8205, (meyer@ncsu.edu).

tions of $\mathbf{Q}$ *and* $\mathbf{P}$, *then then* $\boldsymbol{\pi}^T = \boldsymbol{\phi}^T - \boldsymbol{\epsilon}^T$, *where*

$$(2) \qquad \boldsymbol{\epsilon}^T = \left[\frac{\phi_i}{1 + \boldsymbol{\delta}^T \mathbf{A}_{*i}^{\#}} \right] \boldsymbol{\delta}^T \mathbf{A}^{\#} \qquad (\mathbf{A}_{*i}^{\#} = \text{the } i\text{th column of } \mathbf{A}^{\#}).$$

Multiple row updates to $\mathbf{Q}$ *are accomplished by sequentially applying this formula one row at a time, which means that the group inverse must be sequentially updated. The formula for updating* $(\mathbf{I} - \mathbf{Q})^{\#}$ *to* $(\mathbf{I} - \mathbf{P})^{\#}$ *is as follows:*

$$(3) \qquad (\mathbf{I} - \mathbf{P})^{\#} = \mathbf{A}^{\#} + \mathbf{e}\boldsymbol{\epsilon}^T \left[\mathbf{A}^{\#} - \gamma\mathbf{I} \right] - \frac{\mathbf{A}_{*i}^{\#}\boldsymbol{\epsilon}^T}{\phi_i}, \quad \text{where} \quad \gamma = \frac{\boldsymbol{\epsilon}^T \mathbf{A}_{*i}^{\#}}{\phi_i}.$$

If more than just one or two rows are involved, then Theorem 3.1 is not computationally efficient. If every row needs to be touched, then using (2) together with (3) requires $O(n^3)$ floating point operations, which is comparable to the cost of starting from scratch. Other updating formulas exist [9, 12, 16, 19, 36], but all are variations of a Sherman–Morrison type of formula, and all are $O(n^3)$ algorithms for a general update. Moreover, rank-one updating techniques are not easily adapted to handle the creation or destruction of states.

4. Aggregation. Consider an irreducible n-state Markov chain whose state space has been partitioned into k disjoint groups $\mathcal{S} = G_1 \cup G_2 \cup \cdots \cup G_k$ with associated transition probability matrix

$$(4) \qquad \mathbf{P}_{n \times n} = \begin{array}{c} \\ G_1 \\ G_2 \\ \vdots \\ G_k \end{array} \begin{array}{c} \begin{array}{cccc} G_1 & G_2 & \cdots & G_k \end{array} \\ \left(\begin{array}{cccc} \mathbf{P}_{11} & \mathbf{P}_{12} & \cdots & \mathbf{P}_{1k} \\ \mathbf{P}_{21} & \mathbf{P}_{22} & \cdots & \mathbf{P}_{2k} \\ \vdots & \vdots & \ddots & \vdots \\ \mathbf{P}_{k1} & \mathbf{P}_{k2} & \cdots & \mathbf{P}_{kk} \end{array} \right) \end{array} \quad \text{(square diagonal blocks)}.$$

This *parent* chain induces k smaller Markov chains called *censored chains*. The censored chain associated with G_i is defined to be the Markov process that records the location of the parent chain only when the parent chain visits states in G_i. Visits to states outside of G_i are ignored. The transition probability matrix for the ith censored chain is the ith *stochastic complement* [26] defined by

$$(5) \qquad \mathbf{S}_i = \mathbf{P}_{ii} + \mathbf{P}_{i\star}(\mathbf{I} - \mathbf{P}_i^{\star})^{-1}\mathbf{P}_{\star i},$$

in which $\mathbf{P}_{i\star}$ and $\mathbf{P}_{\star i}$ are, respectively, the ith row and the ith column of blocks with $\mathbf{P}_{ii}$ removed, and $\mathbf{P}_i^{\star}$ is the principal submatrix of $\mathbf{P}$ obtained by deleting the ith row and ith column of blocks. For example, if $\mathcal{S} = G_1 \cup G_2$, then the respective transition matrices for the two censored chains are the two stochastic complements

$$\mathbf{S}_1 = \mathbf{P}_{11} + \mathbf{P}_{12}(\mathbf{I} - \mathbf{P}_{22})^{-1}\mathbf{P}_{21} \quad \text{and} \quad \mathbf{S}_2 = \mathbf{P}_{22} + \mathbf{P}_{21}(\mathbf{I} - \mathbf{P}_{11})^{-1}\mathbf{P}_{12}.$$

In general, if the stationary distribution for $\mathbf{P}$ is $\boldsymbol{\pi}^T = (\, \boldsymbol{\pi}_1^T \,|\, \boldsymbol{\pi}_2^T \,|\, \cdots \,|\, \boldsymbol{\pi}_k^T \,)$ (partitioned conformably with $\mathbf{P}$), then the *ith censored distribution* (the stationary distribution for $\mathbf{S}_i$) is

$$(6) \qquad \mathbf{s}_i^T = \frac{\boldsymbol{\pi}_i^T}{\boldsymbol{\pi}_i^T \mathbf{e}}, \qquad \text{where } \mathbf{e} \text{ is an appropriately sized column of ones [26]}.$$

For aperiodic chains, the jth component of $\mathbf{s}_i^T$ is the limiting conditional probability of being in the jth state of group G_i given that the process is somewhere in G_i.

Each group G_i is compressed into a single state in a smaller k-state aggregated chain by squeezing the original transition matrix $\mathbf{P}$ down to an *aggregated transition matrix*

$$(7) \qquad \mathbf{A}_{k \times k} = \begin{pmatrix} \mathbf{s}_1^T \mathbf{P}_{11} \mathbf{e} & \cdots & \mathbf{s}_1^T \mathbf{P}_{1k} \mathbf{e} \\ \vdots & \ddots & \vdots \\ \mathbf{s}_k^T \mathbf{P}_{k1} \mathbf{e} & \cdots & \mathbf{s}_k^T \mathbf{P}_{kk} \mathbf{e} \end{pmatrix},$$

which is stochastic and irreducible whenever $\mathbf{P}$ is [26]. For aperiodic chains, transitions between states in the aggregated chain defined by $\mathbf{A}$ correspond to transitions between groups G_i in the parent chain when the parent chain is in equilibrium. The remarkable feature of aggregation is that it allows the parent chain to be decomposed into k small censored chains that can be independently solved, and the resulting censored distributions $\mathbf{s}_i^T$ can be combined with the stationary distribution of $\mathbf{A}$ to construct the parent stationary distribution $\boldsymbol{\pi}^T$. This is the aggregation theorem.

THEOREM 4.1. (*the aggregation theorem* [26]) *If* $\mathbf{P}$ *is the block-partitioned transition probability matrix* (4) *for an irreducible n-state Markov chain whose stationary probability distribution is*

$$\boldsymbol{\pi}^T = (\,\boldsymbol{\pi}_1^T \mid \boldsymbol{\pi}_2^T \mid \cdots \mid \boldsymbol{\pi}_k^T\,) \quad \textit{(partitioned conformably with } \mathbf{P}),$$

and if $\boldsymbol{\alpha}^T = (\alpha_1, \alpha_2, \dots, \alpha_k)$ *is the stationary distribution for the aggregated chain defined by the matrix* $\mathbf{A}_{k \times k}$ *in* (7), *then* $\alpha_i = \boldsymbol{\pi}_i^T \mathbf{e}$, *and the stationary distribution for* $\mathbf{P}$ *is*

$$\boldsymbol{\pi}^T = \left(\alpha_1 \mathbf{s}_1^T \mid \alpha_2 \mathbf{s}_2^T \mid \cdots \mid \alpha_k \mathbf{s}_k^T\right),$$

where $\mathbf{s}_i^T$ *is the censored distribution for the stochastic complement* $\mathbf{S}_i$ *in* (5).

5. Approximate Updating by Aggregation. Aggregation as presented in Theorem 4.1 is mathematically elegant but numerically inefficient because costly inversions are embedded in the stochastic complements (5) that are required to produce the censored distributions $\mathbf{s}_i^T$. Consequently, the approach is to derive computationally cheap estimates of the censored distributions as described below.

In many large-scale problems the effects of updating are localized. That is, not all stationary probabilities are equally affected—some changes may be significant while others are hardly perceptible—e.g., this is generally true in applications such as Google's PageRank Problem [21] in which the stationary probabilities obey a power-law distribution (discussed in section 7.2).

Partition the state space of the updated chain as $\mathcal{S} = G \cup \overline{G}$, where G contains the states that are most affected by updating along with any new states created by updating—techniques for determining these states are discussed in section 7. Some nearest neighbors of newly created states might also go into G. Partition the updated (and reordered) transition matrix $\mathbf{P}$ as

$$(8) \qquad \mathbf{P} = \begin{pmatrix} p_{11} & \cdots & p_{1g} & \mathbf{P}_{1\star} \\ \hline \vdots & \ddots & \vdots & \vdots \\ p_{g1} & \cdots & p_{gg} & \mathbf{P}_{g\star} \\ \hline \mathbf{P}_{\star 1} & \cdots & \mathbf{P}_{\star g} & \mathbf{P}_{22} \end{pmatrix} = \begin{array}{c} \\ G \\ \overline{G} \end{array}\!\! \begin{pmatrix} \overset{G}{\mathbf{P}_{11}} & \overset{\overline{G}}{\mathbf{P}_{12}} \\ \mathbf{P}_{21} & \mathbf{P}_{22} \end{pmatrix},$$

where

$$\mathbf{P}_{11} = \begin{pmatrix} p_{11} & \cdots & p_{1g} \\ \vdots & \ddots & \vdots \\ p_{g1} & \cdots & p_{gg} \end{pmatrix}, \quad \mathbf{P}_{12} = \begin{pmatrix} \mathbf{P}_{1\star} \\ \vdots \\ \mathbf{P}_{g\star} \end{pmatrix}, \quad \text{and} \quad \mathbf{P}_{21} = (\mathbf{P}_{\star 1} \cdots \mathbf{P}_{\star g}).$$

Let $\boldsymbol{\phi}^T$ and $\boldsymbol{\pi}^T$ be the respective stationary distributions of the pre- and post-updated chains, and let if $\overline{\boldsymbol{\phi}}^T$ and $\overline{\boldsymbol{\pi}}^T$ contain the respective stationary probabilities from $\boldsymbol{\phi}^T$ and $\boldsymbol{\pi}^T$ that correspond to the states in $\overline{G}$. The stipulation that $\overline{G}$ contains the nearly unaffected states translates to saying that

$$\overline{\boldsymbol{\pi}}^T \approx \overline{\boldsymbol{\phi}}^T.$$

When viewed as a partitioned matrix with $g+1$ diagonal blocks, the first g diagonal blocks in $\mathbf{P}$ are 1×1, and the lower right-hand block is the $(n-g) \times (n-g)$ matrix $\mathbf{P}_{22}$ that is associated with the states in $\overline{G}$. The stochastic complements in $\mathbf{P}$ are

$$\mathbf{S}_1 = \cdots = \mathbf{S}_g = 1, \quad \text{and} \quad \mathbf{S}_{g+1} = \mathbf{P}_{22} + \mathbf{P}_{21}(\mathbf{I} - \mathbf{P}_{11})^{-1}\mathbf{P}_{12}.$$

Consequently, the aggregated transition matrix (7) becomes

(9)
$$\mathbf{A} = \left(\begin{array}{ccc|c} p_{11} & \cdots & p_{1g} & \mathbf{P}_{1\star}\mathbf{e} \\ \hline \vdots & \ddots & \vdots & \vdots \\ \hline p_{g1} & \cdots & p_{gg} & \mathbf{P}_{g\star}\mathbf{e} \\ \hline \mathbf{s}^T\mathbf{P}_{\star 1} & \cdots & \mathbf{s}^T\mathbf{P}_{\star g} & \mathbf{s}^T\mathbf{P}_{22}\mathbf{e} \end{array} \right)_{(g+1)\times(g+1)}$$

$$= \begin{pmatrix} \mathbf{P}_{11} & \mathbf{P}_{12}\mathbf{e} \\ \mathbf{s}^T\mathbf{P}_{21} & \mathbf{s}^T\mathbf{P}_{22}\mathbf{e} \end{pmatrix} = \begin{pmatrix} \mathbf{P}_{11} & \mathbf{P}_{12}\mathbf{e} \\ \mathbf{s}^T\mathbf{P}_{21} & 1 - \mathbf{s}^T\mathbf{P}_{21}\mathbf{e} \end{pmatrix},$$

where $\mathbf{s}^T$ is the censored distribution derived from the only significant stochastic complement $\mathbf{S} = \mathbf{S}_{g+1}$. If the stationary distribution for $\mathbf{A}$ is

$$\boldsymbol{\alpha}^T = (\alpha_1, \ldots, \alpha_g, \alpha_{g+1}),$$

then Theorem 4.1 says that the stationary distribution for $\mathbf{P}$ is

(10)
$$\boldsymbol{\pi}^T = (\pi_1, \ldots \pi_g \mid \pi_{g+1}, \ldots, \pi_n) = (\pi_1, \ldots, \pi_g \mid \overline{\boldsymbol{\pi}}^T) = (\alpha_1, \ldots, \alpha_g \mid \overline{\boldsymbol{\pi}}^T)$$

Since $\mathbf{s}_i^T = \boldsymbol{\pi}_i^T / \boldsymbol{\pi}_i^T \mathbf{e}$, and since $\overline{\boldsymbol{\pi}}^T \approx \overline{\boldsymbol{\phi}}^T$, it follows that

(11)
$$\tilde{\mathbf{s}}^T = \frac{\overline{\boldsymbol{\phi}}^T}{\overline{\boldsymbol{\phi}}^T \mathbf{e}} \approx \mathbf{s}^T$$

is a good approximation to $\mathbf{s}^T$ that is available from the pre-updated distribution. Using this in (9) produces an approximate aggregated transition matrix

(12)
$$\tilde{\mathbf{A}} = \begin{pmatrix} \mathbf{P}_{11} & \mathbf{P}_{12}\mathbf{e} \\ \tilde{\mathbf{s}}^T\mathbf{P}_{21} & 1 - \tilde{\mathbf{s}}^T\mathbf{P}_{21}\mathbf{e} \end{pmatrix}.$$

Notice that

$$\mathbf{A} - \widetilde{\mathbf{A}} = \begin{pmatrix} \mathbf{0} & \mathbf{0} \\ \boldsymbol{\delta}^T \mathbf{P}_{21} & -\boldsymbol{\delta}^T \mathbf{P}_{21} \mathbf{e} \end{pmatrix} = \begin{pmatrix} \mathbf{0} \\ \boldsymbol{\delta}^T \end{pmatrix} \mathbf{P}_{21} (\mathbf{I} \mid -\mathbf{e}), \quad \text{where} \quad \boldsymbol{\delta}^T = \mathbf{s}^T - \tilde{\mathbf{s}}^T.$$

Consequently, $\mathbf{A} - \widetilde{\mathbf{A}}$ and $\mathbf{s}^T - \tilde{\mathbf{s}}^T$ are of the same order of magnitude, so the stationary distribution $\widetilde{\boldsymbol{\alpha}}^T$ of $\widetilde{\mathbf{A}}$ can provide a good approximation to $\boldsymbol{\alpha}^T$, the stationary distribution of $\mathbf{A}$. That is,

$$\widetilde{\boldsymbol{\alpha}}^T = (\tilde{\alpha}_1, \tilde{\alpha}_2, \ldots, \tilde{\alpha}_g, \tilde{\alpha}_{g+1}) \approx (\alpha_1, \ldots, \alpha_g, \alpha_{g+1}) = \boldsymbol{\alpha}^T.$$

Use $\tilde{\alpha}_i \approx \alpha_i$ for $1 \leq i \leq g$ in (10) along with $\overline{\boldsymbol{\pi}}^T \approx \overline{\boldsymbol{\phi}}^T$ to obtain the approximation

$$(13) \qquad\qquad \boldsymbol{\pi}^T \approx \widetilde{\boldsymbol{\pi}}^T = \left(\tilde{\alpha}_1, \tilde{\alpha}_2, \ldots, \tilde{\alpha}_g \mid \overline{\boldsymbol{\phi}}^T \right).$$

Thus an approximate updated distribution is obtained. The degree to which this approximation is accurate clearly depends on the degree to which $\overline{\boldsymbol{\pi}}^T \approx \overline{\boldsymbol{\phi}}^T$. If (13) does not provide the desired accuracy, it can be viewed as the first step in an iterative aggregation scheme described below that performs remarkably well.

6. Updating by Iterative Aggregation. Iterative aggregation as described in [38] is not a general-purpose technique, because it usually does not work for chains that are not nearly uncoupled. However, iterative aggregation can be adapted to the updating problem, and these variations work extremely well, even for chains that are not nearly uncoupled. This is in part due to the fact that the approximate aggregation matrix (12) differs from the exact aggregation matrix (9) in only one row. Our iterative aggregation updating algorithm is described below.

Assume that the stationary distribution $\boldsymbol{\phi}^T = (\phi_1, \phi_2, \ldots, \phi_m)$ for some irreducible Markov chain $\mathcal{C}$ is already known, perhaps from prior computations, and suppose that $\mathcal{C}$ needs to be updated. As in earlier sections, let the transition probability matrix and stationary distribution for the updated chain be denoted by $\mathbf{P}$ and $\boldsymbol{\pi}^T = (\pi_1, \pi_2, \ldots, \pi_n)$, respectively. The updated matrix $\mathbf{P}$ is assumed to be irreducible. It is important to note that m is not necessarily equal to n because the updating process allows for the creation or destruction of states as well as the alteration of transition probabilities.

THE ITERATIVE AGGREGATION UPDATING ALGORITHM

Initialization

    i. Partition the states of the updated chain as $\mathcal{S} = G \cup \overline{G}$ and reorder $\mathbf{P}$ as described in (8)

    ii. $\overline{\boldsymbol{\phi}}^T \longleftarrow$ the components from $\boldsymbol{\phi}^T$ that correspond to the states in $\overline{G}$

    iii. $\mathbf{s}^T \longleftarrow \overline{\boldsymbol{\phi}}^T / (\overline{\boldsymbol{\phi}}^T \mathbf{e})$ (an initial approximate censored distribution)

Iterate until convergence

1. $\mathbf{A} \longleftarrow \begin{pmatrix} \mathbf{P}_{11} & \mathbf{P}_{12}\mathbf{e} \\ \mathbf{s}^T\mathbf{P}_{21} & 1 - \mathbf{s}^T\mathbf{P}_{21}\mathbf{e} \end{pmatrix}_{(g+1)\times(g+1)}$ $(\,g = |G|\,)$

2. $\boldsymbol{\alpha}^T \longleftarrow (\alpha_1, \alpha_2, \ldots, \alpha_g, \alpha_{g+1})$ (the stationary distribution of $\mathbf{A}$)

3. $\boldsymbol{\chi}^T \longleftarrow \left(\alpha_1, \alpha_2, \ldots, \alpha_g \,|\, \alpha_{g+1}\mathbf{s}^T\right)$

4. $\boldsymbol{\psi}^T \longleftarrow \boldsymbol{\chi}^T\mathbf{P}$ (see note following the algorithm)

5. If $\|\boldsymbol{\psi}^T - \boldsymbol{\chi}^T\| < \tau$ for a given tolerance τ, then quit—else $\mathbf{s}^T \longleftarrow \boldsymbol{\psi}^T/\boldsymbol{\psi}^T\mathbf{e}$ and go to step 1

Note concerning step 4. Step 4 is necessary because the vector $\boldsymbol{\chi}^T$ generated in step 3 is a fixed point in the sense that if Step 4 is omitted and the process is restarted using $\boldsymbol{\chi}^T$ instead of $\boldsymbol{\psi}^T$, then the same $\boldsymbol{\chi}^T$ is simply reproduced at Step 3 on each subsequent iteration. Step 4 has two purposes—it moves the iterate off the fixed point while simultaneously contributing to the convergence process. That is, the $\boldsymbol{\psi}^T$ resulting from step 4 can be used to restart the algorithm as well as produce a better approximation because applying a power step makes small progress toward the stationary solution. In the past, some authors [38] have used Gauss–Seidel in place of the power method at Step 4.

While precise rates of convergence for general iterative aggregation algorithms are difficult to articulate, the specialized nature of our iterative aggregation updating algorithm allows us to easily establish its rate of convergence. The following theorem shows that this rate is directly dependent on how fast the powers of the one significant stochastic complement $\mathbf{S} = \mathbf{P}_{22} + \mathbf{P}_{21}(\mathbf{I} - \mathbf{P}_{11})^{-1}\mathbf{P}_{12}$ converge. In other words, since $\mathbf{S}$ is an irreducible stochastic matrix, the rate of convergence is completely dictated by the magnitude and Jordan structure of the largest subdominant eigenvalue of $\mathbf{S}$.

THEOREM 6.1. [22] *The iterative aggregation updating algorithm defined above converges to the stationary distribution $\boldsymbol{\pi}^T$ of $\mathbf{P}$ for all partitions $\mathcal{S} = G \cup \overline{G}$. The rate at which the iterates converge to $\boldsymbol{\pi}^T$ is exactly the rate at which the powers $\mathbf{S}^n$ converge, which is governed by the magnitude and Jordan structure of largest subdominant eigenvalue $\lambda_2(\mathbf{S})$ of $\mathbf{S}$. If $\lambda_2(\mathbf{S})$ is real and simple, then the asymptotic rate of convergence is $R = -\log_{10}|\lambda_2(\mathbf{S})|$.*

7. Determining The Partition. The iterative aggregation updating algorithm is globally convergent, and it never requires more iterations than the power method to attain a given level of convergence [17]. However, iterative aggregation clearly requires more work per iteration than the power method. One iteration of iterative aggregation requires forming the aggregation matrix, solving for its stationary vector, and executing one power iteration. The key to realizing an improvement in iterative aggregation over the power method rests in properly choosing the partition $\mathcal{S} = G \cup \overline{G}$. As Theorem 6.1 shows, good partitions are precisely those that yield a stochastic complement $\mathbf{S} = \mathbf{P}_{22} + \mathbf{P}_{21}(\mathbf{I} - \mathbf{P}_{11})^{-1}\mathbf{P}_{12}$ whose subdominant eigenvalue $\lambda_2(\mathbf{S})$ is small in magnitude.

Experience indicates that as $|G| = g$ (the size of $\mathbf{P}_{11}$) becomes larger, iterative aggregation tends to converge in fewer iterations. But as g becomes larger, each iteration requires more work, so the trick is to strike an acceptable balance. A small g that significantly reduces $|\lambda_2(\mathbf{S})|$ is the ideal situation.

Even for moderately sized problems there is an extremely large number of possible partitions, but there are some useful heuristics that can help guide the choice of G that will produce reasonably good results. For example, a relatively simple approach is to take G to be the set of all states "near" the updates, where "near" might be measured in a graph theoretic sense or else by transient flow [7] (i.e., using the magnitude of entries of $\mathbf{x}_{j+1}^T = \mathbf{x}_j^T \mathbf{P}$ after j iterations, where j is small, say 5 or 10). In the absence of any other information, this naive strategy is at least a good place to start. However, there are usually additional options that lead to even better "G-sets," and some of these are described below.

7.1. Partitioning by differing time scales. In most aperiodic chains, evolution is not at a uniform rate, and consequently most iterative techniques, including the power method, often spend the majority of the time in resolving a small number of components—the slow evolving states. The slow states can be isolated either by monitoring the process for a few iterations or by theoretical means [22]. If the slow states are placed in G while the faster-converging states are lumped into $\overline{G}$, then the iterative aggregation algorithm concentrates its effort on resolving the smaller number of slow-converging states.

In loose terms, the effect of steps 1–3 in the iterative aggregation algorithm is essentially to make progress toward achieving a steady state for a smaller chain consisting of just the slow states in G together with one additional lumped state that accounts for all fast states in $\overline{G}$. The power iteration in step 4 moves the entire process ahead on a global basis, so if the slow states in G are substantially resolved by the relatively cheaper steps 1—3, then not many of the more costly global power steps are required to push the entire chain toward its global equilibrium. This is the essence of the original Simon–Ando idea first proposed in 1961 and explained and analyzed in [26, 37]. As $g = |G|$ becomes smaller relative to n, steps 1–3 become significantly cheaper to execute, and the process converges rapidly in both iteration count and wall-clock time. Examples and reports on experiments are given in [22].

In some applications the slow states are particularly easy to identify because they are the ones having the larger stationary probabilities. This is a particularly nice state of affairs for the updating problem because we have the stationary probabilities from the prior period at our disposal, and thus all we have to do to construct a good G-set is to include the states with prior large stationary probabilities and throw in the states that were added or updated along with a few of their nearest neighbors. Clearly, this is an advantage only when there are just a few "large" states. However, it turns out that this is a characteristic feature of scale-free networks with power-law distributions [1, 2, 5, 10] discussed below.

7.2. Scale-free networks. A scale-free networks with a power-law distribution is a network in which the number of nodes $n(l)$ having l edges (possibly directed) is proportional to l^{-k} where k is a constant that does not change as the network expands (hence the term "scale-free"). In other words, the distribution of nodal degrees obeys a *power-law distribution* in the sense that

$$P[\deg(N) = d] \propto \frac{1}{d^k} \qquad \text{for some } k > 1 \qquad (\propto \text{ means "proportional to"}).$$

A Markov chain with a power-law distribution is a chain in which there are relatively very few states that have a significant stationary probability while the overwhelming majority of states have nearly negligible stationary probabilities. Google's

PageRank application [21, 22] is an important example. Consequently, when the stationary probabilities are plotted in order of decreasing magnitude, the resulting graph has a pronounced "L-shape" with an extremely sharp bend. It is this characteristic "L-shape" that reveals a near optimal partition $\mathcal{S} = G \cup \overline{G}$ for the iterative aggregation updating algorithm presented in section 6. Experiments indicate that the size of the G-set used in our iterative aggregation updating algorithm is nearly optimal around a point that is just to the right-hand side of the pronounced bend in the L-curve. In other words, an apparent method for constructing a reasonably good partition $\mathcal{S} = G \cup \overline{G}$ for the iterative aggregation updating algorithm is as follows.

1. First put all new states and states with newly created or destroyed connections (perhaps along with some of their nearest neighbors) into G.

2. Add other states that remain after the update in order of the magnitude of their prior stationary probabilities up to the point where these stationary probabilities level off.

Of course, there is some subjectiveness to this strategy. However, the leveling-off point is relatively easy to discern in distributions having a sharply defined bend in the L-curve, and only distributions that gradually die away or do not conform to a power law are problematic. If, when ordered by magnitude, the stationary probabilities

$$\pi(1) \geq \pi(2) \geq \cdots \geq \pi(n)$$

for an irreducible chain conform to a power-law distribution so that there are constants $\alpha > 0$ and $k > 0$ such that $\pi(i) \approx \alpha i^{-k}$, then the "leveling-off point" i_{level} can be taken to be the smallest value for which $|d\pi(i)/di| \approx \epsilon$ for some user-defined tolerance ϵ. That is, $i_{level} \approx (k\alpha/\epsilon)^{1/k+1}$. This provides a rough estimate of g_{opt}, the optimal size of G, but empirical evidence suggests that better estimates require a scaling factor $\sigma(n)$ that accounts for the size of the chain; i.e.,

$$g_{opt} \approx \sigma(n) \left(\frac{k\alpha}{\epsilon} \right)^{1/k+1} \approx \sigma(n) \left(\frac{k\pi(1)}{\epsilon} \right)^{1/k+1}.$$

More research and testing is needed to resolve some of these issues.

REFERENCES

[1] Albert-Laszlo Barabasi. *Linked: The New Science of Networks.* Plume, 2003.

[2] Albert-Laszlo Barabasi, Reka Albert, and Hawoong Jeong. Scale-free characteristics of random networks: the topology of the world-wide web. *Physica A*, 281:69–77, 2000.

[3] Sergey Brin and Lawrence Page. The anatomy of a large-scale hypertextual web search engine. *Computer Networks and ISDN Systems*, 33:107–117, 1998.

[4] Sergey Brin, Lawrence Page, R. Motwami, and Terry Winograd. The PageRank citation ranking: bringing order to the web. Technical report, Computer Science Department, Stanford University, 1998.

[5] A. Broder, R. Kumar, F. Maghoul, P. Raghavan, S. Rajagopalan, R. Stata, A. Tomkins, and J. Wiener. Graph structure in the web. In *The Ninth International WWW Conference*, May 2000. http://www9.org/w9cdrom/160/160.html.

[6] Steven Campbell and Carl D. Meyer. *Generalized Inverses of Linear Transformations.* Pitman, San Francisco, 1979.

[7] Steve Chien, Cynthia Dwork, Ravi Kumar, and D. Sivakumar. Towards exploiting link evolution. In *Workshop on algorithms and models for the Web graph*, 2001.

[8] Grace E. Cho and Carl D. Meyer. Markov chain sensitivity measured by mean first passage times. *Linear Algebra and its Applications*, 313:21–28, 2000.

[9] Grace E. Cho and Carl D. Meyer. Comparison of perturbation bounds for the stationary distribution of a Markov chain. *Linear Algebra and its Applications*, 335(1–3):137–150, 2001.

[10] D. Donato, L. Laura, S. Leonardi, and S. Millozzi. Large scale properties of the webgraph. *The European Physical Journal B*, 38:239–243, 2004.

[11] Robert E. Funderlic and Carl D. Meyer. Sensitivity of the stationary distribution vector for an ergodic Markov chain. *Linear Algebra and its Applications*, 76:1–17, 1986.

[12] Robert E. Funderlic and Robert J. Plemmons. Updating **LU** factorizations for computing stationary distributions. *SIAM Journal on Algebraic and Discrete Methods*, 7(1):30–42, 1986.

[13] Gene H. Golub and Charles F. Van Loan. *Matrix Computations*. Johns Hopkins University Press, Baltimore, 1996.

[14] Gene H. Golub and Carl D. Meyer. Using the QR factorization and group inverse to compute, differentiate and estimate the sensitivity of stationary probabilities for Markov chains. *SIAM Journal on Algebraic and Discrete Methods*, 17:273–281, 1986.

[15] Roger A. Horn and Charles R. Johnson. *Matrix Analysis*. Cambridge University Press, 1990.

[16] Jeffrey J. Hunter. Stationary distributions of perturbed Markov chains. *Linear Algebra and its Applications*, 82:201–214, 1986.

[17] Ilse C. F. Ipsen and Steve Kirkland. Convergence analysis of an improved PageRank algorithm. Technical Report CRSC-TR04-02, North Carolina State University, 2004.

[18] Ilse C. F. Ipsen and Carl D. Meyer. Uniform stability of Markov chains. *SIAM Journal on Matrix Analysis and Applications*, 15(4):1061–1074, 1994.

[19] John G. Kemeny and Laurie J. Snell. *Finite Markov Chains*. D. Van Nostrand, New York, 1960.

[20] Amy N. Langville and Carl D. Meyer. A survey of eigenvector methods of web information retrieval. *SIAM Rev.*, 47(1):135–161, 2005.

[21] Amy N. Langville and Carl D. Meyer. *Google's PageRank and Beyond: The Science of Search Engine Rankings*. Princeton University Press, 2006.

[22] Amy N. Langville and Carl D. Meyer. Updating the stationary vector of an irreducible Markov chain with an eye on Google's PageRank. *SIAM Journal on Matrix Analysis and Applications*, 27:968–987, 2006.

[23] Carl D. Meyer. The role of the group generalized inverse in the theory of finite Markov chains. *SIAM Rev.*, 17:443–464, 1975.

[24] Carl D. Meyer. The condition of a finite Markov chain and perturbation bounds for the limiting probabilities. *SIAM Journal on Algebraic and Discrete Methods*, 1:273–283, 1980.

[25] Carl D. Meyer. Analysis of finite Markov chains by group inversion techniques. *Recent Applications of Generalized Inverses, Research Notes in Mathematics, Pitman, Ed. S. L. Campbell*, 66:50–81, 1982.

[26] Carl D. Meyer. Stochastic complementation, uncoupling Markov chains, and the theory of nearly reducible systems. *SIAM Review*, 31(2):240–272, 1989.

[27] Carl D. Meyer. The character of a finite Markov chain. *Linear Algebra, Markov Chains, and Queueing Models, IMA Volumes in Mathematics and its Applications, Ed., Carl D. Meyer and Robert J. Plemmons, Springer-Verlag*, 48:47–58, 1993.

[28] Carl D. Meyer. Sensitivity of the stationary distribution of a Markov chain. *SIAM Journal on Matrix Analysis and Applications*, 15(3):715–728, 1994.

[29] Carl D. Meyer. *Matrix Analysis and Applied Linear Algebra*. SIAM, Philadelphia, 2000.

[30] Carl D. Meyer and Robert J. Plemmons. *Linear Algebra, Markov Chains, and Queueing Models*. Springer-Verlag, 1993.

[31] Carl D. Meyer and James M. Shoaf. Updating finite Markov chains by using techniques of group matrix inversion. *Journal of Statistical Computation and Simulation*, 11:163–181, 1980.

[32] Carl D. Meyer and G. W. Stewart. Derivatives and perturbations of eigenvectors. *SIAM J. Numer. Anal.*, 25:679–691, 1988.

[33] Cleve Moler. The world's largest matrix computation. *Matlab News and Notes*, pages 12–13, October 2002.

[34] Gopal Pandurangan, Prabhakar Raghavan, and Eli Upfal. Using PageRank to Characterize Web Structure. In *8th Annual International Computing and Combinatorics Conference (COCOON)*, 2002.

[35] Eugene Seneta. *Non-negative matrices and Markov chains*. Springer-Verlag, 1981.

[36] Eugene Seneta. Sensitivity analysis, ergodicity coefficients, and rank-one updates for finite Markov chains. In William J. Stewart, editor, *Numerical Solutions of Markov Chains*, pages 121–129, 1991.

[37] Herbert A. Simon and Albert Ando. Aggregation of variables in dynamic systems. *Economet-rica*, 29:111–138, 1961.

[38] William J. Stewart. *Introduction to the Numerical Solution of Markov Chains*. Princeton University Press, 1994.

[39] Twelfth International World Wide Web Conference. *Extrapolation Methods for Accelerating PageRank Computations*, 2003.

PRODUCT PRECONDITIONING FOR
MARKOV CHAIN PROBLEMS

MICHELE BENZI[§][*] AND BORA UÇAR[§][†]

Abstract. We consider preconditioned Krylov subspace methods for computing the stationary probability distribution vector of irreducible Markov chains. We propose preconditioners constructed as the product of two fairly simple preconditioners. Theoretical properties of the proposed product preconditioners are briefly discussed. We use graph partitioning tools to partition the coefficient matrix in order to build the preconditioner matrices, and we investigate the effect of the partitioning on the proposed preconditioners. Numerical experiments with GMRES on various Markov chain problems generated with the MARCA software package demonstrate that the proposed preconditioners are effective in reducing the number of iterations to convergence. Furthermore, the experimental results show that the number of partitions does not severely affect the number of iterations.

Key words. preconditioning, discrete Markov chains, iterative methods, graph partitioning

AMS subject classifications. 05C50, 60J10, 60J22, 65F10, 65F50, 65F35

1. Introduction. Discrete Markov chains with large state spaces arise in many applications, including for instance reliability modeling, queuing network analysis, web-based information retrieval, and computer system performance evaluation [30]. As is well known, the long-run behavior of an ergodic (irreducible) Markov chain is described by the stationary distribution vector of the corresponding matrix of transition probabilities. Recall that the stationary probability distribution vector of a finite, ergodic Markov chain with $N \times N$ transition probability matrix P is the unique $1 \times N$ vector π which satisfies

$$(1.1) \qquad \pi = \pi P, \quad \pi_i > 0 \ \text{for} \ i = 1, \dots N, \quad \sum_{i=1}^{N} \pi_i = 1 \ .$$

Here P is nonnegative ($p_{ij} \geq 0$ for $1 \leq i, j \leq N$), row-stochastic ($\sum_{j=1}^{N} p_{ij} = 1$ for $1 \leq i \leq N$), and due to the ergodicity assumption it is irreducible.

The matrix $A = I - P^T$, where I is the $N \times N$ identity matrix, is called the generator of the Markov process. The matrix A is a singular, irreducible M-matrix of rank $N - 1$. Letting $x = \pi^T$ and hence $x^T = x^T P$, the computation of the stationary vector reduces to finding a nontrivial solution to the homogeneous linear system

$$(1.2) \qquad\qquad Ax = 0 \ ,$$

where $x \in \mathbb{R}^N$, $x_i > 0$ for $i = 1, \dots, N$, and $\sum_{i=1}^{N} x_i = 1$. Perron–Frobenius theory [7] implies that such a vector exists and is unique. We assume that P is large and sparse; hence, so is A. We further assume that P and A are partitioned as

$$(1.3) P = \begin{bmatrix} P_{11} & P_{12} \\ P_{21} & P_{22} \end{bmatrix} \ \text{and} \ \ A = \begin{bmatrix} A_{11} & A_{12} \\ A_{21} & A_{22} \end{bmatrix} = \begin{bmatrix} I_n - P_{11}^T & -P_{21}^T \\ -P_{12}^T & I_m - P_{22}^T \end{bmatrix} \ .$$

---

[*]The work of this author was supported in part by the National Science Foundation grant DMS-0511336.

[†]The work of this author was supported by The Scientific and Technological Research Council of Turkey (TÜBITAK).

[§]Department of Mathematics and Computer Science, Emory University, Atlanta, Georgia 30322, USA (benzi@mathcs.emory.edu, ubora@mathcs.emory.edu).

Here I_n and I_m are identity matrices of size $n \times n$ and $m \times m$, respectively, where $N = n + m$ and typically $n \gg m$.

Our goal is to develop efficient preconditioners for Krylov subspace methods for solving the system (1.2). The gist of the proposed preconditioner is to combine the effects of two simple and inexpensive preconditioners. One of the constituent preconditioners is the well-known block Jacobi preconditioner. The block Jacobi preconditioner is known to deteriorate dramatically with the increasing size of the off-diagonal blocks. The other preconditioner is proposed here to have a corrective effect on the block Jacobi preconditioner. We combine the two preconditioners in a multiplicative fashion and obtain product preconditioners. The resulting algorithm can be regarded as a two-level method where the block Jacobi preconditioner plays the role of a smoothing relaxation with the second preconditioner playing the role of a "coarse grid" correction. Our approach, however, is different from two-level algebraic multigrid or Schwarz methods known in the literature; see, e.g., [19] and the references therein. It is also distinct from (and simpler than) the iterative aggregation-disaggregation (IAD) approach [30].

Due to the very large number N of states typical of many real-world applications, there has been increasing interest in recent years in developing parallel algorithms for Markov chain computations; see [3, 4, 8, 15, 18, 21]. Most of the attention so far has focused on (linear) stationary iterative methods, including block versions of Jacobi and Gauss–Seidel [8, 18, 21], and on (nonlinear) iterative aggregation/disaggregation schemes specifically tailored to stochastic matrices [8, 15]. In contrast, little work has been done with parallel preconditioned Krylov subspace methods. The suitability of preconditioned Krylov subspace methods for solving Markov models has been demonstrated, e.g., in [23, 26], although no discussion of parallelization aspects was given there. Parallel computing aspects can be found in [4], where a symmetrizable stationary iteration (Cimmino's method) was accelerated using the Conjugate Gradients method on a Cray T3D, and in [18], where an out-of-core, parallel implementation of Conjugate Gradient Squared (with no preconditioning) was used to solve very large Markov models with up to 50 million states. We further mention [6], where parallel preconditioners based on sparse approximate pseudoinverses were used to speed-up the convergence of BiCGStab.

The paper is organized as follows. We briefly review background material on M-matrices, stationary iterative methods, matrix splittings, and graph partitioning in Section 2. Then, we discuss two simple preconditioners based on regular splittings and introduce the product preconditioners in Section 3. Section 4 contains materials on partitioning the matrices into the 2×2 block structure (1.3) with an eye to future parallel implementations of the proposed product preconditioner. In Section 5 we investigate the effect of the product preconditioner under various partitionings, the properties of the 2×2 block structure imposed by the graph partitioning, and the performance of the product preconditioner relative to that of some other well-known preconditioners. We present our conclusions in Section 6.

2. Background. Here we borrow some material from [5, 6, 7, 32] to provide the reader with a short summary of the concepts and results that are used in building the proposed preconditioners. We also give a brief description of graph partitioning by vertex separator, which can be used to obtain the 2×2 block structure (1.3).

2.1. Nonnegative matrices and M-matrices. A matrix $A_{N \times N}$ is nonnegative if all of its entries are nonnegative, i.e., $A \geq O$ if $a_{ij} \geq 0$ for all $1 \leq i, j \leq N$.

Any matrix A with nonnegative diagonal entries and nonpositive off-diagonal entries can be written in the form

$$(2.1) \qquad A = sI - B, \quad s > 0, \quad B \geq O.$$

A matrix A of the form (2.1) with $s \geq \rho(B)$ is called an M-matrix. Here, $\rho(B)$ denotes the spectral radius of B. If $s = \rho(B)$ then A is singular, otherwise nonsingular. If A is a nonsingular M-matrix, then $A^{-1} \geq O$.

If A is a singular, irreducible M-matrix, then each $k \times k$ principal square submatrix of A, where $1 \leq k < N$, is a nonsingular M-matrix. If, furthermore, A is the generator of an ergodic Markov chain, then the Schur complement $S_{m \times m} = A_{22} - A_{21} A_{11}^{-1} A_{12}$ of A_{11} (Eq. (1.3)) is a singular, irreducible M-matrix with rank $m - 1$ [6, 20].

2.2. Stationary iterations and matrix splittings. Consider the solution of a linear system of the form $Ax = b$, where A is an $N \times N$ square matrix, possibly singular, and $x, b \in \mathbb{R}^N$. The representation $A = B - C$ is called a splitting if B is nonsingular. A splitting gives rise to the stationary iterative method

$$(2.2) \qquad x^{k+1} = Tx^k + c, \quad k = 0, 1, \dots,$$

where $T = B^{-1}C$ is called the iteration matrix, $c = B^{-1}b$, and $x^0 \in \mathbb{R}^N$ is a given initial guess. The splitting $A = B - C$ is called *regular* if $B^{-1} \geq O$ and $C \geq O$ [32], *weak regular* if $B^{-1} \geq O$ and $T \geq O$ [7], and an *M-splitting* if B is an M-matrix and $C \geq O$ [28].

It is well known that the convergence of the stationary iteration (2.2) depends on the convergence of the sequence T^k as $k \to \infty$; see, e.g., [32]. The matrix T is said to be convergent [22] if the powers T^k converge to a limiting matrix as $k \to \infty$. If the limit is the zero matrix, then T is called zero-convergent. For a nonsingular matrix A, a necessary and sufficient condition for the convergence of (2.2) for any x^0 is that T be zero-convergent, or equivalently, that $\rho(T) < 1$. In the singular case the situation is more involved [7, 31]. In this case, 1 is in the spectrum of T, i.e., $1 \in \sigma(T)$, and a necessary condition for convergence is that $\rho(T) = 1$ be the only eigenvalue on the unit circle, i.e., $\gamma(T) := \max\{|\lambda| : \lambda \in \sigma(T), \lambda \neq 1\} < 1$. If the original system $Ax = b$ is consistent and T is convergent, the iteration (2.2) converges to a solution which depends, in general, on the initial guess x^0.

A related approach is defined by the *alternating iterations*

$$(2.3) \qquad \begin{cases} x^{k+1/2} & = M_1^{-1} N_1 x^k + M_1^{-1} b \\ x^{k+1} & = M_2^{-1} N_2 x^{k+1/2} + M_2^{-1} b, \quad k = 0, 1, \dots, \end{cases}$$

where $A = M_1 - N_1 = M_2 - N_2$ are splittings of A, and x^0 is the initial guess. The convergence of alternating iterations is analyzed by Benzi and Szyld [5]. They construct a single splitting $A = B - C$ associated with the iteration matrix by eliminating $x^{k+1/2}$ from the second equation in (2.3) and obtain

$$(2.4) \qquad x^{k+1} = M_2^{-1} N_2 M_1^{-1} N_1 x^k + M_2^{-1}(N_2 M_1^{-1})b, \quad k = 0, 1, \dots,$$

which is in the form of (2.2) with $T = M_2^{-1} N_2 M_1^{-1} N_1$. Using this formulation, they construct a unique splitting $A = B - C$ with $B^{-1}C = T$. The splitting is defined by (Eq. (10) in [5])

$$(2.5) \qquad B^{-1} = M_2^{-1}(M_1 + M_2 - A)M_1^{-1}.$$

Clearly, the matrix $M_1 + M_2 - A$ must be nonsingular for (2.5) to be well-defined.

2.3. Graph partitioning. Given an undirected graph $G = (V, E)$, the problem of K-way graph partitioning by vertex separator (GPVS) asks for finding a set of vertices V_S of minimum size whose removal decomposes a graph into K disconnected subgraphs with balanced sizes. The problem is NP-hard [9]. Formally, $\Pi = \{V_1, \ldots, V_K; V_S\}$ is a K-way vertex partition by vertex separator V_S if the following conditions hold: $V_k \subset V$ and $V_k \neq \emptyset$ for $1 \leq k \leq K$; $V_k \cap V_\ell = \emptyset$ for $1 \leq k < \ell \leq K$ and $V_k \cap V_S = \emptyset$ for $1 \leq k \leq K$; $\bigcup_k V_k \cup V_S = V$; there is no edge between vertices lying in two different parts V_k and V_ℓ for $1 \leq k < \ell \leq K$; $W_{max}/W_{avg} \leq \epsilon$, where W_{max} is the maximum part size (defined as $\max_k |V_k|$), W_{avg} is the average part size (defined as $(|V| - |V_S|)/K$), and ϵ is a given maximum allowable imbalance ratio. See the works [1, 10, 14, 13, 16] for applications of the GPVS and heuristics for GPVS.

In the weighted GPVS problem, the vertices of the given undirected graph have weights. The weight of the separator or a part is defined as the sum of the weights of the vertices that they contain. The objective of the weighted GPVS problem is to minimize the weight of the separator while maintaining a balance criterion on the part weights.

3. Product splitting preconditioners. We will consider two preconditioners based on regular splittings of A and combine them as in the alternating iterations (2.4) to build an effective preconditioner for Krylov subspace methods.

The first preconditioner is the well-known block Jacobi preconditioner:

$$(3.1) \qquad M_{BJ} = \begin{bmatrix} A_{11} & O \\ O & A_{22} \end{bmatrix} .$$

Note that A_{11} and A_{22} are nonsingular M-matrices, and $A = M_{BJ} - (M_{BJ} - A)$ is a regular splitting (in fact, an M-splitting).

Next, we consider another simple preconditioner:

$$(3.2) \qquad M_{SC} = \begin{bmatrix} D_{11} & A_{12} \\ A_{21} & A_{22} \end{bmatrix} .$$

Here $D_{11} \neq A_{11}$ stands for an approximation of A_{11}; in practice, we take D_{11} to be the diagonal matrix formed with the diagonal entries of A_{11}. More generally, we assume that D_{11} is a matrix obtained from A_{11} by setting off-diagonal entries to zero. Thus, D_{11} is a nonsingular M-matrix [32, Theorem 3.12]. The Schur complement matrix $A_{22} - A_{21} D_{11}^{-1} A_{12}$ is therefore well-defined and under the (very mild) structural conditions given in [6, Theorem 3], it is also a nonsingular M-matrix. These conditions are satisfied for the problems considered in this paper. Therefore M_{SC} is a nonsingular M-matrix and $A = M_{SC} - (M_{SC} - A)$ is an M-splitting (hence, a regular splitting).

Since both M_{BJ} and M_{SC} define regular splittings, the product preconditioner M_{PS} given by

$$(3.3) \qquad M_{PS}^{-1} = M_{SC}^{-1}(M_{BJ} + M_{SC} - A)M_{BJ}^{-1} ,$$

(see (2.5)) implicitly defines a weak regular splitting [5, Theorem 3.4]. Note that since the matrix

$$(3.4) \qquad M_{BJ} + M_{SC} - A = \begin{bmatrix} D_{11} & O \\ O & A_{22} \end{bmatrix}$$

is invertible, M_{PS}^{-1} is well-defined, and so is the corresponding splitting of A. We also have

$$M_{PS} = M_{BJ}(M_{BJ} + M_{SC} - A)^{-1}M_{SC}$$
$$= \begin{bmatrix} A_{11} & O \\ O & A_{22} \end{bmatrix} \begin{bmatrix} D_{11}^{-1} & O \\ O & A_{22}^{-1} \end{bmatrix} \begin{bmatrix} D_{11} & A_{12} \\ A_{21} & A_{22} \end{bmatrix}$$
$$= \begin{bmatrix} A_{11} & A_{11}D_{11}^{-1}A_{12} \\ A_{21} & A_{22} \end{bmatrix},$$

and therefore

$$A - M_{PS} = \begin{bmatrix} O & (A_{12} - A_{11}D_{11}^{-1}A_{12}) \\ O & O \end{bmatrix}.$$

It follows from the identity $M_{PS}^{-1}A = I + M_{PS}^{-1}(A - M_{PS})$ that $M_{PS}^{-1}A$ (or AM_{PS}^{-1}) has at least n eigenvalues all equal to 1. Exactly one eigenvalue is zero; the remaining $m-1$ all have positive real part and lie in a disk of radius $\rho \leq 1$ and center at the point $(1,0)$ in the complex plane, since $A = M_{PS} - (M_{PS} - A)$ is a weak regular splitting [22]. The better D_{11} approximates A_{11}, the smaller ρ is, and the more clustered the nonzero eigenvalues are around the point $(1,0)$. In general, a clustered spectrum near $(1,0)$ implies fast convergence of the preconditioned Krylov subspace iteration.

Since the order of the A_{22} block m is an upper bound on the number of non-unit eigenvalues, it is important to keep this number as small as possible. This is also desirable from the point of view of parallel efficiency; see [6] and the next section.

Application of the M_{BJ} preconditioner requires solving two uncoupled linear systems with the A_{11} and A_{22} blocks:

(3.5)
$$\begin{cases} A_{11}x_1 = b_1 \\ A_{22}x_2 = b_2 \end{cases}.$$

Within a Krylov subspace method, these linear systems can be solved exactly or approximately. When the blocks are large, as they are bound to be in realistic applications, exact solves are inefficient in terms of both time and storage, and inexact solves must be used. Although the use of iterative methods is a possibility (leading to an inner-outer iterative scheme), in this paper we perform inexact solves by means of incomplete factorizations.

Application of the M_{SC} preconditioner requires solving coupled equations of the form

(3.6)
$$\begin{cases} D_{11}x_1 + A_{12}x_2 = b_1 \\ A_{21}x_1 + A_{22}x_2 = b_2 \end{cases}.$$

A convenient way to solve the above system is to eliminate x_1 from the second equation using the first one and to solve the Schur complement system

(3.7)
$$(A_{22} - A_{21}D_{11}^{-1}A_{12})x_2 = b_2 - A_{21}D_{11}^{-1}b_1$$

for x_2. Then substituting x_2 into the first equation in (3.6) results in the system

(3.8)
$$D_{11}x_1 = b_1 - A_{12}x_2,$$

which is easy to solve.

Application of the M_{PS} preconditioner requires the solution of a system of the form (3.5), a matrix-vector multiply with the matrix $M_{BJ} + M_{SC} - A$ given in (3.4), and the solution of two systems of the form (3.7) and (3.8). In analogy with domain decomposition methods for partial differential equations, (3.7) can be interpreted as a kind of "coarse grid" correction, even though there may be no underlying physical grid. Note that x_2 corresponds to the interface unknowns. On the other hand, x_1 corresponds to subdomain unknowns. Note, however, that (3.7) is embedded in the global solve (3.6), which is one of the features that differentiate our approach from standard two-level Schwarz methods.

4. Building the block structure. The first requirement to be met in permuting the matrix A into 2×2 block structure (1.3) is that the permutation should be symmetric. A symmetric permutation on the rows and columns of A guarantees that A_{11} is an $n \times n$ invertible M-matrix, since the transition probability matrix P is irreducible and $A = I - P^T$. Moreover, A_{11} is diagonally dominant by columns.

The second requirement, as already discussed in Section 3, is to keep the order m of A_{22} as small as possible. The requirement is important in order to have fast convergence of the Krylov subspace method, since m is an upper bound on the number of non-unit eigenvalues. Strictly speaking, this is true only if we assume exact solves in the application of the preconditioner. In practice we will use inexact solves, and rather than having n eigenvalues (or more) exactly equal to 1, there will be a cluster of at least n eigenvalues near the point $(1, 0)$. Still, we want this cluster to contain as many eigenvalues as possible.

The second requirement is also desirable from the point of view of parallel implementation. A possible parallelization approach would be constructing and solving the system (3.7) on a single processor and then solving the system (3.8) in parallel. This approach has been taken previously in parallelizing applications of approximate inverse preconditioners [6]. Another possible approach would be parallelizing the solution of (3.7) either by allowing redundancies in the computations (each processor can form the whole system or a part of it) or by running a parallel solver on (3.7) itself. In both cases, the solution with the Schur complement system constitutes a serial bottleneck and requires additional storage space.

The third requirement, also stemming from the discussion in Section 3, is to have D_{11} as close to A_{11} as possible in order to cluster the non-unit eigenvalues of $M_{PS}^{-1}A$. Meeting this requirements would likely help reduce the number of iterations to convergence for the Krylov subspace method.

The fourth requirement, not necessary for the convergence analysis but crucial for an efficient implementation, is that A_{11} should be block diagonal with subblocks of approximately equal size and density. Given K subblocks in the (1,1) block A_{11}, the application of the M_{BJ} preconditioner, i.e., solving the system (3.5), requires $K + 1$ independent solves: one with the A_{22} block and one with each subblock of A_{11}. Similarly, the application of the M_{SC} preconditioner after solving (3.7) requires K independent diagonal solves (scalings) for each subblock of D_{11}. We note that this form of the M_{SC} preconditioner is a special case of a *domain decomposition splitting*, as defined in [33]. Meeting this requirement for a serial implementation will enable solution of very large systems, since the subblocks can be handled one at a time. In any admissible parallelization, each of these subblocks would more likely be assigned to a single processor. Therefore, maintaining balance on the sizes and the densities of the subblocks will relate to maintaining balance on computational loads of the processors. Furthermore, it is desirable that the sizes of these subblocks be larger than the order m of A_{22}, if possible, for the reasons given for the second requirement.

Meeting all of the above four requirements is a very challenging task. Therefore, as a pragmatic approach we totally ignore the third one and apply well established heuristics for addressing the remaining three requirements. As it is common, we adopt the standard undirected graph model to represent a square matrix $A_{N \times N}$. The vertices of the graph $G(A) = (V, E)$ correspond to the rows and columns of A and the edges correspond to the nonzeros of A. The vertex $v_i \in V$ represents the ith row and the ith column of A, and there exists an edge $(v_i, v_j) \in E$ if a_{ij} and a_{ji} are nonzero.

Consider a partitioning $\Pi = \{V_1, \ldots, V_K; V_S\}$ of $G(A)$ with vertex separator V_S. The matrix A can be permuted into the 2×2 block structure (1.3) by permuting the rows and columns associated with the vertices in $\bigcup_k V_k$ before the rows and columns associated with the vertices in V_S. That is, V_S defines the rows and columns of the (2,2) block A_{22}. Notice that the resulting permutation is symmetric, and hence the first requirement is met. Furthermore, since GPVS tries to minimize the size of the separator set V_S, it tries to minimize the order of the block A_{22}. Therefore, the permutation induced by Π meets the second requirement as well.

Consider the A_{11} block defined by the vertices in $\bigcup_k V_k$. The rows and columns that are associated with the vertices in V_k can be permuted before the rows and columns associated with the vertices in V_ℓ for $1 \leq k < \ell \leq K$. Such a permutation of A_{11} gives rise to diagonal subblocks. Since we have already constructed A_{22} using V_S, we end up with the following structure:

$$
A = \begin{bmatrix}
A_1 & & & & B_1 \\
& A_2 & & & B_2 \\
& & \ddots & & \vdots \\
& & & A_K & B_K \\
C_1 & C_2 & \cdots & C_K & A_S
\end{bmatrix}.
$$

The diagonal blocks $A_1, \ldots, A_K$ correspond to the vertex parts $V_1, \ldots, V_K$, and therefore have approximately the same order. The off-diagonal blocks B_i, C_i represent the connections between the subgraphs, and the diagonal block A_S represents the connections between nodes in the separator set. Note that because of the irreducibility assumption, each block A_i is a nonsingular M-matrix. Thus, graph partitioning induces a reordering and block partitioning of the matrix A in the form (1.3) where

$$
A_{11} = \text{diag}(A_1, A_2, \ldots, A_K), \quad A_{22} = A_S
$$

and

$$
A_{12} = [B_1^T \ B_2^T \ \cdots \ B_K^T]^T, \quad A_{21} = [C_1 \ C_2 \ \cdots \ C_K].
$$

Therefore, the permutation induced by the GPVS partially addresses the fourth requirement. Note that the GPVS formulation ignores the requirement of balancing the densities of the diagonal subblocks of A_{11}. In fact, obtaining balance on the densities of the diagonal blocks is a complex partitioning requirement that cannot be met before a partitioning takes place (see [24] for a possible solution) even with a weighted GPVS formulation.

If the matrix is structurally nonsymmetric, which is common for matrices arising from Markov chains, then A cannot be modeled with undirected graphs. In this case, a 2×2 block structure can be obtained by partitioning the graph of $A + A^T$.

TABLE 5.1
Properties of the generator matrices.

Matrix	number of rows/cols	number of nonzeros					
		total	average	row		col	
	N		row/col	min	max	min	max
mutex09	65535	1114079	17.0	16	17	16	17
mutex12	263950	4031310	15.3	9	21	9	21
ncd07	62196	420036	6.8	2	7	2	7
ncd10	176851	1207051	6.8	2	7	2	7
qnatm06	79220	533120	6.7	3	9	4	7
qnatm07	130068	875896	6.7	3	9	4	7
tcomm16	13671	67381	4.9	2	5	2	5
tcomm20	17081	84211	4.9	2	5	2	5
twod08	66177	263425	4.0	2	4	2	4
twod10	263169	1050625	4.0	2	4	2	4

5. Numerical experiments. In this section, we report on experimental results obtained with a Matlab 6 implementation on a 1.2GHz Sun Fire V880 with 2 Gbytes of main memory. The main goal was to test the product splitting preconditioner and to compare it with a few other preconditioners. The Krylov method used was GMRES [27]. For completeness we performed experiments with the stationary iterations corresponding to the various splittings (without GMRES acceleration), but they were found to converge too slowly to be competitive with preconditioned GMRES. Therefore, we do not show these results.

The various methods were tested on the generator matrices of some Markov chain models provided in the MARCA (MARkov Chain Analyzer) collection [29]. The models are discussed in [12, 23, 25] and have been used to compare different solution methods in [6, 11] and elsewhere. These matrices are infinitesimal generators of time-continuous Markov chains, but can be easily converted (as we did) to the form $A = I - P^T$, with P row-stochastic, so that A corresponds to a discrete-time Markov chain, known as the *embedded Markov chain*; see [30, Chapter 1.4.3]. The preconditioning techniques described in this paper can be applied to either form of the generator matrix.

We performed a large number of tests on numerous matrices; here we present a selection of results for a few test matrices, chosen to be representative of our overall findings.

Table 5.1 displays the properties of the test matrices. Each matrix is named by its family followed by its index in the family. For example, mutex09 refers to the 9th matrix in the mutex family. The matrices from the mutex and ncd families are structurally symmetric, the matrices from the qnatm and twod families are structurally nonsymmetric, and the matrices from the tcomm family are very close to being structurally symmetric—the nonzero patterns of tcomm20 and tcomm16 differ from the nonzero patterns of their transposes in only 60 locations.

We compared the product preconditioner (PS) with its factors block Jacobi (BJ) and Schur complement-based (SC) preconditioners. We also compared PS with the block Gauss-Seidel (BGS) and block successive overrelaxation (BSOR) preconditioners, where the preconditioner matrices are

$$M_{BGS} = \begin{bmatrix} A_{11} & A_{12} \\ O & A_{22} \end{bmatrix} \quad \text{or} \quad M_{BGS} = \begin{bmatrix} A_{11} & O \\ A_{21} & A_{22} \end{bmatrix},$$

$$M_{BSOR} = \begin{bmatrix} \frac{1}{\omega}A_{11} & A_{12} \\ O & \frac{1}{\omega}A_{22} \end{bmatrix} \quad \text{or} \quad M_{BSOR} = \begin{bmatrix} \frac{1}{\omega}A_{11} & O \\ A_{21} & \frac{1}{\omega}A_{22} \end{bmatrix}.$$

In agreement with the previously reported results [11] on the MARCA collection, we observed that $\omega = 1.0$ (which reduces the BSOR to BGS) or very close to 1.0 is nearly always the best choice of the relaxation parameter for BSOR. We also observed that the block lower triangular versions of the BGS and BSOR preconditioners are indistinguishable from the block upper triangular versions under either the storage or performance criteria. Therefore, we report only the experiments with the upper triangular BGS preconditioner. Note that application of the BGS preconditioner requires two linear system solutions (one with A_{11} and one with A_{22}) and a matrix-vector multiply with A_{12}.

5.1. Properties of the block structure and the preconditioners. We partitioned the matrix into the 2×2 block structure (1.3) using Metis [17]. In all cases, the partitioning time is negligible compared to the solve time. For the structurally symmetric `mutex` and `ncd` matrices, we used the graph of A, and for the other matrices we used the graph of $A + A^T$ as mentioned in Section 4. As discussed in Section 4, we maintain balance on the size, rather than the densities, of the subblocks of A_{11}. We have conducted experiments with $K = 2, 4, 8, 16$, and 32 subblocks in the (1,1) block. For each K value, K-way partitioning of a test matrix constitutes a partitioning instance. Since Metis incorporates randomized algorithms, it was run 20 times starting from different random seeds for each partitioning instance. The maximum allowable imbalance ratio among the part weights was specified as 25%. In all partitioning instances except the `mutex` matrices, the imbalance ratios among the parts were within the specified limit. Therefore, the partition with the minimum separator size was chosen for those matrices. For the `mutex` matrices, we chose the best among the partitions that satisfies the imbalance ratio, if any. Otherwise, we chose the one with smallest imbalance (this was the case in $K = 16$- and 32-way partitioning of both of the matrices and the resulting imbalance was 35%). The properties of the 2×2 block structures corresponding to these best partitions are given in Table 5.2.

As seen from Table 5.2, only the `mutex` matrices have a large number of rows in the second row block of A, i.e., a large separator set. For these matrices, only for the $K = 2$-way partitioning instances the average part size is larger than the size of the separator set. For the other matrices, the average part size is larger than the size of the second row block in all partitioning instances with $K = 2, 4, 8$, and 16 except in $K = 16$-way partitioning of `ncd07` and `qnatm06`. The average separator sizes for all K values are given at the bottom of the table. These averages contain the `mutex` matrices; without them, the average separator and part sizes for $K = 16$ are 0.043 and 0.060, respectively.

An interesting observation is that in all partitioning instances, the (2,2) block is always very sparse; the minimum and maximum number of nonzeros per row in the (2,2) block are 1.0 (in $K = 2, 4$, and 8-way partitioning of `mutex12` and in almost all `ncd` partitioning instances) and 4.2 (in $K = 16$-way partitioning of `mutex09`) where the overall average is 1.67. All matrices, except `mutex12` and `mutex09` have small numbers of nonzeros per row. Therefore, a very sparse separator is most likely to occur in partitioning these matrices. However, the separator being highly sparse in different Markov chain models is noteworthy, especially from the parallelization perspective. For example, the (2,2) block may be duplicated on all processors to overcome the serial bottleneck of Schur complement solves with only a small storage overhead, or the off-diagonal entries of the (2,2) block may be selectively dropped to

TABLE 5.2
Properties of the partitions and the induced block structures for the test matrices. The column "sep" refers to the number of rows in the 2nd row block of A normalized by the number of rows in A, i.e., m/N; the column "part" refers to the average part size normalized by the number of rows in A, i.e., (n/K)/N; the columns A_{ij} for $i, j = 1, 2$ refer to the number of nonzeros in the (i, j) block normalized by the number of nonzeros in A, i.e., $nnz(A_{ij})/nnz(A)$.

Matrix	K	Partition		Blocks			
		sep	part	A_{11}	A_{12}	A_{21}	A_{22}
mutex09	2	0.205	0.398	0.618	0.177	0.177	0.028
	4	0.337	0.166	0.363	0.300	0.300	0.037
	8	0.415	0.073	0.237	0.348	0.348	0.068
	16	0.469	0.033	0.177	0.354	0.354	0.116
	32	0.473	0.016	0.138	0.389	0.389	0.084
mutex12	2	0.141	0.429	0.621	0.185	0.185	0.009
	4	0.225	0.194	0.397	0.294	0.294	0.015
	8	0.282	0.090	0.243	0.369	0.369	0.018
	16	0.333	0.042	0.162	0.389	0.389	0.060
	32	0.343	0.021	0.124	0.411	0.411	0.053
ncd07	2	0.015	0.493	0.972	0.013	0.013	0.002
	4	0.028	0.243	0.946	0.025	0.025	0.004
	8	0.047	0.119	0.910	0.041	0.041	0.007
	16	0.071	0.058	0.866	0.062	0.062	0.010
	32	0.099	0.028	0.813	0.086	0.086	0.015
ncd10	2	0.012	0.494	0.976	0.011	0.011	0.002
	4	0.023	0.244	0.957	0.020	0.020	0.003
	8	0.036	0.121	0.933	0.031	0.031	0.005
	16	0.057	0.059	0.893	0.049	0.049	0.009
	32	0.076	0.029	0.856	0.066	0.066	0.012
qnatm06	2	0.012	0.494	0.979	0.009	0.009	0.003
	4	0.024	0.244	0.957	0.019	0.019	0.005
	8	0.041	0.120	0.927	0.032	0.032	0.009
	16	0.068	0.058	0.877	0.055	0.055	0.013
	32	0.100	0.028	0.819	0.081	0.081	0.019
qnatm07	2	0.009	0.496	0.986	0.005	0.005	0.004
	4	0.019	0.245	0.966	0.015	0.015	0.004
	8	0.034	0.121	0.940	0.026	0.026	0.008
	16	0.054	0.059	0.903	0.043	0.043	0.011
	32	0.081	0.029	0.854	0.065	0.065	0.017
tcomm16	2	0.002	0.499	0.996	0.002	0.002	0.001
	4	0.007	0.248	0.988	0.005	0.005	0.002
	8	0.016	0.123	0.973	0.011	0.011	0.005
	16	0.034	0.060	0.942	0.024	0.024	0.010
	32	0.059	0.029	0.898	0.042	0.042	0.018
tcomm20	2	0.002	0.499	0.997	0.001	0.001	0.001
	4	0.005	0.249	0.991	0.004	0.004	0.002
	8	0.013	0.123	0.978	0.009	0.009	0.004
	16	0.027	0.061	0.954	0.019	0.019	0.008
	32	0.054	0.030	0.908	0.038	0.038	0.016
twod08	2	0.002	0.499	0.997	0.001	0.001	0.001
	4	0.006	0.249	0.991	0.003	0.003	0.003
	8	0.013	0.123	0.980	0.007	0.007	0.007
	16	0.021	0.061	0.968	0.011	0.011	0.011
	32	0.035	0.030	0.947	0.018	0.018	0.018
twod10	2	0.002	0.499	0.997	0.001	0.001	0.001
	4	0.004	0.249	0.994	0.002	0.002	0.002
	8	0.007	0.124	0.989	0.004	0.004	0.004
	16	0.012	0.062	0.982	0.006	0.006	0.006
	32	0.019	0.031	0.972	0.009	0.009	0.009
Averages							
	2	0.040	0.480	0.914	0.040	0.040	0.005
	4	0.068	0.233	0.855	0.069	0.069	0.008
	8	0.090	0.114	0.811	0.088	0.088	0.013
	16	0.115	0.055	0.772	0.101	0.101	0.026
	32	0.134	0.027	0.733	0.121	0.121	0.026

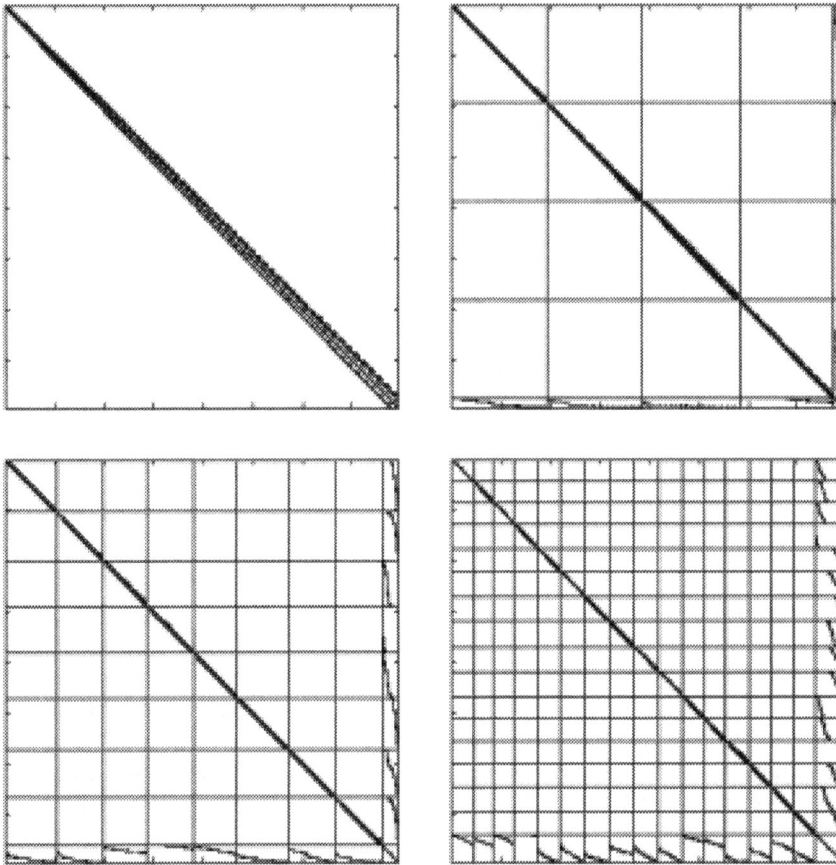

FIG. 5.1. *Sparsity pattern of the matrix* qnatm06 *and structure induced by* K-*way partitioning for* $K = 4, 8, 16$.

yield an approximate Schur complement in order to reduce the storage requirements. As an example of the partitioning outcomes, Figure 5.1 shows the sparsity pattern of the matrix qnatm06 followed by the block structure of the K-way partitionings for $K = 4, 8, 16$.

Each subblock of A_{11} and the (2,2) block A_{22} are factored using the incomplete LU factorization (ILUTH) with threshold parameter $\tau = 0.01$ for the qnatm matrices and $\tau = 0.001$ for the other matrices. The threshold of 0.001 was too small for the qnatm matrices: the resulting preconditioners had 8 times more nonzeros than the generator matrices. The densities of the preconditioners, i.e., the number of nonzeros in the matrices appearing in the preconditioner solve phase divided by the number of nonzeros in the corresponding generator matrices, are given in Table 5.3. The number of nonzeros in the preconditioner solve phase of BJ is the total number of nonzeros in the approximate ILU factors of A_{11} and A_{22}. The number of nonzeros in the preconditioner solve phase of SC is the total number of nonzeros in the approximate ILU factors of the Schur complement matrix in (3.7), plus the number of nonzeros in A_{21} and A_{12}. The number of nonzeros in the preconditioner solve phase of PS is computed by adding the number of nonzeros in $M_{BJ} + M_{SC} - A$ to those of the

BJ and SC preconditioners. The number of nonzeros for the BGS preconditioner is the total number of nonzeros in the ILU factors of A_{11} and A_{22} plus the number of nonzeros in A_{21}.

Since the number of nonzeros in the off-diagonal blocks A_{12} and A_{21} increases for increasing K, the number of nonzeros in the BJ preconditioners decreases for increasing K. By the same token, the number of nonzeros in the SC preconditioners increases for increasing K. Since the proposed preconditioner PS uses those two preconditioners and an additional $n + nnz(A_{22}) \approx N$ nonzeros, the number of nonzeros in the PS preconditioners is not heavily affected by the increasing number of parts. As seen from the averages given in the bottom of the table, the product preconditioners have around 2.5 times the nonzeros of the generator matrices for all K, whereas the BGS preconditioners contain around 1.97 times the nonzeros of the generator matrices, on the average. Thus, on average, the PS preconditioners contain 28% more nonzeros than the BGS preconditioners.

5.2. Performance comparisons. The underlying Krylov subspace method was GMRES, restarted every 50 iterations (if needed). Right preconditioning was used in all the tests. The stopping criterion was set as

$$\|r_k\|_2 / \|r_0\|_2 < 10^{-10} ,$$

where r_k is the residual at the kth iteration and r_0 is the initial residual. We allow at most 250 iterations, i.e., 5 restarts, for the GMRES iteration. Therefore, the number 250 in the following tables marks the cases in which GMRES failed to deliver solutions with the prescribed accuracy within 250 iterations. For each matrix, only one initial solution is chosen at random (with positive entries) and normalized to have an ℓ_1-norm of 1.0. In other words, the initial guess is a random probability distribution vector. The same initial guess is used for all partitioning instances of a matrix with all preconditioners. Iteration counts for GMRES(50) on the various test matrices with no permutation or preconditioning (GMRES) and with the preconditioners BJ, SC, BGS, and PS are given in Table 5.4. The ℓ_1-norms of the residuals corresponding to the approximate solutions returned by GMRES(50) were between 1.9889e-17 and 1.6098e-11 for the converged instances. Note that without preconditioning, GMRES(50) converges only for the `mutex` matrices.

As seen from Table 5.4, the BGS preconditioner is consistently better than the BJ and SC preconditioners. The proposed PS preconditioner, although constructed from the two preconditioners outperformed by BGS, is in turn consistently better than the BGS preconditioner. The last five rows of the table show the performance of the proposed PS preconditioner with respect to the BGS preconditioner. These numbers are computed by first normalizing the number of iterations of BGS and PS by the number of iterations required by BGS preconditioner with $K = 2$-way partitioning for each matrix. Then, the averages of these numbers are displayed in the last five rows. We did not display the averages for the BJ and SC preconditioners because they do not lead to convergence in all instances. As seen from the normalized averages, the proposed PS preconditioner outperforms the BGS one by a factor of two in terms of iteration counts, at the expense of an increase of just 28% (approximately) in the number of nonzeros (see Table 5.3).

Recall from Section 3 that the PS preconditioner is given by $M_{PS}^{-1} = M_{BJ}^{-1}(M_{BJ} + M_{SC} - A)M_{SC}^{-1}$. Consider the matrix in the middle. It is composed of the diagonal of A_{11} and the (2,2) block A_{22}. Since the A_{22} block is very sparse, i.e., close to being a diagonal matrix, the effect of the multiply with $(M_{BJ} + M_{SC} - A)$ is merely a

TABLE 5.3

The densities of the preconditioners, i.e., the total number of nonzeros in the blocks appearing in the preconditioning matrices divided by the number of nonzeros in the corresponding generator matrices.

Matrix	K	Preconditioners			
		BJ	SC	BGS	PS
mutex09	2	0.75	0.67	0.93	1.50
	4	0.47	1.26	0.77	1.81
	8	0.37	1.50	0.72	1.97
	16	0.36	1.56	0.72	2.07
	32	0.28	1.78	0.67	2.17
mutex12	2	0.68	0.62	0.87	1.37
	4	0.47	1.01	0.76	1.55
	8	0.31	1.34	0.68	1.72
	16	0.29	1.51	0.68	1.90
	32	0.24	1.58	0.65	1.92
ncd07	2	0.88	0.18	0.89	1.21
	4	0.86	0.21	0.89	1.22
	8	0.82	0.25	0.86	1.22
	16	0.79	0.30	0.85	1.23
	32	0.75	0.36	0.84	1.25
ncd10	2	0.72	0.17	0.73	1.04
	4	0.71	0.19	0.73	1.05
	8	0.68	0.22	0.72	1.05
	16	0.66	0.26	0.71	1.07
	32	0.64	0.30	0.71	1.09
qnatm06	2	3.43	0.18	3.44	3.77
	4	3.31	0.22	3.33	3.68
	8	3.17	0.27	3.21	3.60
	16	2.93	0.36	2.98	3.44
	32	2.65	0.46	2.73	3.26
qnatm07	2	3.48	0.17	3.48	3.80
	4	3.38	0.20	3.39	3.73
	8	3.25	0.25	3.27	3.65
	16	3.06	0.31	3.11	3.53
	32	2.83	0.40	2.89	3.38
tcomm16	2	2.12	0.21	2.13	2.53
	4	2.10	0.22	2.11	2.52
	8	2.06	0.24	2.07	2.50
	16	1.97	0.27	2.00	2.45
	32	1.88	0.32	1.92	2.41
tcomm20	2	2.19	0.21	2.19	2.60
	4	2.17	0.21	2.18	2.59
	8	2.13	0.23	2.14	2.57
	16	2.06	0.26	2.08	2.52
	32	1.93	0.31	1.97	2.45
twod08	2	2.38	0.25	2.38	2.88
	4	2.36	0.26	2.36	2.87
	8	2.33	0.27	2.34	2.86
	16	2.30	0.29	2.31	2.84
	32	2.24	0.31	2.25	2.81
twod10	2	3.86	0.25	3.86	4.37
	4	3.84	0.26	3.84	4.35
	8	3.80	0.26	3.80	4.31
	16	3.73	0.27	3.74	4.26
	32	3.66	0.28	3.67	4.20
Averages					
	2	2.05	0.29	2.09	2.51
	4	1.97	0.40	2.04	2.54
	8	1.89	0.48	1.98	2.54
	16	1.82	0.54	1.92	2.53
	32	1.71	0.61	1.83	2.50

TABLE 5.4

Number of iterations to reduce the ℓ_2-norm of the initial residual by ten orders of magnitude using GMRES(50) with at most 5 restarts. The number 250 means that the method did not converge in 250 iterations.

Matrix	GMRES	K	Preconditioned GMRES			
			Preconditioners			
			BJ	SC	BGS	PS
mutex09	97	2	25	27	13	8
		4	29	23	15	9
		8	30	20	15	8
		16	26	22	14	9
		32	29	17	15	9
mutex12	91	2	27	23	14	8
		4	29	20	15	7
		8	29	18	15	7
		16	27	17	14	8
		32	28	17	15	8
ncd07	250	2	38	250	25	16
		4	201	250	69	17
		8	250	250	99	19
		16	250	250	99	21
		32	250	250	158	22
ncd10	250	2	42	250	29	19
		4	250	250	181	21
		8	205	250	101	22
		16	250	250	145	25
		32	250	250	188	26
qnatm06	250	2	64	250	41	39
		4	92	250	45	39
		8	120	250	50	42
		16	200	250	62	45
		32	250	250	86	49
qnatm07	250	2	65	250	45	44
		4	94	250	52	46
		8	138	250	71	48
		16	215	250	87	55
		32	245	250	98	68
tcomm16	250	2	33	250	19	15
		4	50	250	27	20
		8	100	250	36	26
		16	250	250	91	42
		32	250	250	79	42
tcomm20	250	2	31	250	19	16
		4	41	250	24	20
		8	101	250	36	26
		16	250	250	73	40
		32	250	250	200	101
twod08	250	2	26	250	14	9
		4	34	250	18	11
		8	43	250	22	16
		16	51	250	26	21
		32	58	250	30	25
twod10	250	2	35	250	18	12
		4	46	250	24	19
		8	52	250	27	22
		16	50	250	26	21
		32	87	250	37	29
Normalized averages with respect to 2-way BGS						
		2	-	-	1.00	0.74
		4	-	-	1.88	0.85
		8	-	-	1.93	0.98
		16	-	-	2.64	1.23
		32	-	-	3.82	1.67

TABLE 5.5
Running times (in seconds) for GMRES(50) without preconditioning (GMRES column) and with BJ, SC, BGS and PS preconditioning for the `mutex` matrices.

Matrix	GMRES		Preconditioned GMRES							
	Total	K	Total time							
	time		Preconditioner construction				Solve			
	Solve		BJ	SC	BGS	PS	BJ	SC	BGS	PS
mutex09	19.6	2	1.8	0.7	2.0	3.2	11.0	8.2	5.8	5.0
		4	0.8	3.2	1.0	4.5	11.8	9.0	6.0	6.2
		8	0.5	4.1	0.7	5.2	11.9	8.8	5.8	6.0
		16	0.4	7.9	0.7	8.9	10.2	10.4	5.4	6.9
		32	0.3	5.2	0.5	6.1	11.2	8.4	5.8	7.0
mutex12	79.1	2	6.5	2.6	7.2	11.7	51.9	26.7	26.7	21.0
		4	3.6	6.5	4.3	12.4	52.0	28.5	26.9	19.8
		8	1.9	12.3	2.7	16.5	49.6	30.1	26.0	21.3
		16	1.7	42.4	2.5	46.4	46.4	30.7	23.9	25.3
		32	1.2	35.5	2.1	38.8	47.0	31.6	25.3	25.2

scaling. Therefore, we conducted experiments with the PS preconditioner without the matrix-vector multiplies with $(M_{BJ} + M_{SC} - A)$ and observed that in 41 partitioning instances (even with the `mutex` matrices that have relatively denser A_{22}) the number of iterations did not change at all, in 3 of the cases omitting the multiply decreased the number of iterations by 1, in 4 cases it increased the number of iterations by 1, and only in $K = 32$-way partitioning of `tcomm20` it increased the number of iterations by 5.

We close this section by discussing running times for GMRES. The timings are obtained using Matlab 6's `cputime` command and are in seconds. In Table 5.5, the total time for GMRES without preconditioning is the total time spent in performing the GMRES iterations. In Table 5.5 and 5.6, the total time for the preconditioned GMRES is dissected into the preconditioner construction and the solve phases. We first discuss the case of `mutex` matrices, since all the preconditioners lead to convergence for these matrices. In all partitioning instances of the `mutex` matrices, the proposed PS preconditioner solve phase time is less than the solve phase times of its factors BJ and SC preconditioners. Furthermore, in half of the instances its solve phase time is smaller than the BGS preconditioner solve phase time. On the other hand, the total running time of BGS preconditioner is always the minimum except in $K = 2$-way partitioning of `mutex12` in which case BJ gives the minimum total running time. We observe that the `mutex` matrices are the worst case for the construction of the PS preconditioner, since the size of the separator set is very large already for $K = 2$, thus forming the Schur complement is very time-consuming.

Table 5.6 contains the running times of the preconditioned GMRES with the BGS and PS preconditioners for the larger matrices in each matrix family. As seen from the table, for these matrices (whose partitions have small separators) the preconditioner construction phases' time are always smaller than the solve phases' time. Furthermore, the construction times for the BGS and PS preconditioners are almost the same, and PS is faster than BGS (due to smaller number of iterations) in all instances except for $K = 2$-way partitioning of `qnatm07`. Note that the ith iteration of GMRES after a restart requires i inner product computations with vectors of length N [2]. Therefore, the performance gains in the solve phase with the PS preconditioners are not only due to the savings in preconditioner solves and matrix-vector multiplies, but also to the savings in the inner product computations. What is important in this

TABLE 5.6

Running times (in seconds) for GMRES(50) with BGS and PS preconditioners for the larger matrices in each family.

Matrix	K	Total time			
		Precond const		Solve	
		BGS	PS	BGS	PS
ncd10	2	1.3	1.7	28.1	19.8
	4	1.3	1.7	193.0	22.0
	8	1.2	1.6	108.6	23.4
	16	1.2	1.6	153.7	27.0
	32	1.1	1.6	198.4	28.5
qnatm07	2	20.9	21.2	46.5	49.6
	4	14.9	15.2	53.7	50.8
	8	9.6	10.0	70.3	53.4
	16	6.5	6.9	85.8	61.1
	32	4.9	5.3	93.6	70.9
tcomm20	2	0.2	0.2	1.3	1.2
	4	0.2	0.2	1.7	1.5
	8	0.2	0.2	2.6	2.0
	16	0.2	0.2	5.6	3.3
	32	0.1	0.2	15.8	8.8
twod10	2	17.3	17.6	31.1	22.6
	4	14.5	14.8	42.4	35.9
	8	6.6	6.9	47.3	42.0
	16	6.1	6.4	45.4	39.6
	32	5.2	5.5	67.1	56.4

table is how the total time with the PS preconditioner increases as the number of parts increases. These increases are fairly modest compared to the number of parts. Except for `tcommm20`, which is rather small, the solution times with 32 parts are at most 3 times larger than the solution times with 2 parts. This heralds considerable speed-ups in the total solution time in a parallel computing environment. Additionally, this also gives leeway in parallelizing the iterations.

6. Conclusions. We have described and investigated a new preconditioning technique for Markov chain problems. The idea is to combine two simple preconditioners to get a new, more effective one. This "product splitting" preconditioner relies on a block 2×2 structure of the generator matrix, which is obtained through graph partitioning. Our approach is somewhat similar to a two-level, non-overlapping additive Schwarz (block Jacobi) preconditioner with a multiplicative "coarse grid correction" represented by an approximate Schur complement solve. Numerical experiments with preconditioned GMRES using test matrices from MARCA indicate that the product preconditioner is much more effective than the two constituent preconditioners, at the expense of only a slight increase in storage requirements. The product splitting was also found to be competitive in most cases with an appropriate block Gauss–Seidel preconditioner. Furthermore, the numerical experiments indicate that, for most problems, the number of iterations grows slowly with the number of parts (subdomains). This suggests that the product splitting preconditioner should perform very well in a parallel implementation.

Acknowledgment. We thank Billy Stewart for making the MARCA software available.

REFERENCES

[1] C. AYKANAT, A. PINAR, AND U. V. ÇATALYÜREK, *Permuting sparse rectangular matrices into block-diagonal form*, SIAM J. Sci. Comput., 25 (2004), pp. 1860–1879.

[2] R. BARRETT, M. BERRY, T. F. CHAN, J. DEMMEL, J. DONATO, J. DONGARRA, V. EIJKHOUT, R. POZO, C. ROMINE, AND H. A. VAN DER VORST, *Templates for the Solution of Linear Systems: Building Blocks for Iterative Methods*, SIAM, Philadelphia, 1994.

[3] M. BENZI AND T. DAYAR, *The arithmetic mean method for finding the stationary vector of Markov chains*, Parallel Algorithms Appl., 6 (1995), pp. 25–37.

[4] M. BENZI, F. SGALLARI, AND G. SPALETTA, *A parallel block projection method of the Cimmino type for finite Markov chains*, in Computations with Markov Chains, W. J. Stewart, ed., Kluwer Academic Publishers, Boston/London/Dordrecht, 1995, pp. 65–80.

[5] M. BENZI AND D. B. SZYLD, *Existence and uniqueness of splittings for stationary iterative methods with applications to alternating methods*, Numer. Math., 76 (1997), pp. 309–321.

[6] M. BENZI AND M. TŮMA, *A parallel solver for large-scale Markov chains*, Appl. Numer. Math., 41 (2002), pp. 135–153.

[7] A. BERMAN AND R. J. PLEMMONS, *Nonnegative Matrices in the Mathematical Sciences*, Academic Press, New York, 1979. Reprinted by SIAM, Philadelphia, PA, 1994.

[8] P. BUCHHOLZ, M. FISCHER, AND P. KEMPER, *Distributed steady state analysis using Kronecker algebra*, in Numerical Solutions of Markov Chains (NSMC'99), B. Plateau, W. J. Stewart, and M. Silva, eds., Prensas Universitarias de Zaragoza, Zaragoza, Spain, 1999, pp. 76–95.

[9] T. N. BUI AND C. JONES, *Finding good approximate vertex and edge partitions is NP hard*, Inform. Process. Lett., 42 (1992), pp. 153–159.

[10] T. N. BUI AND C. JONES, *A heuristic for reducing fill in sparse matrix factorization*, in Proc. 6th SIAM Conf. Parallel Processing for Scientific Computing, SIAM, Philadelphia, 1993, pp. 445–452.

[11] T. DAYAR AND W. J. STEWART, *Comparison of partitioning techniques for two-level iterative solvers of large, sparse Markov chains*, SIAM J. Sci. Comput., 21 (2000), pp. 1691–1705.

[12] P. FERNANDES, B. PLATEAU, AND W. J. STEWART, *Efficient descriptor-vector multiplications in stochastic automata networks*, J. ACM, 45 (1998), pp. 381–414.

[13] B. HENDRICKSON AND R. LELAND, *A multilevel algorithm for partitioning graphs*, in Supercomputing'95: Proceedings of the 1995 ACM/IEEE conference on Supercomputing (CDROM), ACM Press, New York, NY, USA, 1995, p. 28.

[14] B. HENDRICKSON AND E. ROTHBERG, *Improving the run time and quality of nested dissection ordering*, SIAM J. Sci. Comput., 20 (1998), pp. 468–489.

[15] M. JARRAYA AND D. EL BAZ, *Asynchronous iterations for the solution of Markov systems*, in Numerical Solutions of Markov Chains (NSMC'99), B. Plateau, W. J. Stewart, and M. Silva, eds., Prensas Universitarias de Zaragoza, Zaragoza, Spain, 1999, pp. 335–338.

[16] G. KARYPIS AND V. KUMAR, *A fast and high quality multilevel scheme for partitioning irregular graphs*, SIAM J. Sci. Comput., 20 (1998), pp. 359–392.

[17] G. KARYPIS AND V. KUMAR, *MeTiS: A software package for partitioning unstructured graphs, partitioning meshes, and computing fill-reducing orderings of sparse matrices version 4.0*, University of Minnesota, Department of Computer Science / Army HPC Research Center, Minneapolis, MN 55455, September 1998.

[18] W. J. KNOTTENBELT AND P. G. HARRISON, *Distributed disk-based solution techniques for large Markov models*, in Numerical Solutions of Markov Chains (NSMC'99), B. Plateau, W. J. Stewart, and M. Silva, eds., Prensas Universitarias de Zaragoza, Zaragoza, Spain, 1999, pp. 58–75.

[19] I. MAREK AND D. B. SZYLD, *Algebraic Schwarz methods for the numerical solution of Markov chains*, Linear Algebra Appl., 386 (2004), pp. 67–81.

[20] C. D. MEYER, *Stochastic complementation, uncoupling Markov chains, and the theory of nearly reducible systems*, SIAM Rev., 31 (1989), pp. 240–272.

[21] V. MIGALLÓN, J. PENADÉS, AND D. B. SZYLD, *Experimental studies of parallel iterative solutions of Markov chains with block partitions*, in Numerical Solutions of Markov Chains (NSMC'99), B. Plateau, W. J. Stewart, and M. Silva, eds., Prensas Universitarias de Zaragoza, Zaragoza, Spain, 1999, pp. 96–110.

[22] M. NEUMANN AND R. J. PLEMMONS, *Convergent nonnegative matrices and iterative methods for consistent linear systems*, Numer. Math., 31 (1978), pp. 265–279.

[23] B. PHILIPPE, Y. SAAD, AND W. J. STEWART, *Numerical methods in Markov chain modeling*, Oper. Res., 40 (1992), pp. 1156–1179.

[24] A. PINAR AND B. HENDRICKSON, *Partitioning for complex objectives*, Proceedings of the 15th International Parallel and Distributed Processing Symposium (CDROM), IEEE Computer

Society Washington, DC, USA, 2001, p. 121.

[25] P. K. Pollet and S. E. Stewart, *An efficient procedure for computing quasi-stationary distributions of Markov chains with sparse transition structure*, Adv. Appl. Probab., 26 (1994), pp. 68–79.

[26] Y. Saad, *Preconditioned Krylov subspace methods for the numerical solution of Markov chains*, in Computations with Markov Chains, W. J. Stewart, ed., Kluwer Academic Publishers, Boston/London/Dordrecht, 1995, pp. 49–64.

[27] Y. Saad and M. H. Schultz, *GMRES: A generalized minimal residual algorithm for solving nonsymmetric linear systems*, SIAM J. Sci. Stat. Comput., 7 (1986), pp. 856–869.

[28] H. Schneider, *Theorems on M-splittings of a singular M-matrix which depend on graph structure*, Linear Algebra Appl., 58 (1984), pp. 407–424.

[29] W. J. Stewart, *MARCA Models: A collection of Markov chain models.* URL http://www.csc.ncsu.edu/faculty/stewart/MARCA_Models/MARCA_Models.html.

[30] W. J. Stewart, *Introduction to the Numerical Solution of Markov Chains*, Princeton University Press, Princeton, NJ, 1994.

[31] D. B. Szyld, *Equivalence of convergence conditions for iterative methods for singular equations*, Numer. Linear Algebra Appl., 1 (1994), pp. 151–154.

[32] R. S. Varga, *Matrix Iterative Analysis*, Prentice-Hall, Englewood Cliffs, New Jersey, 1962. Second Edition, revised and expanded, Springer, Berlin, Heidelberg, New York, 2000.

[33] R. E. White, *Domain decomposition splittings*, Linear Algebra Appl., 316 (2000), pp. 105–112.

THE BOUNDING DISCRETE PHASE–TYPE METHOD

JEAN–SÉBASTIEN TANCREZ[†‡] AND PIERRE SEMAL[†]

Abstract. Models of production systems have always been essential. They are needed at a strategic level in order to guide the design of production systems but also at an operational level when, for example, the daily load and staffing have to be chosen.

Models can be classified into three categories: analytical, simulative and approximate. In this paper, we propose an approximation approach that works as follows. Each arrival or service distribution is discretized using the same time step. The evolution of the production system can then be described by a Markov chain. The performances of the production system can then be estimated from the analysis of the Markov chain.

The way the discretization is carried on determines the properties of the results. In this paper, we investigate the "grouping at the end" discretization method and, in order to fix ideas, in the context of production lines. In this case, upper and lower bounds on the throughput can be derived. Furthermore, the distance between these bounds is proved to be related to the time step used in the discretization. They are thus refinable and their precision can be evaluated a priori. All these results are proved using the concept of critical path of a production run.

Beside the conceptual contribution of this paper, the method has been successfully applied to a line with three stations in which three buffer spaces have to be allocated. Nevertheless, the complexity and solution aspects will require further attention before making the method eligible for real large scale problems.

Key words. Production line, Discretization, Bounds, Critical path.

AMS subject classifications. 60J20, 60K25, 90B22, 90B30

1. Introduction. Production systems follow various types of organization, among which the job–shop and the flow line are most typical [5, 12]. In this paper, we focus on production lines although we strongly believe the approach and the method presented here can be applied to any production system. The details of the production line we study are described below in §1.1.

Models of production systems have always been essential. They are needed during the design phases, when the following elements have to be selected: the type of production equipments and their power, the layout of the different stations and the mechanisms of synchronization (equipment, transfer lot size, buffer size, ...). At a more operational level, models are also needed to provide support for daily production decisions like the load, the sequence of jobs or the necessary staffing, and for customer oriented decisions like accepting a new job or promising some delivery time. Refer to [4] and [2] for an overview of manufacturing systems models. A systematic review of the literature for models of production lines is given below in §1.2.

The model presented in this paper is approximate in the sense that it simplifies the system to be studied, without changing its general structure, in order to facilitate the evaluation of its performance. The model has two peculiar features: it provides not only an approximation of the performance of interest but bounds on it, and these lower and upper bounds can be tightened. This will be illustrated by determining bounds on the throughput of a production line, which are valid both in the transient and in steady–state. The approach seems to be applicable to other performances like the buffer utilization, the job flow time or the work in progress.

[†]Université Catholique de Louvain, Place des Doyens, 1, 1348 Louvain–la–Neuve, Belgium, {`tancrez,semal@poms.ucl.ac.be`}.
[‡]Facultés Universitaires Catholiques de Mons.

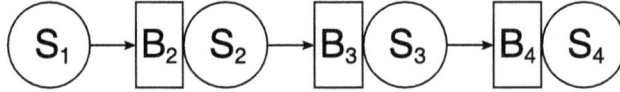

FIG. 1. *Production line including four stations and three buffers.*

Practically, the paper is organized as follows. Here below, §1.1 details the type of production lines we focus on and §1.2 provides a short literature review of models used for such lines. Section 2 and 3 are the heart of the paper. Section 2 describes the BDPH method and §3 states its properties. Section 4 provides an example. Finally, §5 focuses on possible extensions of the method.

1.1. Production Lines. This paper focuses on production lines with asynchronous part transfer, in other terms tandem queueing systems. As shown on Figure 1, production lines have a very special structure. It is a linear network of m service stations $(S_1, S_2, \ldots, S_m)$ separated by $m - 1$ buffers storages $(B_2, B_3, \ldots, B_m)$. The manufacturing of one item consists in the sequential processing of this item by the stations S_1 to S_m. The item enters the system at station S_1 and leaves at S_m. After its processing by a station, let us say S_i, the unit is stored in the buffer B_{i+1} if a space is available. There it waits until the next station, S_{i+1}, finishes its job on the previous item and gets rid of it. At this moment, the processing of the unit on station S_{i+1} starts. This is repeated until the item gets its last processing at station S_m and then leaves the system.

Such lines experience productivity losses due to blocking and starving. First, a station is said to be blocked when it cannot get rid of an item because the next buffer is full. Second, a station is to be starved when it cannot begin to work on a new item because the previous buffer is empty. Increasing buffer sizes allows to limit these productivity losses.

Here is the list of assumptions we make on the lines we analyze, in order to fix ideas. None of these assumptions is restrictive.

- *General finite service time distributions.* The service times are generally distributed but finite. Successive processing times are independent and identically distributed. The finiteness assumption is not restrictive since it is always the case in practice.
- *Finite buffer sizes.* We do not make any assumption on the buffers sizes except their finiteness.
- *Infinite arrival.* With this assumption, the first station is never starved. This assumption can be relaxed by using an initial station that models the arrival process.
- *Infinite demand.* With this assumption, the last station is never blocked. Again, this assumption can be relaxed by using a last station that models the real demand or the real storage space.
- *Blocking after service.* This means that if the next station is full, a job is blocked in its current station after having been processed. A blocking before service policy can be modeled by adapting the buffer size.

1.2. Models for Production Lines. Several good comprehensive reviews of models for production lines are available (see for example [2], [4], [17], [7], [10], and [16]). In order to situate precisely our method in the state of the art, let us review the main approaches used for the evaluation of production line performance. As already said, there are three kinds of models: exact analytical models, approximate analytical models and simulations.

Exact analytical models are the richest since they allow a direct and exact understanding of the influence of a decision variable on the performance of interest. Unfortunately, most production systems are too complex to be modeled analytically. In our case, exact models can be used for simple production lines only. Three methods are significant:

- *Closed–form models.* Closed–form results are available for very simple configurations, e.g. two stations lines and exponentially (or sometimes Erlang) distributed service times.
- *State models.* These models build continuous (rarely discrete) Markov chains to analyze lines with exponential or phase–type distributions. Based on the identified state space, a transition matrix is derived and the stationary equations are solved numerically to obtain the steady–state probabilities. The main difficulty lies in the explosion of the state space size. A first paper [13] dates from 1967.
- *Holding time models.* Introduced by Muth, this method aims to be computationally more efficient than state models for tandem queues without intermediate buffers. It considers the sequence of holding times (blocking time added to processing time) for successive jobs at each station, and constructs recursive relationships. For an overview, see [15].

Simulation is at the other end of the continuum. It does not rely on any assumption and is therefore very general. Almost all systems can be simulated. The weakness of the simulation approach mainly lies in its development cost. An example of simulation for production lines is given in [8].

Approximate analytical models are in between: the system to be analyzed is simplified in order to be analytically modeled. This keeps the development costs low. However, the uncertainty about the results is the weakness of the approach. The method presented in this paper is approximate but tries to give certainties about the results, by proving bounds. Here are the two main approximate methods:

- *Decomposition.* The idea is to decompose a system into smaller subsystems. Solving more, but much easier, subproblems, allows to approximately analyze the global system much more quickly. A set of equations that determines the unknown parameters and the links between subsystems is first derived. An iterative procedure is then used to solve the equations. This method has initially been created for exponentially distributed production lines, see [9] and [3] for examples or [6] for a good review. Some authors then extended it to lines with phase–type processing times, see [1] for continuous PH and [11] for discrete PH.
- *Expansion.* The Generalized Expansion Method (GEM) is also based on the idea of decomposing the system, but it adds the concept of an artificial node that registers the blocked jobs. For a description and an example of application, see [14].

2. The Bounding Discrete Phase–Type Method. The method presented in this paper works as follows. First, each distribution is discretized using the same time step. The evolution of the production system can then be described by a Markov chain whose states describe the stages of the different service centers and the current utilizations of the various storage areas (buffers). The performance of the production system can then be estimated from the analysis of the Markov chain. The production

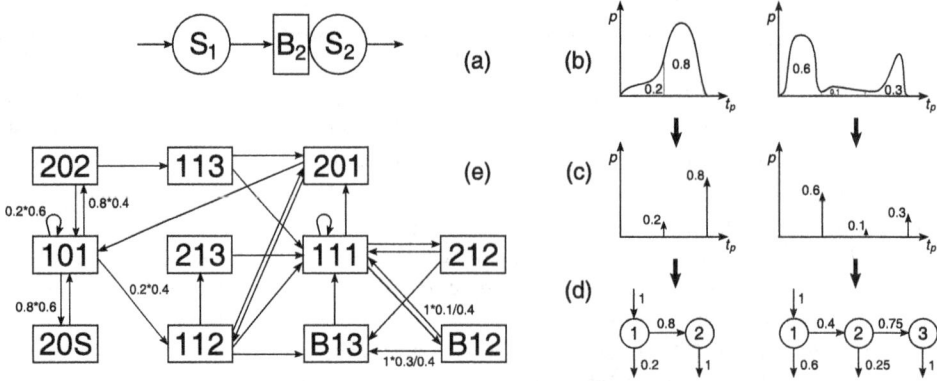

FIG. 2. *Stages of the BDPH method applied to a two station line: discretization of the original service time distributions by "grouping at the end", PH representation and Markov Chain.*

rate and the buffer utilization, for example, can be derived from the steady–state probabilities. Transient characteristics can also be determined.

The originality comes from the discretization method we use. Using a discretization step τ, we transform the original distributions into a discrete one by concentrating the probability mass distributed in the interval $[k\tau, (k+1)\tau]$ on a single value. The choice of this value is open. Here, we chose a "grouping at the end" principle, that is, the probability mass is carried forward on the point $(k+1)\tau$. This discretization creates a bias, as each job length is increased. However, it has the advantage to keep an intelligible link with the original distributions.

The method is best illustrated on an example. Let us look at a simple line made of two stations separated by a buffer of size one depicted on Figure 2.a. Figure 2.b shows the exact service time distributions for the two servers. The discretized distributions by "grouping at the end" are shown in Figure 2.c and their phase–type representation in Figure 2.d.

The behavior of the complete system can now be modeled by a Markov Chain, i.e., using a state model. The Markov chain given in Figure 2.e lists all the possible recurrent states of the system (the first symbol refers to the first station, the second to the buffer and the third to the second station) and the possible transitions between these states. Each station can be starved (S), blocked (B) or in some stage of service (1 means, for example, that the station already spent one time step working on the current job). Each buffer is described by its utilization (0 or 1 with a buffer of size one). For example, state B12 means that the first station is blocked, that the buffer is full and that the second station already worked during two time steps on the current job. Two transitions are possible from B12, depending if the second station continues to work on the same job or ends. In the first case, the new state will be B13. The probability of this transition is 0.3/0.4, the probability that the processing time is greater or equal to three knowing that it is greater or equal to two. It is easily deduced from the discrete distribution given in Figure 2.c. In the second case, the second station ends his job, picks up the next item in the buffer and begin to work on it. The first station can thus get rid of its blocking item and begins a new job. The new state is thus 111.

Transient performances, like the throughput at some time t, can be derived from the matrix of transition probabilities. The steady–state performances can be computed from the steady–state probabilities derived from the Markov chain. Steady–

state productivity, work in progress or buffer utilization, for example, can be approximated. The size of the Markov chain[1] and the complexity problems are not addressed in this paper. They require attention in the future.

The method is approximate. However, it has the advantage to offer some theoretical control. Indeed, it is proven here below that the time needed to achieve any production target will be overestimated by the method. Furthermore, the amount by which this time is overestimated can also be bounded. Finally, these bounds can be tightened, by reducing the time step τ.

The method is named "Bounding Discrete Phase–type" (BDPH) since it relies on a discrete phase type approximation of the various distributions of the system to be studied and since it leads to bounds.

3. Properties. In this section, we try to better understand the behavior of the production line and the effect of the discretization by "grouping at the end". We indeed have the intuition that the BDPH method leads to a pessimistic estimation of the productivity. In the next two subsections, we will lay the foundations that allow this result to be formally proved. These foundations rely on the concept of critical path.

3.1. Structural Properties. Let us consider a random infinite real time production run. To construct such a run, we only need a sequence of random processing times drawn according to the original service distributions. The jobs then find their places in the run according to the structure of the production system. We simply denote this infinite real time production run by r. We have:

$$\{ l^r(W_{i,k}) \} \overset{\Delta}{\longmapsto} r,$$

where $W_{i,k}$ denotes the job k at station i and $l^r(W_{i,k})$ the time it takes in this particular run r.

If the original service times $l^r(W_{i,k})$ are discretized, giving the discretized service times $\overline{l^r(W_{i,k})}$, we get the discretized production run, denoted $\overline{r}$:

$$\{ l^r(W_{i,k}) \} \overset{\text{disc.}}{\longmapsto} \{ \overline{l^r(W_{i,k})} \} \overset{\Delta}{\longmapsto} \overline{r}.$$

When using the "grouping at the end" discretization, we have, $\forall \, r, i, k$:

$$(1) \qquad \overline{l^r(W_{i,k})} \overset{\Delta}{\equiv} \left\lceil \frac{l^r(W_{i,k})}{\tau} \right\rceil \tau,$$

with, by construction, the following property, $\forall \, r, i, k$:

$$(2) \qquad \overline{l^r(W_{i,k})} - \tau \leq l^r(W_{i,k}) \leq \overline{l^r(W_{i,k})}.$$

Before stating a first result, we define the moment a job is started and the moment it is ended. Obviously, a job length, in a particular run r, is given by the difference between these moments:

$$l^r(W_{i,k}) = t^r_{end}(W_{i,k}) - t^r_{start}(W_{i,k}).$$

Since the production system has a definite structure, i.e., a line in our case, these moments are subject to structural constraints, stated in the following lemma.

---

[1]The size of the Markov chain is, in first approximation, proportional to $(a+2)^m(b+1)^{m-1}$, with m the number of machines, a the number of steps and b the buffer size.

LEMMA 1 (Structural properties of a production line). *Given a production line including a buffer of size b_i before each station i, the jobs verify the following inequalities, $\forall r, k$:*

$$(3) \quad t_{start}^r (W_{i,k}) \geq t_{end}^r (W_{i-1,k}) \qquad \forall i \geq 2$$

$$(4) \qquad\qquad\qquad \geq t_{end}^r (W_{i,k-1}) \qquad \forall i$$

$$(5) \qquad\qquad\qquad \geq t_{end}^r \left(W \begin{matrix} i+1, & k-b_{i+1}-2 \\ i+2, & k-b_{i+1}-b_{i+2}-3 \\ i+3, & k-b_{i+1}-b_{i+2}-b_{i+3}-4 \\ \vdots & \vdots \end{matrix} \right) \begin{matrix} \forall i \leq m-1 \\ \forall i \leq m-2 \\ \forall i \leq m-3 \\ \vdots \end{matrix}$$

with at least one equality.

Intuitively, the proof works as follows. Because of the line structure, a job k can only be started on station i: (3) if its processing on the previous station $i-1$ is ended, (4) if the processing of the previous job $k-1$ in station i is ended and (5) if this previous job $k-1$ is not blocked in station i by some unfinished jobs downstream. Furthermore, since there is no reason to wait once all these conditions are satisfied, $t_{start}^r (W_{i,k})$ will be given by the maximum of the right hand sides of Lemma 1. The formal proof is given in the Appendix.

Note that the inequalities of Lemma 1 are valid for any run: the job $W_{i,k}$ cannot be started before all the jobs on the right hand sides are finished. These are just static structural properties of the line, independent of the run. However, which precise job end will trigger the start of job $W_{i,k}$ depends on the processing times and thus on the particular run we consider.

3.2. Productivity and Critical Path. In this subsection, we are interested in the time needed to produce p units in a particular run r. This time is given by $t_{end}^r (W_{m,p})$ if we fix, without loss of generality, that the run has been started at time 0^2.

For the determination of $t_{end}^r (W_{m,p})$, it is first clear that not all the events of the run r are relevant. From Lemma 1, we see that $t_{end}^r (W_{m,p})$ only depends on job p or on previous jobs (by any station). We can thus restrict our attention to the following part of the run r, called its p-part and denoted r_p:

$$\{ l^r(W_{i,k}) \mid 1 \leq i \leq m, 1 \leq k \leq p\} \overset{\Delta}{\longmapsto} r_p.$$

The length of r_p, denoted $l(r_p)$, equals $t_{end}^r (W_{m,p})$, i.e., the time needed to produce p units in r. We thus focus on determining the length of this part of the run.

Second, the structural properties given in Lemma 1 allow us to introduce a useful concept. We define the *critical path* of r_p, $cp(r_p)$, as the sequence of jobs that covers r_p. It can be built quite easily. Starting with the last job that leaves the system, job $W_{m,p}$, we can look which job end, in this precise run, has triggered its start, in other words which inequality of Lemma 1 is satisfied at equality^3. Repeating this process, we can proceed backward in time until the start time of the run. It is obvious by Lemma 1 that every run r_p has at least one critical path.

---

^{2}The start time may correspond to various situations. In most cases, it will be given by the start of the first job on the first workstation, $t_{start}^r (W_{1,1})$. However, arbitrary loaded lines at start time may also be considered. For simplicity, we will just assume that the run is started with no job being partially processed.

^{3}It can happen that several equalities are satisfied at the same time. In this case, one of them is chosen arbitrarily (let us say the one corresponding to the preceding state in the same station, see Figure 5).

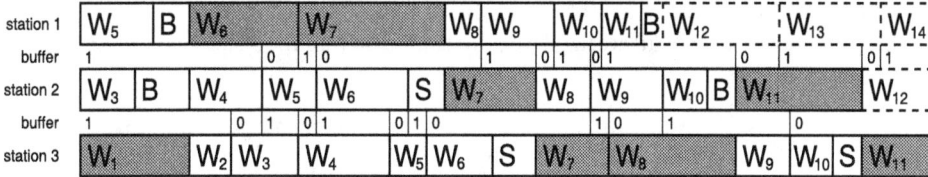

FIG. 3. *Gantt chart of the 11–part of a run on a production line with three stations and buffers of size one. The critical path is given in gray.*

This critical path $cp(r_p)$ has some nice properties. It is a set of jobs $W_{i,k}$ which covers the time of this part of the production run r_p without overlap and without gap. In other words, the length of r_p equals the length of its critical path, denoted $l(cp(r_p))$. It is given by:

$$(6) \qquad l(r_p) = t^r_{end}(W_{m,p}) = l(cp(r_p)) = \sum_{W_{i,k} \in cp(r_p)} l^r(W_{i,k}) = \sum_{j \in cp(r_p)} l^r(j).$$

As already said, the inequalities of Lemma 1 are valid for any run. The absence of overlap in the critical path $cp(r_p)$ is thus a property independent of the considered run. Therefore, in another run, the sequence of jobs $cp(r_p)$ will just be one non–overlapping path (maybe with gaps) whose total length is shorter or equal to the length of the p–part of this other run.

The notion of critical part can be best illustrated on the Gantt chart (see Figure 3) associated to a particular run. In this chart, the time goes from left to right. The state of a station at a given time is represented either by a letter (B for blocked, S for starved) or by the job currently processed. The state of a buffer is represented by the number of jobs waiting inside. For the run depicted, the line is fully loaded at time 0, with one job waiting in each buffer. The critical path (in gray) of the 11–part is made of the following backward sequence:

$$cp(r_{11}) = \{\, W_{3,11}, W_{2,11}, W_{3,8}, W_{3,7}, W_{2,7}, W_{1,7}, W_{1,6}, W_{1,3} \,\}$$

It can be checked that each couple of successive jobs satisfies one of the inequalities of Lemma 1 at equality.

The concept of critical path offers a useful tool to understand what is happening in a production system. Its ability to cover the time allows to express the time to produce p units in a run r as the sum of job lengths. When an intelligible transformation is operated on the job lengths, i.e., on their distribution, it allows to relate this transformation and its effect on the global length of a production run. We believe the concept of critical path can be generalized to other production systems and used for other transformations of the service distributions.

In the case of production lines and discretization by "grouping at the end", the critical path leads us to the following result, where the p–part of $\bar{r}$ is defined similarly :

$$\{\, \overline{l^r(W_{i,k})} \mid 1 \le i \le m, 1 \le k \le p \} \overset{\Delta}{\longmapsto} \bar{r}_p.$$

LEMMA 2. *The time an m–station line takes to produce p units in a random real time production run r can be bounded as follows:*

$$l(\bar{r}_p) - \tau(p + m - 1) \le l(r_p) \le l(\bar{r}_p).$$

Proof. Using equations (6) and (2) and the fact that $cp(r_p)$ is just a non–overlapping path in the discretized production run (smaller than the critical path, $cp(\overline{r}_p)$), we can write:

$$l(r_p) = \sum_{j \in cp(r_p)} l^r(j) \leq \sum_{j \in cp(r_p)} \overline{l^r(j)} \leq \sum_{j \in cp(\overline{r}_p)} \overline{l^r(j)} = l(\overline{r}_p),$$

that states the right inequality of the lemma. For the left inequality, using the same equations and the fact that $cp(\overline{r}_p)$ is non–overlapping in the original run, we get:

$$l(\overline{r}_p) - \tau|cp(\overline{r}_p)| = \sum_{j \in cp(\overline{r}_p)} (\overline{l^r(j)} - \tau) \leq \sum_{j \in cp(\overline{r}_p)} l^r(j) \leq \sum_{j \in cp(r_p)} l^r(j) = l(r_p),$$

where $|cp(\overline{r}_p)|$ denotes the cardinality of the critical path $cp(\overline{r}_p)$, i.e., the number of jobs making it up. The proof ends by showing that this cardinality is smaller than $p + m - 1$ (see Lemma 8 in the Appendix).   □

Lemma 2 provides a major result. Considering any random production run, we have upper and lower bounds on the time it would take to produce a given production. Unfortunately, this result cannot yet be directly used since it refers to a given random production run. For the results to be useful, we need to be able to say something about an average production run. This point is tackled in the next subsection.

3.3. Bounds on the Throughput. Let us first consider the mean time T_P necessary to reach a given production P. By definition,

$$T_P = \int f(r_P) l(r_P) dr_P,$$

where $f(r_P)$ is the density function of the p–part of the production runs. This time can be bounded as follows.

THEOREM 3. *The mean time T_P an m–station line takes to produce P units can be bounded on the basis of the information computed by the BDPH method. If $\overline{T}_P$ is the mean time to produce P units using the "grouping at the end" discretized times, we have:*

$$\overline{T}_P - \tau(P + m - 1) \leq T_P \leq \overline{T}_P.$$

The proof detailed in the Appendix relies on Lemma 2 and on the fact that the probabilities of r_P and $\overline{r}_P$ are the same since they are both derived from the same run r.

If we are interested in a fixed time instead of a fixed production, bounds can quite easily be derived from the previous theorem.

THEOREM 4. *The mean production P_T produced by an m–station line during a fixed time T can be bounded on the basis of the information computed by the BDPH method. If $\overline{P}_T$ is the mean production during time T using the "grouping at the end" discretized times, and $\overline{P}^*$ is the mean production during discrete time $\overline{T}^*$, where chosen minimal such that $\overline{T}^* - \tau(\overline{P}^* + m - 1) \geq T$, we have:*

$$\overline{P}_T \leq P_T \leq \overline{P}^*.$$

The formal proof given in the Appendix relies on the following simple argument. Since the average time to produce a given production is longer with discretized times than with the original times, the quantity produced in a given time is smaller with discretized times than with the original times.

When focusing on the steady–state productivity, the results get even simpler. Indeed, in this case, simple bounds are derived from Theorem 3. The average time between the completion of two units, in steady-state, is called the cycle time c where $c = \lim_{P \to \infty} T_P / P$. If $\bar{c}$ denotes the cycle time measured using the "grouping at the end" discretized times (BDPH method), we have the following result.

COROLLARY 5 (BDPH bounds for a production line). *When it measures the productivity of a production line in steady–state, the BDPH method is pessimistic, i.e., the cycle time is overvalued. Moreover, the error is smaller than the discretization time step:*

$$(7) \qquad\qquad \bar{c} - \tau \leq c \leq \bar{c}.$$

The proof is given in the Appendix.

Theorems 3 and 4 and Corollary 5 show that our method allows to bound the productivity, in transient or in steady–state, from below and from above. Moreover, these bounds become tighter and converge to the exact productivity when the discretization step is decreased.

These results also show another feature of the method: the accuracy of the bounds is directly related to the selected discretization step. Of course every accuracy improvement will require additional computational efforts caused by the increase of the state space size.

The BDPH bounds lead to two simple approximations of the cycle time. More precise approximations are goals for future research. Inequality (7) leads quite intuitively to a first approximation.

APPROXIMATION 6. *The cycle time of a line can be approximated by:*

$$c \approx \bar{c} - \frac{\tau}{2}.$$

This approximation can be seen as an approximation of the cycle time we would obtain by grouping "at the middle", i.e., by concentrating the probability mass at the middle of the step instead of at the end. More rigorously, it can be seen as a converging approximation of the following better approximation.

APPROXIMATION 7. *The cycle time of a line can be approximated, $\forall i$, by:*

$$c \approx \bar{c} - e_W(i),$$

with $e_W(i) = E[\overline{l^r(W_{i,k})}] - E[l^r(W_{i,k})]$, the discretization bias on the service distribution of station i.

This result comes from the fact that the cycle time can be divided up into two components: the processing time and the blocking/starving time. A good approximation of the cycle time can thus be obtained by removing the known discretization bias on the service time distribution. When the discretization step τ decreases, the bias tends to $\tau/2$ and Approximation 7 tends to Approximation 6. Note that both approximations converge to the exact cycle time as they are between the converging bounds.

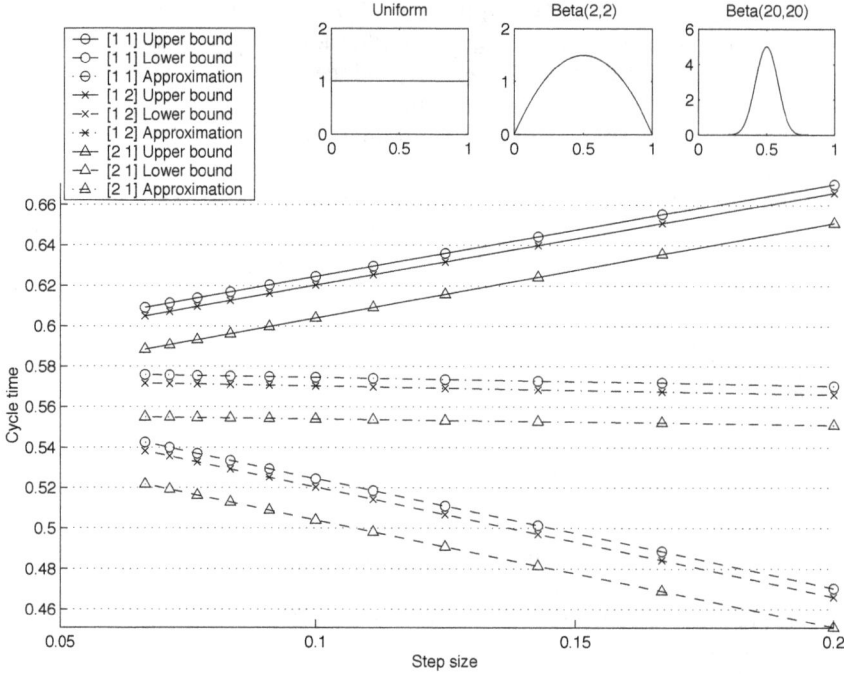

FIG. 4. *Bounds and approximations on the cycle time for a three station line with various buffers configurations. The upper part shows the service distributions, with first station on the left.*

The cycle time in discretized time, $\bar{c}$, can easily be computed from the steady–state probabilities given by the BDPH method. Let us define p_1 as the steady–state probability that a station, the last one for example, is in first stage of service. As every job on this station pass through this first stage during one time step, p_1 equals one step length divided by the cycle time. We get: $\bar{c} = \tau/p_1$.

4. An Example. In this section, we briefly show how the bounding discrete phase–type method performs on a simple example. More realistic examples are not in the scope of this paper.

Let us consider a three station line with the processing times depicted on top of Figure 4. Different buffer configurations are being studied : [1 1], [1 2] and [2 1]. The minimization of the cycle time is the objective here. The chart of Figure 4 gives the bounds on the cycle time for the original buffer configuration and for the two configurations with one more buffer space. The Approximation 7 is also given (with $i = 3$). We see that the bounds and the approximation regularly converge when the step size decreases. Moreover, the accuracy of the approximation can be assessed by comparing it to a simulation result. The [1 1] buffer configuration leads to a cycle time of 0.5786, with a 99% confidence interval given by [0.5780, 0.5792]. In this case, the approximation makes an error of 1% with seven steps and 0.47% with fifteen steps. The other buffer configurations lead to similar accuracy.

As expected, the configuration with a buffer of size two in first position turns out to be better, as the beginning of the line is more variable. The benefit in term of productivity can be estimated. Moreover, as the BDPH method offers a quite complete modeling, other performances (like the work–in–progress, i.e., the average number of items in the system, and the buffer occupancy for example) can be estimated.

5. Conclusion and Future Work. In this paper, we presented a new method, called BDPH method, to determine bounds on the performance of a production line. The method relies on a discrete phase–type approximation of the service time distributions with a "grouping at the end" approach. The study of the critical path of a part of a production run allowed the main results to be stated.

In this paper, we sticked to simple production lines and to simple performance in order to present rigorous results. It is clear that the BDPH approach calls for various extensions and future research.

The method has been used to compute bounds on the throughput and on the productivity of a production line. These bounds are valid both in the transient and in the steady–state. It should now be investigated how other performances of interest like the buffer utilization or the average job flow time can be bounded by the BDPH method.

A second direction for future research is related to the "approximation" methods. Indeed, on the basis of the stated BDPH bounds, very obvious approximation methods have been derived (Approximation 6 and 7). A more thorough analysis of the critical path could open the door to more subtle approximation approaches.

In terms of production systems, more complex organization can be studied. Indeed, Lemma 1 states the structural conditions of a line system. In case of an assembly tree, i.e., a set of production lines converging to a unique final workstation, these structural conditions can be very easily updated so that similar conclusions can be drawn. For job–shops, the door is still open. However, the notion of critical path will still constitute the heart of the proofs.

Finally, a large field of research is related to the solution methods to be implemented in order to solve the generated discrete time Markov Chain. It is clear that iterative methods [18] that take advantage of the sparsity of the transition matrix constitute promising ways in that respect. Moreover, the decomposition methods presented in §1.2 offer another way to accelerate the solution.

Appendix. Here are the formal proofs of the results presented in the paper.

Proof of Lemma 1. We first show the three inequalities, simply giving their practical significance:

(3) A station i cannot begin to work on an item k before the preceding station $i - 1$ has done his job on this item k.

(4) A station i cannot begin to work on an item k before it has done his job on item $k - 1$.

(5) To begin to work on an item k, station i has to finish its job on item $k - 1$ (inequality (4)) *and* get rid of it. There is space in the next buffer if station $i + 1$ already began to work on item $k - b_{i+1} - 1$:

$$t^r_{start}(W_{i,k}) \geq t^r_{start}(W_{i+1,k-b_{i+1}-1}).$$

Combining this and inequality (4), we get inequalities (5).

The fact that one of the inequalities is always satisfied at equality comes from the fact that a job always begins due to the end of another job. As each possible state preceding $W_{i,k}$ on the same station corresponds to one of the inequalities (satisfied at equality), the inequalities give all the possibilities:

(3) Before $W_{i,k}$, the station was starved, (3) is thus satisfied at equality: $W_{i,k}$ begins when the previous station pass the item on (see Figure 5.b).

(4) The station was already working previously, (4) is thus satisfied at equality: the station begin to work on item k directly when it ends on $k - 1$ (see Figure 5.a).

(5) The station was blocked, it had thus to pass the blocking item on, what is possible when the start of a job in a next station vacate a space in the buffers between. According to the number of blocked states one after the other, it corresponds to one of the inequalities (5) satisfied at equality (see Figure 5.c). □

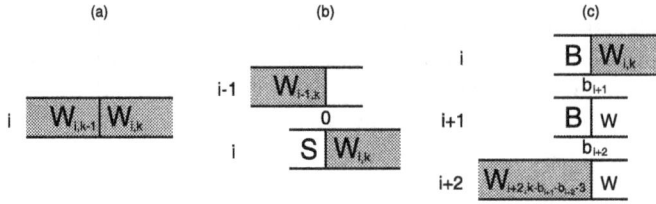

FIG. 5. *Predecessors of $W_{i,k}$ in the critical path, according to the preceding state in station i and corresponding to inequalities (4), (3) and (5).*

Lemma 8 and Proof. Here, we give the result needed in proof of Lemma 2.

LEMMA 8. *The number of jobs in the critical path is smaller than the number of productions plus the number of workstations in the line, m, minus one:*

$$|cp(r_p)| \leq p + m - 1.$$

Proof. The construction of the critical path begins on job $W_{m,p}$. To count the number of jobs in $cp(r_p)$, we relate them to the underlying items. Let us call δ the difference between item indexes of the first and the last job of $cp(r_p)$ plus one. We study $d = |cp(r_p)| - \delta$ which counts the number of items counted twice minus those omitted when the critical path is constructed. Three possibilities exist, corresponding to the inequalities of Lemma 1.

(3) The predecessor is a job on the same item, by the previous station (see Figure 5.b). The same item is thus counted twice.

(4) The predecessor is a job by the same station, on the previous item (see Figure 5.a). No item is thus omitted or counted twice.

(5) The predecessor is on a next station (see Figure 5.c). Depending on the number of stations skipped, the number of omitted items follow from the inequality (5).

The value of d is thus maximal when the critical path jumps upward a lot without jumping downward. We thus get $d \leq m - 1$ and $|cp(r_p)| \leq \delta + m - 1$. As the first and last items of the critical path are part of the production, as each item between, we have $\delta \leq p$ and the lemma is proved. $\square$

Proof of Theorem 3. To get the mean time to produce P, we consider each possible P–part r_P of the possible runs and weight its length by its probability, giving:

$$T_P = \int f(r_P) l(r_P) dr_P,$$

where $f(r_P)$ is the density function of the P–parts. Aiming to use Lemma 2, we have to relate r_P to its discrete correspondent, $\bar{r}_P$ (which also produce P). Let us note $\gamma(\bar{r}_P)$ the set of continuous P–parts (in infinite number) which have the same discrete correspondent $\bar{r}_P$. We can decompose the previous integral:

$$(8) \qquad\qquad T_P = \sum_{\bar{r}_P} \int_{r_P \in \gamma(\bar{r}_P)} f(r_P) l(r_P) dr_P.$$

By Lemma 2, we get:

$$T_P \leq \sum_{\bar{r}_P} l(\bar{r}_P) \int_{r_P \in \gamma(\bar{r}_P)} f(r_P) dr_P.$$

As the discretization by "grouping at the end" simply concentrates the probability masses in intervals, the integrals in the last equation give the probabilities of the P–parts in discrete time. We get:

$$T_P \leq \sum_{\bar{r}_P} l(\bar{r}_P) P[\bar{r}_P] = \bar{T}_P.$$

The way to the lower bound is very similar. By Lemma 2, (8) becomes:

$$T_P \geq \sum_{\bar{r}_P} \left(l(\bar{r}_P) - \tau(P + m - 1) \right) \int_{r_P \in \gamma(\bar{r}_P)} f(r_P) dr_P.$$

For the same reasons as previously and as $\sum_{\bar{r}_P} \int_{r_P \in \gamma(\bar{r}_P)} f(r_P) dr_P = 1$, we get the lower bound. $\square$

Proof of Theorem 4. The lower bound, $\overline{P}_T \leq P_T$, follows from the upper bound in Theorem 3, $T_P \leq \overline{T}_P$. As the mean time to produce a given production is longer in discretized time, the mean quantity produced in a given time T is smaller in discretized time.

Similarly, the upper bound, $P_T \leq \overline{P}^*$, comes from the lower bound in Theorem 3. The time $\overline{T}^*$ to produce $\overline{P}^*$ respects $\overline{T}^* - \tau(\overline{P}^* + m - 1) \leq T^*$, where T^* is the mean time, in continuous time, to produce $\overline{P}^*$. By definition of $\overline{P}^*$, we get $T \leq T^*$. Consequently, the production in T, P_T, is smaller than the one in T^*, $\overline{P}^*$. □

Proof of Corollary 5. Let us divide the equation of Theorem 3 by the production P, in steady–state case (when $P \to \infty$). First, the last term becomes the cycle time computed by the BDPH method. Second, the middle term becomes the exact cycle time. Finally, the first term simplifies to the cycle time in discretized time minus the step length, as the term $(m - 1)$ disappears when the time becomes infinite. □

Acknowledgments. The work of J.S. Tancrez has been supported by the European Social Fund and the Walloon Region of Belgium and is made in collaboration with Fontainunion SA. Authors are also grateful to Prof. P. Chevalier for his helpful comments on earlier versions of the work.

REFERENCES

[1] T. ALTIOK, *Approximate analysis of queues in series with phase-type service times and blocking*, Oper. Res., 37 (1989), pp. 601–610.

[2] R.G. ASKIN AND C.R. STANDRIDGE, *Modeling and analysis of manufacturing systems*, John Wiley & Sons, Inc., New York, 1993.

[3] A. BRANDWAJN AND Y.L. JOW, *An approximation method for tandem queues with blocking*, Oper. Res., 36 (1988), pp. 73–83.

[4] J.A. BUZACOTT AND J.G. SHANTHIKUMAR, *Stochastic models of manufacturing systems*, Prentice-Hall, Englewood Cliffs, New Jersey, 1993.

[5] R.B. CHASE, N.J. AQUILANO, AND F.R. JACOBS, *Operations management for competitive advantage*, McGraw-Hill/Irwin, New York, 2001.

[6] Y. DALLERY AND Y. FREIN, *On decomposition methods for tandem queueing networks with blocking*, Oper. Res., 41 (1993), pp. 386–399.

[7] Y. DALLERY AND S.B. GERSHWIN, *Manufacturing flow line systems: a review of models and analytical results*, Queueing Syst., 12.

[8] C. DINÇER AND B. DELER, *On the distribution of throughput of transfer lines*, Journal of the Operational Research Society, 52 (2000), pp. 1170–1178.

[9] S.B. GERSHWIN, *An efficient decomposition method for the approximate evaluation of tandem queues with finite storage space and blocking*, Oper. Res., 35 (1987), pp. 291–305.

[10] M.K. GOVIL AND M.C. FU, *Queueing theory in manufacturing : A survey*, Journal of Manufacturing Systems, 18 (1999), pp. 214–240.

[11] L. GUN AND A.M. MAKOWSKI, *An approximation method for general tandem queueing sytems subject to blocking*, src technical report 87-209-r1, Electrical Engineering Departement and Systems Research Center, University of Maryland, 1987.

[12] T. HILL, *Manufacturing strategy*, Irwin, Burr Ridge, Illinois, 1994.

[13] F.S. HILLIER AND R.W. BOLING, *Finite queues in series with exponential or erlang service times—a numerical approach*, Oper. Res., 15 (1967), pp. 286–303.

[14] L. KERBACHE AND J. MACGREGOR SMITH, *Multi-objective routing within large scale facilities using open finite queueing networks*, European J. Oper. Res., 121 (2000), pp. 105–123.

[15] H.T. PAPADOPOULOS, *The throughput of multistation production lines with no intermediate buffers*, Oper. Res., 43 (1995), pp. 712–715.

[16] H.T. PAPADOPOULOS AND C. HEAVEY, *Queueing theory in manufacturing systems analysis and design : A classification of models for production and transfer lines*, European J. Oper. Res., 92 (1996), pp. 1–27.

[17] H.G. PERROS, *Queueing networks with blocking : exact and approximate solutions*, Oxford University Press, 1994.

[18] W.J. STEWART, *Introduction to the numerical solution of Markov chains*, Princeton University Press, 1994.

ENTRYWISE CONDITIONING OF THE STATIONARY VECTOR FOR A GOOGLE-TYPE MATRIX[*]

S. KIRKLAND[†]

Abstract. We give conditioning bounds for the stationary vector π^T of a stochastic matrix of the form $cA + (1 - c)B$, where $c \in (0, 1)$ is a scalar, and A and B are stochastic matrices, the latter being rank one. Such matrices and their stationary vectors arise as a key component in Google's PageRank algorithm. The conditioning bounds considered include absolute componentwise, and relative componentwise, and the bounds depend on c, and on quantities such as the number of dangling nodes (which correspond to rows of A having all entries equal), or the lengths of certain cycles in the directed graph associated with A. In particular, we find that if vertex j is on only long cycles in that directed graph, then the corresponding entry in π^T exhibits better conditioning properties, and that for dangling nodes, the sensitivity of the corresponding entries in π^T decreases as the number of dangling nodes increases. Conditions are given that are sufficient to ensure that an iterate of the power method accurately reflects the relative ordering of two entries in π^T.

Key words. Stochastic Matrix, Stationary Vector, Condition Number, PageRank.

AMS subject classifications. 15A18, 15A51.

1. Introduction and preliminaries. Suppose that S is a row-stochastic matrix having 1 as an algebraically simple eigenvalue, and having stationary distribution vector σ^T (normalized so that $\sigma^T \mathbf{1} = 1$, where $\mathbf{1}$ is the all ones vector of the appropriate order). The stationary distribution vector is a one of the key parameters of the Markov chain associated with S, and there is a wealth of literature on the subject. (See [6] for more background on stochastic matrices and stationary vectors.)

One of the most compelling applications of the stationary distribution vector arises in the context of the internet search engine Google. Specifically, Google's algorithm PageRank involves a stochastic matrix of massive size; the stationary vector of that matrix is estimated numerically, and the results are used to provide a ranking of the pages on the internet. Google has reported using the power method in order to obtain its estimate for the desired stationary distribution vector [1].

The stochastic matrix used in the PageRank algorithm has a particularly special structure. First, a directed graph D is constructed whose vertices correspond to web pages, with a directed arc from vertex i to vertex j if and only if page i has a link out to page j. Next, a stochastic matrix A is constructed from the directed graph as follows. For each i, j, we have $a_{ij} = 1/d(i)$ if the outdegree of vertex $i, d(i)$, is positive and $i \to j$ in the directed graph D, and $a_{ij} = 0$ if $d(i) > 0$ but there is no arc from i to j in D. Finally, if vertex i has outdegree zero, we have $a_{ij} = 1/n$ for all j, where n is the order of the matrix. (There are several possible ways of dealing with rows of A corresponding to vertices of outdegree zero, but here we focus only on the convention noted above.) We note that because of the highly disconnected nature of the web graph, the matrix A has a block triangular structure, with several diagonal blocks that are stochastic matrices of smaller order. Next, a positive row vector v^T is selected, normalized so that $v^T \mathbf{1} = 1$. Finally a parameter $c \in (0, 1)$

---

[*]The results presented here have been extracted from a forthcoming paper. Readers seeking further details on these results are invited to consult [5], which is available from the author upon request.

[†]Department of Mathematics and Statistics, University of Regina, Regina, Saskatchewan, S4S 0A2, CANADA. (kirkland@math.uregina.ca) Research supported in part by an NSERC Research Grant.

is chosen (Google reports that c is approximately 0.85), and the *Google matrix* G is constructed as follows:

$$(1.1) \qquad\qquad G = cA + (1 - c)\mathbf{1}v^T.$$

It is the stationary distribution vector of G that is estimated, and the results are then used in Google's ranking of the pages on the web.

Given that Google produces an estimate for the stationary vector of G, it is natural to wonder how close that estimate is to the actual stationary distribution vector of G. Our goal is to investigate that issue. In order to do so, we focus our attention on stochastic matrices of the following type:

$$(1.2) \qquad\qquad M = cA + (1 - c)\mathbf{1}v^T,$$

where A is an $n \times n$ stochastic matrix, $c \in (0, 1)$ and v^T is a nonnegative row vector such that $v^T\mathbf{1} = 1$. We remark that any matrix M of the form (1.2) has a column with all positive entries (in fact, the converse also holds), and so has a single essential class of indices (see [6]); hence M has 1 as a simple eigenvalue. Throughout, we impose the additional hypothesis that for index $1 \leq i \leq n$, the principal submatrix of $I - M$ formed by deleting row and column i is invertible. We note that there is no real loss of generality in this assumption, for if that principal submatrix is singular, it is straightforward to determine that necessarily, the i-th entry of the stationary vector for M is 0, and so the problem can be reduced to one of smaller order. Observe that in the special case that v^T is a positive vector and A is block triangular with at least two diagonal blocks that are stochastic, a matrix of the form (1.2) coincides with the Google matrix G of (1.1).

To fix the notation, henceforth we take A, n, c, v^T, M and G to be as described above, and we denote the stationary distribution vector for M by π^T. For any column vector z, we define $||z^T||_1 = \sum_i |z_i|$, while for any $n \times n$ matrix P, we take $||P||_\infty = max\{||e_i^T P||_1 | i = 1, \ldots, n\}$. Henceforth it will be convenient to take A, π^T and v^T to be partitioned as follows:

$$(1.3) \qquad A = \left[\begin{array}{c|c} A_n & 1 - A_n\mathbf{1} \\ \hline a^T & 1 - a^T\mathbf{1} \end{array} \right], \pi^T = \left[\overline{\pi}^T | \pi_n \right], v^T = \left[\overline{v}^T | v_n \right].$$

Much of the discussion proceeds with reference to the directed graph associated with A, which we denote by $\Delta(A)$ (see [2] for basics on the relationship between a matrix and its directed graph). Consequently, the results give some insight into the qualitative features of $\Delta(A)$ that influence the estimation of π^T.

2. Componentwise absolute bounds. Suppose that p^T is a vector whose entries sum to 1, and let $r^T = p^T(I - M)$. Given that the entries in the stationary vector π^T for the Google matrix G are used to rank web pages, it is natural to ask for more detailed information on this point. In this section, we consider, for each index $j = 1, \ldots, n$, questions of the following type:
1. Given a vector p^T whose entries sum to 1, how close is p_j to π_j?
2. If p^T is an estimate of π^T and we know that $p_i \geq p_j$, under what circumstances can we conclude that $\pi_i \geq \pi_j$?
To fix the notation, throughout the remainder of the paper, we take p^T, r^T and $\overline{v}^T$ to be as described above.

For each $j = 1, \ldots, n$, let $\kappa_j(M) = \frac{1}{2}\pi_j||(I - M_j)^{-1}||_\infty$. The following result indicates the role played by κ_j in discussing the questions posed at the beginning of this section.

THEOREM 2.1. *a) Suppose that p^T is an n-vector whose entries sum to 1. Then for each $j = 1, \ldots, n$, we have $|p_j - \pi_j| \leq ||r^T||_1 \kappa_j(M)$.*
b) Fix an index j between 1 and n. For each sufficiently small $\epsilon > 0$, there is a positive vector p^T whose entries sum to 1 such that $||r^T||_1 = \epsilon$ and $|p_j - \pi_j| = ||r^T||_1 \kappa_j(M)$.

We see from Theorem 2.1 that the quantity $\kappa_j(M)$ provides a precise measure of the difference between p_j and π_j. Further, if $\tilde{M} = M + E$ is another stochastic matrix having 1 as an algebraically simple eigenvalue, and stationary vector $\tilde{\pi}^T$, then $|\tilde{\pi}_j - \pi_j| \leq \kappa_j(M)||E||_\infty$. Consequently, $\kappa_j(M)$ also measures the conditioning of π_j under perturbation of M (and so κ_j is known as a *condition number*). In what follows, we present bounds on $\kappa_j(M)$ for any matrix M satisfying (1.2). Those bounds rely on the closed form expression in the following result.

THEOREM 2.2. *Suppose that the matrix A is partitioned as in (1.3). Then*
$$\kappa_n(M) = \max\{\frac{e_i^T(I-cA_n)^{-1}\mathbf{1}}{2(1+ca^T(I-cA_n)^{-1}\mathbf{1})}|i = 1, \ldots, n-1\}.$$
Next is one of the main results.

THEOREM 2.3. *a) Suppose that vertex j is on a cycle of length at least 2 in $\Delta(A)$, and let g be the length of a shortest such cycle. Then $\kappa_j(M) \leq \frac{1}{2(1-c^g-ca_{jj}(1-c^{g-1}))}$. Equality holds if and only if there is some i such that there is no path from vertex i to vertex j in $\Delta(A)$, and there is a principal submatrix of A of the following form*

(2.1)
$$S = \begin{bmatrix} 0 & S_{g-1} & \cdots & 0 & 0 \\ 0 & 0 & S_{g-2} & \cdots & 0 \\ & & & \ddots & \vdots \\ 0 & 0 & \cdots & 0 & 1 \\ b^T & 0 & \cdots & 0 & 1-b^T\mathbf{1} \end{bmatrix},$$

where the last row and column of (2.1) corresponds to index j.
b) If vertex j is on no cycle of length at least 2 in $\Delta(A)$ and $a_{jj} \neq 1$, then $\kappa_j(M) = \frac{1}{2(1-ca_{jj})}$.
c) If $a_{jj} = 1$, then $\kappa_j(M) \leq \frac{1}{2(1-c)}$, with equality if and only if there is a vertex i such that there is no path from vertex i to vertex j in $\Delta(A)$.

The following is immediate.

COROLLARY 2.4. *a) If j is on a cycle of length at least 2 and g is the length of the shortest such cycle, then $|p_j - \pi_j| \leq \frac{||r^T||_1}{2(1-c^g-ca_{jj}(1-c^{g-1}))}$.*
b) Suppose that vertex j is on no cycle of length 2 or more in $\Delta(A)$. Then $|p_j - \pi_j| \leq \frac{||r^T||_1}{2(1-ca_{jj})}$.

REMARK 2.1. Observe that the upper bound of Theorem 2.3 a) on κ_j is readily seen to be decreasing in g. We can interpret this bound as implying that if vertex j of $\Delta(A)$ is only on long cycles, then π_j will exhibit good conditioning properties.

As noted in the introduction, the construction of the Google matrix G of (1.1) typically involves a matrix A having a number of rows equal to $\frac{1}{n}\mathbf{1}^T$. Such rows correspond to so-called "dangling nodes" in the directed graph associated with the internet, and arise from pages from which there are no links out or from pages whose links out have not yet been crawled. The results of a web crawl in 2001 are reported in [4], in which, out of a total of 290 million pages crawled, about 220 million pages corresponded to dangling nodes. Our next result leads to a bound on κ_j where j is associated with a row of A that is equal to $\frac{1}{n}\mathbf{1}^T$.

COROLLARY 2.5. *Suppose that A has $m \geq 2$ rows equal to $\frac{1}{n}\mathbf{1}^T$, and that row j is one of those rows. Then $\kappa_j(M) \leq \frac{n-c(m-1)}{2((1-c^2)n-c(1-c)m)}$.*

REMARK 2.2. Suppose that A has m rows that are equal to $\frac{1}{n}\mathbf{1}^T$, and let $\mu = m/n$. For large values of n, we see that if $\mu > 0$, then the upper bound of Corollary 2.5 is roughly $\frac{1-c\mu}{2(1-c)(1+c-c\mu)}$, which is readily seen to be decreasing in μ. Thus, as the proportion of dangling nodes increases, there is likely to be a decrease in the sensitivity of the entries in π^T corresponding to those dangling nodes. Note that for $c = .85, \mu = \frac{22}{29}$ and large n, the bound of Corollary 2.5 is approximately .9824.

We have the following additional result for the Google matrix G.

THEOREM 2.6. *Consider the Google matrix G of (1.1), and fix an index $j = 1,\ldots,n$. For all sufficiently small $\epsilon > 0$, there is a positive vector p^T whose entries sum to 1 such that $\|r^T\|_1 = \epsilon$ and $|p_j - \pi_j| \geq \frac{\|r^T\|_1}{2(1-ca_{jj})}$.*

The following result helps to address the second question posed at the beginning of this section. It follows immediately from Corollary 2.4.

COROLLARY 2.7. *a) Suppose that vertices i and j of $\Delta(A)$ are on cycles of length two or more, and let g_i and g_j denote the lengths of the shortest such cycles, respectively. If $p_i \geq p_j + \|r^T\|_1 \left(\frac{1}{2(1-c^{g_i}-ca_{ii}(1-c^{g_i-1}))} + \frac{1}{2(1-c^{g_j}-ca_{jj}(1-c^{g_j-1}))} \right)$, then $\pi_i \geq \pi_j$.*

b) Suppose that vertex i of $\Delta(A)$ is on a cycle of length two or more, and let g_i denote the length of the shortest such cycle. Suppose that vertex j is on no cycle of length two or more. If $p_i \geq p_j + \|r^T\|_1 \left(\frac{1}{2(1-c^{g_i}-ca_{ii}(1-c^{g_i-1}))} + \frac{1}{2(1-ca_{jj})} \right)$, then $\pi_i \geq \pi_j$.

c) Suppose that neither of vertices i and j of $\Delta(A)$ are on a cycle of length two or more. If $p_i \geq p_j + \|r^T\|_1 \left(\frac{1}{2(1-ca_{ii})} + \frac{1}{2(1-ca_{jj})} \right)$, then $\pi_i \geq \pi_j$.

We have the following analogue of Corollary 2.7 for rows of M that correspond to dangling nodes.

COROLLARY 2.8. *Suppose that A has $m \geq 2$ rows equal to $\frac{1}{n}\mathbf{1}^T$, one of which is row j.*

a) Suppose that vertex i of $\Delta(A)$ is on a cycle of length two or more, and let g_i be the length of a shortest such cycle. If

$$p_i \geq p_j + \|r^T\|_1 \left(\frac{1}{2(1 - c^{g_i} - ca_{ii}(1 - c^{g_i-1}))} + \frac{n - c(m - 1)}{2((1 - c^2)n - c(1 - c)m)} \right),$$

then $\pi_i \geq \pi_j$.

b) Suppose that vertex i is on no cycle of length two or more. If

$$p_i \geq p_j + \|r^T\|_1 \left(\frac{1}{2(1 - ca_{ii})} + \frac{n - c(m - 1)}{2((1 - c^2)n - c(1 - c)m)} \right),$$

then $\pi_i \geq \pi_j$.

c) Suppose that row i of A is equal to $\frac{1}{n}\mathbf{1}^T$. If $p_i \geq p_j + \|r^T\|_1 \left(\frac{n-c(m-1)}{((1-c^2)n-c(1-c)m)} \right)$, then $\pi_i \geq \pi_j$.

Our next result considers the accuracy of the iterates arising from the power method in estimating a particular entry of π^T.

COROLLARY 2.9. *Suppose that $x(0)^T \geq 0^T$, with $x(0)^T\mathbf{1} = 1$, and that for each $k \in \mathbb{N}$, $x(k)^T$ is the k-th vector in the sequence of iterates generated by applying the power method to $x(0)^T$ with the matrix M. If vertex j is on no cycle of length at least 2 in $\Delta(A)$, then for each $k \in \mathbb{N}$, $|x(k)^T e_j - \pi_j| \leq \frac{c^k\|\{x(1)^T-x(0)^T\}A^k\|_1}{2(1-ca_{jj})} \leq \frac{c^k\|x(1)^T-x(0)^T\|_1}{2(1-ca_{jj})}$. On the other hand, if vertex j is on a cycle of length at least 2 and*

g is the length of the shortest such cycle, then for each $k \in \mathbb{N}$, $|x(k)^T e_j - \pi_j| \leq \frac{c^k \|\{x(1)^T - x(0)^T\}A^k\|_1}{2(1 - c^g - ca_{jj}(1 - c^{g-1}))} \leq \frac{c^k \|x(1)^T - x(0)^T\|_1}{2(1 - c^g - ca_{jj}(1 - c^{g-1}))}$.

Our final two results of this section are applications of Corollaries 2.7 and 2.8 to the sequence of iterates arising from the power method.

COROLLARY 2.10. *Suppose that $x(0)^T \geq 0^T$, with $x(0)^T \mathbf{1} = 1$, and that for each $k \in \mathbb{N}$, $\mathrm{x(k)}^T$ is the $k-$th vector in the sequence of iterates generated by applying the power method to $x(0)^T$ with the matrix M.*

a) Suppose that vertices i and j of $\Delta(A)$ are on cycles of length two or more, and let g_i and g_j denote the lengths of the shortest such cycles, respectively. If $x(k)_i \geq x(k)_j + c^k \|x(1)^T - x(0)^T\|_1 \left(\frac{1}{2(1 - c^{g_i} - ca_{ii}(1 - c^{g_i - 1}))} + \frac{1}{2(1 - c^{g_j} - ca_{jj}(1 - c^{g_j - 1}))} \right)$, then $\pi_i \geq \pi_j$.

b) Suppose that vertex i of $\Delta(A)$ is on a cycle of length two or more, and let g_i denote the length of the shortest such cycle. Suppose that vertex j is on no cycle of length two or more. If $x(k)_i \geq x(k)_j + c^k \|x(1)^T - x(0)^T\|_1 \left(\frac{1}{2(1 - c^{g_i} - ca_{ii}(1 - c^{g_i - 1}))} + \frac{1}{2(1 - ca_{jj})} \right)$, then $\pi_i \geq \pi_j$.

c) Suppose that neither of vertices i and j of $\Delta(A)$ are on a cycle of length two or more. If $x(k)_i \geq x(k)_j + c^k \|x(1)^T - x(0)^T\|_1 \left(\frac{1}{2(1 - ca_{ii})} + \frac{1}{2(1 - ca_{jj})} \right)$, then $\pi_i \geq \pi_j$.

COROLLARY 2.11. *Suppose that A has $m \geq 2$ rows equal to $\frac{1}{n}\mathbf{1}^T$, one of which is row j. Suppose that $x(0)^T \geq 0^T$, with $x(0)^T \mathbf{1} = 1$, and that for each $k \in \mathbb{N}$, $x(k)^T$ is the $k-$th vector in the sequence of iterates generated by applying the power method to $x(0)^T$ with the matrix M.*

a) Suppose that vertex i of $\Delta(A)$ is on a cycle of length two or more, and let g_i be the length of a shortest such cycle. If $x(k)_i \geq x(k)_j + c^k \|x(1)^T - x(0)^T\|_1 \left(\frac{1}{2(1 - c^{g_i} - ca_{ii}(1 - c^{g_i - 1}))} + \frac{n - c(m-1)}{2((1 - c^2)n - c(1 - c)m)} \right)$, then $\pi_i \geq \pi_j$.

b) Suppose that vertex i is on no cycle of length two or more. If

$$x(k)_i \geq x(k)_j + c^k \|x(1)^T - x(0)^T\|_1 \left(\frac{1}{2(1 - ca_{ii})} + \frac{n - c(m-1)}{2((1 - c^2)n - c(1 - c)m)} \right),$$

then $\pi_i \geq \pi_j$.

c) Suppose that row i of A is equal to $\frac{1}{n}\mathbf{1}^T$. If

$$x(k)_i \geq x(k)_j + c^k \|x(1)^T - x(0)^T\|_1 \left(\frac{n - c(m-1)}{((1 - c^2)n - c(1 - c)m)} \right),$$

then $\pi_i \geq \pi_j$.

3. Componentwise relative error bounds. In the preceding section, we considered the absolute error $|p_j - \pi_j|$; this section focuses on the corresponding relative errors, $\frac{|p_j - \pi_j|}{\pi_j}$. In [3], a method designed to speed up the convergence of the sequence of iterates $x(k)^T$, $k \in \mathbb{N}$ arising from the power method (applied with the matrix M) is presented. A key aspect of that method involves the monitoring of a quantity that can be thought of as an estimate of the relative error associated with the j-th entry of $x(k)^T$, and this provides some of the motivation for a discussion of relative errors. In order to simplify notation in this section, we consider, without loss of generality, the problem of bounding the relative error for π_n.

Let $\hat{S}$ be the set of vertices in $\Delta(A)$ for which there is no path to vertex n. For each vertex $j \notin \hat{S}$, let $d(j, n)$ be the distance from vertex j to vertex n, and let $d = max\{d(j, n) | j \notin \hat{S}\}$. For each $i = 0, \ldots, d$, let $S_i = \{j \notin \hat{S} | d(j, n) = i\}$ (evidently

$S_0 = \{n\}$ here). Suppose also that $\bar{v}^T$ is partitioned accordingly into sub-vectors $\overline{v_i}^T, i = 0, \ldots, d$, and $\hat{v}^T$. Finally, for each $i = 1, \ldots, d$, let α_i be the minimum row sum of $A[S_i, S_{i-1}]$, the submatrix of A on rows S_i and columns S_{i-1}.

We use this notation in our next result.

COROLLARY 3.1. *We have*

$$\kappa_n(M) \leq \frac{\pi_n}{2(1-c)(v_n + \sum_{i=1}^d c^i \alpha_1 \ldots \alpha_i \overline{v_i}^T \mathbf{1})},$$

so that in particular,

$$\frac{|p_n - \pi_n|}{\pi_n} \leq \frac{||r^T||_1}{2(1-c)(v_n + \sum_{i=1}^d c^i \alpha_1 \ldots \alpha_i \overline{v_i}^T \mathbf{1})}.$$

If $\hat{S} \neq \emptyset$, then

$$\frac{\pi_n}{2(1-c)(v_n + \sum_{i=1}^d c^i \overline{v_i}^T \mathbf{1})} \leq \kappa_n(M).$$

In particular, for each $\epsilon > 0$, there is a positive vector p^T whose entries sum to 1 such that $||r^T||_1 = \epsilon$ and

$$\frac{|p_n - \pi_n|}{\pi_n} \geq \frac{||r^T||_1}{2(1-c)(v_n + \sum_{i=1}^d c^i \overline{v_i}^T \mathbf{1})}.$$

REMARK 3.1. From Corollary 3.1, we see that the vector v^T is influential on the relative conditioning of π_n. Specifically, if v^T places more weight on vertices in S_i for small values of i (i.e. on vertices at short distance from vertex n), then that has the effect of improving the relative conditioning properties of π_n.

As in Section 2, we treat the rows of M that correspond to dangling nodes as a special case. Again, we adhere to the notation above.

THEOREM 3.2. *Suppose that A has m rows equal to $\frac{1}{n}\mathbf{1}^T$, one of which is row n. Write $\bar{v}^T$ as $\bar{v}^T = [u_1^T | u_2^T]$, where the partitioning conforms with that of A_n. Then*

$$\kappa_n(M) \leq \pi_n \frac{n - c(m-1)}{2(1-c)(v_n(n - c(m-1)) + cu_2^T \mathbf{1})},$$

with equality holding if and only if $\hat{A}_1$ is stochastic. In particular,

$$\frac{|p_n - \pi_n|}{\pi_n} \leq \frac{(n - c(m-1))||r^T||_1}{2(1-c)(v_n(n - c(m-1)) + cu_2^T \mathbf{1})}.$$

REMARK 3.2. In the notation of Theorem 3.2, we note that in the case that $v^T = \frac{1}{n}\mathbf{1}^T$ and $\frac{m}{n} = \mu$, we find that the upper bound of Theorem 3.2 is roughly $\frac{n\pi_n(1-c\mu)}{2(1-c)}$. For instance if $c = .85$ and $\mu = \frac{22}{29}$, that quantity is approximately $1.184n\pi_n$.

REFERENCES

[1] S. BRIN, L. PAGE, R. MOTWAMI AND T. WINOGRAD, *The PageRank citation index: bringing order to the web*, Technical Report 1999-66, Computer Science Department, Stanford University, 1999.

[2] R. BRUALDI AND H. RYSER, *Combinatorial Matrix Theory*, Cambridge University Press, Cambridge, UK, 1991.

[3] S. KAMVAR, T. HAVELIWALA AND G. GOLUB, *Adaptive methods for the computation of PageRank*, Linear Algebra and its Applications, 386 (2004), pp. 51–65.

[4] S. KAMVAR, T. HAVELIWALA, C. MANNING AND G. GOLUB, *Exploiting the block structure of the web for computing PageRank*, Preprint, 13 pages, 2003.

[5] S. KIRKLAND, *Conditioning of the Entries in the Stationary Vector of a Google-Type Matrix*, Linear Algebra and its Applications, to appear.

[6] E. SENETA, *Non-negative Matrices and Markov Chains*, Second ed., Springer-Verlag, New York, NY, 1981.

ANALYZING MARKOV CHAINS
BASED ON KRONECKER PRODUCTS[*]

TUĞRUL DAYAR[†]

Abstract. Kronecker products are used to define the underlying Markov chain (MC) in various modeling formalisms, including compositional Markovian models, hierarchical Markovian models, and stochastic process algebras. The motivation behind using a Kronecker structured representation rather than a flat one is to alleviate the storage requirements associated with the MC. With this approach, systems that are an order of magnitude larger can be analyzed on the same platform. The developments in the solution of such MCs are reviewed from an algebraic point of view and possible areas for further research are indicated with an emphasis on preprocessing using reordering, grouping, and lumping and numerical analysis using block iterative, multilevel, and preconditioned projection methods.

Key words. Markov chains, Kronecker products, reordering, grouping, lumping, block iterative methods, multilevel methods, preconditioned projection methods

AMS subject classifications. 60J27, 65F50, 15A72, 65F10, 65B99

1. Introduction. We consider discrete state space, continuous-time Markovian processes, namely *continuous-time Markov chains* (CTMCs). A CTMC having n states may be represented by an $(n \times n)$ square matrix $Q \in \mathbb{R}^{n \times n}$ having nonnegative off-diagonal elements indicating the rates of the exponentially distributed transition times between pairs of states and a diagonal formed by the negated row sums of its off-diagonal elements. Therefore, Q has a nonpositive diagonal and row sums of zero. This matrix is also known as the *infinitesimal generator* of the associated Markovian process [62].

Let $\pi_0 \in \mathbb{R}^{1 \times n}$ denote the *initial* probability distribution row vector of Q, where $\pi_0 \geq 0$, $\pi_0 e = 1$, and e is a column vector of ones. Then the *transient* probability distribution row vector at time $t \geq 0$ is given by

$$(1.1) \qquad \pi_t = \pi_0 \exp(Qt) = \pi_0 \sum_{i=0}^{\infty} \frac{(Qt)^i}{i!}.$$

Whenever the *steady-state* (or *limiting, long-run*) probability distribution row vector $\pi = \lim_{t \to \infty} \pi_t$ exists, it satisfies

$$(1.2) \qquad \pi Q = 0, \quad \pi e = 1.$$

Hence, the steady-state distribution is also a *stationary* distribution [62].

In the Kronecker based approach, Q is represented using Kronecker products [30, 65] of smaller matrices and is never explicitly generated. The implementation of transient solvers for (1.1) and steady-state solvers for (1.2) can rest on this compact Kronecker representation, thanks to the existence of an efficient vector-Kronecker product multiplication algorithm known as the *shuffle* algorithm [30].

In practice, the representation of Q based on Kronecker products is obtained using various modeling formalisms. These include compositional Markovian models such as stochastic automata networks (SANs) [55, 56, 57, 62] and different classes of

---

[*]This work is partly supported by the Alexander von Humboldt Foundation.
[†]Department of Computer Engineering, Bilkent University, TR-06800 Bilkent, Ankara, Turkey; Tel: (+90) (312) 290-1981; Fax: (+90) (312) 266-4047 (`tugrul@cs.bilkent.edu.tr`).

superposed stochastic Petri nets [34, 47], hierarchical Markovian models (HMMs) of queueing networks [6], generalized stochastic Petri nets (GSPNs) [22], or systems of asynchronously communicating stochastic modules [24], and stochastic process algebras such as the performance evaluation process algebra (PEPA) [44]. These modeling formalisms are integrated to various software packages such as the Abstract Petri Net Notation (APNN) toolbox [1, 2], the Performance Evaluation of Parallel Systems (PEPS) software tool [4, 54], the PEPA Workbench [29, 53], and the Stochastic Model checking Analyzer for Reliability and Timing (SMART) [28, 61].

An advantage of HMMs is their ability to represent Q using Kronecker products without introducing unreachable states. Matrix diagrams [27] and representations for specific models as in [43] can also be used to achieve the same effect when state spaces are expressed compositionally. There are other approaches that can be used to deal with unreachable states as discussed in [5, 11, 17]. Throughout our discussion, we make the assumption that the MC at hand does not have unreachable states and is irreducible. Yet, in many practical applications, it is very large and has many nonzeros necessitating its storage in memory using Kronecker products. In order to analyze Markovian models based on Kronecker products efficiently, various algorithms for vector-Kronecker product multiplication based on the shuffle algorithm are devised [5, 17, 35, 36, 57, 58] and used as kernels in iterative solution techniques proposed for different modeling formalisms. The transient distribution in (1.1) can be computed through uniformization using vector-Kronecker product multiplications as in [6]. The steady-state distribution in (1.2) also needs to be computed using vector-Kronecker product multiplications, since direct methods based on complete factorizations, such as Gaussian elimination, normally introduce new nonzeros which cannot be accommodated. The two papers [12, 63] provide good overviews of iterative solution techniques for the analysis of MCs based on Kronecker products. Issues related to reachability analysis, vector-Kronecker product multiplication, hierarchical state space generation in Kronecker based matrix representations for large Markov models are surveyed in [23]. A comparison of the merits of the SAN and GSPN modeling formalisms using the PEPS and SMART software packages can be found in [26].

Although Kronecker representations for CTMCs underlying many models of practical applications have been considered, so far only a handful of discrete-time MCs (DTMCs) based on Kronecker products appeared in the literature. For instance, the one in [59] is a model of synchronization via message passing in a distributed system, the one in [56] is a model of the mutual exclusion algorithm of Lamport, those in [37, 38] are models of buffer admission mechanisms for asynchronous transfer mode (ATM) networks from telecommunications, and the one in [66] is a multiservices resource allocation policy for wireless ATM networks. The model in [42] is a larger, scalable, and extended version of that in [66]. It serves as a good example showing that the underlying MC of a discrete-time model based on Kronecker products can be relatively dense and numerically difficult to analyze. These case studies are based on the SAN modeling formalism, whereas [60] extends the Kronecker representation to stochastic Petri nets with discrete phase-type distributions. Clearly, the area of DTMCs based on Kronecker products can use other case studies and formalisms.

Here, we take an algebraic view and discuss recent results related to the analysis of MCs based on Kronecker products independently from modeling formalisms. In section 2, we provide background material on the Kronecker representation of a CTMC, show that it has a rich structure which is nested and recursive, and introduce a small CTMC expressed as a sum of Kronecker products; this CTMC is used as a running example throughout the discussion. In section 3, we consider preprocessing of the Kronecker representation so as to expedite numerical analysis. We discuss permuting the nonzero structure of the underlying CTMC symmetrically by reordering, changing the orders of the nested blocks by grouping, and reducing the size of the state space by lumping. Sections 4, 5, and 6 are devoted to the steady-state analysis of CTMCs based on Kronecker products with block iterative methods, multilevel methods, and preconditioned projection methods, respectively. In section 7, we conclude. The results can be extended to DTMCs based on Kronecker products with minor modifications. Areas that need further research are mentioned within the sections as they are discussed. In passing, we remark that parallel implementations exploiting the Kronecker representation are beyond the scope of this paper and form an open area for research.

2. Background. Recall that the *Kronecker* (or *tensor*) *product* [30, 65] of two (rectangular) matrices $A \in \mathbb{R}^{n_A \times m_A}$ and $B \in \mathbb{R}^{n_B \times m_B}$ is written as $A \otimes B$ and yields the (rectangular) matrix $C \in \mathbb{R}^{n_A n_B \times m_A m_B}$ whose elements satisfy

$$c(i_C, j_C) = a(i_A, j_A)b(i_B, j_B) \quad \text{with} \quad i_C = (i_A - 1)n_B + i_B \quad \text{and} \quad j_C = (j_A - 1)m_B + j_B$$

for

$$(i_A, j_A) \in \{1, 2, \ldots, n_A\} \times \{1, 2, \ldots, m_A\},$$
$$(i_B, j_B) \in \{1, 2, \ldots, n_B\} \times \{1, 2, \ldots, m_B\},$$

where $\times$ is the Cartesian product operator. Note that in a 2-dimensional representation, the row indices of C are in $\{1, 2, \ldots, n_A\} \times \{1, 2, \ldots, n_B\}$, whereas its column indices are in $\{1, 2, \ldots, m_A\} \times \{1, 2, \ldots, m_B\}$. Hence, the ordering of rows and columns of C with respect to this 2-dimensional representation is lexicographical, since

$$c(i_C, j_C) = c((i_A, i_B), (j_A, j_B)) = c((i_A - 1)n_B + i_B, (j_A - 1)m_B + j_B).$$

The Kronecker product is associative and defined for more than two matrices. To explain this further for a MC setting, let us consider the Kronecker product of H square matrices as in

$$X = X^{(1)} \otimes X^{(2)} \otimes \cdots \otimes X^{(H)} = \otimes_{h=1}^{H} X^{(h)},$$

where $X^{(h)} \in \mathbb{R}^{n_h \times n_h}$ and row/column indices of $X^{(h)}$ are in $\mathcal{S}^{(h)} = \{1, 2, \ldots, n_h\}$ for $h = 1, 2, \ldots, H$. Therefore, $X \in \mathbb{R}^{n \times n}$ with $n = \prod_{h=1}^{H} n_h$, and the ordered H-dimensional tuples $(i_1, i_2, \ldots, i_H) \in \times_{h=1}^{H} \mathcal{S}^{(h)}$ and $(j_1, j_2, \ldots, j_H) \in \times_{h=1}^{H} \mathcal{S}^{(h)}$ may be used to represent the row and column indices of X, respectively. Hence, the Kronecker product of H square matrices implies a one-to-one onto mapping between an H-dimensional state space and a one-dimensional state space that are lexicographically ordered, and naturally the Kronecker product has been used to define MCs having multi-dimensional state spaces.

2.1. Kronecker representation of Q. Without going into detail, we assume that the H-dimensional CTMC at hand is represented as a sum of Kronecker products plus a diagonal matrix. Specifically, we have

$$(2.1) \qquad Q = Q_O + Q_D, \qquad Q_O = \sum_{k=1}^{K} \bigotimes_{h=1}^{H} Q_k^{(h)}, \qquad Q_D = \text{diag}(-Q_O e),$$

where Q_O and Q_D correspond respectively to the off-diagonal and diagonal parts of Q, K is the number of Kronecker products (or terms) forming Q_O, H is the number of factors in each Kronecker product, $Q_k^{(h)} \in \mathbb{R}^{n_h \times n_h}$ and satisfies $Q_k^{(h)} \geq 0$ for $k = 1, 2, \ldots, K$ and $h = 1, 2, \ldots, H$, and diag is used to denote a diagonal matrix which has its vector argument along its diagonal. Observe that $Q_O \geq 0$ and $Q_D \leq 0$. If row/column indices of $Q_k^{(h)}$ are in $\mathcal{S}^{(h)} = \{1, \ldots, n_h\}$ for $k = 1, 2, \ldots, K$ and $h = 1, 2, \ldots, H$, then the H-dimensional state space of Q is given by $\mathcal{S} = \times_{h=1}^{H} \mathcal{S}^{(h)}$. Observe that $|\mathcal{S}| = \prod_{h=1}^{H} |\mathcal{S}^{(h)}| = \prod_{h=1}^{H} n_h = n$. Furthermore, the one-dimensional value of state $s \in \mathcal{S}$ corresponding to $(s_1, s_2, \ldots, s_H)$, where $s_h \in \mathcal{S}^{(h)}$ for $h = 1, 2, \ldots, H$, is given by

$$s = 1 + \sum_{h=1}^{H} (s_h - 1) \prod_{i=h+1}^{H} n_i.$$

Throughout the text, we will be using the one-dimensional and multi-dimensional representations of states interchangeably.

One needs space for the diagonal matrix Q_D and the matrices in the Kronecker representation of Q_O, meaning a floating-point vector of length $\prod_{h=1}^{H} n_h$ and at most K (sparse) floating-point matrices of order n_h are stored for $h = 1, 2, \ldots, H$. In the worst case, this amounts to a storage space of $n + \sum_{h=1}^{H} nz_{Q^{(h)}}$ floating-point values, where $nz_{Q^{(h)}}$ is the sum of the number of nonzeros in $Q_k^{(h)}$ for $k = 1, 2, \ldots, K$, compared to nz nonzeros required by the flat representation. We remark that Q_D can also be expressed as a sum of Kronecker products:

$$Q_D = -\sum_{k=1}^{K} \bigotimes_{h=1}^{H} \text{diag}(Q_k^{(h)} e).$$

However, in order to enable the efficient implementation of numerical solvers, most of the time Q_D is precomputed and stored explicitly.

The complexity of a vector multiplication with Q_O, which consists of K Kronecker product terms, amounts to

$$K \prod_{h=1}^{H} n_h + 2 \sum_{k=1}^{K} \prod_{h=1}^{H} n_h \sum_{l=1}^{H} nz_{Q_k^{(l)}} / n_l = K \prod_{h=1}^{H} n_h + 2 \prod_{h=1}^{H} n_h \sum_{l=1}^{H} \left(\sum_{k=1}^{K} nz_{Q_k^{(l)}} \right) / n_l$$

$$= n \left(K + 2 \sum_{h=1}^{H} nz_{Q^{(h)}} / n_h \right)$$

floating-point arithmetic operations [35], where $nz_{Q_k^{(l)}}$ is the number of nonzeros in $Q_k^{(l)}$ for $k = 1, 2, \ldots, K$ and $l = 1, 2, \ldots, H$.

Now, observe that each nonzero element of the matrix $Q_k^{(h)}$ in (2.1) is located by its row and column indices, which are members of $\mathcal{S}^{(h)}$. In a more general setting, a nonzero element of $Q_k^{(h)}$ may be a function of states in state spaces other than $\mathcal{S}^{(h)}$, thus a function of non-local states. This phenomenon is a by-product of the modeling process and has been utilized in the SAN modeling formalism. These nonzero elements are referred to as *functional transitions* and the corresponding Kronecker products are said to be *generalized* [55]. Although it is possible to remove functional transitions from a sum of generalized Kronecker products by introducing new terms [57] and/or factors [5], functional transitions enable a more compact Kronecker representation with fuller factor matrices [26]. We do not consider functional transitions in this discussion, but indicate that the results extend to generalized Kronecker products whereever appropriate.

There is a rich structure associated with the Kronecker representation in (2.1). This structure is nested and recursive [18, 19, 20, 32, 39, 41, 42, 64]. Let level 0 denote the highest level at which Q is perceived as a single block of order $n = \prod_{h=1}^{H} n_h$. At the next level, which we call level 1, Q is an $(n_1 \times n_1)$ block matrix with blocks of order $\prod_{h=2}^{H} n_h$. At level 2, Q is an $(n_1 n_2 \times n_1 n_2)$ block matrix with blocks of order $\prod_{h=3}^{H} n_h$. Continuing in this manner, at level H, Q is a $(\prod_{h=1}^{H} n_h \times \prod_{h=1}^{H} n_h)$, in other words, $(n \times n)$ block matrix with blocks of order 1. More formally, we have

$$(2.2) \quad b_l = \begin{cases} 1, & l = 0 \\ n_l^2 b_{l-1}, & l = 1, 2, \ldots, H \end{cases} \quad \text{and} \quad o_l = \begin{cases} n, & l = 0 \\ o_{l-1}/n_l, & l = 1, 2, \ldots, H \end{cases},$$

where b_l and o_l denote the number and order of blocks at level $l = 0, 1, \ldots, H$, respectively. Unrolling the recurrences, we obtain

$$b_l = \prod_{h=1}^{l} n_h^2 \quad \text{and} \quad o_l = \prod_{h=l+1}^{H} n_h.$$

Note that at level l there are $\sqrt{b_l}$ blocks each of order o_l along the diagonal of Q. Furthermore, block $((i_1, i_2, \ldots, i_l), (j_1, j_2, \ldots, j_l))$ of Q at level l is given by

$$Q((i_1, i_2, \ldots, i_l), (j_1, j_2, \ldots, j_l))$$

$$(2.3) \quad = \sum_{k=1}^{K} \left(\prod_{h=1}^{l} q_k^{(h)}(i_h, j_h) \right) \left(\bigotimes_{h=l+1}^{H} Q_k^{(h)} \right) + Q_D((i_1, i_2, \ldots, i_l), (j_1, j_2, \ldots, j_l))$$

$$\text{for} \quad l = 0, 1, \ldots, H,$$

where $Q_D((i_1, i_2, \ldots, i_l), (j_1, j_2, \ldots, j_l))$ is block $((i_1, i_2, \ldots, i_l), (j_1, j_2, \ldots, j_l))$ of Q_D with the understanding that $l = 0$ yields Q and $l = H$ yields the scalar

$$q((i_1, i_2, \ldots, i_H), (j_1, j_2, \ldots, j_H)) = \sum_{k=1}^{K} \prod_{h=1}^{H} q_k^{(h)}(i_h, j_h)$$

$$+ q_D((i_1, i_2, \ldots, i_H), (j_1, j_2, \ldots, j_H)).$$

Observe that $Q_D((i_1, i_2, \ldots, i_l), (j_1, j_2, \ldots, j_l)) = 0$ if $(i_1, i_2, \ldots, i_l) \neq (j_1, j_2, \ldots, j_l)$, meaning it is an off-diagonal block at level l. The nested structure associated with (2.1) is also valid in the presence of functional transitions.

Now, we introduce a small example to illustrate the Kronecker representation of a CTMC.

2.2. An example. Consider the following matrices for a 4-dimensional problem (each dimension with 2 states) having 7 terms of Kronecker products:

$$Q_1^{(1)} = \begin{pmatrix} & 1 \\ & \end{pmatrix}, \ Q_2^{(1)} = Q_3^{(1)} = Q_4^{(1)} = Q_5^{(1)} = I, \ Q_6^{(1)} = \begin{pmatrix} & 1 \\ & \end{pmatrix}, \ Q_7^{(1)} = \begin{pmatrix} & \\ 10 & \end{pmatrix},$$

$$Q_1^{(2)} = I, \ Q_2^{(2)} = \begin{pmatrix} & \\ 1 & \end{pmatrix}, \ Q_3^{(2)} = Q_4^{(2)} = Q_5^{(2)} = Q_6^{(2)} = I, \ Q_7^{(2)} = \begin{pmatrix} & 1 \\ & \end{pmatrix},$$

$$Q_1^{(3)} = Q_2^{(3)} = I, \quad Q_3^{(3)} = \begin{pmatrix} & 1 \\ & \end{pmatrix}, \quad Q_4^{(3)} = I, \quad Q_5^{(3)} = \begin{pmatrix} & 1 \\ & \end{pmatrix}, \quad Q_6^{(3)} = \begin{pmatrix} & \\ 10 & \end{pmatrix}, \quad Q_7^{(3)} = I,$$

$$Q_1^{(4)} = Q_2^{(4)} = Q_3^{(4)} = I, \quad Q_4^{(4)} = \begin{pmatrix} & 1 \\ & \end{pmatrix}, \quad Q_5^{(4)} = \begin{pmatrix} & \\ 10 & \end{pmatrix}, \quad Q_6^{(4)} = I, \quad Q_7^{(4)} = I.$$

Then, from equation (2.1)

$$Q = \sum_{k=1}^{7} \bigotimes_{h=1}^{4} Q_k^{(h)} + Q_D,$$

and is given by

```
                1  1  1  1  1  1  1  1  2  2  2  2  2  2  2  2
                1  1  1  2  2  2  2  2  1  1  1  2  2  2  2  2
                1  1  2  2  1  1  2  2  1  1  2  2  1  1  2  2
                1  2  1  2  1  2  1  2  1  2  1  2  1  2  1  2

       ( -3    1    1                            1                              )
       (     -12   10    1                            1                        )
       (          -12    1                       10        1                   )
       (               -11                            10        1             )
       (   1               -4    1    1                        1              )
       (        1              -13   10    1                        1         )
Q =    (             1              -13    1                  10        1      )
       (                  1              -12                      10        1  )
       (                       10              -12    1    1                   )
       (                            10              -21   10    1             )
       (                                 10              -11    1             )
       (                                      10              -10             )
       (                                           1              -3    1    1 )
       (                                                1              -12   10    1 )
       (                                                     1              -2    1 )
       (                                                          1              -1 )
```

Now, if we assume the absence of a matrix in the Kronecker representation indicates that it is identity, then it is possible to do without storing identity matrices. With this understanding, the number of floating-point values stored in the Kronecker representation of Q is 10 for the matrices and 16 for the diagonal elements, thus totaling 26; whereas, it is 60 for the flat representation. The discrepancy between the Kronecker and the flat representations becomes substantial for larger values of the state space size, n. In passing, we remark that it is also possible to take advantage of identity matrices in the vector-Kronecker product multiplication algorithm.

Since $n = 16$ and $n_1 = n_2 = n_3 = n_4 = 2$, the recursive definition in (2.2) reveals the nested block structure of Q as follows. At level 0, we have an order 16 matrix; at level 1, we have a (2×2) block matrix with blocks of order 8; at level 2, we have a (4×4) block matrix with blocks of order 4, at level 3, we have an (8×8) block matrix with blocks of order 2; and finally, at level 4, we have a (16×16) block matrix with blocks of order 1.

In the next section, we discuss preprocessing techniques to expedite the analysis of MCs based on Kronecker products.

3. Preprocessing. There are three techniques that can be used to put the Kronecker representation into a more favorable form before solvers take over. These are reordering, grouping, and lumping.

3.1. Reordering and grouping. We assume that Q is a time-homogeneous CTMC. Therefore, the left-hand side of (2.1) is constant up to a symmetric permutation, that is, up to a reordering of the state space, $\mathcal{S}$. Yet, there may be multiple ways in which the number of Kronecker product terms, K, and the number of factors in each Kronecker product, H, forming Q_O on the right-hand side of (2.1) are chosen. Obviously, the choice $(K, H) = (1, 1)$ indicates a flat representation and is assumed to be impossible due to memory limitations. As H decreases towards 1, the Kronecker representation becomes flatter, implying increased storage requirements. On the other hand, if K were 1, then Q could be analyzed along each dimension independently. Hence, K is normally assumed to be larger than 1. Observe that it would be advantageous to be able to make K as small as possible without changing H, since then we would be decreasing the number of terms in the Kronecker representation of Q_O and making the matrices $Q_k^{(h)}$ fuller.

Reordering in MCs based on Kronecker products refers to either permuting the factors of Kronecker products or renumbering the states in the state spaces of factors. We remark that the latter corresponds to a symmetric permutation of the factor matrices $Q_k^{(h)}$ for $k = 1, 2, \ldots, K$ associated with the renumbered state space $\mathcal{S}^{(h)}$. As indicated in [5, 36], reordering of the first kind may be used to reduce the overhead associated with vector-Kronecker product multiplication in the presence of functional transitions. Furthermore, reordering of both kinds can change the nonzero structure of the underlying MC, and thereby can have an effect on the convergence of iterative methods sensitive to the nonzero structure [31]. Hence, by the help of reordering, it may be possible to symmetrically permute the nonzero structure of the underlying MC to a more favorable form for the iterative method of choice. In doing this, we can use the nonzero structure of $\sum_{k=1}^{K} Q_k^{(h)}$, which indicates how the factor h contributes to the nonzero structure of Q_O for $h = 1, 2, \ldots, H$.

Grouping in MCs based on Kronecker products refers to collapsing the same adjacent factors in each Kronecker product. Consequently, the factors in each Kronecker product are reduced by the same number and the state space sizes of the factors are increased. The effect of grouping factors in Kronecker products forming Q_O is investigated in a sequence of papers [5, 35, 36] under functional transitions. The objective is to reduce the number of factors and thereby the overhead associated with evaluating functional transitions. Results show that in some cases grouping may help reduce the state space if it had unreachable states, may decrease the overhead associated with functional transitions, and may even decrease the number of terms in the Kronecker representation. When there are functional transitions, the best approach seems to group those factors which have functional dependencies among each other. In the absence of functional transitions, it is recommended to group as many factors as possible given available memory starting from the highest indexed factor. This ensures a flatter representation for diagonal blocks at a particular level, which is a useful feature in certain iterative methods.

The effects of reordering and grouping of factors of Kronecker products on the convergence and space requirements of iterative methods have been investigated in a number of papers [18, 20, 32, 39, 64], but a broad, systematic study seems to be lacking.

In the next section, we recall the concept and types of lumpability, and provide a brief summary of existing work associated with lumping on MCs based on Kronecker products.

3.2. Lumping. *Lumpability* [46] is a property possessed by some MCs which, if conditions are met, may be used to reduce a large state space to a smaller one. The idea is to find a partitioning of the original state space such that, when the states in each partition are lumped (or aggregated) to form a single state, the resulting MC described by the lumped states has equivalent behavior to the original chain. It is therefore important to consider lumping to reduce the size of the state space, $\mathcal{S}$, before moving to the solution phase.

In this work we refer to two kinds of lumpability: ordinary lumpability and exact lumpability. Here we give definitions for CTMCs. Equivalent definitions can be stated for DTMCs. A CTMC Q is said to be *ordinarily lumpable* with respect to a partitioning of its state space $\mathcal{S} = \cup_i \mathcal{S}_i$ and $\mathcal{S}_i \cap \mathcal{S}_j = \emptyset$ for all $i \neq j$ if for all $\mathcal{S}_i \subset \mathcal{S}$ and all $s_i, s_i' \in \mathcal{S}_i$

$$(3.1) \qquad \sum_{s_j \in \mathcal{S}_j} q(s_i, s_j) = \sum_{s_j \in \mathcal{S}_j} q(s_i', s_j) \text{ for all } \mathcal{S}_j \subset \mathcal{S}.$$

A CTMC Q is said to be *exactly lumpable* with respect to a partitioning of its state space $\mathcal{S} = \cup_i \mathcal{S}_i$ and $\mathcal{S}_i \cap \mathcal{S}_j = \emptyset$ for all $i \neq j$ if for all $\mathcal{S}_i \subset \mathcal{S}$ and all $s_i, s_i' \in \mathcal{S}_i$

$$(3.2) \qquad \sum_{s_j \in \mathcal{S}_j} q(s_j, s_i) = \sum_{s_j \in \mathcal{S}_j} q(s_j, s_i') \text{ for all } \mathcal{S}_j \subset \mathcal{S}.$$

Ordinary lumpability refers to a partitioning of the state space in which the sums of transition rates from each state in a partition to a(nother) partition are the same. On the other hand, exact lumpability refers to a partitioning of the state space in which the sums of transition rates from all states in a partition into each state of a(nother) partition are the same.

Let $\mathcal{S}_{lumped}$ denote the lumped state space. On the ordinarily lumped MC one can only compute performance measures defined over $\mathcal{S}_{lumped}$. On the exactly lumped MC one can compute steady-state performance measures defined over $\mathcal{S}$, transient performance measures defined over $\mathcal{S}_{lumped}$, and transient performance measures defined over $\mathcal{S}$ if the states in the exactly lumpable partitions have the same initial probabilities. Since MCs satisfy a row sum property rather than a column sum property, the exact lumpability condition in (3.2) is more difficult to be satisfied than the ordinary lumpability condition in (3.1). See [7] and the references therein for more information regarding the concept of lumpability and its implications.

Lumpability can be investigated on the flat representation of the MC. Detection of ordinary and exact lumpability on Q through partition refinement would imply a time complexity of $O(nz \log n)$ and a space complexity of $O(nz)$ [52]. Since this is expensive in terms of time and storage, techniques that investigate lumpability on the Kronecker representation have been considered.

Lumpability can be investigated within each of the state spaces $\mathcal{S}^{(h)}$ that define the Kronecker representation of Q_O in (2.1) for $h = 1, 2, \ldots, H$ independently. For the state space $\mathcal{S}^{(h)}$, detection of ordinary and exact lumpability through partition refinement as in [16] requires a time complexity of $O(nz_{Q^{(h)}} \log n_h)$ and a space complexity of $O(nz_{Q^{(h)}})$. Then the lumped Kronecker representation may be obtained by replacing each of the state spaces $\mathcal{S}^{(h)}$ and its corresponding matrices $Q_k^{(h)}$ for $k = 1, 2, \ldots, K$ with equivalent lumped ones. Lumpability can also be investigated among the state spaces $\mathcal{S}^{(h)}$ that are replicated (or identical) with respect to the Kronecker representation of Q_O as in [3]. Therein ordinary lumpability of replicated state spaces is shown in the presence of functional transitions. Note that replication refers to a very specific kind of symmetry in the Kronecker representation, and with ordinary lumpability, only performance measures of interest over $\mathcal{S}_{lumped}$ can be computed. Lumpability can also be investigated among the state spaces $\mathcal{S}^{(h)}$ by considering dependencies and matrix properties in the Kronecker representation as in [41, 42]. Therein sufficient conditions that satisfy ordinary lumpability are specified and an iterative steady-state solution method which is able to compute performance measures over $\mathcal{S}$ is given for CTMCs and DTMCs in the presence of functional transitions. The work identifies lumpable partitionings on the underlying MC induced by the nested block structure of the Kronecker representation in (2.2). Although the particular approach of lumping one or more state spaces $\mathcal{S}^{(h)}$ totally as in [41, 42] is a very specific kind of performance equivalence and lumping considered in [8, 10], due to its accommodation of functional transitions it also enables the detection of certain ordinarily lumpable partitionings in which blocks are composed of multiple (non-identical) state spaces but the individual state spaces cannot be lumped by themselves. This is not possible with the approaches in [3, 8, 10].

We remark that neither of the two approaches in [3] and [41, 42] that investigate lumping among the state spaces $\mathcal{S}^{(h)}$ for $h = 1, 2, \ldots, H$ is completely automated, use a Kronecker representation for the lumped MC, and possess a proper complexity analysis. Furthermore, since the Kronecker representation is rich in structure and the three approaches presented in this section do not work on the flat representation, there can very well be other symmetries in the Kronecker representation which also lead to lumpability. This may be worthwhile investigating.

In the next section, we consider block iterative methods based on splittings for MCs that are in the form of (2.1).

4. Block iterative methods.

We begin by splitting the smaller matrices that form the Kronecker products as in [64].

4.1. Splitting the smaller matrices.

Let

$$(4.1) \quad Q_k^{(h)} = D_k^{(h)} + U_k^{(h)} + L_k^{(h)} \quad \text{for} \quad k = 1, 2, \ldots, K \quad \text{and} \quad h = 1, 2, \ldots, H,$$

where $D_k^{(h)}$, $U_k^{(h)}$, and $L_k^{(h)}$ are respectively the diagonal, strictly upper-triangular, and strictly lower-triangular parts of $Q_k^{(h)}$. Observe that $D_k^{(h)} \geq 0$, $U_k^{(h)} \geq 0$, and $L_k^{(h)} \geq 0$ since $Q_k^{(h)} \geq 0$. Then using Lemma A.8 in [64, p. 183], which rests on the associativity of Kronecker product and the distributivity of Kronecker product over matrix addition, it is possible to express Q_O of Q in (2.1) at level $l = 0, 1, \ldots, H$ using (4.1) as

$$(4.2) \quad Q_O = Q_{U(l)} + Q_{L(l)} + Q_{DU(l)} + Q_{DL(l)},$$

where

$$(4.3) \quad Q_{U(l)} = \sum_{k=1}^{K} \sum_{h=1}^{l} \left(\bigotimes_{f=1}^{h-1} D_k^{(f)} \right) \otimes U_k^{(h)} \otimes \left(\bigotimes_{f=h+1}^{H} Q_k^{(f)} \right),$$

$$(4.4) \quad Q_{L(l)} = \sum_{k=1}^{K} \sum_{h=1}^{l} \left(\bigotimes_{f=1}^{h-1} D_k^{(f)} \right) \otimes L_k^{(h)} \otimes \left(\bigotimes_{f=h+1}^{H} Q_k^{(f)} \right)$$

correspond respectively to the strictly block upper- and lower-triangular parts of Q_O at level l, and

$$(4.5) \quad Q_{DU(l)} = \sum_{k=1}^{K} \sum_{h=l+1}^{H} \left(\bigotimes_{f=1}^{h-1} D_k^{(f)} \right) \otimes U_k^{(h)} \otimes \left(\bigotimes_{f=h+1}^{H} Q_k^{(f)} \right),$$

$$(4.6) \qquad Q_{DL(l)} = \sum_{k=1}^{K} \sum_{h=l+1}^{H} \left(\bigotimes_{f=1}^{h-1} D_k^{(f)} \right) \otimes L_k^{(h)} \otimes \left(\bigotimes_{f=h+1}^{H} Q_k^{(f)} \right).$$

correspond respectively to the strictly upper- and lower-triangular parts of the block diagonal of Q_O at level l. Observe that $Q_{U(l)} \geq 0$, $Q_{L(l)} \geq 0$, $Q_{DU(l)} \geq 0$, and $Q_{DL(l)} \geq 0$. Furthermore, we remark that $l = 0$ implies Q_O is a single block for which $Q_{U(0)} = Q_{L(0)} = 0$, whereas $l = H$ corresponds to a point-wise partitioning of Q_O for which $Q_{DU(H)} = Q_{DL(H)} = 0$. Hence, for iterative methods based on block partitionings $l = 1, 2, \ldots, H - 1$ should be used.

4.2. Example (continued). Consider the block partitioning of the 4-dimensional problem at level 1 for which $l = 1$, $b_1 = 4$, and Q is viewed as a (2×2) block matrix with blocks of order $o_1 = 8$ (see (2.2)). Then, from (4.3) and (4.4), we have

$$Q_{U(1)} = \sum_{k=1}^{7} U_k^{(1)} \otimes Q_k^{(2)} \otimes Q_k^{(3)} \otimes Q_k^{(4)} \quad \text{and} \quad Q_{L(1)} = \sum_{k=1}^{7} L_k^{(1)} \otimes Q_k^{(2)} \otimes Q_k^{(3)} \otimes Q_k^{(4)},$$

implying

$$Q_{U(1)} + Q_{L(1)} = \left(\begin{array}{cccccccc|cccccccc}
 & & & & & & & & 1 & & & & & & & \\
 & & & & & & & & & 1 & & & & & & \\
 & & & & & & & & 10 & & 1 & & & & & \\
 & & & & & & & & & 10 & & 1 & & & & \\
 & & & & & & & & & & & & 1 & & & \\
 & & & & & & & & & & & & & 1 & & \\
 & & & & & & & & & & & & 10 & & 1 & \\
 & & & & & & & & & & & & & 10 & & 1 \\ \hline
 & & 10 & & & & & & & & & & & & & \\
 & & & 10 & & & & & & & & & & & & \\
 & & & & 10 & & & & & & & & & & & \\
 & & & & & 10 & & & & & & & & & & \\
 & & & & & & & & & & & & & & & \\
 & & & & & & & & & & & & & & & \\
 & & & & & & & & & & & & & & & \\
 & & & & & & & & & & & & & & &
\end{array} \right),$$

whereas from (4.5) and (4.6), we have

$$Q_{DU(1)} = \sum_{k=1}^{7} D_k^{(1)} \otimes U_k^{(2)} \otimes Q_k^{(3)} \otimes Q_k^{(4)} + \sum_{k=1}^{7} D_k^{(1)} \otimes D_k^{(2)} \otimes U_k^{(3)} \otimes Q_k^{(4)}$$
$$+ \sum_{k=1}^{7} D_k^{(1)} \otimes D_k^{(2)} \otimes D_k^{(3)} \otimes U_k^{(4)},$$

$$Q_{DL(1)} = \sum_{k=1}^{7} D_k^{(1)} \otimes L_k^{(2)} \otimes Q_k^{(3)} \otimes Q_k^{(4)} + \sum_{k=1}^{7} D_k^{(1)} \otimes D_k^{(2)} \otimes L_k^{(3)} \otimes Q_k^{(4)}$$
$$+ \sum_{k=1}^{7} D_k^{(1)} \otimes D_k^{(2)} \otimes D_k^{(3)} \otimes L_k^{(4)},$$

implying

$$Q_{DU(1)} + Q_{DL(1)} = \left(\begin{array}{cccccccc|cccccccc}
 & 1 & 1 & & & & & & & & & & & & & \\
 & & 10 & 1 & & & & & & & & & & & & \\
 & & & 1 & & & & & & & & & & & & \\
1 & & & & 1 & 1 & & & & & & & & & & \\
 & 1 & & & & 10 & 1 & & & & & & & & & \\
 & & 1 & & & & 1 & & & & & & & & & \\
 & & & & & & & & & & & & & & & \\
 & & & & & & & & & & & & & & & \\ \hline
 & & & & & & & & & 1 & 1 & & & & & \\
 & & & & & & & & & & 10 & 1 & & & & \\
 & & & & & & & & & & & 1 & & & & \\
 & & & & & & & 1 & & & & & 1 & 1 & & \\
 & & & & & & & & 1 & & & & & 10 & 1 & \\
 & & & & & & & & & 1 & & & & & 1 & \\
 & & & & & & & & & & & & & & &
\end{array} \right).$$

Note that there are the $\sqrt{b_1} = 2$ blocks along the diagonal.

4.3. Block iterative methods for Kronecker products. Now, let Q in (2.1) be irreducible and split at level l using (4.2) as

$$(4.7) \qquad Q = Q_O + Q_D = Q_{U(l)} + Q_{L(l)} + Q_{DU(l)} + Q_{DL(l)} + Q_D = M - N,$$

where M is nonsingular (i.e., M^{-1} exists). Then, the power, block Jacobi over-relaxation (BJOR), and block successive over-relaxation (BSOR) methods are based on different splittings of Q, and each satisfies

$$\pi_{(m+1)}M = \pi_{(m)}N \quad \text{for} \quad m = 0, 1, \ldots$$

with the sequence of approximations $\pi_{(m+1)}$ to the steady-state distribution in (1.2), where $\pi_{(0)} > 0$ is the initial approximation such that $\pi_{(0)}e = 1$ and $T = NM^{-1}$ is the iteration matrix. Note that T does not change from iteration to iteration and only the current approximation is used to compute the new approximation. Hence, these methods, which are based on splittings of the coefficient matrix, are also known as stationary iterative methods. Since Q is a singular matrix and assumed to be irreducible, the largest eigenvalue of T in magnitude is one. In order to ensure convergence, T should not have other eigenvalues with magnitude one. For converging approximations, the magnitude of the eigenvalue of T closest to one determines the rate of convergence [62].

The particular splittings corresponding to the power, BJOR, and (forward) BSOR methods are

$$M^{Power} = -\alpha I,$$
$$N^{Power} = -\alpha(I + Q/\alpha),$$
$$M^{BJOR} = (Q_D + Q_{DU(l)} + Q_{DL(l)})/\omega,$$
$$N^{BJOR} = (1 - \omega)(Q_D + Q_{DU(l)} + Q_{DL(l)})/\omega - Q_{U(l)} - Q_{L(l)},$$
$$M^{BSOR} = (Q_D + Q_{DU(l)} + Q_{DL(l)})/\omega + Q_{U(l)},$$
$$N^{BSOR} = (1 - \omega)(Q_D + Q_{DU(l)} + Q_{DL(l)})/\omega - Q_{L(l)},$$

where $\alpha \in [\max_{s \in \mathcal{S}} |q_D(s, s)|, \infty)$ is the uniformization parameter of the power method and $\omega \in (0, 2)$ is the relaxation parameter of the BJOR and BSOR methods. The power method works at level $l = H$ since it is a point method. Furthermore, the BJOR and BSOR methods reduce to the block Jacobi (BJacobi) and block Gauss-Seidel (BGS) methods for $\omega = 1$, and the BJOR and BSOR methods become point JOR and point SOR methods for $l = H$. We remark that [40] shows how one can find $\max_{s \in \mathcal{S}} |q_D(s, s)|$ in the presence of functional transitions when Q_D is given as a sum of Kronecker products.

Since $Q = Q_O + Q_D$, it is possible to express the power method at iteration m as

$$(4.8) \qquad \pi_{(m+1)} = \pi_{(m)} + \pi_{(m)}Q_D/\alpha + \pi_{(m)}Q_O/\alpha.$$

Observe that the second term in (4.8) poses no problem from a computational point of view since Q_D is diagonal, and the third term can be efficiently implemented using the vector-Kronecker product multiplication algorithm since Q_O is a sum of Kronecker products (see (2.1)).

Regarding the BJOR method with a level l block partitioning, at iteration m we have

$$\pi_{(m+1)}(Q_D + Q_{DU(l)} + Q_{DL(l)})$$

(4.9) $= (1 - \omega)\pi_{(m)}Q_D + (1 - \omega)\pi_{(m)}Q_{DU(l)} + (1 - \omega)\pi_{(m)}Q_{DL(l)}$

$$-\omega\pi_{(m)}Q_{U(l)} - \omega\pi_{(m)}Q_{L(l)}.$$

This is a block diagonal linear system with $\sqrt{b_l}$ blocks of order o_l along the diagonal of the nonsingular coefficient matrix $(Q_D + Q_{DU(l)} + Q_{DL(l)})$ and a nonzero right-hand side which can be efficiently computed using the vector-Kronecker product multiplication algorithm, since $Q_{U(l)}, Q_{L(l)}, Q_{DU(l)}$, and $Q_{DL(l)}$ are sums of Kronecker products (see (4.3), (4.4), (4.5), and (4.6)). Hence, (4.9) is equivalent to $\sqrt{b_l}$ independent, nonsingular linear systems each of order o_l and a nonzero right-hand side. If there is space, one can generate and factorize in sparse storage the nonsingular blocks of the form

$$Q((i_1, i_2, \ldots, i_l), (i_1, i_2, \ldots, i_l)) = \sum_{k=1}^{K} \left(\prod_{h=1}^{l} q_k^{(h)}(i_h, i_h) \right) \left(\bigotimes_{h=l+1}^{H} Q_k^{(h)} \right)$$

(4.10) $+ Q_D((i_1, i_2, \ldots, i_l), (i_1, i_2, \ldots, i_l))$ for $(i_1, i_2, \ldots, i_l) \in \times_{h=1}^{l} \mathcal{S}^{(h)}$

along the diagonal (see (2.3)) of $(Q_D + Q_{DU(l)} + Q_{DL(l)})$ at the outset and solve the $| \times_{h=1}^{l} \mathcal{S}^{(h)} | = \sqrt{b_l}$ systems directly at each iteration. Otherwise, one can use an iterative method; indeed, it is even possible to use a block iterative method, such as BJOR, since the off-diagonal parts of the diagonal blocks given by

$$\sum_{k=1}^{K} \left(\prod_{h=1}^{l} q_k^{(h)}(i_h, i_h) \right) \left(\bigotimes_{h=l+1}^{H} Q_k^{(h)} \right)$$

are sums of Kronecker products.

The situation with the BSOR method is not very different from that of BJOR. For BSOR with a level l block partitioning, at iteration m we have

$$\pi_{(m+1)}(Q_D + Q_{DU(l)} + Q_{DL(l)} + \omega Q_{U(l)})$$

(4.11) $= (1 - \omega)\pi_{(m)}Q_D + (1 - \omega)\pi_{(m)}Q_{DU(l)} + (1 - \omega)\pi_{(m)}Q_{DL(l)}$

$$-\omega\pi_{(m)}Q_{L(l)}.$$

This is a block upper-triangular linear system with $\sqrt{b_l}$ blocks of order o_l along the diagonal of the nonsingular coefficient matrix $(Q_D + Q_{DU(l)} + Q_{DL(l)} + \omega Q_{U(l)})$ and a nonzero right-hand side which can be efficiently computed using the vector-Kronecker product multiplication algorithm, since $Q_{L(l)}, Q_{DU(l)}$, and $Q_{DL(l)}$ are sums of Kronecker products. In [64], a recursive algorithm is given for a nonsingular linear system with a lower-triangular coefficient matrix in the form of a sum of Kronecker products and a nonzero right-hand side. Such a system arises in backward point SOR. Therein, a version of the same algorithm for backward BSOR is also discussed. Here we remark that an iterative block upper-triangular solution algorithm for (4.11) is also possible [18] and a block row-oriented version is preferable in the presence of functional transitions:

ALGORITHM 1. *Iterative block upper-triangular solution at level l for MCs based on Kronecker products*

$b = (1 - \omega)\pi_{(m)}Q_D + (1 - \omega)\pi_{(m)}Q_{DU(l)} + (1 - \omega)\pi_{(m)}Q_{DL(l)} - \omega\pi_{(m)}Q_{L(l)}$;

For row of blocks $(i_1, i_2, \ldots, i_l) = (1, 1, \ldots, 1)$ to $(n_1, n_2, \ldots, n_l)$ lexicographically,

> Solve $\pi_{(m+1)}((i_1, i_2, \ldots, i_l))Q((i_1, i_2, \ldots, i_l), (i_1, i_2, \ldots, i_l)) = b((i_1, i_2, \ldots, i_l))$;
>
> > For column of blocks $(j_1, j_2, \ldots, j_l) > (i_1, i_2, \ldots, i_l)$,
> >
> > $b((j_1, j_2, \ldots, j_l)) = b((j_1, j_2, \ldots, j_l))$
> > $\quad - \omega\pi_{(m+1)}((i_1, i_2, \ldots, i_l))Q_{U(l)}((i_1, i_2, \ldots, i_l), (j_1, j_2, \ldots, j_l))$.

Observe in Algorithm 1 that initially the nonzero right-hand side vector b can be efficiently computed using the vector-Kronecker product multiplication algorithm, since $Q_{L(l)}$, $Q_{DU(l)}$, and $Q_{DL(l)}$ are sums of Kronecker products. Furthermore, $Q((i_1, i_2, \ldots, i_l), (i_1, i_2, \ldots, i_l))$ is given in (4.10) in terms of a sum of Kronecker products, and $Q_{U(l)}((i_1, i_2, \ldots, i_l), (j_1, j_2, \ldots, j_l))$ for $(j_1, j_2, \ldots, j_l) > (i_1, i_2, \ldots, i_l)$ can be expressed in terms of a sum of Kronecker products using (4.3) as

$$Q((i_1, i_2, \ldots, i_l), (j_1, j_2, \ldots, j_l))$$

$$= \sum_{k=1}^{K} \sum_{h=1}^{l} \left(\prod_{f=1}^{h-1} d_k^{(f)}(i_f, j_f) \right) u_k^{(h)}(i_h, j_h) \left(\prod_{f=h+1}^{l} q_k^{(f)}(i_f, j_f) \right) \left(\bigotimes_{f=l+1}^{H} Q_k^{(f)} \right).$$

To the contrary of BJOR, the nonsingular diagonal blocks $Q((i_1, i_2, \ldots, i_l), (i_1, i_2, \ldots, i_l))$ in BSOR must be solved in lexicographical order. If there is space, one can generate and factorize in sparse storage these blocks as in BJOR at the outset and solve the $\sqrt{b_l}$ systems directly at each iteration. Otherwise, one can use an iterative method such as BSOR, since the off-diagonal parts of the diagonal blocks are also sums of Kronecker products. After each block is solved for the unknown subvector $\pi_{(m+1)}((i_1, i_2, \ldots, i_l))$, b is updated by multiplying the computed subvector with the corresponding row of blocks above the diagonal. Finally, we emphasize that BSOR at level l reduces to point SOR if $Q_{DL(l)} = 0$ (see Remark 4.1 in [64, p. 176]).

It is quite surprising to notice that block iterative solvers, which are sometimes called two-level iterative solvers, have still not been incorporated into most analysis packages based on Kronecker representations although they are shown to be more effective than point solvers on many test cases [18, 64]. Furthermore, to the contrary of block partitionings considered in [33] for sparse MCs, block partitionings of Kronecker products are nested and recursive due to the lexicographical ordering of states. Therefore, there tends to be more common structure among the diagonal blocks of a MC expressed as a sum of Kronecker products. Diagonal blocks having identical off-diagonal parts and diagonals which differ by a multiple of the identity are exploited in [18]. Therein, it is shown that such diagonal blocks can share and work with the factorization of only one diagonal block. This approach saves not only from time spent for factorization of diagonal blocks at the outset, but also from space. The same paper also discusses a three-level version of BSOR for MCs based on Kronecker products in which the diagonal blocks that are too large to be factorized are solved using BSOR. Similar results also appear in [39]. Finally, we remark that it is possible to alter the nonzero structure of the underlying MC of a Kronecker representation by reordering factors and states of factors so as to make it more suitable for block iterative methods. Obviously, the power and point JOR methods will not benefit from such reordering.

In the next section, we introduce a simple version of the *multilevel* (ML) method [15, 19] for irreducible MCs based on Kronecker products which happens to be a generalization of the well-known *iterative aggregation-disaggregation* (IAD) method [62] to more than two levels.

5. Multilevel methods. In [6, 11, 12, 13], aggregation-disaggregation steps are coupled with various iterative methods for MCs based on Kronecker products to accelerate convergence. An IAD method for MCs based on Kronecker products and its adaptive version, which analyzes aggregated systems for those parts where the error is estimated to be high, are proposed in [9] and [14], respectively. The adaptive IAD method in [14] is improved in [15] through a recursive definition and called ML.

5.1. The simple multilevel method for Kronecker products. Let $\mathcal{S}_{(l)} = \times_{h=l+1}^{H}\mathcal{S}^{(h)}$ for $l = 0, 1, \ldots, H$ and the mapping $f_{(l)} : \mathcal{S}_{(l)} \longrightarrow \mathcal{S}_{(l+1)}$ represent the aggregation of dimension $(l+1)$ (i.e., the state space $S^{(l+1)}$) so that the states in $\mathcal{S}_{(l)}$ are mapped to the states in $\mathcal{S}_{(l+1)}$. Note that $\mathcal{S}_{(0)} = \mathcal{S}$ and $\mathcal{S}_{(H)} = \{1\}$. Furthermore, let the aggregated CTMCs $\tilde{Q}_{(m,l)}$ with state spaces $\mathcal{S}_{(l)}$ be defined at levels $l = 1, 2, \ldots, H$ with $\tilde{Q}_{(m,0)} = Q$ for iteration m. Finally, let the power method be used as a *smoother* (or *accelerator*) before aggregation $\eta_{(m,l)}$ times and after disaggregation $\nu_{(m,l)}$ times with $\alpha_{(m,l)} \in [\max_{s_{(l)} \in \mathcal{S}_{(l)}} |\tilde{q}_{(m,l)}(s_{(l)}, s_{(l)})|, \infty)$ at level l for iteration m. Then the ML iteration matrix at level l for iteration m is given by

$$(5.1) \quad T_{(m,l)}^{ML} = (I + \tilde{Q}_{(m,l)}/\alpha_{(m,l)})^{\eta_{(m,l)}} R_{(l)} T_{(m,l+1)}^{ML} P_{x_{(m,l)}} (I + \tilde{Q}_{(m,l)}/\alpha_{(m,l)})^{\nu_{(m,l)}},$$

and satisfies

$$\pi_{(m+1,l)} = \pi_{(m,l)} T_{(m,l)}^{ML} \quad \text{for} \quad m = 0, 1, \ldots,$$

where

$$(5.2) \quad x_{(m,l)} = \pi_{(m,l)}(I + \tilde{Q}_{(m,l)}/\alpha_{(m,l)})^{\eta_{(m,l)}},$$

$$r_{(l)}(s_{(l)}, s_{(l+1)}) = \begin{cases} 1 & \text{if } f_{(l)}(s_{(l)}) = s_{(l+1)} \\ 0 & \text{otherwise} \end{cases}$$

$$(5.3) \quad \text{for } s_{(l)} \in \mathcal{S}_{(l)} \text{ and } s_{(l+1)} \in \mathcal{S}_{(l+1)},$$

$$p_{x_{(m,l)}}(s_{(l+1)}, s_{(l)}) = \begin{cases} \dfrac{x_{(m,l)}(s_{(l)})}{\sum_{s_{(l)} \in \mathcal{S}_{(l)}, f_{(l)}(s_{(l)})=s_{(l+1)}} x_{(m,l)}(s_{(l)})} & \text{if } f_{(l)}(s_{(l)}) = s_{(l+1)} \\ 0 & \text{otherwise} \end{cases}$$

$$(5.4) \quad \text{for } s_{(l+1)} \in \mathcal{S}_{(l+1)} \text{ and } s_{(l)} \in \mathcal{S}_{(l)},$$

$$(5.5) \quad \pi_{(m,l+1)} = x_{(m,l)} R_{(l)} \quad \text{and} \quad \tilde{Q}_{(m,l+1)} = P_{x_{(m,l)}} \tilde{Q}_{(m,l)} R_{(l)}.$$

At iteration m, the recursion ends and backtracking starts when $\tilde{Q}_{(m,l+1)}$ in (5.5) is the last aggregated CTMC and solved exactly to give $T_{(m,l+1)} = e\pi_{(m+1,l+1)}$, where $\pi_{(m+1,l+1)}\tilde{Q}_{(m,l+1)} = 0$ and $\pi_{(m+1,l+1)}e = 1$. The level to end recursion depends on available memory since there must be space to store and factorize the aggregated CTMC at that level. When the initial approximation is positive (i.e., $\pi_{(0,0)} > 0$), the aggregated CTMCs $\tilde{Q}_{(m,l+1)}$ are irreducible [15, p. 348], and the ML method has been observed to converge if a sufficient number of smoothings are performed to improve the approximate solution vector, $\pi_{(m,l)}$, at each level.

Observe that to the contrary of block iterative methods, the ML iteration matrix in (5.1) changes from iteration to iteration, and hence, the method is non-stationary. Nevertheless, the $(|\mathcal{S}_{(l)}| \times |\mathcal{S}_{(l+1)}|)$ *aggregation* operator, $R_{(l)}$, in (5.3) is constant and need not be stored since it is defined by $f_{(l)}$. At level l, the $|\mathcal{S}_{(l)}| = \prod_{h=l+1}^{H} n_l$ states represented by $(H - l)$-tuples are mapped to the $|\mathcal{S}_{(l+1)}| = \prod_{h=l+2}^{H} n_l$ states represented by $(H - l - 1)$-tuples by aggregating the leading dimension $\mathcal{S}^{(l+1)}$ in $\mathcal{S}_{(l)}$. We remark that this corresponds to an aggregation based on a contiguous and non-interleaved block partitioning if the states in $\mathcal{S}_{(l)}$ were ordered anti-lexicographically. On the other hand, the $(|\mathcal{S}_{(l+1)}| \times |\mathcal{S}_{(l)}|)$ *disaggregation* operator, $P_{x_{(m,l)}}$, in (5.4) depends on the smoothed vector $x_{(m,l)}$ in (5.2) and has the nonzero structure of $R_{(l)}^T$. Therefore, $P_{x_{(m,l)}}$ can be stored in a vector of length $|\mathcal{S}_{(l)}|$ since it has one nonzero per column by definition. These vectors amount to a total storage of $\sum_{l=0}^{H-1} \prod_{h=l+1}^{H} n_h$ floating-point values if the recursion terminates at level H.

In [15, p. 347], it is shown that $\tilde{Q}_{(m,l+1)}$ can be expressed as a sum of Kronecker products using at most K vectors of length $|\mathcal{S}_{(l+1)}|$ and the matrices corresponding to the factors $(l + 2)$ through H. More specifically, the $s_{(l+1)}$st element of the vector corresponding to the kth term in the Kronecker representation at level $(l + 1)$ for iteration m is defined as

$$a_{(m,l+1),k}(s_{(l+1)})$$

$$(5.6) \quad = \frac{\left(\sum_{s_{(l)} \in \mathcal{S}_{(l)}, f_{(l)}(s_{(l)})=s_{(l+1)}} x_{(m,l)}(s_{(l)}) \, a_{(m,l),k}(s_{(l)}) \, (e_{s_{(l)}(l+1)}^T Q_k^{(l+1)} e) \right)}{\pi_{(m,l+1)}(s_{(l+1)})}$$

$$\text{for } s_{(l+1)} \in \mathcal{S}_{(l+1)} \text{ and } k = 1, 2, \ldots, K,$$

where $a_{(m,0),k} = e$, $s_{(l)}(l + 1) \in \mathcal{S}^{(l+1)}$, and $e_{s_{(l)}(l+1)}$ is the $s_{(l)}(l + 1)$st column of I. Then

$$\tilde{Q}_{(m,l+1)} = \sum_{k=1}^{K} \text{diag}(a_{(m,l+1),k}) \bigotimes_{h=l+2}^{H} Q_k^{(h)}$$

$$(5.7) \qquad\qquad - \sum_{k=1}^{K} \text{diag}(a_{(m,l+1),k}) \bigotimes_{h=l+2}^{H} \text{diag}(Q_k^{(h)} e).$$

Observe that the second summation in (5.7) returns a diagonal matrix which sums the rows of $\tilde{Q}_{(m,l+1)}$ to zero. Furthermore, the vectors $a_{(m,0),k}$ for $k = 1, 2, \ldots, K$ at level 0 consist of all ones, and therefore need not be stored. If the recursion ends at level H, then $\tilde{Q}_{(m,H)}$ is a (1×1) CTMC equal to zero, and need not be stored since its steady-state vector is one. We remark that $a_{(m,l+1),k} = e$ for those k which either have a single $Q_k^{(h)} \neq I$ for $h = 1, 2, \ldots, H$, or have all $Q_k^{(h)} = I$ for $h = l + 2, \ldots, H$. Such vectors need not be stored either. The K vectors at a particular level have the same length, but vary in length from $\prod_{h=2}^{H} n_h$ at level 1 to n_H at level $(H - 1)$, implying a storage requirement of at most $K \sum_{l=1}^{H-1} \prod_{h=l+1}^{H} n_h$ floating-point values to facilitate the Kronecker representation of the aggregated CTMCs. We remark that grouping of factors will further reduce the storage requirement for vectors.

5.2. Example (continued). Consider the 4-dimensional problem with the initial distribution $\pi_{(0,0)} = e/16$, $\alpha_{(0,0)} = 21$, and $\eta_{(0,0)} = \nu_{(0,0)} = 1$. Then, $x_{(0,0)} = \pi_{(0,0)}(I + \tilde{Q}_{(0,0)}/21)$ from (5.2) yields

$$x_{(0,0)} = (19\ 11\ 21\ 13\ 27\ 19\ 29\ 21\ 21\ 13\ 23\ 15\ 29\ 21\ 31\ 23)/336.$$

Furthermore,

$$R_{(0)} = \begin{pmatrix}
1 & & & & & & & \\
& 1 & & & & & & \\
& & 1 & & & & & \\
& & & 1 & & & & \\
& & & & 1 & & & \\
& & & & & 1 & & \\
& & & & & & 1 & \\
& & & & & & & 1 \\
1 & & & & & & & \\
& 1 & & & & & & \\
& & 1 & & & & & \\
& & & 1 & & & & \\
& & & & 1 & & & \\
& & & & & 1 & & \\
& & & & & & 1 & \\
& & & & & & & 1
\end{pmatrix}$$

and

$$P_{x_{(0,0)}} = \begin{pmatrix}
\frac{19}{40} & & & & & & & & \frac{21}{40} & & & & & & & \\
& \frac{11}{24} & & & & & & & & \frac{13}{24} & & & & & & \\
& & \frac{21}{44} & & & & & & & & \frac{23}{44} & & & & & \\
& & & \frac{13}{28} & & & & & & & & \frac{15}{28} & & & & \\
& & & & \frac{27}{56} & & & & & & & & \frac{29}{56} & & & \\
& & & & & \frac{19}{40} & & & & & & & & \frac{21}{40} & & \\
& & & & & & \frac{29}{60} & & & & & & & & \frac{31}{60} & \\
& & & & & & & \frac{21}{44} & & & & & & & & \frac{23}{44}
\end{pmatrix}$$

from (5.3) and (5.4), respectively. Hence, the 16 states represented by 4-tuples in $\mathcal{S}_{(0)} = \mathcal{S}$ are mapped to the 8 states represented by 3-tuples in $\mathcal{S}_{(1)}$. For instance, states $(1,1,1,1)$ and $(2,1,1,1)$ are mapped to $(1,1,1)$, whereas states $(1,1,1,2)$ and $(2,1,1,2)$ are mapped to $(1,1,2)$. Using $R_{(0)}$ in (5.5), we obtain the starting approximation at level 1 as

$$\pi_{(0,1)} = (40\ 24\ 44\ 28\ 56\ 40\ 60\ 44)/336.$$

Through (5.6), the 7 vectors used to represent the aggregated CTMC at level 1 are computed as

$$a_{(0,1),1} = (19/40\ 11/24\ 21/44\ 13/28\ 27/56\ 19/40\ 29/60\ 21/44),$$
$$a_{(0,1),2} = a_{(0,1),3} = a_{(0,1),4} = a_{(0,1),5} = e,$$
$$a_{(0,1),6} = (19/40\ 11/24\ 21/44\ 13/28\ 27/56\ 19/40\ 29/60\ 21/44),$$
$$a_{(0,1),7} = (210/44\ 130/24\ 230/44\ 150/28\ 290/56\ 210/40\ 310/60\ 230/44).$$

and the aggregated CTMC is expressed as

$$\tilde{Q}_{0,1} = P_{x_{(0,0)}} \tilde{Q}_{(0,0)} R_{(0)}$$

$$= \sum_{k=1}^{7} \mathrm{diag}(a_{(0,1),k}) \bigotimes_{h=2}^{4} Q_k^{(h)} - \sum_{k=1}^{7} \mathrm{diag}(a_{(0,1),k}) \bigotimes_{h=2}^{4} \mathrm{diag}(Q_k^{(h)} e).$$

Observe that the effect of $a_{(0,1),1}$ in the first term of the first summation will be to the diagonal of $\tilde{Q}_{0,1}$ since $Q_1^{(2)} = Q_1^{(3)} = Q_2^{(4)} = I$. But this effect will be cancelled by the first term of the second summation simply because $\mathrm{diag}(Q_1^{(2)} e) = \mathrm{diag}(Q_1^{(3)} e) = \mathrm{diag}(Q_2^{(4)} e) = I$. Hence, we may very well set $a_{(0,1),1} = e$ as suggested before. Furthermore, $a_{(0,1),2} = a_{(0,1),3} = a_{(0,1),4} = a_{(0,1),5} = e$ since $Q_2^{(1)} = Q_3^{(1)} = Q_4^{(1)} = Q_5^{(1)} = I$ and $a_{(0,0),k} = e$ for $k = 1, 2, \ldots, 7$. Therefore, we implicitly have

$$
\tilde{Q}_{(0,1)} =
\begin{array}{cccccccc}
1 & 1 & 1 & 1 & 2 & 2 & 2 & 2 \\
1 & 1 & 2 & 2 & 1 & 1 & 2 & 2 \\
1 & 2 & 1 & 2 & 1 & 2 & 1 & 2 \\
\end{array}
$$

$$
\left(
\begin{array}{c|c|c|c|c|c|c|c}
-\frac{290}{40} & \frac{40}{40} & \frac{40}{40} & & \frac{210}{40} & & & \\
\hline
& -\frac{394}{24} & \frac{240}{24} & \frac{24}{24} & & \frac{130}{24} & & \\
\hline
\frac{210}{44} & & -\frac{484}{44} & \frac{44}{44} & & & \frac{230}{44} & \\
\hline
& \frac{130}{28} & & -\frac{280}{28} & & & & \frac{150}{28} \\
\hline
\frac{56}{56} & & & & -\frac{168}{56} & \frac{56}{56} & \frac{56}{56} & \\
\hline
& \frac{40}{40} & & & & -\frac{480}{40} & \frac{400}{40} & \frac{40}{40} \\
\hline
& & \frac{60}{60} & & \frac{290}{60} & & -\frac{410}{60} & \frac{60}{60} \\
\hline
& & & \frac{44}{44} & & \frac{210}{44} & & -\frac{254}{44}
\end{array}
\right).
$$

In the next step, similar operations will be carried out at level 1 unless the aggregated CTMC is solved exactly, upon which backtracking from recursion starts for iteration m.

5.3. A class of multilevel methods for Kronecker products. The ML method we discussed follows a V-cycle [48] at each iteration. That is, starting from the finest level, at each step it smoothes the current approximation and moves to a coarser level by aggregation until it reaches a level at which the aggregated CTMC can be solved exactly. Once the exact solution is obtained at the coarsest level, the method starts moving in the opposite direction. At each step on the way to the finest level, the method disaggregates the current approximation passed by the coarser level and smoothes it. Furthermore, the state spaces $\mathcal{S}^{(h)}$ are aggregated according to the fixed ordering $h = 1, 2, \ldots, H$. However, to the contrary of the ML method for sparse MCs in [45], the definition of the aggregated state spaces follows naturally from the Kronecker representation in (2.1) and the aggregated CTMCs can also be represented using Kronecker products as shown in (5.7). In [19], a sophisticated class of ML methods is given. The methods therein are capable of using JOR and SOR as smoothers, performing the W- and F-cycles inspired by multigrid [67], and aggregating the state spaces in cyclic and adaptive orderings. Numerical experiments in [19] proved ML methods to be very strong, robust, and scalable solvers for MCs based on Kronecker products.

The convergence properties of the class of ML methods in [19] are discussed in [21]; however, it is not clear how its behavior would be affected if block iterative methods, such as BJOR and BSOR, are used as smoothers rather than power, JOR, and SOR. Note that BJOR and BSOR should normally not use a direct method for the solution of the diagonal blocks when employed as smoothers with the ML method, since the aggregated CTMC at each level changes from iteration to iteration and the factorization may be too time consuming to offset. In [40], an efficient algorithm that finds a nearly completely decomposable (NCD) [33, 62] partitioning of $\mathcal{S}$ in the presence of functional transitions for a user specified decomposability parameter is given. Since IAD using NCD partitionings has certain rate of convergence guarantees, the algorithm may be useful in the context of ML methods to determine the loosely coupled dimensions to be aggregated first in a given iteration.

The next section discusses various preconditioners to be used with projection methods for MCs based on Kronecker products.

6. Preconditioned projection methods. *Projection* (or *Krylov subspace*) methods for MCs based on Kronecker products are non-stationary iterative methods in which approximate solutions satisfying various constraints are extracted from small dimensional subspaces [62]. Being iterative, their basic operation is vector-Kronecker product multiplication. However, compared to block iterative methods, they require a larger number of supplementary vectors as long as the state space size, n. But, more importantly they need to be used with preconditioners to result in effective solvers.

At each iteration of a preconditioned projection method, the row residual vector, r, is used as the right-hand side of the linear system

$$(6.1) \qquad\qquad\qquad\qquad zM = r$$

to compute the preconditioned row residual vector, z. The objective of this preconditioning step is to improve the error in the approximate solution vector at that iteration. Note that if M were a multiple of I (as in (4.8)), the preconditioned residual would be a multiple of the residual computed at that iteration, implying no improvement. Hence, the preconditioner should approximate the coefficient matrix of the original system in a better way, yet the solution of linear systems as in (6.1) involving the *preconditioner* matrix, M, should be cheap. It is shown in [33] through a large number of numerical experiments on benchmark problems that, to result as effective solvers, projection methods for sparse MCs should be used with preconditioners, such as those based on incomplete LU (ILU) factorizations. However, it is still not clear how one can devise ILU-type preconditioners for MCs that are in the form of (2.1).

So far, various preconditioners are proposed for Kronecker structured representations such as those based on truncated Neumann series [62, 63], the cheap and separable preconditioner [12], and circulant preconditioners for a specific class of problems [25]. The Kronecker product approximate preconditioner for MCs based on Kronecker products developed in a sequence of papers [49, 50, 51], although encouraging, is in the form of a prototype implementation. Numerical experiments in [12, 13, 50, 51, 63] indicate that there is still room for research regarding the development of effective preconditioners for MCs based on Kronecker products.

In introducing another class of preconditioners, we remark that each of the block iterative methods introduced in this work is actually a preconditioned power method for which the preconditioning matrix is M in (4.7). Since M is based on Kronecker products, a BSOR preconditioner exploiting this property is proposed in [20]. To the contrary of the BSOR preconditioner entertained for sparse MCs in [33], the BSOR preconditioner for MCs based on Kronecker products has a rich structure induced by the lexicographical ordering of states. Through numerical experiments, it is shown in

[20] that two-level BSOR preconditioned projection methods in which the diagonal blocks are solved exactly emerge as effective solvers that are competitive with block iterative methods and ML methods.

It will be interesting to compare point JOR, BJOR, and point SOR preconditioners as defined in (4.9) and (4.11) with the existing preconditoners for MCs based on Kronecker products. Clearly, the class of ML methods proposed in [19] is another candidate for preconditioning projection methods.

7. Conclusion. MCs based on Kronecker products have a rich structure, which is nested and recursive. Preprocessing techniques that take advantage of this rich structure to expedite analysis are reordering, grouping, and lumping. Block iterative methods based on splittings, multilevel methods, and projection methods preconditioned with block iterative methods come across as a strong set of solvers which should be integrated to software packages working with Kronecker products. However, all of this is easier said, than done. Implementation of these solvers requires intricate programming with dynamically allocated, relatively complex data structures, which needs time, careful testing, and tuning.

REFERENCES

[1] APNN-Toolbox. Available at: http://www4.cs.uni-dortmund.de/APNN-TOOLBOX/ Last accessed February 14, 2006.

[2] F. BAUSE, P. BUCHHOLZ, AND P. KEMPER, *A toolbox for functional and quantitative analysis of DEDS*, in Quantitative Evaluation of Computing and Communication Systems, Lecture Notes in Computer Science 1469, R. Puigjaner, N. N. Savino, and B. Serra, eds., Springer Verlag, 1998, pp. 356–359.

[3] A. BENOIT, L. BRENNER, P. FERNANDES, AND B. PLATEAU, *Aggregation of stochastic automata networks with replicas*, Linear Algebra and Its Applications, 386 (2004), pp. 111–136.

[4] A. BENOIT, L. BRENNER, P. FERNANDES, B. PLATEAU, AND W. J. STEWART, *The PEPS software tool*, in Computer Performance Evaluation: Modelling Techniques and Tools, Lecture Notes in Computer Science 2794, P. Kemper and W. H. Sanders, eds., Springer Verlag, 2003, pp. 98–115.

[5] A. BENOIT, P. FERNANDES, B. PLATEAU, AND W. J. STEWART, *On the benefits of using functional transitions and Kronecker algebra*, Performance Evaluation, 58 (2004), pp. 367–390.

[6] P. BUCHHOLZ, *A class of hierarchical queueing networks and their analysis*, Queueing Systems, 15 (1994), pp. 59–80.

[7] P. BUCHHOLZ, *Exact and ordinary lumpability in finite Markov chains*, Journal of Applied Probability, 31 (1994), pp. 59–75.

[8] P. BUCHHOLZ, *Hierarchical Markovian models: symmetries and reduction*, Performance Evaluation, 22 (1995), pp. 93–110.

[9] P. BUCHHOLZ, *An aggregation\disaggregation algorithm for stochastic automata networks*, Probability in the Engineering and Informational Sciences, 11 (1997), pp. 229–253.

[10] P. BUCHHOLZ, *Exact performance equivalence: An equivalence relation for stochastic automata*, Theoretical Computer Science, 215 (1999), pp. 263–287.

[11] P. BUCHHOLZ, *Hierarchical structuring of superposed GSPNs*, IEEE Transactions on Software Engineering, 25 (1999), pp. 166–181.

[12] P. BUCHHOLZ, *Structured analysis approaches for large Markov chains*, Applied Numerical Mathematics, 31 (1999), pp. 375–404.

[13] P. BUCHHOLZ, *Projection methods for the analysis of stochastic automata networks*, in

Numerical Solution of Markov Chains, B. Plateau, W. J. Stewart, and M. Silva, eds., Prensas Universitarias de Zaragoza, Zaragoza, Spain, 1999, pp. 149–168.

[14] P. BUCHHOLZ, *An adaptive aggregation/disaggregation algorithm for hierarchical Markovian models*, European Journal of Operational Research, 116 (1999), pp. 545–564.

[15] P. BUCHHOLZ, *Multilevel solutions for structured Markov chains*, SIAM Journal on Matrix Analysis and Applications, 22 (2000), pp. 342–357.

[16] P. BUCHHOLZ, *Efficient computation of equivalent and reduced representations for stochastic automata*, Computer Systems Science & Engineering, 15 (2000), pp. 93–103.

[17] P. BUCHHOLZ, G. CIARDO, S. DONATELLI, AND P. KEMPER, *Complexity of memory-efficient Kronecker operations with applications to the solution of Markov models*, INFORMS Journal on Computing, 12 (2000), pp. 203–222.

[18] P. BUCHHOLZ AND T. DAYAR, *Block SOR for Kronecker structured Markovian representations*, Linear Algebra and Its Applications, 386 (2004), pp. 83–109.

[19] P. BUCHHOLZ AND T. DAYAR, *Comparison of multilevel methods for Kronecker structured Markovian representations*, Computing, 73 (2004), pp. 349–371.

[20] P. BUCHHOLZ AND T. DAYAR, *Block SOR preconditioned projection methods for Kronecker structured Markovian representations*, SIAM Journal on Scientific Computing, 26 (2005), pp. 1289–1313.

[21] P. BUCHHOLZ AND T. DAYAR, *On the convergence of a class of multilevel methods for large, sparse Markov chains*, Technical Report BU-CE-0601, Department of Computer Engineering, Bilkent University, Ankara, Turkey, 2006. Available at: http://www.cs.bilkent.edu.tr/tech-reports/2006/BU-CE-0601.pdf Last accessed February 14, 2006.

[22] P. BUCHHOLZ AND P. KEMPER, *On generating a hierarchy for GSPN analysis*, Performance Evaluation Review, 26 (1998), pp. 5–14.

[23] P. BUCHHOLZ AND P. KEMPER, *Kronecker based representations of large Markov chains*, in Validation of Stochastic Systems, Lecture Notes in Computer Science 2925, B. Haverkort, H. Hermanns, and M. Siegle, eds., Springer Verlag, 2004, pp. 256–295.

[24] J. CAMPOS, S. DONATELLI, AND M. SILVA, *Structured solution of asynchronously communicating stochastic models*, IEEE Transactions on Software Engineering, 25 (1999), pp. 147–165.

[25] R. H. CHAN AND W. K. CHING, *Circulant preconditioners for stochastic automata networks*, Numerische Mathematik, 87 (2000), pp. 35–57.

[26] M.-Y. CHUNG, G. CIARDO, S. DONATELLI, N. HE, B. PLATEAU, W. STEWART, E. SULAIMAN, AND J. YU, *A comparison of structural formalisms for modeling large Markov models*, in Proceedings of the 18th International Parallel and Distributed Processing Symposium, IEEE CS-Press, 2004, pp. 196b.

[27] G. CIARDO AND A. S. MINER, *A data structure for the efficient Kronecker solution of GSPNs*. in Proceedings of the 8th International Workshop on Petri Nets and Performance Models, P. Buchholz and M. Silva, eds., IEEE CS-Press, 1999, pp. 22–31.

[28] G. CIARDO, R. L. JONES, A. S. MINER, AND R. SIMINICEANU, *Logical and stochastic modeling with SMART*, in Computer Performance Evaluation: Modelling Techniques and Tools, Lecture Notes in Computer Science 2794, P. Kemper and W. H. Sanders, eds., Springer Verlag, 2003, pp. 78–97.

[29] G. CLARK, S. GILMORE, J. HILLSTON, AND N. THOMAS, *Experiences with the PEPA performance modelling tools*, IEE Software, 146 (1999), pp. 11–19.

[30] M. DAVIO, *Kronecker products and shuffle algebra*, IEEE Transactions on Computers, C-30 (1981), pp. 116–125.

[31] T. DAYAR, *State space orderings for Gauss-Seidel in Markov chains revisited*, SIAM Journal on Scientific Computing, 19 (1998), pp. 148–154.

[32] T. DAYAR, *Effects of reordering and lumping in the analysis of discrete-time SANs*, in

Mathematics and Computer Science: Algorithms, Trees, Combinatorics and Probabilities, D. Gardy and A. Mokkadem, eds., Birkhauser Verlag, 2000, pp. 209–220.

[33] T. DAYAR AND W. J. STEWART, *Comparison of partitioning techniques for two-level iterative solvers on large, sparse Markov chains*, SIAM Journal on Scientific Computing, 21 (2000), pp. 1691–1705.

[34] S. DONATELLI, *Superposed stochastic automata: a class of stochastic Petri nets with parallel solution and distributed state space*, Performance Evaluation, 18 (1993), pp. 21–26.

[35] P. FERNANDES, B. PLATEAU, AND W. J. STEWART, *Efficient descriptor-vector multiplications in stochastic automata networks*, Journal of the ACM, 45 (1998), pp. 381–414.

[36] P. FERNANDES, B. PLATEAU, AND W. J. STEWART, *Optimizing tensor product computations in stochastic automata networks*, RAIRO Operations Research, 32 (1998), pp. 325–351.

[37] J.-M. FOURNEAU, L. KLOUL, N. PEKERGIN, F. QUESSETTE, AND V. VÉQUE *Modelling buffer admission mechanisms using stochastic automata networks*, Revue Annales des Télécommunications, 49 (1994), pp. 337–349.

[38] J.-M. FOURNEAU, H. MAISONNIAUX, N. PEKERGIN, AND V. VÉQUE, *Performance evaluation of a buffer policy with stochastic automata networks*, in IFIP Workshop on Modelling and Performance Evaluation of ATM Technology, vol. C-15, La Martinique, IFIP Transactions North-Holland, Amsterdam, Holland, 1993, pp. 433–451.

[39] O. GUSAK AND T. DAYAR, *Iterative aggregation-disaggregation versus block Gauss-Seidel on continuous-time stochastic automata networks with unfavorable partitionings*, in Proceedings of the 2001 International Symposium on Performance Evaluation of Computer and Telecommunication Systems, M. S. Obaidat and F. Davoli, eds., Orlando, Florida, 2001, pp. 617–623.

[40] O. GUSAK, T. DAYAR, AND J.-M. FOURNEAU, Stochastic automata networks and near complete decomposability, SIAM Journal on Matrix Analysis and Applications, 23 (2001), pp. 581–599.

[41] O. GUSAK, T. DAYAR, AND J.-M. FOURNEAU, *Lumpable continuous-time stochastic automata networks*, European Journal of Operational Research, 148 (2003), pp. 436–451.

[42] O. GUSAK, T. DAYAR, AND J.-M. FOURNEAU, *Iterative disaggregation for a class of lumpable discrete-time stochastic automata networks*, Performance Evaluation, 53 (2003), pp. 43–69.

[43] S. HADDAD AND P. MOREAUX, *Asynchronous composition of high-level Petri nets: a quantitative approach*, in Proceedings of the 17th International Conference on Application and Theory of Petri Nets, Lecture Notes in Computer Science 1091, J. Billington and W. Reisig, eds., Springer Verlag, 1996, pp. 192–211.

[44] J. HILLSTON AND L. KLOUL, *An efficient Kronecker representation for PEPA models*, in Proceedings of the 1st Process Algebras and Performance Modeling, Probabilistic Methods in Verification Workshop, Lecture Notes in Computer Science 2165, L. de Alfaro and S. Gilmore, eds., Springer Verlag, 2001, pp. 120–135.

[45] G. HORTON AND S. LEUTENEGGER, *A multi-level solution algorithm for steady state Markov chains*, Performance Evaluation Review, 22 (1994), pp. 191–200.

[46] J. G. KEMENY AND J. L. SNELL, *Finite Markov Chains*, Springer-Verlag, New York, 1983.

[47] P. KEMPER, *Numerical analysis of superposed GSPNs*, IEEE Transactions on Software Engineering, 22 (1996), pp. 615–628.

[48] U. KRIEGER, *Numerical solution of large finite Markov chains by algebraic multigrid techniques*, in Computations with Markov Chains, W. J. Stewart, ed., Kluwer Academic Publishers, Boston, Massachusetts, 1995, pp. 403-424.

[49] A. N. LANGVILLE AND W. J. STEWART, *The Kronecker product and stochastic automata networks*, Journal of Computational and Applied Mathematics, 167 (2004), pp. 429–

447.

[50] A. N. LANGVILLE AND W. J. STEWART, *Testing the nearest Kronecker product precon- ditioner on Markov chains and stochastic automata networks*, Informs Journal on Computing, 16 (2004), pp. 300–315.

[51] A. N. LANGVILLE AND W. J. STEWART, *A Kronecker product approximate precon- ditioner for SANs*, Numerical Linear Algebra with Applications, 11 (2004), pp. 723–752.

[52] R. PAIGE AND R. E. TARJAN, *Three partition refinement algorithms*, SIAM Journal on Computing, 16 (1987), pp. 973–989.

[53] PEPA Home Page. Available at: http://www.dcs.ed.ac.uk/pepa/tools/ Last accessed February 14, 2006.

[54] PEPS Home Page. Available at: http://www-id.imag.fr/Logiciels/peps Last accessed February 14, 2006.

[55] B. PLATEAU, *On the stochastic structure of parallelism and synchronization models for distributed algorithms*, in Proceedings of the ACM SIGMETRICS Conference on Measurement and Modelling of Computer Systems, Austin, Texas, 1985, pp. 147– 154.

[56] B. PLATEAU AND K. ATIF, *Stochastic automata network for modeling parallel systems*, IEEE Transactions on Software Engineering, 17 (1991), pp. 1093–1108.

[57] B. PLATEAU AND J.-M. FOURNEAU, *A methodology for solving Markov models of paral- lel systems*, Journal of Parallel and Distributed Computing, 12 (1991), pp. 370–387.

[58] B. PLATEAU, J.-M. FOURNEAU, AND K.-H. LEE, *PEPS: A package for solving complex Markov models of parallel systems*, in Modeling Techniques and Tools for Computer Performance Evaluation, R. Puigjaner and D. Ptier, eds., Palma de Mallorca, Spain, 1988, pp. 291–305.

[59] B. PLATEAU AND K. TRIPATHI, *Performance analysis of synchronization for two com- municating processes*, Performance Evaluation, 8 (1988), pp. 305–320.

[60] M. SCARPA AND A. BOBBIO, *Kronecker representation of stochastic Petri nets with discrete PH distributions*, in Proceedings of the IEEE International Computer Per- formance and Dependability Symposium, IEEE CS-Press, 1998, pp. 52–61.

[61] SMART Project Home page. Available at: http://www.cs.ucr.edu/~ciardo/SMART Last accessed February 14, 2006.

[62] W. J. STEWART, *Introduction to the Numerical Solution of Markov Chains*, Princeton University Press, Princeton, New Jersey, 1994.

[63] W. J. STEWART, K. ATIF, AND B. PLATEAU, *The numerical solution of stochastic automata networks*, European Journal of Operational Research, 86 (1995), pp. 503– 525.

[64] E. UYSAL AND T. DAYAR, *Iterative methods based on splittings for stochastic automata networks*, European Journal of Operational Research, 110 (1998), pp. 166–186.

[65] C. F. VAN LOAN, *The ubiquitous Kronecker product*, Journal of Computational and Applied Mathematics, 123 (2000), pp. 85–100.

[66] V. VÈQUE AND J. BEN-OTHMAN, *MRAP: A multiservices resource allocation policy for wireless ATM network*, Computer Networks and ISDN Systems, 29 (1998), pp. 2187–2200.

[67] P. WESSELING, *An Introduction to Multigrid Methods*, Wiley, Chichester, New York, 1992.

STRUCTURED STOCHASTIC MODELING AND PERFORMANCE ANALYSIS OF A MULTIPROCESSOR SYSTEM[*]

I. SBEITY[†] AND B. PLATEAU [‡]

Abstract. With excellent cost/performance tradeoff and good scalability, multiprocessor systems are becoming attractive alternatives when high performance, reliability and availability are needed. They are now more popular in universities, research labs and industries. In these communities, life-critical applications requiring high degrees of precision and performance are executed and controlled. Thus, it is important to the developers of such applications to analyze during the design phase how hardware, software and performance related failures affect the quality of service delivered to the users. This analysis can be conducted using modeling techniques such as transition systems. However, the high complexity of such systems (large state space) makes them difficult to analyze. This work presents an efficient way to model and analyze multiprocessor system in a structured and compact manner using a Stochastic Automata Network (SAN). A SAN is a high-level formalism for modeling very large and complex Markov chains. The formalism permits complete systems to be represented as a collection of interacting sub-systems. The basic concept which renders SAN powerful is the use of tensor algebra for its representation and analysis. Furthermore, a new modeling alternative has been recently incorporated into SANs: the use of phase-type distributions, which remains a desirable objective for the more accurately modeling of numerous real phenomena such as the repair and service time in multiprocessor systems.

Key words. Multiprocessor System, Stochastic Automata Networks, Phase-Type Distribution.

1. Introduction. Over the last decade, cluster systems, distributed and multiprocessor systems have emerged as popular computing platforms for computationally intensive applications with diverse computing needs [2]. In constructing such systems, QoS (Quality of Service) control mechanisms are incorporated, but the evaluation of systems with QoS control has several problems that must be solved. One of the problems is to build and analyze stochastic models for the availability of the system which incorporate task deadlines, system processing time, and so on. Almost all traditional evaluations use exponential probability distributions functions. Such distribution functions facilitate the evaluation. However, several producible events in multiprocessor systems, e.g., repair and service time, cannot be represented sufficiently accurately by the exponential distribution. These events follow a general non-Markovian distribution, and the exponential approximation of these general distributions may not be sufficiently accurate. An efficient way to handle this problem is to approximate a general distribution by a phase-type (PH) distribution. The exponential distribution is widely used in performance modeling. One of the most obvious reasons is the exceptional mathematical tractability that flows from the memoryless property of this distribution. But sometimes, mathematical tractability is not sufficient to overcome the need to model processes for which the exponential distribution is simply not adequate [13]. This leads us to explore ways in which we can model more general distributions while maintaining some of the mathematical advantages of the exponential. This is precisely what PH distributions allow us to do. Additionally, PH distributions prove to be very useful when it is necessary to form a distribution having some given expectation and variance.

---

[*]This work is supported by ACI Sure-Paths project.

[†]Informatique et Distribution - 51, Av. Jean Kuntzman - 38330 Montbonnot, France (`ihab.sbeity@imag.fr`).

[‡]Informatique et Distribution - 51, Av. Jean Kuntzman - 38330 Montbonnot, France (`brigitte.plateau@imag.fr`).

Even though PH distribution modeling methods work well in theory, they partly suffer from the problem that the state space of the models can grow exponentially with the number of phases, a behavior called state-space explosion. Most of the previous research in PH distributions has concentrated chiefly on the mathematical part of the approximation (PH fitting), but only marginally on the performance, or practical feasibility of the results. The research with PH performance modeling gave rise to what has been called Markovian Arrival Processes (MAPs) [13, 10]. However, the analyzing procedure is still limited by the size of the state space. PH fitting consists of adjusting the parameters of a PH distribution until it best matches the required general distribution, where best is often defined by maximum likelihood. Several methods for fitting the parameters of a PH distribution have been presented. Asmussen, Nerman, and Olsson proposed a fitting method for general PH distributions based on the expectation-maximization (EM) algorithm [1]. Based on earlier work from [8], the authors of [7] developed an algorithm to fit the parameters of a hyperexponential distribution to values of a given general distribution. The resulting approach is very efficient and yields good fitting results for heavy tailed distributions with monotonically decreasing density functions. However, the use of hyperexponential distributions restricts the class of distributions which can be represented, since hyperexponential distributions cannot adequately capture general distributions with increasing and decreasing densities or with a coefficient of variation less than one. However, PH fitting is not the objective of our work. In this paper, general distribution will be approximated by an Erlang distribution which can more efficiently approximate a tightly tailed distribution [12] such as the service and repair time distribution in our model. Nevertheless, other PH distributions can be also easily modeled.

To handle the state-space explosion problem, we suggest the use of Stochastic Automata Networks (SANs) [14] which may be used to develop performance models of parallel and distributed computer systems [3, 11]. The SAN formalism is usually quite attractive when modeling a system with several parallel cooperative activities. An important advantage of the SAN formalism is that efficient numerical algorithms have been developed to compute stationary and transient measures [17, 16, 9]. These algorithms take advantage of structured and modular definitions which allow the treatment of considerably larger models. Another important advantage of the SAN formalism is the recent possibility of modeling and analysing systems with PH distribution [15]. For these reasons, we believe SANs are more than adequate for describing the cooperative activities and general distributions of multiprocessor systems.

This paper describes how SANs may be used to model and analyse multiprocessor systems using different probability distribution functions. It also presents the impact of the use of PH distribution, such as the Erlang distribution, instead of exponential distribution when describing the repair and service time of such systems. In Section 2, we briefly describe the Stochastic Automata Network formalism. Section 3 presents the multiprocessor systems which will be modeled and analyzed. Section 4 presents the description of the system using SANs and analyzes the results. Finaly, we conclude our paper in Section 5.

2. Stochastic Automata Network. Stochastic Automata Networks, SANs, were first proposed by Plateau in 1985 [14]. The SAN formalism enables a complete system to be represented as a collection of interacting subsystems. Each subsystem is represented by an automaton which is simply a directed and labeled graph whose states are referred to as *local states*, being local to that subsystem, and whose edges, relating local states to one another, are labeled with probabilistic and event information. The different subsystems apply this label information to enable them to interact with each other and to coordinate their behavior.

The states of a SAN are defined by the cartesian product of the *local states* of the automata and are called the *global states* of the SAN. Thus, a global state may be described by a vector whose i^{th} component denotes the local state occupied by the ith automaton. The global state of a SAN is altered by the occurrence (referred to as the *firing*) of an *event*. Each event has a unique *identifier* and a *firing rate*. At any moment, multiple events may be enabled to fire (we shall also use the word *fireable* to describe events that are enabled): the one which actually fires is determined in a Markovian fashion, i.e., from the relative firing rates of those which are enabled. The firing of an event changes a given global *source* state into a global *destination* state. An event may be one of two different types. A *local event* causes a change in one automaton only, so that the global source and destination states differ in one component (local state) only. A *synchronizing event*, on the other hand, can cause more than one automaton to simultaneously change its state with the result that the global source and destination states may differ in multiple components. Indeed, each synchronizing event is associated with multiple automata and the occurrence of a synchronizing event forces all automata associated with the event to simultaneously change state in accordance with the dictates of this synchronizing event on each implicated automata. Naturally, a synchronizing event must be enabled in *all* of the automata on which it is defined before it can fire.

Transitions from one local state to another within a given automaton are not necessarily in one-to-one correspondence with events: several different events may occasion the same local transition. Furthermore, the firing of a single event may give rise to several possible destinations on leaving a local source state. In this case, *routing probabilities* must be associated with the different possible destinations. Routing probabilities may be omitted only if the firing of an event gives rise to a transition having a single destination. Also, automata may interact with one another by means of *functional rates*: the firing rate of any event may be expressed, not only as a constant value (a positive real number), but also as a function of the state of other automata. Functional rates are defined within a single automaton, even though their parameters involve the states of other automata.

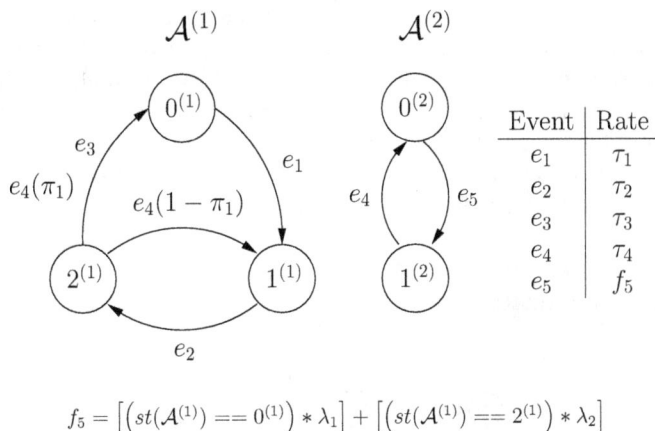

$$f_5 = \left[\left(st(\mathcal{A}^{(1)}) == 0^{(1)}\right) * \lambda_1\right] + \left[\left(st(\mathcal{A}^{(1)}) == 2^{(1)}\right) * \lambda_2\right]$$

FIG. 2.1. *Example of a* SAN *model*

As an example of the previous discussion, Figure 2.1 presents a SAN model with two automata, $\mathcal{A}^{(1)}$ and $\mathcal{A}^{(2)}$, the first with 3 local states, $0^{(1)}$, $1^{(1)}$ and $2^{(1)}$ and the second with two local states, $0^{(2)}$ and $1^{(2)}$. The model contains four local events, e_1, e_2, e_3 and e_5 and one synchronizing event, e_4. When automaton $\mathcal{A}^{(1)}$ is in local state $0^{(1)}$, and $\mathcal{A}^{(2)}$ is in local state $0^{(2)}$, (global state $[0,0]$), two events are eligible to fire, namely e_1 and e_5. The event e_1 fires at rate τ_1. This is taken to mean that the random variable which describes the time t from the moment that automaton $\mathcal{A}^{(1)}$ moves into state $0^{(1)}$ until the event e_1 fires, taking it into state $1^{(1)}$, is exponentially distributed with probability density function given by $\tau_1 e^{-\tau_1 t}$. Similar remarks hold for the firing rate of the other events. The firing of e_1 when the system is in global state $[0,0]$ moves it to global state $[1,0]$ in which e_5 is still eligible to fire, along now with event e_2. The event e_1 cannot fire from this new state. The synchronizing event e_4 is enabled in global state $[2,1]$ and when it fires it changes automaton $\mathcal{A}^{(2)}$ from state $1^{(2)}$ to state $0^{(2)}$ while simultaneously changing automaton $\mathcal{A}^{(1)}$ from state $2^{(1)}$ to either state $0^{(1)}$, with probability π_1, or to state $1^{(1)}$ with probability $1 - \pi_1$. Observe that two events are associated with the same edge in automaton $\mathcal{A}^{(1)}$, namely e_3 and e_4. If event e_3 fires, then the first automaton will change from state $2^{(1)}$ to state $0^{(1)}$; if event e_4 fires, then the first automaton can change from state $2^{(1)}$ to either state $0^{(1)}$ or state $1^{(1)}$ as previously described. There is one functional rate, f_5, the rate at which event e_5 fires, defined as

$$
f_5 = \begin{cases} \lambda_1 & \text{if } \mathcal{A}^{(1)} \text{ is in state } 0^{(1)} \\[2mm] 0 & \text{if } \mathcal{A}^{(1)} \text{ is in state } 1^{(1)} \\[2mm] \lambda_2 & \text{if } \mathcal{A}^{(1)} \text{ is in state } 2^{(1)} \end{cases}
$$

Thus event e_5, which changes the state of automaton $\mathcal{A}^{(2)}$ from $0^{(2)}$ to $1^{(2)}$, fires at rate λ_1 if the first automaton is in state $0^{(1)}$ or at rate λ_2 if the first automaton is in state $2^{(1)}$. The event e_5 is prohibited from firing if the first automaton is in state $1^{(1)}$. Functional transitions are written more compactly, e.g.,

$$
f_5 = \left[(st(\mathcal{A}^{(1)}) == 0^{(1)}) * \lambda_1 \right] + \left[(st(\mathcal{A}^{(1)}) == 2^{(1)}) * \lambda_2 \right]
$$

in which conditions such as $st(\mathcal{A}^{(1)} == 2^{(1)})$ (which means "the state of $\mathcal{A}^{(1)}$ is $2^{(1)}$") have the value 1 if the condition is true and are equal to 0 otherwise. This is the notation used to describe functions in the PEPS software tool [4]. In this setting, the interpretation of a function may be viewed as the evaluation of an expression in the C programming language. The use of functional expressions in SANs is not limited to the rates at which events occur; indeed, probabilities also may be expressed as functions. Figure 2.2 shows the equivalent Markov chain transition rate diagram for this example.

The transition matrix of the Markov chain described by a SAN is given in the following tensor [5] format:

$$
(2.1) \qquad Q = \bigoplus_{i=1}^{N} Q_l^{(i)} + \sum_{e \in \mathcal{ES}} \left(\bigotimes_{i=1}^{N} Q_{e+}^{(i)} + \bigotimes_{i=1}^{N} Q_{e-}^{(i)} \right).
$$

In this equation, each matrix $Q_l^{(i)}$ corresponds to the local transitions of automaton $\mathcal{A}^{(i)}$. $\mathcal{ES}$ is the set of synchronizing events and matrices $Q_{e+}^{(i)}$ correspond to the

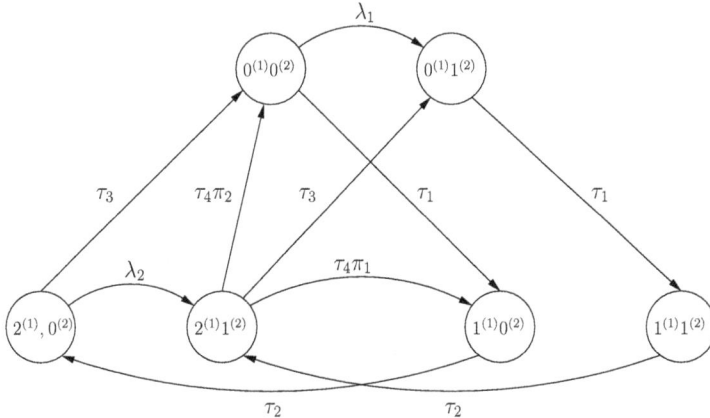

FIG. 2.2. *Transition rate diagram of corresponding Markov chain.*

synchronizing transitions of automaton $\mathcal{A}^{(i)}$ which may be fired by event e. The matrix $Q_{e^-}^{(i)}$ is used for the normalization of the matrix $Q_{e^+}^{(i)}$. This tensor formula is referred to as the *descriptor* of the stochastic automata network [14, 17, 16, 9]. The real interest in developing stochastic automata networks lies, not so much in their specification, although this can be extremely useful in certain circumstances, but rather in the fact that the transition matrix of the underlying Markov chain of a SAN can be represented compactly by storing the elementary matrices of the descriptor and never the transition matrix itself.

Furthemore, a new methodology has been recently incorporated into SANs: the use of phase-type distributions [15]. With the exception of some rare research such as [6], the exponential distribution has been the only distribution used to model the passage of time in the evolution of the different SAN components. In [15], it is shown how *phase-type* distributions may be incorporated into SANs thereby providing the wherewithal by which arbitrary distributions can be used, which in turn leads to an improved ability for more accurately modeling numerous real phenomena. A tensor representation of a SAN model with *phase-type* distributions has been also provided.

3. System architecture. The multiprocessor system architecture which we consider, is a queueing model of an n-site distributed system in which n sites are connected via a network to process independent tasks submitted by m users. The system model, depicted in Figure 3.1, is composed of a task schedule queue of size K, a task scheduler, and n sites $S_1, S_2, \cdots, S_n$ (processors). We assume that the function of the scheduler is to make appropriate task allocation decisions for each arriving task that satisfies a mean average of service time.

In this system, a major feature is the possibility of a site to fail. More precisely, we consider a *crash fault* assumption [2] of sites where a failed site cannot continue to carry out its tasks until it is repaired. Furthermore, faults that may affect a site during its active state are of two types: *transient* or *permanent*. A transient fault can be simply treated by rebooting the failed site. However, a permanent fault implies a relatively longer repair phase of the failed site. The occurrence process of faults is considered to be exponential, and depends on the number of active sites. Moreover, we assume that when the number of active sites decreases, the probability of failure of a given active site increases because of the high workload on this site. Thus the failure

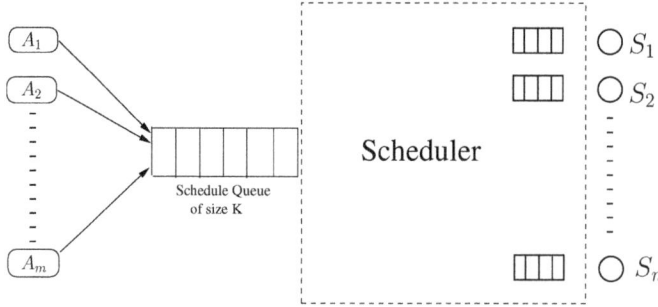

FIG. 3.1. *The model*

rate is given by $flt \times 1/(\# \ active \ sites)$, where $1/flt$ is the mean of an exponential distribution. With probability p, a failure is transient while with probability $1 - p$ the failure is permanent. Subsequent to a transient failure, a site becomes active after a rebooting delay that is considered also to be exponential with mean $1/reb$. However, the repair time is not exponential, but rather it follows a general non-Markovian distribution with mean $1/rep$.

In addition, we assume a Poisson arrival process of tasks to the schedule queue with rate λ. The service time follows a general distribution and depends on the number of active sites. Moreover, we consider the variation of service time to be small, meaning that service times of each task are very close. Thus, the service rate is given as $\mu \times (\# \ active \ sites)$, where $1/\mu$ is the mean of the general distribution of service time whose coefficient of variation is very small (< 1).

The use of general distributions remains a desirable objective for the more accurate modeling of numerous real phenomena. However, such distributions are difficult to analyze. One possible approach is to replace a general distribution by an exponential which can be easily analyzed mathematically. However, the use of the exponential distribution is not appropriate in all cases. Specifically, the exponential approximation of a general distribution with a very small/large coefficient of variation (cv^2) may be not robust. Such distributions are called heavy-tailed distributions (with large coefficient of variation) and tightly-tailed distributions (with small coefficient of variation). The coefficient of variation is given as:

$$cv^2 = \frac{Var(X)}{E[X]^2}$$

where X is the random variable of the distribution, $Var(X)$ its variance and $E[X]$ its mean.

Another possibility for including non-Markovian distributions into a model, is to approximate them by a phase-type distribution PH (PH fitting). Popular examples of phase-type distributions include the Erlang distribution, the hyperexponential distribution and the Coxian distribution. Indeed, any distribution having a rational Laplace transform can be modeled arbitrarily closely with a Coxian distribution. In our experiments, we shall replace general distributions by an Erlang distribution (Figure 3.2) which can more efficiently represent a tightly tailed distribution [12] such as the service and repair time distributions in our model. In this Figure, state i ($i = 1$, ..., N) represents the ith phase of the Erlang distribution. State **A** represents the absorption state of the distribution.

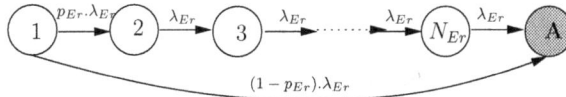

FIG. 3.2. *Representation of an Erlang-N_{Er} distribution*

More precisely, the coefficient of variation of an Erlang distribution decreases when the number of Erlang phases increases, as illustrated in the following equation:

$$cv_{Er}^2 = \frac{N_{Er}/\lambda_{Er}^2}{(N_{Er}/\lambda_{Er})^2} = \frac{1}{N_{Er}} \xrightarrow{N_{Er} \to \infty} 0$$

This implies that an Erlang distribution with a large value of N_{Er} ($N_{Er} \to \infty$) can represent a deterministic distribution whose coefficient of variation is zero ($cv^2 = 0$).

As mentioned in the introduction, PH distribution are not commonly used in stochastic modeling because of the well-known problem of state-space explosion generated by the use of this distribution. One of the main goals of the work presented here is to show the, almost *positive*, impact of the use of PH distribution to model the service and repair time of our system. Performance and result quality will be analyzed using the SAN formalism, which can easily handle the problem of state-space explosion to model the system. For our model, we will use Erlang distribution to approximate general distributions. However, other PH distributions can be also easily modeled [15].

4. SAN model. This section describes the manner used to translate the system presented above into a SAN model. First, we consider a sizing based on a pure availability (failure/repair) model. Later the performance model is presented. The advantage of this two step modeling process will become apparent in the following sections.

4.1. Availability model. As presented, a site of the system may be in any one of three different states. Figure 4.1 represents the SAN model of the site S_i in the system. The model is composed of one automata **S[i]** which models the three previous states of the site S_i, and a table of events. Indeed, the automaton **S[i]** is composed of the three states: *active* which presents the state when the site S_i works normally – it processes tasks delivered from the scheduler; *Perm_Failure* which models the situation where the site is in a permanent failure; and *Tran_Failure* which presents the transient failure state. Suppose that automaton **S[i]** is in its *active* state. After an exponentially distributed amount of time spent in this state, event $fault[i]$ fires and automaton **S[i]** has the possibility of moving either to state *Tran_Failure* or to state *Perm_Failure* with probabilities p and $1 - p$ respectively. If the automaton moves to state *Tran_Failure*, meaning that a transient failure has occurred, it will move back to the *active* state after an exponentially distributed amount of time spent in the *Tran_Failure* state. This is presented by the occurrence of event $reboot[i]$, meaning that the site has been rebooted. On the other hand, if event $fault[i]$ fires and automaton **S[i]** moves to the permanent failure state *Perm_Failure*, it only moves back to its *active* state after an arbitrary sejourn time in state *Perm_Failure*. This sojourn time follows a general distribution with mean $1/rep$.

In the case of an exponential repair time distribution, the global SAN availability model is composed of n automata **S[1]**, **S[2]**, $\cdots$, **S[n]**, which model the sites of the

$$S[i]$$

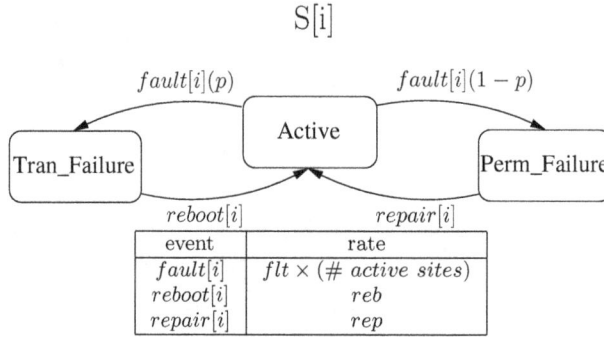

event	rate
$fault[i]$	$flt \times (\# \ active \ sites)$
$reboot[i]$	reb
$repair[i]$	rep

FIG. 4.1. *The SAN model - site i*

system. However, when the repair time distribution is an Erlang-N_{Er}, we need to add n other automata for modeling the Erlang-N_{Er} representation [15].

Er[i]

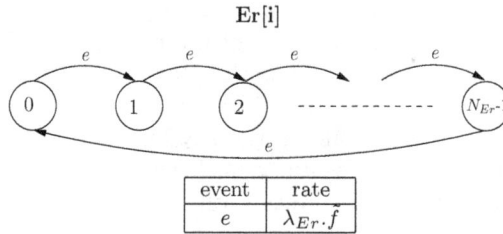

event	rate
e	$\lambda_{Er}.\tilde{f}$

With :

- $\lambda_{Er} = \frac{rep}{N_{Er}}$, $p_{Er} = 1$

- $\tilde{f} = \begin{cases} 1 & \text{if the site is in Perm\_Failure state} \\ 0 & \text{otherwise} \end{cases}$

FIG. 4.2. *The SAN model - Erlang representation*

Figure 4.2 represents the SAN automaton, **Er[i]**, which models the Erlang-N_{Er} repair time distribution of the site S_i. The automaton **Er[i]** is composed of N_{Er} states, denoted 0, 1, ..., N_{Er}. In this automaton, state 0 models initially the state where the distribution is not activated, because the site S_i is not in its *Perm_Faiure* state: the booleen function $\tilde{f}$ associated with the only possible event e, is equal to 0 in this situation. When the site (automaton **S[i]**) moves to the permanent failure state *Perm_Faiure*, the value of the function $\tilde{f}$ is now equal to 1. This means that the Erlang delay in the state *Perm_Faiure* is also activated. The state 0 of the automaton **Er[i]** moves then to represent the first phase of the distribution. After an exponential amount of time with rate λ_{Er}, event e fires and automaton **Er[i]** moves to state 1 which represents the second phase of the distribution, and so on. Finally, when the exponential sojourn time in the last phase of the Erlang distribution, i.e. state $N_{Er}-1$, is finished, automaton **Er[i]** moves back to state 0. This also implies the end of the Erlang sojourn in state *Perm_Faiure* of automatan **S[i]** which goes to its *active* state as described previously. Intuitively, this means that transitions from state $N_{Er} - 1$ to state 0 inside the automaton **Er[i]** must be synchronized with transitions from state

Perm_Faiure to state *active* of automaton **S[i]**. In [15], we introduced a systematic procedure for applying the connection between basic automata (such as automatan **S[i]**) and automata which model a PH distribution (such as automatan **Er[i]**), while considering different preemption policies of the PH distribution. This connection is transparent to the SAN modeler who just has to describe the PH distributions of a model separately, by simple automata, and without establishing the relation with the base model. This last action is done automatically in the PEPS software tool [4].

In addition, we consider three different definitions of the global failure of the system. The system is said to be in a DOWN state in the following cases:

Case 1 All-failed : system is down when all sites are not *active*

Case 2 Half-failed : if at least half the sites are not *active*

Case 3 One-failed : if at least one site is not *active*

These three assumptions are realistic. In fact, we may consider that *Case* 1 is the case of persistent systems, *Case* 2 is that of systems with data duplication and *Case* 3 is that of highly sensitive systems.

To analyze the availability of our system, we calculate the probability of global failure of the system under these three definitions of global failure. Figure 4.3 shows the results of plotting the probability of global failure against the number of sites. Table 4.1 presents the value of the parameters of the model. Moreover, for each global failure hypothesis, we show three curves which correspond to the results obtained using respectively, an exponential distribution, an Erlang-3 distribution and an Erlang-8 distribution to replace the general distribution of the repair time. In all three cases, the mean of the substitute distribution is taken as $1/rep$. This permits us to study the influence of the diminuation of the coefficient of the variation on the availability of the system.

TABLE 4.1

Values of parameters of the availability model. The experimental results are given in Figure 4.3.

parameter	value	comment
flt	10^{-5}	the mean time to failure of each site is
		$10^5 \times$ (# *active sites*) minutes – at least 10^5 minutes
reb	1	the mean reboot time is one minute
rep	0.033	the mean repair time is about 30 minutes
p	0.9	a transient failure is more probable
		than a permanent failure.

An important result which can be observed from Figure 4.3 is that all three curves lie on top of each other. This means that the difference in variation of the distributions does not have any impact whatsoever on the stationary solution of the model, if they have the same mean. In other words, the system is not sensitive to the variation of the distribution. Additionally, it is obvious that the probability of global failure is larger when we consider the "One-failed" hypothesis. This probability decreases relatively when assuming the "Half-failed" hypothesis, and it is smallest under the "All-failed" hypothesis.

4.2. Performance model. Our observations in the previous section permit us to conclude that the exponential distribution is robust when representing the repair time distributions. This is taken into account in our performance model presented in this section. The state space of the model is now smaller because the product factor originating from the Erlang distribution of the availability model has been eliminated.

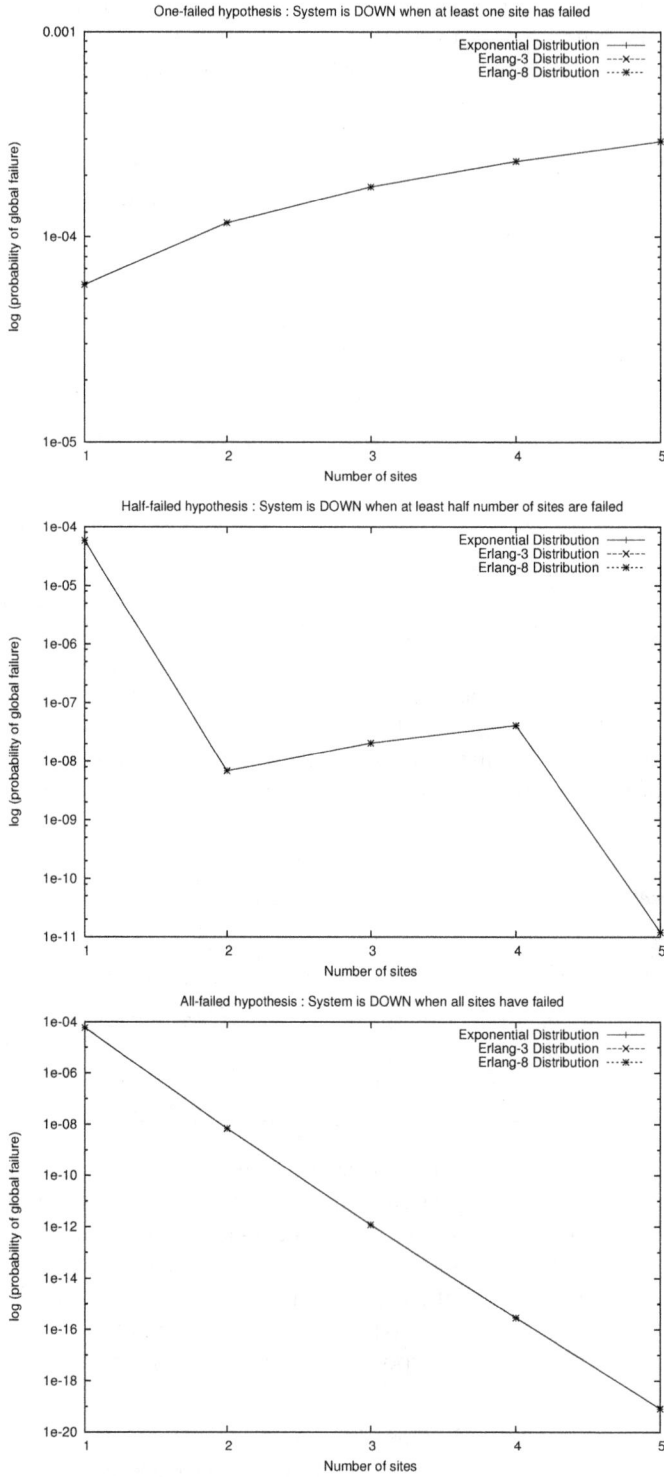

FIG. 4.3. *Availability Analysis: different distributions for the three global failure hypothesis*

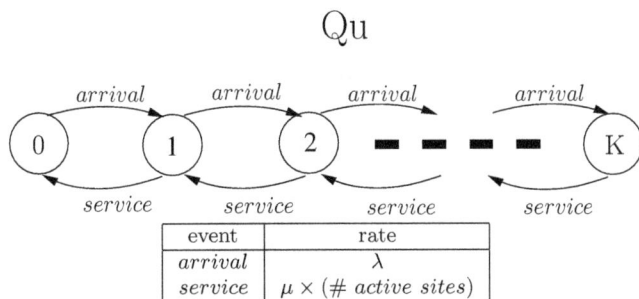

event	rate
arrival	λ
service	$\mu \times (\# \ active \ sites)$

FIG. 4.4. *The SAN model - task schedule queue*

Figure 4.4 presents the SAN model which corresponds to the task schedule queue. This model is composed of one automaton **Qu** and a table of events. Automaton **Qu** is composed of K states where state i, $i = 0, 1, \cdots, K$, models the situation when there are i waiting tasks in the queue. An *arrival* to the queue, with an exponential rate λ increases the number of tasks in the queue. The *service* event has a rate $\mu \times (\# \ active \ sites)$, where $1/\mu$ is the mean of a general distribution.

Thus, the total SAN performance model is composed of the n automata **S[i]** of the availability model and the automaton **Qu**. As mentioned previously, we assume that the repair time distributions (in automata **S[i]**) are exponential because other approximate distribution gives rise to nearly identical results. However, to model the service time distribution, we proceed to replace it respectively by exponential and Erlang distributions as in the availability model. In the case of Erlang service time distributions, we need to add one automaton to model this Erlang representation, as it was the case for the repair time distribution of the availablity model.

In Figure 4.5, we present the curves corresponding to the mean average number of waiting tasks in the queue (a) and the probability of overflow of the system (b), by varying the number of sites. Table 4.2 gives the value of parameters of the model in our experiments. The Figure shows the influence of the variation on the distribution on the results. Indeed, the exponential distribution is not adequate when it is used to replace a service time distribution with a small coefficient of variation. The Erlang distribution is a better approximation than the exponential when the general service time distribution has small variation.

In addition, Figure 4.5 (a) shows that starting from a number of sites greater than 4, Erlang-3 and Erlang-8 curves approach each other when computing the mean number of waiting tasks. This observation suggests the use of Erlang-3 to analyse the situation when the number of sites is greater than 4. This has the obvious impact of reducing the state space. This is not the case when analyzing the probability of overflow in the system as seen in Figure 4.5 (b). Even for five sites, Erlang-3 is still far from identical with Erlang-8. The use of Erlang-3 in this case does not produce an efficient approximation when used to approximate a distribution with a small coefficient of variation. Recall that the use of an Erlang-8 has the disadvantage of increasing the size of the state-space. However, PH distributions other than the Erlang and having a smaller number of phases, such as the Coxian distribution may be a better fit. This leads us to exploit and take advantage of several methods for fitting PH parameters, as suggested in the introduction.

In [18], the authors investigated the value of the minimal number of sites (processors) which guarantees some given QoS, while using only the exponential distributions

TABLE 4.2

Values of parameters of the performance model. The experimental results are given in Figure 4.5.

parameter	value	comment
λ	5	
μ	3	
ρ	$\frac{\lambda}{\mu} \times \frac{1}{\# \ active \ sites}$	ρ is the workload of the system

(a)

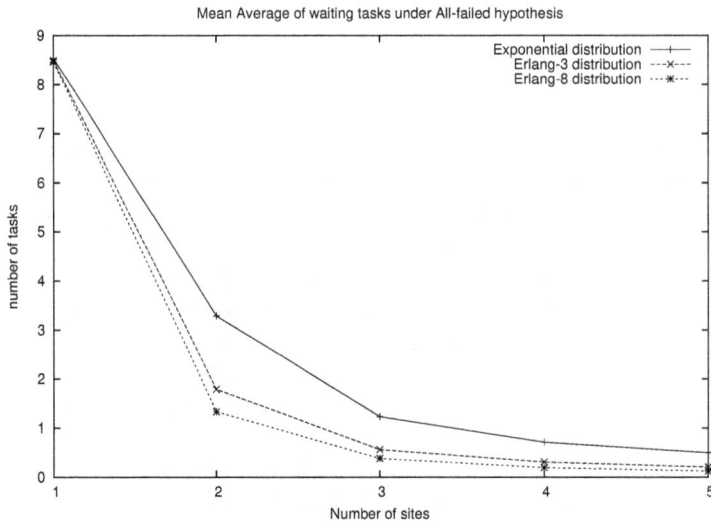

(b)

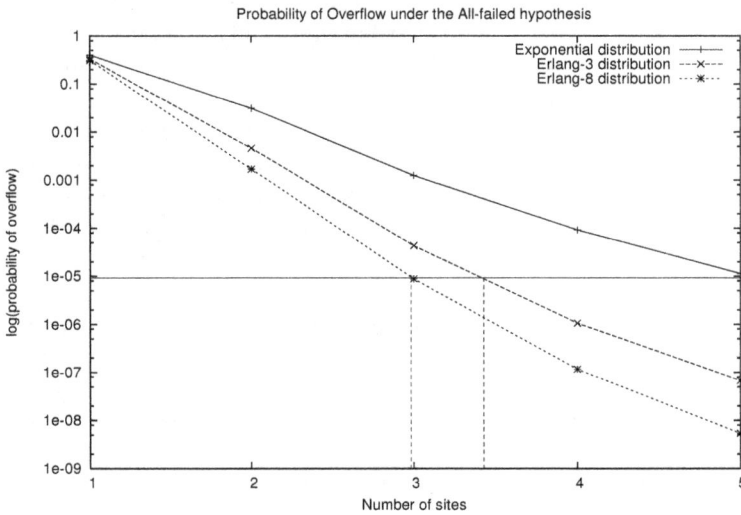

FIG. 4.5. *Performance Analysis: different distributions under the All-failed hypothesis*

functions. Assume that this QoS is given as: "the logarithmic value of the probability of overflow is less than 10^{-5}". According to Figure 4.5 (b), this minimal site number has the value 3 for Erlang-8 distribution, 4 for the Erlang-3 distribution and 6 for the exponential distribution which is not a significant value compared to 3 or even 4. Therefore, the exponential distribution does not give us a precise prediction in the case of a tightly-tailed service time distribution for our multiprocessor system. On the contrary, the use of Erlang distribution allows us to obtain more precise results, when it represents a distribution having a small coefficient of variation.

5. Conclusion. In this paper we used the formalism of Stochastic Automata Networks to model multiprocessor systems with general service and repair time distributions, and to generate some performance predictions. General distributions were substituted by exponential and Erlang distributions. The basic idea has been to use SANs to handle the large state space of the model. Morever, SANs allows us to easily use PH distributions, such as the Erlang distribution. We have shown first the robustess of the exponential distribution in replacing the general repair time distribution. This has the advantage of decreasing the size of the state space of the model. However, the exponential distribution does not approximate the service time distribution with a small coefficient of variation sufficiently accurately. We have shown how an Erlang distribution is more appropriate in this case. The use of the Erlang distribution permits us to obtain more precise predictions and analyses.

We intend to analyze this multiprocessor system with other PH distributions, such as the Coxian distribution. This requires us to fit the parameters of the PH distribution and we shall examine several methods that are already available. We shall also evaluate the system with other non-exponential events, such as the arrival process, using SANs, naturally.

REFERENCES

[1] S. Asmussen, O. Nerman, and M.Olsson, "Fitting Phase-type Distributions via the EM algorithm", Scandinavian Journal of Statistics 23, 419-441, 1996.
<www.maths.lth.se/matstat/staff/asmus/pspapers.html>

[2] A. Avizienis, J.C. Laprie , B. Randell, "Dependability and its threats: a taxonomy", Rapport LAAS No04386, 18th IFIP World Computer Congress. Building the Information Society, Toulouse (France), 22-27 Juin 2004.

[3] C. Bertolini, A.G. Farina, P. Fernandes and F.M. Oliveira, "Test Case Generation using Stochastic Automata Networks: Quantative Analysis", In proceedings of the 2nd IEEE International Conference on Software Engineering and Formal Methods, Beijing, China, September 2004.

[4] A. Benoit, L. Brenner, P. Fernandes, B. Plateau and W.J. Stewart. The PEPS Software Tool. In *Performance TOOLS 2003*, pages 98–115, Urbana, Illinois, USA, 2003. Springer-Verlag.

[5] M. Davio. Kronecker products and shuffle algebra. IEEE Transactions on Computers, 30(2) : 116-125, 1981.

[6] T. Dayar, O. I. Pentakalos, and A. B. Stephens, "Analytical modeling of robotic tape libraries using stochastic automata", IASTED International Conference on Applied Modelling and Simulation, 27 July-1 August 1997, Banff, Canada.

[7] R. El Abdouni Khayri, R. Sadre, and B.R. Haverkort, "Fitting world-wide web request traces with the EM-algorithm", Performance Evaluation 52, 175-191, 2003.

[8] A. Feldmann and W. Whitt, "Fitting mixtures of exponentials to long-tail distributes to analyze network performance models", Performance Evaluation 31, 245-258, 1998.

[9] P. Fernandes, B. Plateau, and W.J. Stewart, "Efficient descriptor-Vector multiplication in Stochastic Automata Networks", Journal of the ACM, 45(3): 381-414, 1998.

[10] M. Malhorta and A. Reibman, "Selecting and implementing phase approximations for semi-Markov models", Communications in Statistics, Stochastic Models 9, (1993) pp. 473-506.

[11] R. Marculescu and A. Nandi, "Probabilistic Application Modeling for System-Level Performance Analysis", In Design Automation & Test in Europe (DATE), pp 572-579, Munich, Germany, March 2001.

[12] R. Marie, "Calculating equilibrium probabilities for queues", In proceedings of the 7th IFIP W.G.7.3 International Symposium on Computer Performance Modelling, Measurement and Evaluation, ACM SIGMETRICS, 117-122, Toronto, Canada, 1980.

[13] M.F. Neuts and K. S. Meier. On the Use of Phase Type Distributions in Reliability Modeling of Systems with a Small Number of Components, OR Spectrum, 2, 1981, pp. 227-234

[14] B. Plateau, "On the stochastic structure of parallelism and synchronization models for distributed algorithms", Proc. of the SIGMETRICS Conference, Texas, 1985, 147–154.

[15] I. Sbeity, L. Brenner, B. Plateau and W. J. Stewart, "Phase Type Distributions in Stochastic Automata Networks", NCSU Technical Report number TR-2005-49, December 20, 2005. <http://www.csc.ncsu.edu/research/tech/reports.php>

[16] W.J. Stewart, "Introduction to the Numerical Solution of Markov Chains", Princeton University Press, 1994.

[17] W.J. Stewart, K. Atif, B. Plateau, "The numerical solution of stochastic automata networks", European Journal of Operational Research, 86 (1995), 503–525.

[18] K. S. Trivedi, A. S. Sathaye, O. C. Ibe, R. C. Howe and A. Aggarwal, "Availability and Performance-Based Sizing of Multiprocessor Systems", Communications in Reliability, Maintainability and Serviceability: An International Journal published by SAE International, 1996.

MATRIX AND RANDOM WALK METHODS IN WEB ANALYSIS

RAVI KUMAR *

Abstract. We discuss the use of matrix and random walk methods in web analysis. The last few years have witnessed a surge of activity in applying tools and techniques from linear algebra to search and retrieval on the web. We cover some of the key results in this area, including the use of hyperlinks to enhance the quality of web information retrieval and the use of Markov chains and random walks to aid web analysis.

Key words. Hyperlinks, matrices, eigenvectors, Markov chains, random walks, stationary distribution, information retrieval, HITS, PageRank, decay

AMS subject classifications. 60G50, 60J10, 60J20, 65C40, 68P20

A set of web pages is conceptually richer than a collection of pure textual documents, the distinguishing feature being the presence of hyperlinks in the former. A hyperlink is an explicit pointer from one web page to another, placed to aid the browsing and navigation of web pages. Hyperlinks can be interpreted in one of several ways. A hyperlink from a page p to a page q, denoted $p \to q$, can be thought of as an implicit endorsement of the page q by the page p. It can also be interpreted as the page p conferring authority on the page q with respect to p's topic of discourse, thereby bringing q closer to p. The most natural way to model the hyperlink relationship between pages is by a matrix $M = \{M_{pq}\}$ where $M_{pq} > 0$ if and only if there is a hyperlink from page p to page q. In this case, the value of M_{pq} can be use to encode the significance of this hyperlink, possibly taking into account other aspects such as the words surrounding the hyperlink (i.e., the anchortext); in the simplest case, $M_{pq} \in \{0, 1\}$. Exploiting hyperlinks has paid rich dividends in terms of developing novel and effective methods to analyze the web, especially in the context of web information retrieval.

In a seminal piece of work, Kleinberg [15] used the existence of hyperlinks to construct a method (called *HITS*) to find the most authoritative sources in a given collection of web pages. In his model, each page p has two attributes: a hub value $h(p)$ and an authority value $a(p)$. The authority value of a page is determined by the hub values of the pages that point to it and the hub value of a page is determined by the authority value of the pages it points to. This mutual reinforcement can be mathematically written as $a(p) = \sum_{q|q \to p} M_{qp} h(q)$ and $h(p) = \sum_{q|p \to q} M_{pq} a(q)$. These equations suggest an iterative scheme to compute the authority and hub scores of all pages: $\vec{a} \leftarrow M^T \vec{h}$ and $\vec{h} \leftarrow M \vec{a}$. It follows that $\vec{a}$ is the principal eigenvector of $M^T M$ and $\vec{h}$ is the principal eigenvector of MM^T. These can be computed, for instance, by a simple power iteration. The most authoritative pages are given by the ones with the highest authority scores. The HITS method can be used to construct a web search engine — the search engine Teoma (`teoma.com`) is based on these ideas.

There have been several extensions to Kleinberg's original work; see, for instance, [4, 8, 9, 17]. Chakrabarti et al. [8] used the anchortext to fill the entries of the matrix M. Bharat and Henzinger [4] proposed several heuristics, including ones to address the issue of a page receiving unduly high authority score by virtue of many pages from the *same* web site pointing to it. Lempel and Moran [17] proposed a HITS-like method with slightly different update rules for authority and hub values.

---

*Yahoo! Research, 701 First Ave., Sunnyvale, CA 94089, (ravikumar@yahoo-inc.com.).

Brin and Page [7] proposed a slightly different way to make use of hyperlinks. In their method, called *PageRank*, each page p has a value $\pi(p)$. This value is the limiting value of the fraction of the time spent at page p by the following process. At each step, with probability α, the process jumps to a page chosen uniformly at random from the entire set of pages. With probability $1 - \alpha$, the process jumps to a page q chosen uniformly at random from the set of pages to which page p has a hyperlink; if p has no hyperlinks, then the process jumps to a page chosen uniformly at random from the entire set of pages. This process can be abstracted by the update rule $\vec{\pi} \leftarrow P^T \vec{\pi}$ with $P = (1 - \alpha)M' + \alpha J$, where M' is a stochastic matrix corresponding to the above process, J is the matrix of all ones, and α is the teleportation probability, usually chosen around 0.15. Thus, the vector $\vec{\pi}$ is the principal eigenvector of P and a global ordering of pages can be obtained using the values in $\vec{\pi}$. PageRank has been a successful method to construct web search engines and is the brainchild behind Google (`google.com`).

There have been several enhancement, modifications, and extensions to the vanilla PageRank method. Computational issues of the PageRank method have been discussed in several work; see, for instance, [16, 14]. Haveliwala [11] proposed a topic-sensitive PageRank method, where the process jumps to a page chosen uniformly at random from a subset of pages on a specific topic instead of the entire set of web pages. Jeh and Widom [13] propose an efficient algorithm to compute this personalized version of the PageRank method. Boldi, Santini, and Vigna [5] studied the PageRank method as a function of the teleportation probability. Dwork et al. [10] studied the effect on $\pi(q)$ of adding a new hyperlink $p \rightarrow q$ and proposed an incremental PageRank method to recompute $\vec{\pi}$ after this hyperlink addition.

For a comparative account of these methods, the readers are referred to a survey by Borodin et al. [6].

Web pages together with their hyperlinks suggest a directed graph (*web graph*) whose nodes are the web pages and directed edges are the hyperlinks. There has been lot of work on using this web graph for analysis and interpretation of the web. One particularly interesting phenomenon studied on the web graph is random walks. In fact, PageRank itself has a random-walk interpretation: the stochastic process can be modeled by a random surfer. In a recent work, Broder et al. [2] proposed a random walk with absorbing states to compute the *decay* score $d(p)$ of a web page p. Informally, the decay score of a page measures how close the page is to being *non-existent*. If p itself is non-existent, then it is fully decayed and if p exists, then its decay score measures how easily a fully decayed page can be reached from p by a random walk following hyperlinks. This can be modeled by a random walk with absorption states according to the following update rule: if p is non-existent, then $d(p) = 1$ and if p exists, then $d(p) = (1 - \alpha) \sum_{q \in \Gamma(p)} d(q)/|\Gamma(p)|$, where $\Gamma(p) = \{q \mid p \rightarrow q\}$, the set of hyperlinks of p, and $0 < \alpha < 1$ is a parameter.

Random walks on the web graph have also been used to generate uniform random sample web pages. One particularly interesting application of this method is the measurement of the size of the web. See Henzinger et al. [12], Bar-Yossef et al. [1], and the very recent work by Bar-Yossef and Gurevich [3]. Rafiei and Mendelzon [18] used random walks and PageRank to compute the topics for which a given web page has a good reputation.

REFERENCES

[1] Z. BAR-YOSSEF, A. BERG, S. CHIEN, J. FAKCHAROENPHOL, AND D. WEITZ, *Approximate aggregate queries about web pages via random walks*, in Proc. 26th International Conference on Very Lage Data Bases, 2000, pp. 535–544.

[2] Z. BAR-YOSSEF, A. BRODER, R. KUMAR, AND A. TOMKINS, *Sic transit gloria telae: Towards and understanding of the web's decay*, in Proc. 13th International World Wide Web Conference, 2004, pp. 328–337.

[3] Z. BAR-YOSSEF AND M. GUREVICH, *Random sampling from a search engine's index*, in Proc. 15th International World Wide Web Conference, 2006.

[4] K. BHARAT AND M. HENZINGER, *Improved algorithms for topic distillation in a hyperlinked environment*, in Proc. of the 21st Annual International ACM SIGIR Conference on Research and Development in Information Retrieval, ACM Press, 1998, pp. 104–111.

[5] P. BOLDI, M. SANTINI, AND S. VIGNA, *PageRank as a function of the damping factor*, in Proc. 14th International World Wide Web Conference, 2005, pp. 557–566.

[6] A. BORODIN, G. O. ROBERTS, J. S. ROSENTHAL, AND P. TSAPARAS, *Link analysis ranking algorithms, theory, and experiments*, ACM Transactions on Internet Technology, 5 (2005), pp. 231–297.

[7] S. BRIN AND L. PAGE, *The anatomy of a large-scale hypertextual Web search engine*, WWW7/Computer Networks, 30 (1998), pp. 107–117.

[8] S. CHAKRABARTI, B. DOM, D. GIBSON, J. KLEINBERG, P. RAGHAVAN, AND S. RAJAGOPALAN, *Automatic resource compilation by analyzing hyperlink structure and associated text*, WWW7/Computer Networks, 30 (1998), pp. 65–74.

[9] S. CHAKRABARTI, B. DOM, D. GIBSON, R. KUMAR, P. RAGHAVAN, S. RAJAGOPALAN, AND A. TOMKINS, *Spectral filtering for resource discovery*, in Proc. of the ACM SIGIR Workshop on Hypertext Analysis, ACM Press, 1998, pp. 13–21.

[10] S. CHIEN, C. DWORK, R. KUMAR, D. SIMON, AND D. SIVAKUMAR, *Link evolution: Analysis and algorithms*, Internet Mathematics, 1 (2004), pp. 277–304.

[11] T. H. HAVELIWALA, *Topic-sensitive PageRank: A context-sensitive ranking algorithm for web search*, IEEE Transactions on Knowledge and Data Engineering, 15 (2003), pp. 784–796.

[12] M. R. HENZINGER, A. HEYDON, M. MITZENMACHER, AND M. NAJORK, *Measuring index quality using random walks on the web*, WWW8/Computer Networks, 31 (1999), pp. 1291–1303.

[13] G. JEH AND J. WIDOM, *Scaling personalized web search*, in Proc. 12th International World Wide Web Conference, 2003, pp. 271–279.

[14] S. KAMVAR, T. H. HAVELIWALA, C. D. MANNING, AND G. H. GOLUB, *Extrapolation methods for accelerating pagerank computations*, in Proc. 12th International World Wide Web Conference, 2003, pp. 261–270.

[15] J. KLEINBERG, *Authoritative sources in a hyperlinked environment*, Journal of the ACM, 46 (2000), pp. 604–632.

[16] A. N. LANGVILLE AND C. D. MEYER, *Deeper inside PageRank*, Internet Mathematics, 1 (2005), pp. 335–380.

[17] R. LEMPEL AND S. MORAN, *SALSA: The stochastic approach for link-structure analysis*, ACM Transactions on Information Systems, 19 (2001), pp. 131–160.

[18] D. RAFIEI AND A. O. MENDELZON, *What is this page known for? Computing web page reputations*, WWW10/Computer Networks, 33 (2000), pp. 823–835.

APPLICATIONS OF MARKOV CHAINS IN DEMOGRAPHY[*]

HAL CASWELL[†]

Abstract. The life cycle of an individual organism can be described by an absorbing finite-state Markov chain. The transient states of the chain are life cycle stages (size classes, age classes, developmental stages, spatial locations, occupational categories, etc.). The absorbing states include death. Other absorbing states may be added. I show how to analyze longevity, competing risks, the net reproductive rate, the birth interval, and age at maturity directly from the Markov chain. Perturbation analysis of the fundamental matrix gives the sensitivity of demographic characteristics to changes in the vital rates. The model is extended to include spatial or temporal variation. The time-varying model permits calculation of both period and cohort rates.

Key words. birth interval, demography, *Diomedea exulans*, *Eubalaena glacialis*, matrix population models, net reproductive rate, North Atlantic right whale, population dynamics, fundamental matrix, life expectancy, vec-permutation matrix, wandering albatross.

AMS subject classifications. 60J20, 60J10, 92D25, 91D20

1. Introduction. As an individual organism develops through its life cycle it increases in size, changes its morphology, develops new physiological functions, exhibits new behaviors, and moves to different locations. It may change sex and/or change its reproductive status. These changes can be dramatic. This developmental process, and its attendant risks of death and opportunities for reproduction determine the rates of birth and death that, in turn, determine population growth or decline. Human demographers originally focused on age as a way to describe the state of an individual in its life cycle (e.g., Lotka 1939). They still largely do so, although it is now widely appreciated that many other individual properties (marriage, occupation, location, parity, disease state) influence birth and death (e.g., Land and Rogers 1982). Plant and animal demographers, confronting a wider array of taxa with an enormous diversity of life histories, have long viewed age as only one among many measures of individual state (e.g., Lefkovitch 1965, Harper 1977, Werner and Caswell 1977).

Regardless of the species under consideration, demographic analysis requires differentiation among individuals and a way to infer the consequences of those differences. Thus, in this paper, I focus on stage-classified matrix population models (discrete in time and in which individuals are classified into discrete states). Age-classified models follow as a special case. To develop such a model, one must define stages, choose a projection interval (i.e., a time scale), and assign transitions among the stages on the time scale of the projection interval. These can be indicated on a directed graph, the *life cycle graph* of the species. For example, Figure 1 shows a life cycle graph used by Caswell and Fujiwara (2004) for the North Atlantic right whale (*Eubalaena glacialis*). Figure 2 shows a graph for the life cycle of the wandering albatross (*Diomedea exulans*).

What follows is a biologist's tour of some of the demographic results that emerge from treating these life cycle descriptions as absorbing finite-state Markov chains. Little or nothing is mathematically new, but I will present some new demographic results and hopefully suggest problems that might lead to interesting new mathematics.

---

[*]This work was supported by the National Science Foundation grants DEB-0235692 and DEB-0343820 and NOAA grant NA03NMF4720491.

[†]Biology Department MS-34, Woods Hole Oceanographic Institution, Woods Hole MA 02543 (hcaswell@whoi.edu).

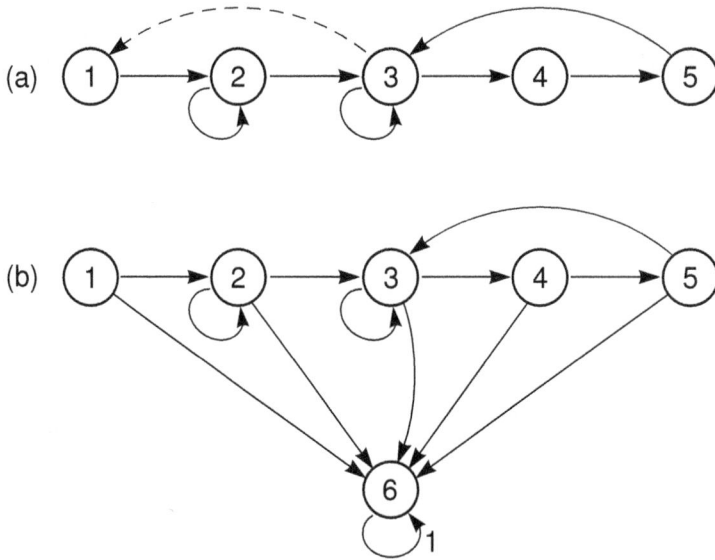

FIG. 1. *Life cycle graphs for females of the North Atlantic right whale (*Eubalaena glacialis*). Projection interval is one year. Nodes correspond to stages; 1=calf, 2=immature, 3=mature, 4=mother (reproducing female), 5=post-breeding female. (a) The graph corresponding to the population projection matrix* **A**. *Solid arcs indicate transitions, dashed arcs reproduction. Every time a mature female enters stage 4, she produces a calf. See Caswell and Fujiwara (2004) for explanation and parameter estimates. (b) The graph corresponding to the transition matrix* **P***; it shows transitions of extant individuals only. Stage 6=death is an absorbing state.*

The approach I present here was pioneered by Gustav Feichtinger (1971, 1973). Much later, Cochran and Ellner (1992) independently explored some of the same territory, but without the convenience of matrix formulations. Caswell (2001, Chapter 5) and Keyfitz and Caswell (2005, Chapter 11) focused on stage-classified populations and addressed some problems complementary to those presented here.

Section 2 will show how to embed the life cycle graph in an absorbing Markov chain. Following sections will explore demographic implications. Section 3 will focus on longevity its perturbation analysis. Section 4 will consider competing risks in stage-classified models. Section 5 will combine transitions and reproduction. Section 6 will focus on intervals between demographically important events, including the inter-birth interval and the time to recruitment. Finally, §7 will develop models including spatial and temporal variation, and use these to examine period and cohort effects in changing environments. Biological examples will be sprinkled throughout.

2. The life cycle as an absorbing Markov chain. The directed graph in Figure 1(a) corresponds to a projection matrix **A**, in which a_{ij} is the expected number of stage i individuals at $t+1$ per stage j individual at time t, and

$$(2.1) \qquad \mathbf{n}(t+1) = \mathbf{A}\mathbf{n}(t)$$

where $\mathbf{n}(t)$ is a vector of the abundances of the stages at time t. We partition **A** into a matrix **U** describing transitions of individuals already existing at t and a matrix **F** describing the production of new individuals by reproduction

$$(2.2) \qquad \mathbf{A} = \mathbf{U} + \mathbf{F}$$

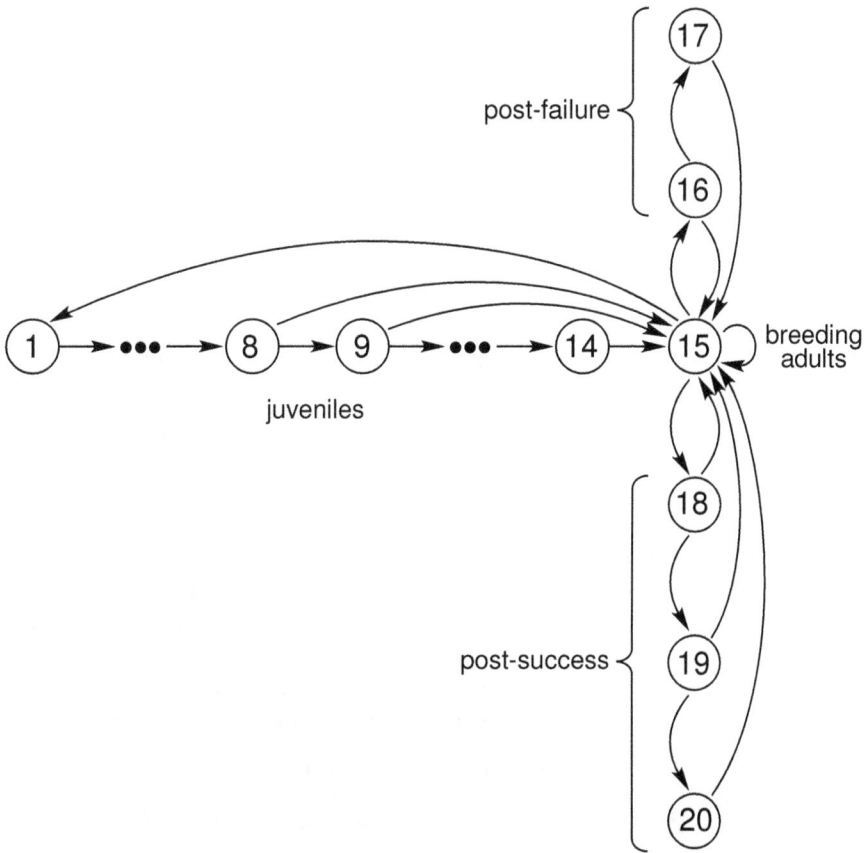

FIG. 2. *Life cycle graph for females of the wandering albatross* (Diomedea exulans). *Projection interval is one year. Stages 1–14 are age classes of pre-breeding immatures. Stage 15 is breeding adults. Stages 16 and 17 are adults who are not breeding following a failed breeding attempt. Stages 18–20 are adults not breeding following a successful breeding episode. Because incubation and fledging take more than a year, adults breed in successive years only after an (early) failure. Death is not shown in this graph.*

The matrix $\mathbf{U}$ is column sub-stochastic; it describes only the transitions of survivors. The graph corresponding to $\mathbf{U}$ is obtained by removing all arcs representing reproduction.

Death (an absorbing state) is implicit in Figure 1(a). If we make it explicit (Figure 1(b)), we obtain the graph corresponding to the transition matrix for an absorbing Markov chain

$$(2.3) \qquad \mathbf{P} = \left(\begin{array}{c|c} \mathbf{U} & 0 \\ \hline \mathbf{m} & 1 \end{array} \right)$$

The element m_j of the vector $\mathbf{m}$ is the probability of mortality of an individual in stage j. If there are multiple absorbing states (as there will be in some of the applications shown later), then

$$(2.4) \qquad \mathbf{P} = \left(\begin{array}{c|c} \mathbf{U} & 0 \\ \hline \mathbf{M} & \mathbf{I} \end{array} \right)$$

The element m_{ij} of $\mathbf{M}$ is the probability that an individual in transient state j enters absorbing state i. I assume throughout that at least one absorbing state is accessible from any state represented in $\mathbf{U}$, and that the spectral radius of $\mathbf{U}$ is strictly less than 1.

Note that in this paper, the transition matrix $\mathbf{P}$ is column-stochastic, and operates on column vectors, rather than the row-stochasticity and row vectors common in much of the literature on Markov chains.

3. Lifespan and longevity. All men are mortal. Absorbtion is certain. The question is, how long does it take and what happens en route? From a demographic perspective, this is asking about the lifespan of an individual, starting from a specified stage in the life cycle. Longevity can be calculated directly from age-classified models using life tables (e.g., Keyfitz and Caswell 2005), but well-known results on absorbing Markov chains (e.g., Kemeny and Snell 1976, Iosifescu 1980) permit easy generalization to stage classified models.

Let ν_{ij} denote the number of occurrences of transient state i of an individual starting in transient state j. The moments of ν_{ij} depend on $\mathbf{U}$. The expectation is

$$(3.1) \qquad \left(\, E(\nu_{ij}) \, \right) = \mathbf{N} = (\mathbf{I} - \mathbf{U})^{-1}.$$

$\mathbf{N}$ is the *fundamental matrix* of the chain. The second moments are

$$(3.2) \qquad \left(\, E\left(\nu_{ij}^2\right) \, \right) = \mathbf{N}_2 = (2\mathbf{N}_{\mathrm{diag}} - \mathbf{I})\,\mathbf{N}$$

where $\mathbf{N}_{\mathrm{diag}} = \mathbf{I} \circ \mathbf{N}$ and $\circ$ denotes the Hadamard product. The variance is

$$(3.3) \qquad \left(\, V\left(\nu_{ij}\right) \, \right) = (2\mathbf{N}_{\mathrm{diag}} - \mathbf{I})\,\mathbf{N} - \mathbf{N} \circ \mathbf{N}.$$

Let η_j be the time to absorbtion for an individual starting in stage j;

$$(3.4) \qquad \left(\, \eta_j \, \right) = \sum_i \nu_{ij}$$

The expected value, second moment, and variance of η are

$$(3.5) \qquad E\left(\, \eta_j \, \right) = \bar{\eta} = \sum_i \nu_{ij} = \mathbf{e}^{\mathsf{T}}\mathbf{N}$$

$$(3.6) \qquad E\left(\, \eta_j^2 \, \right) = \mathbf{e}^{\mathsf{T}}\mathbf{N}\,(2\mathbf{N} - \mathbf{I})$$

$$(3.7) \qquad V\left(\, \eta_j \, \right) = \mathbf{e}^{\mathsf{T}}\mathbf{N}\,(2\mathbf{N} - \mathbf{I}) - (\mathbf{e}^{\mathsf{T}}\mathbf{N}) \circ (\mathbf{e}^{\mathsf{T}}\mathbf{N})$$

where $\mathbf{e}$ is a column vector of ones. The probability distribution of ν_j is

$$(3.8) \qquad P(\nu_j = t) = \mathbf{e}^{\mathsf{T}}\left(\mathbf{U}^{t-1} - \mathbf{U}^t\right).$$

EXAMPLE 1. **Longevity of the North Atlantic right whale** Caswell and Fujiwara (2004) estimated $\mathbf{U}$ by applying multi-stage mark-recapture methods to photographic identification data, compiled by the New England Aquarium (Crone and Kraus 1990), on the North Atlantic right whale. The best model, out of a large number evaluated, included significant time variation in survival and birth rates. I will return to this later. Here is the transition matrix estimated for 1980–1981, a period when the environment supported positive population growth:

$$(3.9) \qquad \mathbf{P} = \left(\begin{array}{ccccc|c} 0 & 0 & 0 & 0 & 0 & 0 \\ 0.9281 & 0.8461 & 0 & 0 & 0 & 0 \\ 0 & 0.1245 & 0.5930 & 0 & 0.9895 & 0 \\ 0 & 0 & 0.3964 & 0 & 0 & 0 \\ 0 & 0 & 0 & 0.9683 & 0 & 0 \\ \hline 0.0719 & 0.0295 & 0.0105 & 0.0317 & 0.0105 & 1 \end{array}\right).$$

The fundamental matrix,

$$(3.10) \qquad \mathbf{N} = \left(\begin{array}{ccccc} 1.00 & 0.00 & 0.00 & 0.00 & 0.00 \\ 6.03 & 6.50 & 0.00 & 0.00 & 0.00 \\ 27.62 & 29.76 & 36.81 & 35.27 & 36.42 \\ 10.95 & 11.80 & 14.59 & 14.98 & 14.44 \\ 10.60 & 11.43 & 14.13 & 14.51 & 14.98 \end{array}\right),$$

the matrix of variances in ν,

$$(3.11) \quad V(\nu_{ij}) = \left(\begin{array}{ccccc} 0.00 & 0.00 & 0.00 & 0.00 & 0.00 \\ 35.95 & 35.70 & 0.00 & 0.00 & 0.00 \\ 1243.09 & 1275.68 & 1318.32 & 1317.48 & 1318.56 \\ 197.26 & 202.53 & 209.71 & 209.47 & 209.72 \\ 194.68 & 200.38 & 209.60 & 209.72 & 209.47 \end{array}\right),$$

and the corresponding standard deviations

$$(3.12) \qquad (SD(\nu_{ij})) = \left(\begin{array}{ccccc} 0.00 & 0.00 & 0.00 & 0.00 & 0.00 \\ 6.00 & 5.97 & 0.00 & 0.00 & 0.00 \\ 35.26 & 35.72 & 36.31 & 36.30 & 36.31 \\ 14.04 & 14.23 & 14.48 & 14.47 & 14.48 \\ 13.95 & 14.16 & 14.48 & 14.48 & 14.47 \end{array}\right)$$

are obtained directly from $\mathbf{U}$.

Row 4 of $\mathbf{N}$ is of particular interest; it is the expected number of calving events that a female will experience during her remaining lifetime. In 1980 a newborn calf could expect to give birth about 11 times, with a standard deviation of 14. By the time she became mature, she could expect to give birth 14.6 times; the difference reflects the likelihood of mortality between birth and maturity. (Drastic changes took place after 1980; the picture now is not so rosy.)

The expectation, variance, and standard deviation of the times to absorbtion are

$$(3.13) \qquad E(\eta) = (\ 56.21 \quad 59.48 \quad 65.54 \quad 64.76 \quad 65.85\ )$$

$$(3.14) \qquad V(\eta) = (\ 4049.66 \quad 4108.95 \quad 4215.49 \quad 4215.29 \quad 4215.85\ )$$

$$(3.15) \qquad SD(\eta) = (\ 63.64 \quad 64.10 \quad 64.93 \quad 64.93 \quad 64.93\ )$$

Thus life expectancy for a calf was about 56 years, with a standard deviation of 64 years. Note that assuming that 2.5% of a cohort would live longer than 56.21 + 1.96 × 63.34 = 180.9 years would *not* be a good idea. The standard deviation reflects the self-loops in the life cycle graph, which render mortality stage-dependent but age-independent. Thus survival eventually falls off exponentially; the tail of that distribution is not biologically realistic.   △

3.1. Perturbation analysis of longevity. Perturbation analysis plays an important role in demography, especially perturbation analysis of the population growth rate λ, calculated as the dominant eigenvalue of $\mathbf{A}$ (e.g., Keyfitz 1971, Caswell 1978, 2000, 2001). Let θ be a parameter that affects $\mathbf{A}$. Knowledge of $d\lambda/d\theta$ is useful as a way to project the effects of future environmental changes, to evaluate possible population management tactics, to quantify the effects of past environmental differences, to determine the direction and intensity of natural selection, and to calculate confidence intervals.

Perturbation analysis of longevity is equally interesting. Keyfitz (see Keyfitz and Caswell 2005, Chapter 4) and Vaupel and Romo (2003) have explored this in detail for age-classified models. Here I present a new analysis for stage-classified models.

3.1.1. Matrix calculus conventions. I use the vector-rearrangement formalism of matrix calculus (Magnus and Neudecker 1988); the term comes from a useful review paper of Nel (1980). In this formalism, the derivative of a $q \times 1$ vector $\mathbf{y}$ with respect to a $r \times 1$ vector $\mathbf{x}$ is the $q \times r$ Jacobian matrix

$$(3.16) \qquad \frac{d\mathbf{y}}{d\mathbf{x}^{\mathsf{T}}} = \left(\frac{dy_i}{dx_j} \right)$$

The derivative of a matrix $\mathbf{Y}$ with respect to a matrix $\mathbf{X}$ is obtained by turning $\mathbf{Y}$ and $\mathbf{X}$ into vectors using the vec operator, (which stacks the columns one above the next), to write the derivative as

$$(3.17) \qquad \frac{d\mathrm{vec}\,\mathbf{Y}}{d\mathrm{vec}\,^{\mathsf{T}}\mathbf{X}},$$

where vec $^{\mathsf{T}}\mathbf{X}$ is written for $(\mathrm{vec}\,\mathbf{X})^{\mathsf{T}}$.

Life expectancy depends on the transient matrix $\mathbf{U}$, which I suppose is a function of a $p \times 1$ vector $\boldsymbol{\theta}$ of parameters.

We begin with the derivative of the fundamental matrix. Write $\mathbf{I} - \mathbf{U} = \mathbf{Z}$, so that $\mathbf{N} = \mathbf{Z}^{-1}$. Since

$$(3.18) \qquad\qquad \mathbf{I} = \mathbf{Z}^{-1}\mathbf{Z},$$

taking the differential of both sides yields

$$(3.19) \qquad\qquad 0 = (d\mathbf{Z}^{-1})\mathbf{Z} + \mathbf{Z}^{-1}(d\mathbf{Z}).$$

Multiply by an identity matrix of order s

$$(3.20) \qquad\qquad 0 = \mathbf{I}_s(d\mathbf{Z}^{-1})\mathbf{Z} + \mathbf{Z}^{-1}(d\mathbf{Z})\mathbf{I}_s$$

and apply the vec operator to both sides, using the result (Roth 1934) that

$$(3.21) \qquad\qquad \mathrm{vec}\,\mathbf{ABC} = (\mathbf{C}^{\mathsf{T}} \otimes \mathbf{A})\mathrm{vec}\,\mathbf{B},$$

to obtain

(3.22) $$0 = (\mathbf{Z}^\mathsf{T} \otimes \mathbf{I}_s) \, d\text{vec} \, (\mathbf{Z}^{-1}) + (\mathbf{I}_s \otimes \mathbf{Z}^{-1}) \, d\text{vec} \, \mathbf{Z}.$$

Solving for $d\text{vec} \, \mathbf{Z}^{-1}$ and setting $\mathbf{Z}^{-1} = \mathbf{N}$, gives

(3.23) $$d\text{vec} \, \mathbf{N} = (\mathbf{N}^\mathsf{T} \otimes \mathbf{N}) \, d\text{vec} \, \mathbf{U}$$

from which

(3.24) $$\frac{d\text{vec} \, \mathbf{N}}{d\boldsymbol{\theta}^\mathsf{T}} = (\mathbf{N}^\mathsf{T} \otimes \mathbf{N}) \frac{d\text{vec} \, \mathbf{U}}{d\boldsymbol{\theta}^\mathsf{T}}.$$

The sensitivity of life expectancy is obtained by taking the differential of

(3.25) $$d\bar{\eta} = \mathbf{e}^\mathsf{T} d\mathbf{N}.$$

Multiplying on the right by $\mathbf{I}_s$ and applying the vec operator gives

(3.26) $$d\text{vec} \, \bar{\eta} = (\mathbf{I}_s \otimes \mathbf{e}^\mathsf{T}) \, d\text{vec} \, \mathbf{N},$$

so that

(3.27) $$\frac{d\text{vec} \, \bar{\eta}}{d\boldsymbol{\theta}^\mathsf{T}} = (\mathbf{I}_s \otimes \mathbf{e}^\mathsf{T}) (\mathbf{N}^\mathsf{T} \otimes \mathbf{N}) \frac{d\text{vec} \, \mathbf{U}}{d\boldsymbol{\theta}^\mathsf{T}}.$$

EXAMPLE 2. **Sensitivity of life expectancy in the right whale.** Life expectancy in the right whale as of 1980 was strikingly sensitive to changes in $\mathbf{U}$. Focusing on $\bar{\eta}_1$, the life expectancy of a calf, we obtain

(3.28) $$\left(\frac{d\eta_1}{du_{ij}} \right) = \begin{pmatrix} 56.21 & 338.87 & 1552.72 & 615.56 & 596.03 \\ \mathbf{59.48} & \mathbf{358.61} & 1643.19 & 651.43 & 630.76 \\ 65.54 & \mathbf{395.11} & \mathbf{1810.40} & 717.72 & \mathbf{694.94} \\ 64.76 & 390.40 & \mathbf{1788.85} & 709.18 & 686.67 \\ 65.85 & 396.97 & 1818.95 & \mathbf{721.10} & 698.22 \end{pmatrix}$$

The boldface entries correspond to the transitions that actually occur in the life cycle, and are thus biologically relevant. Life expectancy is particularly sensitive to changes in the fate of mature females (stage 3). This no doubt reflects the fact (see $\mathbf{N}$ in (3.10) above) that a calf can expect to spend about three times as long in this stage as any other. △

4. Competing risks. Sometimes more than one absorbing state exists. Two applications are of particular demographic interest. In one, multiple causes of death are distinguished, and the goal is to obtain the probability of death due to each cause, and the results of eliminating one or more causes. In the other, multiple absorbing states are added to a model in order to study the occurrence of certain events in the life cycle (see §6).

Let a be the number of absorbing states. The row vector of mortality probabilities $\mathbf{m}$ is replaced by an $a \times s$ matrix $\mathbf{M}$, so that

(4.1) $$\mathbf{P} = \left(\begin{array}{c|c} \mathbf{U} & 0 \\ \hline \mathbf{M} & \mathbf{I} \end{array} \right)$$

The probability of absorbtion in each of the absorbing states is given by

(4.2) $$\mathbf{B} = \mathbf{MN}$$

where b_{ij} is the probability of eventual absorbtion in absorbing state i for an individual in transient state j (Kemeny and Snell 1976, Iosifescu 1980). This result has many applications in demography. Here are two.

EXAMPLE 3. **The distribution of stage at death in the wandering albatross.** The probability distribution of the age at death is a standard entry in an age-classified life table. The corresponding quantity for a stage-classified model has never (as far as I know) been calculated. It can be done easily by defining a separate absorbing state for death from each stage. This is equivalent to defining $\mathbf{M} = \text{diag}(\mathbf{m})$. Column j of $\mathbf{B}$ gives the probability distribution of the *stage* at death for an individual in stage j.

When applied to the wandering albatross life cycle (Figure 2), using parameter values extracted from Croxall et al. (1990), the results shown in Figure 3 are obtained, for newly fledged chicks and for breeding adults. A chick has about a 65% chance of dying as a juvenile (stages 1–14), and a 35% chance of dying as an adult (stages 15–20). Because the connection between juveniles and adults in this life cycle is one-way (adults never return to become juveniles), the probability of dying in any of the adult stages for a breeding adult is the same as the probability for a chick, conditional on surviving to breeding. △

EXAMPLE 4. **The probability of reproduction.** What is the probability that a right whale will survive to reproduce (stage 4)? This question can be answered by modifying the life cycle graph (Figure 1) to include a new absorbing state ("reproduced before dying") that will trap all individuals entering stage 4. To do this, set row 4 of $\mathbf{U}$ to zero, and set row 2 of $\mathbf{M}$ to row 4 of $\mathbf{U}$ (Figure 4). The two absorbing states are now competing risks; an individual may die before reproducing, or reproduce before dying, but not both. The probability of reproducing at least once is obtained from the fundamental matrix from the modified life cycle by multiplying by $\mathbf{M}$ to obtain

$$(4.3) \qquad \mathbf{B} = \begin{pmatrix} 0.27 & 0.21 & 0.03 & 0.07 & 0.04 \\ 0.73 & 0.79 & 0.97 & 0.93 & 0.96 \end{pmatrix}.$$

The second row is the probability of reproducing at least once before death. A newborn calf has a 73% chance of reproducing at least once; once reaching maturity, the probability increases to 97%. △

5. The net reproductive rate. As individuals move through their life cycle, they also (if they are lucky) reproduce. The combination of transitions, survival, and reproduction determines such quantities as the population growth rate, the net reproductive rate, and the generation time. The net reproductive rate R_0 is the expected number of offspring produced by an individual over its lifetime. It is also the rate of population growth per generation. Recall that $\mathbf{F}$ is a matrix whose (i, j) entry is the expected number of offspring of stage i produced by an individual of stage j. It can be shown (Cushing and Yicang 1994, Caswell 2001) that

$$(5.1) \qquad R_0 = \lambda(\mathbf{FN}),$$

where $\lambda(\cdot)$ denotes the dominant eigenvalue.

The perturbation analysis of the fundamental matrix makes it easy to calculate the sensitivity of R_0 to changes in $\mathbf{F}$ and $\mathbf{U}$. Let $\mathbf{w}$ and $\mathbf{v}$ be the right and left eigenvectors of $\mathbf{FN}$ corresponding to λ, scaled so that $\mathbf{v}^*\mathbf{w} = 1$. Then

$$(5.2) \qquad dR_0 = (\mathbf{w}^\mathsf{T} \otimes \mathbf{v}^\mathsf{T})\, d\text{vec}\,(\mathbf{FN}).$$

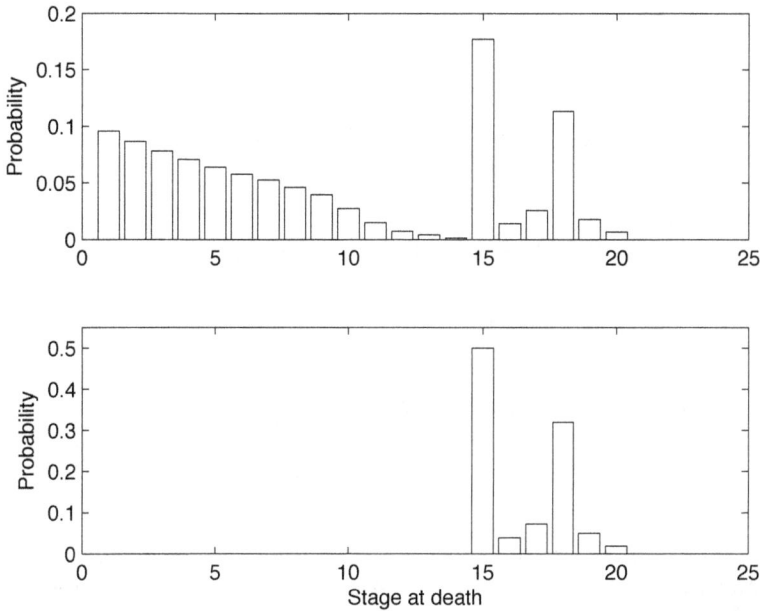

FIG. 3. *The probability distribution of the stage at death for the wandering albatross life cycle of Figure 2. Results are shown for newly fledged chicks (stage 1) and breeding adults (stage 15).*

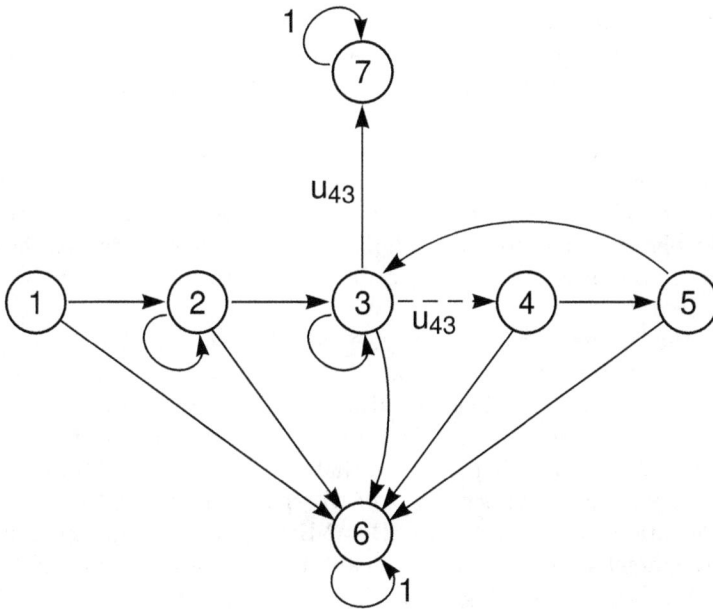

FIG. 4. *The modification of the right whale life cycle graph to make reproduction an absorbing state. The target state is stage 4; all transitions into stage 4 are removed (the dashed arc for u_{43}). A new absorbing state 7, "reproduced before dying" is added. The transitions into this stage consist of all transitions formerly into stage 4.*

But

$$(5.3) \qquad d(\mathbf{FN}) = \mathbf{I}_s(d\mathbf{F})\mathbf{N} + \mathbf{F}(d\mathbf{N})\mathbf{I}_s$$

so

$$(5.4) \qquad d\mathrm{vec}\,(\mathbf{FN}) = (\mathbf{N}^\mathsf{T} \otimes \mathbf{I}_s)\,d\mathrm{vec}\,\mathbf{F} + (\mathbf{I}_s \otimes \mathbf{F})\,d\mathrm{vec}\,\mathbf{N}.$$

Thus

$$(5.5) \qquad \frac{dR_0}{d\boldsymbol{\theta}^\mathsf{T}} = (\mathbf{w}^\mathsf{T} \otimes \mathbf{v}^\mathsf{T}) \left[(\mathbf{N}^\mathsf{T} \otimes \mathbf{I}_s)\frac{d\mathrm{vec}\,\mathbf{F}}{d\boldsymbol{\theta}^\mathsf{T}} + (\mathbf{I}_s \otimes \mathbf{F})(\mathbf{N}^\mathsf{T} \otimes \mathbf{N})\frac{d\mathrm{vec}\,\mathbf{U}}{d\boldsymbol{\theta}^\mathsf{T}} \right].$$

When applied to the right whale life cycle for 1980, the result is

$$(5.6) \qquad \left(\frac{dR_0}{du_{ij}} \right) = \begin{pmatrix} 5.42 & 32.69 & 149.78 & 59.38 & 57.49 \\ \mathbf{5.84} & \mathbf{35.22} & 161.38 & 63.98 & 61.95 \\ 7.23 & \mathbf{43.56} & \mathbf{199.59} & 79.13 & \mathbf{76.62} \\ 6.92 & 41.73 & \mathbf{191.22} & 75.81 & 73.40 \\ 7.15 & 43.10 & 197.49 & \mathbf{78.29} & 75.81 \end{pmatrix}$$

where the entries in boldface correspond to the extant transitions in the right whale life cycle. As with life expectancy, R_0 is most sensitive to changes in rates for mature females.

6. Intervals between events. The intervals between one demographic event and another can be important. The interval between births is particularly important for animals (e.g., whales, seabirds, humans) that produce only a single offspring at a time. Such animals cannot respond to environmental conditions by adjusting the number of their offspring, but they can and do respond by changing the interval between births. Similarly, the interval between birth and maturity (the latter event termed "recruitment" in many branches of animal ecology) is often plastic and reflects the environmental conditions in which the immature animal is developing.

The methods of §3 provide the distribution and the moments of the time required for one particular event (death). These can be applied to the time required for other events by creating multiple absorbing states and conditioning on absorbtion in the state of interest. This corresponds to the way that the interval would be estimated from data, by observing a cohort until all had either died or experienced the event, and then calculating the mean, variance, etc. of the time to the event for the sub-cohort that experienced it.

Suppose that we begin with a life cycle with a single absorbing state (death).

1. Define a target state J; arrival in this state is the event whose timing we wish to calculate. The calculations can also be done for a set of target states.

2. Create a new transient matrix $\mathbf{U}_2 = \mathbf{U}$; then set row J of $\mathbf{U}_2$ to 0.

3. Create a new absorbing state, so that the vector of absorbtion probabilities $\mathbf{m}$ becomes a $2 \times s$ matrix $\mathbf{M}$; set row 2 of $\mathbf{M}$ equal to row J of $\mathbf{U}$.

4. Calculate $\mathbf{N}_2 = (\mathbf{I} - \mathbf{U}_2)^{-1}$ and $\mathbf{B} = \mathbf{M}\mathbf{N}_2$. Row 2 of $\mathbf{B}$ gives the probabilities of reaching the target state (and thus being absorbed in absorbing state 2) before dying (and being absorbed in absorbing state 1).

5. Create a new chain, conditional on absorbtion in absorbing state 2:

$$(6.1) \qquad \mathbf{U}_c = \mathrm{diag}\,(b_2.)\ \mathbf{U}_2\ \mathrm{diag}\,(b_2.)^{-1}$$

where $b_2.$ is row 2 of $\mathbf{B}$.

6. From the fundamental matrix $\mathbf{N}_c = (\mathbf{I} - \mathbf{U}_c)^{-1}$, calculate the moments of the conditional time to absorbtion η_c.

EXAMPLE 5. **Birth interval in the right whale.** The fertility of the right whale is influenced by the dynamics of their food supply (the copepod *Calanus finmarchicus*), which in turn is influenced in any given year by oceanographic circulation patterns, particularly those related to the North Atlantic Oscillation (NAO); see Greene et al. (2003). Thus it is of interest to estimate the birth interval from mark-recapture data. Setting $J = 4$, and applying the algorithm here to the 1980 right whale transition matrix, gives

$$(6.2) \qquad \bar{\eta}_c = (\ \ 9.95 \quad 8.95 \quad 2.46 \quad 4.46 \quad 3.46 \ \ )$$

measured in years. Thus the birth interval, η_4, was about 4.5 years. The standard deviation was 1.89 years, and the distribution of birth intervals is given in Figure 5. △

EXAMPLE 6. **Age at recruitment in the wandering albatross.** Albatrosses are known for their very long times to recruitment, during which the immature birds remain continuously at sea. Setting $J = 15$ and applying the algorithm here to the life cycle graph of Figure 2 gives the mean times to breeding for each stage (Figure 6). For the immature stages 1–14, these are times to recruitment. The mean time to breeding for a newly fledged chick is just over 10 years. Recall that stage 15 is the breeding adult stage; thus the mean time to next breeding for this stage is the birth interval; it is $\bar{\eta}_{15} = 2.01$ years, with a standard deviation of 1.21 years.   △

7. Spatial and temporal variation. The calculations to this point have assumed that the transition probabilities are time-invariant. This is often not the case, and sometimes estimates are available of the temporal variation in $\mathbf{P}$. For example, Fujiwara and Caswell (2001) and Caswell and Fujiwara (2004) documented changes in the right whale transition probabilities between 1980 and 1998. Such data raise questions about how quantities like life expectancy respond to the changing rates. To approach this, first consider an apparently unrelated problem: spatial heterogeneity.

7.1. Spatial heterogeneity. Consider a population with s transient stages and a absorbing states living in a set of p regions or patches. Each patch has its own, patch-specific transition matrix

$$(7.1) \qquad \mathbf{P}_i = \left(\begin{array}{c|c} \mathbf{U}_i & 0 \\ \hline \mathbf{M}_i & \mathbf{I} \end{array} \right) \qquad i = 1, \dots, p$$

Individuals disperse among patches at rates that vary from stage to stage. Let $\mathbf{D}_i$, $i = 1, \dots, s + a$, be a $p \times p$ transition matrix describing the probabilities of movement among patches for stage i.

Using the method of Hunter and Caswell (2005), we develop a model for the entire set of patches, using the transition and dispersal matrices. The critical assumption is that the processes of transition and dispersal can be described sequentially; with sufficient cleverness at defining time intervals this can usually be done.

Let $\mathbf{X}$ be a matrix, of dimension $(s + a) \times p$, where x_{ij} is the probability that the individual is in stage i and in patch j. The probability distribution of the state of the individual at time t is

$$(7.2) \qquad \mathbf{x}(t) = \operatorname{vec} \mathbf{X}^{\mathsf{T}}(t)$$

$$(7.3) \qquad = (\ \ x_{11} \quad \cdots \quad x_{1p} \ | \ \cdots \ | \ x_{s+a,1} \quad \cdots \quad x_{s+a,p} \ \ )^{\mathsf{T}}.$$

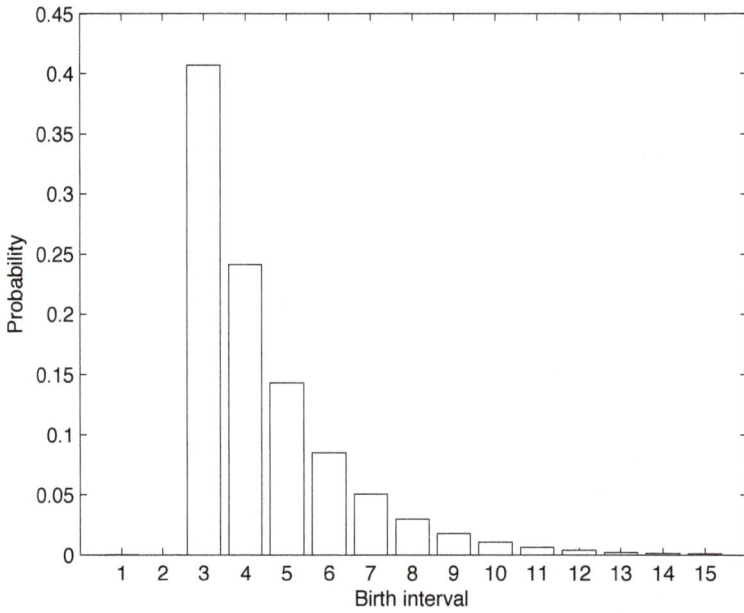

FIG. 5. *The probability distribution of the birth interval for the North Atlantic right whale. This is the interval between one birth and the next, conditional on surviving to the second birth.*

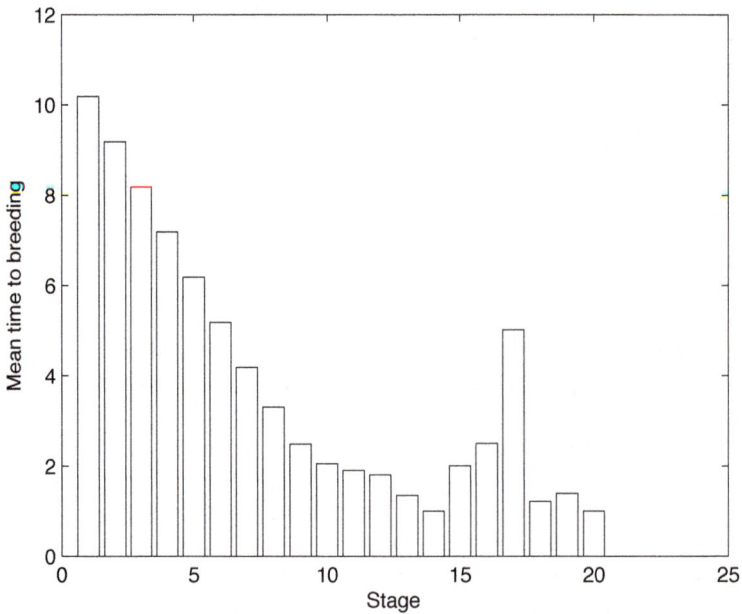

FIG. 6. *The distribution of the times to breeding, conditional on surviving to breed, for the wandering albatross. Parameters for the life cycle graph of Figure 2 were extracted from data in Croxall et al. (1990).*

This choice of arrangement groups together stage 1 from all patches, then stage 2 from all patches, and finally stage $s + a$ from all patches.

Define the block diagonal matrices

$$(7.4) \qquad \mathbb{P} = \begin{pmatrix} \mathbf{P}_1 & \cdots & 0 \\ 0 & \ddots & 0 \\ 0 & \cdots & \mathbf{P}_p \end{pmatrix}$$

$$(7.5) \qquad \mathbb{D} = \begin{pmatrix} \mathbf{D}_1 & \cdots & 0 \\ 0 & \ddots & 0 \\ 0 & \cdots & \mathbf{D}_{s+a} \end{pmatrix}.$$

If demographic transitions occur first, followed by dispersal, we can write

$$(7.6) \qquad \mathbf{x}(t+1) = \mathbb{D}\, \mathbf{V}\, \mathbb{P}\, \mathbf{V}^\mathsf{T}\, \mathbf{x}(t)$$

where $\mathbf{V} = \mathbf{V}(s + a, p)$ is the vec-permutation matrix (Henderson and Searle 1981) given by

$$(7.7) \qquad \mathbf{V}(s + a, p) = \sum_{i=1}^{s+a} \sum_{j=1}^{p} \mathbf{E}_{ij} \otimes \mathbf{E}_{ij}^\mathsf{T}$$

and $\mathbf{E}_{ij}$ has a 1 in the i, j position and zeros elsewhere. The vec-permutation matrix (also called the commutation matrix and many other names) satisfies $\operatorname{vec} \mathbf{X}^\mathsf{T} = \mathbf{V}\operatorname{vec}\mathbf{X}$; its use here permits the organization of $\mathbb{P}$ and $\mathbb{D}$ in block-diagonal form.

The matrix

$$(7.8) \qquad \mathbf{\Phi} = \mathbb{D}\, \mathbf{V}\, \mathbb{P}\, \mathbf{V}^\mathsf{T}$$

is the multiregional transition matrix. It is column stochastic, and can be partitioned into the familiar form

$$(7.9) \qquad \mathbf{\Phi} = \left(\begin{array}{c|c} \mathbb{U} & 0 \\ \hline \mathbb{M} & \mathbf{Y} \end{array} \right)$$

The transition matrix $\mathbf{\Phi}$ is of dimension $(s + a)p \times (s + a)p$; the submatrix $\mathbb{U}$ is of dimension $sp \times sp$ and the submatrix $\mathbb{M}$ is of dimension $ap \times sp$. The nature of $\mathbf{Y}$ depends on the assumptions made about the movement of individuals in the absorbing states. I will assume that $\mathbf{D}_i = \mathbf{I}$ for $i = s + 1, \ldots, s + a$. That is, individuals in the absorbing states do not move among patches; in this case $\mathbf{Y} = \mathbf{I}$. In some cases it might be of interest to allow individuals in absorbing states to move among patches.

Analysis of the transient matrix $\mathbb{U}$ gives the times to absorbtion and the probabilities of absorbtion, as in the single-patch case.

7.2. Time variation. The spatial analysis also solves the temporal problem, because temporal variation can be treated as a special case of spatial variation, in which . "dispersal" has a very restricted form. Consider the case where $p = 4$. Then

$$(7.10) \qquad \mathbf{D}_i = \begin{pmatrix} 0 & 0 & 0 & 0 \\ 1 & 0 & 0 & 0 \\ 0 & 1 & 0 & 0 \\ 0 & 0 & 1 & 1 \end{pmatrix} \qquad i = 1, \ldots, s$$

Each time is followed, with probability 1, by the next. Setting $d_{4,4} = 1$ makes the sequence repeat itself indefinitely in its last observed state. This is a hypothetical, and other patterns might be of interest in specific cases.

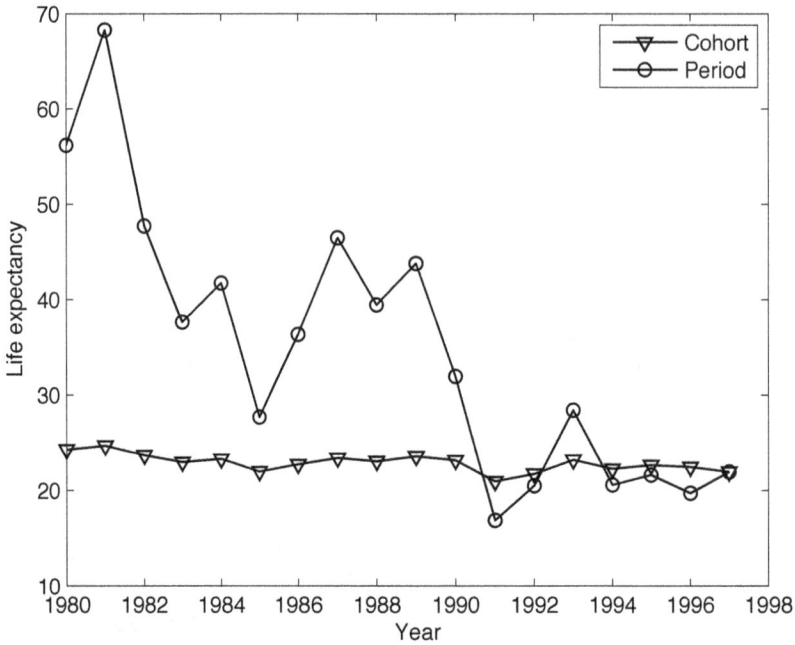

FIG. 7. *The period and cohort life expectancy for the North Atlantic right whale between 1980 and 1998. Results are shown for a model including a time trend and dependence on the North Atlantic Oscillation, rated as the best in a model selection procedure using Akaike's Information Criterion (Caswell and Fujiwara 2004).*

7.2.1. Describing temporal variation: period and cohort effects. Consider a population living in an environment that is changing over time. There are two ways to describe the demographic effects of this change on, say, life expectancy. One is to calculate $\bar{\eta}(t)$ from the vital rates operational at $t = 1, \ldots, p$. These values give a trajectory of life expectancy on the assumption that the rates at time t will remain fixed forever. Although this assumption is known to be false, the calculations are useful because they summarize the demographic consequences of the conditions in effect at time t. They are called *period* calculations, because they give results specific to the period for which they are computed.

Alternatively, one can calculate the demographic fate of a cohort of individuals born at t and living for one interval under the conditions of period t, one interval under the conditions of period $t + 1$, and so on. These are called *cohort* calculations because they apply to a cohort living through a sequence of periods, rather than to any particular period.

Period calculations are simple, since they require only the machinery for time-invariant calculations, applied at each time. Cohort calculations are more complicated, and have not been assayed for stage-classified models before. The Markov chain formulation, however, makes it possible to do so, in the following sequence.

    1. Obtain the sequence of time-specific projection matrices $\mathbf{P}_i$.

    2. Construct the "dispersal" matrices $\mathbf{D}_i$ according to (7.10)

    3. Construct $\boldsymbol{\Phi}$ from (7.8) and extract the $sp \times sp$ submatrix $\mathbb{U}$

4. Calculate the fundamental matrix $\mathbb{N} = (\mathbf{I} - \mathbb{U})^{-1}$

5. The $1 \times sp$ vector $\mathbf{e}^\mathsf{T}\mathbb{N} = \bar{\eta}$ gives the life expectancy of individuals at any stage at any time, projected forward as a cohort through the sequence of p environments. The first p entries of $\bar{\eta}$ are the life expectancies of stage 1 individuals at times $1, \ldots, p$. The next p entries are the life expectancies of stage 2 individuals, and so on.

EXAMPLE 7. **Period and cohort life expectancy in the right whale.** A model including time trends and effects of the NAO (on calf survival, mother survival, and the probability of reproduction) revealed a dramatic decline in survival of mothers between 1980 and 1998. The decline was sufficient to reduce population growth rate below replacement, and thus is of major conservation concern (Fujiwara and Caswell 2001, Caswell and Fujiwara 2004). The results for period and cohort life expectancy are shown in Figure 7. The period estimates reveal how the environment changed between 1980 and 1998, in terms of its effects on longevity. They show that the changes in survival had a major impact — enough to reduce life expectancy by 65%, from 54 years to 19 years.

However, no individual actually experienced such an environment. The cohort calculations show how the environment was experienced by individuals living through the period. The cohort life expectancy decreases only slightly, from 21.2 to 19.1 years. This is partly because the 18-year time period is short for such a long-lived animal as the right whale. Even early cohorts spend most of their lives in the conditions of the latter part of the sequence. Later cohorts actually spend most of their lives living under 1997 conditions, which are assumed fixed. $\triangle$

8. Conclusion. The ability to describe demography on a stage-specific basis is one of the great advances in the population sciences in the last 50 years. The matrix population model framework introduced by Leslie (1945) and reviatalized by Keyfitz (1964), Rogers (1966), and Lefkovitch (1967), has proven to be a powerful tool (Caswell 2001). The connection of the life cycle to Markov chains (Feichtinger 1971, 1973) is still being developed. I hope the examples presented here will contribute to this development.

As a general rule, new demographic calculations become more relevant as examples accumulate and patterns emerge in comparative studies. Some of the quantities reported here have never been calculated before, so there is as yet no basis for comparison. One of the advantages of the Markov chain approach is that computations are easy, regardless of the life cycle structure. Hopefully this will lead to more examples and new insights into the structure of the life cycle.

Acknowledgements. This work has benefited from discussions with Masami Fujiwara, Christine Hunter, Mike Neubert, James Vaupel, and the participants in the WHOI lecture series on Markov chains in demography. I am grateful to the New England Aquarium for access to data on the North Atlantic right whale.

REFERENCES

[1] CASWELL, H. 1978. *A general formula for the sensitivity of population growth rate to changes in life history parameters.* Theoretical Population Biology 14:215–230.

[2] CASWELL, H. 2000. *Prospective and retrospective perturbation analyses and their use in conservation biology.* Ecology 81:619–627.

[3] CASWELL, H. 2001. *Matrix Population Models: Construction, Analysis, and Interpretation.* 2nd ed. Sinauer, Sunderland, Massachusetts, USA.

[4] CASWELL, H. AND M. FUJIWARA. 2004. *Beyond survival estimation: mark-recapture, ma-*

trix population models, and population dynamics. Animal Biodiversity and Conservation 27:471–488.

[5] COCHRAN, M.E. AND S. ELLNER. 1992. *Simple methods for calculating age-based life history parameters for stage-structured populations.* Ecological Monographs 62:345–364.

[6] CRONE, M.J. AND S.D. KRAUS. 1990. *Right whale (*Eubalaena glacialis*) in the western North Atlantic: a catalog of identied individuals.* New England Aquarium, Boston, Massachusetts.

[7] CROXALL, J.P., P. ROTHERY, S.P.C. PICKERING AND P.A. PRINCE. 1990. *Reproductive performance, recruitment and survival of wandering albatrosses* Diomedia exulans *at Bird Island, South Georgia.* Journal of Animal Ecology 59:775–796.

[8] CUSHING, J.M. AND Z. YICANG. 1994. *The net reproductive value and stability in matrix population models.* Natural Resources Modeling 8:297–333.

[9] FEICHTINGER, G. 1971. *Stochastische Modelle Demographischer Prozesse.* Lecture notes in operations research and mathematical systems. Springer-Verlag, Berlin.

[10] FEICHTINGER, G. 1973. *Markovian models for some demographic processes.* Statistiche Hefte 14:310–334.

[11] FUJIWARA, M. AND H. CASWELL. 2001. *Demography of the endangered North Atlantic right whale.* Nature 414:537-541.

[12] GREENE, C.H., A.J. PERSHING, R.D. KENNEY, AND J.W. JOSSI. 2003. *Impact of climate variability on the recovery of endangered North Atlantic right whales.* Oceanography 16:96–101.

[13] HARPER, J.L. 1977. *Population Biology of Plants.* Academic Press, New York, New York, USA.

[14] HENDERSON, H.V. AND S.R. SEARLE. 1981. *The vec-permutation, the vec operator and Kronecker products: a review.* Linear and Multilinear Algebra 9:271–288.

[15] HUNTER, C.M. AND H. CASWELL. 2005. *The use of the vec-permutation matrix in spatial matrix population models.* Ecological Modelling 188:15–21.

[16] IOSIFESCU, M. 1980. *Finite Markov Processes and Their Applications.* Wiley, New York, USA.

[17] KEMENY, J.G. AND J.L. SNELL. 1976. *Finite Markov Chains.* Springer-Verlag, New York, New York, USA.

[18] KEYFITZ, N. 1964. *The population projection as a matrix operator.* Demography 1:56–73.

[19] KEYFITZ, N. 1971. *Linkages of intrinsic to age-specific rates.* Journal of the American Statistical Association 66:275–281.

[20] KEYFITZ, N. AND H. CASWELL. 2005. *Applied Mathematical Demography.* Third edition. Springer-Verlag, New York.

[21] LAND, K.C. AND A. ROGERS. 1982. *Multidimensional Mathematical Demography.* Academic Press, New York, New York, USA.

[22] LEFKOVITCH, L.P. 1965a. *The study of population growth in organisms grouped by stages.* Biometrics 21:1–18

[23] LESLIE, P.H. 1945. *On the use of matrices in certain population mathematics.* Biometrika 33:183–212.

[24] LOTKA, A.J. 1939. *Théorie analytique des associations biologiques. Part II. Analyse démographique avec application particulière à l'espèce humaine.* Actualités Scientifiques et Industrielles, No. 780. Hermann et Cie, Paris, France. (Published in translation as: *Analytical Theory of Biological Populations,* translated by D.P. Smith and H. Rossert. Plenum Press, New York, 1998)

[25] MAGNUS, J.R. AND H. NEUDECKER. 1988. *Matrix Differential Calculus with Applications in Statistics and Econometrics.* John Wiley & Sons, New York.

[26] NEL, D.G. 1980. *On matrix differentiation in statistics.* South African Statistical Journal 14:137–193.

[27] ROGERS, A. 1966. *The multiregional matrix growth operator and the stable interregional age structure.* Demography 3:537–544.

[28] VAUPEL, J.W. AND V.C. ROMO. 2003. *Decomposing change in life expectancy: a bouquet of formulas in honor of Nathan Keyfitz's 90th birthday.* Demography 40:201–216.

[29] WERNER, P.A. AND H. CASWELL. 1977. *Population growth rates and age versus stage-distribution models for teasel (*Dipsacus sylvestris *Huds.).* Ecology 58:1103–1111.